LARGE HADRON COLLIDER
PHENOMENOLOGY

LARGE HADRON COLLIDER PHENOMENOLOGY

Proceedings of the Fifty Seventh Scottish
Universities Summer School in Physics
St Andrews, 17 August to 29 August 2003

Co-sponsored by
The Institute for Particle Physics Phenomenology (IPPP),
University of Durham, England

Edited by

M Krämer — University of Edinburgh
F J P Soler — University of Glasgow and CCLRC, Rutherford
Appleton Laboratory

Series Editor

J F Cornwell — University of St Andrews

CRC Press
Taylor & Francis Group
Boca Raton London New York

CRC Press is an imprint of the
Taylor & Francis Group, an **informa** business

CRC Press
Taylor & Francis Group
6000 Broken Sound Parkway NW, Suite 300
Boca Raton, FL 33487-2742

First issued in paperback 2019

© 2004 by Taylor & Francis Group, LLC
CRC Press is an imprint of Taylor & Francis Group, an Informa business

No claim to original U.S. Government works

ISBN-13: 978-0-7503-0986-8 (hbk)
ISBN-13: 978-0-367-39379-3 (pbk)

Visit the Taylor & Francis Web site at
http://www.taylorandfrancis.com

and the CRC Press Web site at
http://www.crcpress.com

SUSSP Proceedings

/continued

SUSSP Proceedings (continued)

Lecturers

John Ellis	CERN, Geneva, Switzerland.
R. Keith Ellis	Fermilab, Batavia (IL), USA.
Valerie Gibson	University of Cambridge, UK.
Hans Hoffmann	CERN, Geneva, Switzerland.
Berndt Müller	Duke University, Durham (NC), USA.
Michael Andrew Parker	University of Cambridge, UK.
Albert de Roeck	CERN, Geneva, Switzerland.
Douglas Ross	University of Southampton, UK.
Rüdiger Schmidt	CERN, Geneva, Switzerland.
Tejinder S. Virdee	CERN, Geneva, Switzerland, and Imperial College, London, UK.

Key to photograph of participants

1	Emilie BURTON	Université Catholique de Louvain, Belgium
2	Keith ELLIS	Fermilab, IL, USA
3	Marie-Hélène GENEST	University of Montreal, Canada
4	Berndt MÜLLER	Duke University, Durham, NC, USA
5	Douglas ROSS	University of Southampton, UK
6	Ivan REID	Brunel University, UK
7	James STIRLING	Durham University, UK
8	Tony DOYLE	University of Glasgow, UK
9	Colin FROGGATT	University of Glasgow, UK
10	Leanne O'DONNELL	University of Edinburgh, UK
11	Elena GININA	Institute for Nuclear Research and Nuclear Energy, Sofia, Bulgaria
12	Malin SJODAHL	Lund University, Sweden
13	Kurt RINNERT	Karlsruhe University, Germany
14	Heuijin LIM	DESY, Hamburg, Germany
15	Otilia MILITARU	Université Catholique de Louvain, Belgium
16	James HAESTIER	Durham University, UK
17	Andrew MACGREGOR	University of Glasgow, UK
18	Reinhardt CHAMONAL	University of Edinburgh, UK
19	Jean-Philippe LANSBERG	Liège University, Belgium
20	Francesco SPANO	University of Chicago, IL, USA
21	Mariano CICCOLINI	University of Edinburgh, UK
22	Jean-François ARGUIN	University of Toronto, Canada
23	Torsten SOLDNER	Institut Laue-Langevin, France
24	Nicolas ARNAUD	CERN, Switzerland
25	Silvia AMERIO	University of Trento, Italy
26	Laura EDERA	University of Milan, Italy
27	Emanuela CAVALLO	Universita degli Studi di Bari, Italy
28	Maddelena ANTONELLO	L'Aquila University, Italy
29	Fabio FERRI	L'Aquila University, Italy
30	Marco ROVERE	Universita degli Studi dell'Insurbia, Italy
31	Alexei ILLARIONOV	Joint Institute for Nuclear Research, Russia
32	Irina TITKOVA	Joint Institute for Nuclear Research, Russia
33	Anne-Sylvie NICOLLERAT	ETH Zürich, Switzerland
34	Maria MARTISIKOVA	DESY, Hamburg, Germany
35	Paul SOLER	University of Glasgow and Rutherford Appleton Laboratory, UK
36	Alejandro DALEO	Universidad Nacional de la Plata, Argentina
37	Tejinder VIRDEE	Imperial College, London, UK and CERN, Switzerland
38	Mark TOBIN	University of Liverpool, UK
39	Michael KRÄMER	University of Edinburgh, UK

/continued

Key to photograph of participants (continued)

40	Albert DE ROECK	CERN, Switzerland
41	Bryan FIELD	SUNY Stony Brook, NY, USA
42	John ELLIS	CERN, Switzerland
43	Hans HOFFMAN	CERN, Switzerland
44	Rüdiger SCHMIDT	CERN, Switzerland
45	Chris MAXWELL	Durham University, UK
46	Matthew PEARSON	Rutherford Appleton Laboratory, UK
47	Andy PARKER	University of Cambridge, UK
48	Felix NAGEL	Heidelberg University, Germany
49	Ernest JANKOWSKI	University of Alberta, Canada
50	David ERIKSSON	Uppsala University, Sweden
51	Val GIBSON	University of Cambridge, UK
52	Mónica VÁZQUEZ ACOSTA	Universidad Autónoma de Madrid, Spain
53	Nicholas KERSTING	Tsinghua University, China
54	Jozef DUDEK	University of Oxford, UK
55	Partha KONAR	Harish-Chandra Research Institute, India
56	Davide RASPINO	Cagliari University, Italy
57	Enrico CATTARUZZA	University of Trieste, Italy
58	Massimiliano FIORINI	University of Ferrara, Italy
59	Pietro Antonio VERGINE	University of Ferrara, Italy
60	Kim GIOLO	Purdue University, USA
61	Simon SABIK	University of Toronto, Canada
	James LAWRENCE	University of Edinburgh, UK

Organising Committee

Prof A.T. Doyle	University of Glasgow	*Director*
Prof W.J. Stirling	IPPP, University of Durham	*Director*
Dr C.J. Maxwell	IPPP, University of Durham	*Secretary*
Prof C.D. Froggatt	University of Glasgow	*Treasurer*
Dr M. Krämer	University of Edinburgh	*Editor*
Dr F.J.P. Soler	University of Glasgow and RAL	*Steward and Editor*
Ms L. O'Donnell	University of Edinburgh	*Administration*
Ms L. Wilkinson	IPPP, University of Durham	*Administration*

Directors' Preface

With the Large Hadron Collider (LHC) under construction, and due to come online in 2007, it was timely for a summer school in 2003 to focus on LHC phenomenology. At a time when most of the experimental effort is directed to detector construction and software development, it was vitally important to focus young members of the experimental community on the physics that the LHC will deliver. At the same time, there is a continuing need to bring more young theorists into phenomenology, and in particular to inform them about the basic properties and capabilities of the machine, detectors and software required for physics analysis. It was with this in mind that we set about organising the summer school. Senior postgraduate students and postdocs from all over the world were attracted to the lectures provided by the leading figures in their respective fields.

The lectures covered many aspects of LHC phenomenology and the research and development necessary to understand future data. Douglas Ross provided the foundation in his overview of the Standard Model with Keith Ellis emphasising the phenomenological development necessary to truly understand LHC data. John Ellis introduced the key concepts and ideas in going beyond the Standard Model and Andy Parker showed how these are being translated into search methods for the general purpose detectors. Jim Virdee described the design and construction of these detectors and Rüdiger Schmidt gave a conceptual overview of the collider itself at a level that was appreciated by theorists and experimentalists alike, while Hans Hoffmann described the LHC computing challenge and how this was being met by Grid technology. The LHC provides a rich seam of physics: Val Gibson showed how current measurements would be extended on CP-violation in the b sector, Albert de Roeck presented the concepts necessary to understand forward and diffractive physics at the LHC, and Berndt Müller described how ALICE would explore the quark-gluon plasma when the collider runs with heavy ions.

We believe that this series of lectures will provide a thorough introduction to the phenomenology of LHC, not only for those currently working on this outstanding endeavour, but also for those inspired by the breadth of physics covered but, perhaps, overawed by the scale and complexity of the detectors and the detailed understanding necessary to develop the phenomenology. While the wealth of data from the Tevatron will lead to new discoveries and, we hope, clues to the physics beyond the Standard Model, it is already clear that the model provides a very good description of the data, and a solid platform from which to analyse the processes that will underpin the physics at the LHC. These lectures will be a useful guide to these processes, and will enable the student, whether theorist or experimentalist, to judge the significance of these developments as they unfold.

Summer schools are not just about science: they are also about dialogue, discussion, meeting people and making friends. The discussion sessions in the evenings were spent probing the lecturers further, and each ended with a welcome pint. From the initial whisky tasting, sponsored by Chivas Regal, or the challenge of asking John Ellis an impossible question at the evening discussion sessions, or the guest appearance of Peter

Higgs at the school dinner, or the ne'er-to-be-forgotten sight of Keith Ellis in a kilt addressing the haggis, students, lecturers and organisers all appreciated the event. The school succeeded in its secondary aim, aided by a full social programme and a friendly environment provided by the staff of the John Burnet Hall. Within this, the scientific discussions and personal interactions flourished. Finally, the organising committee wish to acknowledge the guidance received from Ken Bowler, Peter Negus, David Saxon and Alan Walker at SUSSP and thank Linda Wilkinson, Emma Durrant and Leanne O'Donnell who all contributed to the school's overall success. We also wish to thank Colin Scott and Keith Geddes from Chivas Regal, who gave up their valuable time to provide a memorable whisky tasting, and to Ivan Reid, who became the unofficial school photographer.

The Organizing Committee acknowledge the support of the Scottish Universities Summer Schools in Physics, the Institute for Particle Physics Phenomenology at the University of Durham, and the Physics and Astronomy Departments of the Universities of Edinburgh, Glasgow and St. Andrews, without which the school would not have been possible.

We would also like to thank all of the lecturers and participants for their enthusiasm, both for physics and for life, which helped make this a truly memorable school.

Tony Doyle and James Stirling

Directors, October 2003

Contents

Standard model and beyond

Large Hadron Collider technology

Physics at the Large Hadron Collider

Foundations of the standard model

Douglas Ross

University of Southampton, UK

1   Introduction

1.1   Quantum field theory

Particle physics is the study of particle interactions at the smallest possible scales. This immediately tells us that we need a quantum description of these particle interactions. In addition, we know from Heisenberg's uncertainty principle that in order to probe such short distances we need sufficiently high energies, so that most or all of the particles involved in a particular scattering event will be moving relativistically.

Quantum field theory is the consistent synthesis of quantum mechanics and special relativity applied to point particles. The goal of a particle theorist is to construct a quantum field theory, which accounts for the different interactions between particles and which can be used to make predictions for scattering cross-sections, decay rates and other measurable quantities, and it it is the task of particle experimentalists to perform the measurements in order to prove or disprove that a particular quantum field theory is consistent with experiment.

The techniques used to calculate measurable quantities from a quantum field theory are

(a) perturbation theory in which an expansion is performed in a power series of one or more coupling constants which are measures of the strength of interactions, and

(b) non-perturbative techniques such as lattice theories or numerical solutions of various equations derived from quantum field theory.

1.2   Renormalization

An additional difficulty that arises in quantum field theory but not in ordinary quantum mechanics is the occurrence of ultraviolet divergences. Higher order perturbative calculations involve summing over all possible intermediate states. In the language of Feynman diagrams this is represented by a loop and the Feynman rules require an integration over all possible momenta of particles inside the loop.

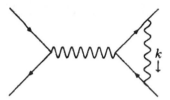

In general, this integral diverges for large momenta. Nowadays we do not interpret this divergence literally (as a genuine infinity) but assume that some new physics enters at some sufficiently high scale Λ to cutoff. The most successful new physics which does this job is string theory in which the point particle description is regarded as a low energy limit of a theory of extended objects (short strings). At 'low' energies the strings are in their ground state and can be described in terms of point-particle theories, but at some energy Λ, string excitations can be excited and in certain string theories these have been shown to conspire to cancel these ultraviolet divergences. From the 'low' energy view-point, the scale Λ of the string excitations acts as a cutoff for these divergences.

Nevertheless, these higher order perturbative corrections are very large since they depend (usually logarithmically) on this very large cutoff. This would render the perturbative expansion uncontrollable unless one can absorb the cutoff dependences by rescaling the quantum fields, masses, and coupling constants, which are the ingredients of the quantum field theory. This process of rescaling is called 'renormalization' and the goal is to rescale these parameters in such a way that in terms of rescaled parameters all physical quantities are finite (i.e. cutoff independent) in all perturbative orders.

Quantum field theories for which this is possible are called 'renormalizable theories'. The criteria for a theory to be renormalizable are:

(a) The interaction terms must not have dimension greater than four. Recalling that a bosonic field, ϕ, A_μ, ... has dimension 1 whereas a fermionic field, ψ, has dimension $\frac{3}{2}$, this means that interactions such as

$$\bar{\psi}\phi\psi, \quad \bar{\psi}\gamma^\mu\psi A_\mu, \quad \phi^4, \quad \text{and} \quad \phi^2 A^\mu A_\mu$$

are renormalizable, whereas interactions such as

$$\phi^6, \quad \bar{\psi}\gamma^\mu\gamma^\nu\psi\partial_\nu A_\mu, \quad \text{and} \quad \bar{\psi}\psi\bar{\psi}\psi$$

are not renormalizable.

(b) All propagators must vanish as the momentum of the propagating particle becomes infinite. This is usually the case, e.g. the propagator for a scalar particle of mass m and momentum p is

$$\frac{i}{(p^2 - m^2)}.$$

There is, however, one important exception, namely a massive vector (spin-one) particle, whose propagator is

$$-i\frac{\left(g^{\mu\nu} - \frac{p^\mu p^\nu}{m^2}\right)}{(p^2 - m^2)},$$

which goes to a constant $\sim 1/m^2$ as the momentum p becomes infinite Note, however, that for a massless vector particle the propagator can be written as

$$-i\frac{g^{\mu\nu}}{p^2},$$

which *does* vanish in the ultraviolet limit as required.

1.3 Early quantum field theories

Until the 1970's the only successful renormalizable quantum field theory was QED – the quantum field theory describing the electromagnetic interactions. The interaction term between a charged fermion and a photon is

$$e\bar{\psi}\gamma^\mu\psi A_\mu,$$

which has dimension four as required, and for which photon and fermion propagators both vanish as the momentum p tends to infinity. Furthermore, the coupling (charge) e is sufficiently small that a perturbative expansion in the fine structure constant

$$\alpha = \frac{e^2}{4\pi} = \frac{1}{137.06\ldots}$$

is reasonable.

Attempts to write quantum field theories in which nucleons interacted with pions via strong interactions through a (Yukawa) interaction term of the form

$$g\bar{\psi}_N\gamma^5\psi_N\phi_\pi$$

were unhelpful because the strong interactions require a large coupling constant, g. In fact a comparison of a leading order (tree-level) calculation of pion-nucleon scattering with measured cross-sections suggested a value of g such that

$$\frac{g^2}{4\pi} \approx 14,$$

rendering the perturbative expansion meaningless. At that time, no progress had been made on calculational techniques beyond perturbation theory such as lattice theories.

The weak interactions have small coupling, but were known to be described by an interaction which was the product of two currents. In terms of fermion fields the weak interaction is of the form

$$\mathcal{H}_{WK} \sim \bar{\psi}\gamma^\mu(1-\gamma^5)\psi\,\bar{\psi}\gamma_\mu(1-\gamma^5)\psi,$$

which is of dimension six and therefore non-renormalizable. Attempts were made to ameliorate this by introducing an 'intermediate vector boson', which was the progenitor of the $W^\pm$. In this theory the interaction term was of the form

$$\bar{\psi}\gamma^\mu(1-\gamma^5)\psi W_\mu,$$

and the effective weak interaction Hamiltonian $\mathcal{H}_{WK}$ arose by the propagation of this vector boson between one current and the other. Now the interaction term was renormalizable, but it was known that the currents were vector and axial-vector so the intermediate boson had to be a vector particle and furthermore it had to be massive as the interactions are short-range. Thus once again the intermediate vector boson theory was not renormalizable.

1.4   Gauge theories

The feature of the successful theory QED that led to the development of the standard model, which is a renormalizable quantum field theory that describes strong, weak, and electromagnetic interactions, is the fact that it is an example of a 'gauge theory'.

A gauge theory is a theory which is invariant under a set of *local* transformations, i.e. transformations described by parameters that can vary in space-time. In the case of QED, the transformation is a phase transformation acting on the fields representing charged particles, but in general it can be a set of transformations (usually forming a group) such as isospin transformations on fields transforming as an isodoublet (or indeed any other isospin multiplet).

For each generator of the local transformations the invariance of the theory requires the introduction of a massless vector boson known as a 'gauge boson'. In the case of QED where there is only one transformation, there is one such massless vector particle which is identified as the photon. In the case of local isospin transformations one needs three such massless gauge bosons corresponding to the three possible isospin transformations.

1.5   A renormalizable theory of weak interactions

In weak interaction processes a particle undergoes an isospin transformation. For example, a neutron decaying into a proton and leptons can be viewed as an isospin transformation of a nucleon.

The extension of the idea of a gauge theory is therefore very appealing as a quantum field theory to describe weak interactions, in which the gauge bosons would be interpreted as the intermediate vector bosons.

However, gauge theories require that the gauge bosons be massless. This is suitable for electromagnetic interactions which are long-range (and possibly also gravitational interactions, which will not be discussed in these lectures), but unsuitable for weak interactions which are known to be short range and therefore involve the exchange of a massive particle.

We could simply break the gauge invariance explicitly by introducing a mass term for the gauge bosons. This would certainly not give a renormalizable theory and it would destroy the gauge invariance. There is a far more elegant way of breaking the gauge invariance known as 'spontaneous symmetry breaking'. In this scenario the invariance is maintained at the level of the Lagrangian, but the ground state of the system (the vacuum) is *not* invariant under the transformations. When spontaneous symmetry breaking is applied in a gauge theory, a mass is automatically generated for the gauge bosons. Such models were developed in the 1960's (independently) by Glashow, Weinberg, and Salam – but these authors did not address the problem of renormalization. Their model predicts the existence of a massive scalar particle – the elusive Higgs boson.

In 1971, Tini Veltman's graduate student, Gerard 't Hooft, showed that it was possible to exploit gauge invariance to cast the theory into a form in which the propagators of the massive particles did vanish as their momenta became infinite – thus fulfilling the condition for renormalizability which was absent in the intermediate vector boson theory. Thus we had a renormalizable quantum field theory for weak interactions.

1.6 A quantum field theory for strong interactions

In the 1960's Gell-Mann and Low, and independently Callan and Symanzik, showed that as a result of higher order perturbative corrections, a coupling 'constant' was not constant at all but that the effective coupling between particles varies with the energy at which the interaction takes place.

This variation is usually positive, i.e. the effective coupling increases with increasing energy. However, in 1973 Sidney Coleman's graduate student, David Politzer, showed that for non-Abelian gauge theories (a gauge theory in which the different transformations do not commute with each other – unlike the case of QED) the variation of the effective (or 'running') coupling *decreases* with energy.

This paved the way for the development of QCD – a gauge theory describing interactions involving quarks and gluons – as a model for the strong interactions. Although it would be the case that at the low energies at which experiments were performed in the 1960's the effective couplings are far too large for meaningful perturbative expansions, at the energy scales at which experiments are carried out today, the running coupling is sufficiently small that perturbative calculations can be carried out and the results compared with experiment.

2 QED as an Abelian gauge theory

Consider the Lagrangian density for a free Dirac field ψ:

$$\mathcal{L} = \overline{\psi} \left(i\gamma^\mu \partial_\mu - m \right) \psi. \tag{1}$$

Now this Lagrangian density is invariant under a phase transformation of the fermion field

$$\psi \rightarrow e^{i\omega}\psi,$$

since the conjugate field $\overline{\psi}$ transforms as

$$\overline{\psi} \rightarrow e^{-i\omega}\overline{\psi}.$$

The set of all such phase transformations is called the 'group U(1)' and it is said to be 'Abelian' which means that any two elements of the group commute. This just means that

$$e^{i\omega_1} e^{i\omega_2} = e^{i\omega_2} e^{i\omega_1}.$$

For the purposes of these lectures it will usually be sufficient to consider infinitesimal group transformations, i.e. we assume that the parameter ω is sufficiently small that we can expand in ω and neglect all but the linear term. Thus we write

$$e^{i\omega} = 1 + i\omega + \mathcal{O}(\omega^2).$$

Under such infinitesimal phase transformations the field ψ changes by $\delta\psi$, where

$$\delta\psi = i\omega\psi,$$

and the conjugate field $\overline{\psi}$ by $\delta\overline{\psi}$, where

$$\delta\overline{\psi} = -i\omega\overline{\psi},$$

such that the Lagrangian density remains unchanged (to order ω).

Now suppose that we wish to allow the parameter ω to depend on space-time. In that case, for infinitesimal transformations, we have

$$\delta\psi(x) = i\omega(x)\psi(x), \tag{2}$$

$$\delta\overline{\psi}(x) = -i\omega(x)\overline{\psi}(x). \tag{3}$$

Such local (i.e. space-time dependent) transformations are called 'gauge transformations'. Note now that the Lagrangian density (1) is *no longer* invariant under these transformations, because of the partial derivative that is interposed between $\overline{\psi}$ and ψ, which will act on the space-time dependent parameter $\omega(x)$, such that the Lagrangian density changes by an amount $\delta\mathcal{L}$, where

$$\delta\mathcal{L} = -\overline{\psi}(x)\gamma^\mu(\partial_\mu\omega(x))\psi(x). \tag{4}$$

It turns out that we can repair the damage if we assume that the fermion field interacts with a vector field A_μ, called a 'gauge field', with an interaction term

$$-e\overline{\psi}\gamma^\mu A_\mu\psi$$

added to the Lagrangian density, which now becomes

$$\mathcal{L} = \overline{\psi}(i\gamma^\mu(\partial_\mu + ieA_\mu) - m)\psi. \tag{5}$$

In order for this to work we must also assume that apart from the fermion field transforming under a gauge transformation according to (2, 3), the gauge field, A_μ, also changes by δA_μ where

$$\delta A_\mu(x) = -\frac{1}{e}\partial_\mu\omega(x). \tag{6}$$

This change exactly cancels with eq.(4), so that once this interaction term has been added the gauge invariance is restored.

We recognize eq.(5) as being the fermionic part of the Lagrangian density for QED, where e is the electric charge of the fermion and A_μ is the photon field.

In order to have a proper quantum field theory, in which we can expand the photon field, A_μ, in terms of creation and annihilation operators for photons, we need a kinetic term for the field, A_μ, i.e. a term which is quadratic in the derivative of the field. We need to ensure that in introducing such a term we do not spoil the invariance under gauge transformations. This is achieved by defining the field strength, $F_{\mu\nu}$, as

$$F_{\mu\nu} = \partial_\mu A_\nu - \partial_\nu A_\mu. \tag{7}$$

It is easy to see that under the gauge transformation (6) each of the two terms on the right hand side of eq.(7) changes, but the changes cancel out. Thus we may add to the Lagrangian any term which depends on $F_{\mu\nu}$ (and which is Lorentz invariant – so we must contract all Lorentz indices). Such a term is $aF_{\mu\nu}F^{\mu\nu}$, which gives the desired term which is quadratic in the derivative of the field A_μ. Furthermore, if we choose the constant a to be $-\frac{1}{4}$ then the Lagrange equations of motion match exactly the (relativistic formulation) of Maxwell's equations. (The determination of this constant a is the *only* place that a match to QED has been used. The rest of the Lagrangian density is obtained purely from the requirement of local U(1) invariance.)

We have thus arrived at the Lagrangian density for QED, but from the viewpoint of demanding invariance under U(1) gauge transformations rather than starting with Maxwell's equations and formulating the equivalent quantum field theory.

The Lagrangian density is:

$$\mathcal{L} = -\frac{1}{4}F_{\mu\nu}F^{\mu\nu} + \overline{\psi}\left(i\gamma^\mu\left(\partial_\mu + ieA_\mu\right) - m\right)\psi. \tag{8}$$

Note that we are *not* allowed to add a mass term for the photon. A term such as $M^2 A_\mu A^\mu$ added to the Lagrangian density is not invariant under gauge transformations, but would give us a transformation

$$\delta\mathcal{L} = -\frac{2M^2}{e}A^\mu(x)\partial_\mu\omega(x).$$

Thus the masslessness of the photon can be understood in terms of the requirement that the Lagrangian be gauge invariant.

2.1   Covariant Derivatives

It is useful to introduce the concept of a 'covariant derivative'. This is not essential for Abelian gauge theories, but will be an invaluable tool when we extend these ideas to non-Abelian gauge theories.

The covariant derivative D_μ is defined to be

$$D_\mu = \partial_\mu + ieA_\mu. \tag{9}$$

This has the property that given the transformations of the fermion field (2) and the gauge field (6) the quantity

$$D_\mu\psi$$

is covariant under gauge transformations, i.e. it transforms by a phase rotation under gauge transformations.

We may thus rewrite the QED Lagrangian density as

$$\mathcal{L} = -\frac{1}{4}F_{\mu\nu}F^{\mu\nu} + \overline{\psi}\left(i\gamma^\mu D_\mu - m\right)\psi. \tag{10}$$

Furthermore the field strength, $F_{\mu\nu}$, can be expressed in terms of the commutator of two covariant derivatives, i.e.

$$F_{\mu\nu} = -\frac{i}{e}[D_\mu, D_\nu] = -\frac{i}{e}[\partial_\mu, \partial_\nu] + [\partial_\mu, A_\nu] + [A_\mu, \partial_\nu] + ie[A_\mu, A_\nu] = \partial_\mu A_\nu - \partial_\nu A_\mu. \tag{11}$$

2.2   Gauge fixing

There is a small difficulty that arises when we wish to quantize this theory. The Lagrange equation of motion for the photon field is

$$\left(-g^{\mu\nu}\partial^2 + \partial^\mu\partial^\nu\right) A_\nu = 0. \tag{12}$$

The propagator $G_{\mu\nu}(x-y)$ for the photon field is given by a solution to the Green function equation

$$\left(-g^{\mu\nu}\partial^2 + \partial^\mu\partial^\nu\right)_x G_{\nu\rho}(x-y) = i\delta^4(x-y)\delta^\mu_\rho. \tag{13}$$

Unfortunately this equation has no solution. What has gone wrong is that, because of the gauge invariance, a given value of the physical fields of the tensor $F_{\mu\nu}$ (the electric and magnetic fields) can be described by an infinite set of photon fields, all related by a gauge transformation.

It is therefore necessary to select a particular gauge in which to carry out the programme of quantization. This is done by imposing some constraint on the gauge field. There is an infinite range of possibilities, but the most convenient and most often used is the constraint or 'gauge condition'

$$\partial^\mu A_\mu = 0.$$

We impose this condition by adding the term

$$\frac{1}{(1-\xi)}\frac{1}{2}(\partial \cdot A)^2$$

to the Lagrangian density. Here ξ is a Lagrange multiplier called the 'gauge parameter'. After imposing this condition, the equation of motion, eq.(12), becomes

$$\left(-g^{\mu\nu}\partial^2 - \frac{\xi}{(1-\xi)}\partial^\mu\partial^\nu\right) A_\nu = 0, \tag{14}$$

and the Green function equation, eq.(13), changes to

$$\left(-g^{\mu\nu}\partial^2 - \frac{\xi}{(1-\xi)}\partial^\mu\partial^\nu\right)_x G_{\nu\rho}(x-y) = i\delta^4(x-y)\delta^\mu_\rho. \tag{15}$$

This has a simple solution whose Fourier transform gives the propagator of a photon with momentum p:

$$\tilde{G}^{\mu\nu}(p) = -i\frac{g^{\mu\nu} - \xi p^\mu p^\nu / k^2}{p^2}.$$

The most convenient choice of gauge parameter is $\xi = 0$. This is called the 'Feynman gauge' for which the photon propagator is simply

$$\tilde{G}^{\mu\nu}(p) = -i\frac{g^{\mu\nu}}{p^2}.$$

3   Non-Abelian gauge theories

3.1   Non-Abelian gauge transformations

We now move on to apply the ideas of the previous lecture to the case where the transformations are 'non-Abelian', i.e. different elements of the group do not commute with each other. This was originally considered by Yang and Mills (Yang and Mills 1954). As an example, we shall use isospin. An isodoublet such as a quark doublet can be either u-type or d-type depending on whether the third component of isospin is $+\frac{1}{2}$ or $-\frac{1}{2}$ respectively. The set of isospin transformations forms the group SU(2), which is mathematically almost identical to the group of rotations.

The quark isodoublet fermion field, ψ_i, now carries an index i, which takes the value 1 if the fermion is a u-type quark and 2 if the fermion is a d-type quark. The conjugate field is written $\overline{\psi}^i$.

The Lagrangian density for a free isodoublet is

$$\mathcal{L} = \overline{\psi}^i \left(i\gamma^\mu \partial_\mu - m \right) \psi_i. \tag{16}$$

A general isospin rotation requires three parameters ω^a, $a = 1\ldots 3$ (in the same way that a rotation is specified by three parameters which indicate the angle of the rotation and the axis about which the rotation is performed). Under such an isospin transformation the field ψ_i transforms as

$$\psi_i \rightarrow \left(e^{i\omega^a \mathbf{T}^a} \right)_i^j \psi_j,$$

where $\mathbf{T}^a$, $a = 1\ldots 3$ are the generators of isospin transformations in the isospin one-half representation. As in the case of the generators of rotations for a spin one-half particle these are given by $\frac{1}{2}$ times the Pauli spin matrices, i.e.

$$\mathbf{T}^1 = \frac{1}{2}\begin{pmatrix} 0 & 1 \\ 1 & 0 \end{pmatrix}, \quad \mathbf{T}^2 = \frac{1}{2}\begin{pmatrix} 0 & -i \\ i & 0 \end{pmatrix}, \quad \mathbf{T}^3 = \frac{1}{2}\begin{pmatrix} 1 & 0 \\ 0 & -1 \end{pmatrix}, \tag{17}$$

and obey the commutation relations

$$\left[\mathbf{T}^a, \mathbf{T}^b \right] = i\,\epsilon_{abc}\mathbf{T}^c. \tag{18}$$

This means that two such isospin transformations do *not* commute:

$$\left(e^{i\omega_1^a \mathbf{T}^a} \right)_i^k \left(e^{i\omega_2^b \mathbf{T}^b} \right)_k^j \psi_j, \neq \left(e^{i\omega_2^b \mathbf{T}^b} \right)_i^k \left(e^{i\omega_1^a \mathbf{T}^a} \right)_k^j \psi_j.$$

Groups of such transformations are called 'non-Abelian groups'.

Once again, it is convenient to consider only infinitesimal transformations under which the field ψ_i changes by an infinitesimal amount $\delta\psi_i$, where

$$\delta\psi_i = i\omega^a \left(\mathbf{T}^a \right)_i^j \psi_j, \tag{19}$$

and the conjugate field $\overline{\psi}^i$ changes by $\delta\overline{\psi}^i$, where

$$\delta\overline{\psi}^i = -i\omega^a \overline{\psi}^j \left(\mathbf{T}^a \right)_j^i. \tag{20}$$

We see that these two changes cancel each other out in the Lagrangian (16), provided that the parameters, ω_a, are constant. (Note that the conjugate field has a superscript i because strictly it transforms as the $\bar{2}$ representation of SU(2). For SU(2) these two representations are equivalent, but this will not be the case when we consider other groups.)

If we allow these parameters to depend on space-time, $\omega^a(x)$, then the Lagrangian density changes by $\delta\mathcal{L}$ under this 'non-Abelian gauge transformation', where

$$\delta\mathcal{L} = -\overline{\psi}^i \, (\mathbf{T}^a)^j_i \, \gamma^\mu \, (\partial_\mu \omega^a(x)) \, \psi_j.$$

3.2 Non-Abelian gauge fields

The symmetry can once again be restored by introducing interactions with vector (spin-one) gauge bosons. In this case we need three such gauge bosons, A^a_μ – one for each generator of SU(2). Under an infinitesimal gauge transformation these gauge bosons transform as

$$\delta A^a_\mu(x) = \epsilon_{abc} A^b_\mu(x) \omega^c(x) - \frac{1}{g}\partial_\mu \omega^a(x). \tag{21}$$

The first term on the right hand side of eq.(21) is the transformation that one would expect if the gauge bosons transformed as a usual triplet (isospin one), and this is indeed the case for constant ω^a. The second term (which is not linear in the field A^a_μ) is an extra term which needs to be added for the case of space-time dependent ω^a.

The interaction with these gauge bosons is again encoded by replacing the ordinary partial derivative in the Lagrangian density (16) with a covariant derivative, which in this case is a 2×2 matrix defined by

$$\mathbf{D}_\mu = \left(\partial_\mu \mathbf{I} + i\,g\,\mathbf{T}^a A^a_\mu\right), \tag{22}$$

where I is the unit matrix.

The Lagrangian density thus becomes

$$\mathcal{L} = \overline{\psi}^i \, (i\gamma^\mu \mathbf{D}_\mu - m\mathbf{I})^j_i \, \psi_j. \tag{23}$$

The quantity $\mathbf{D}_\mu\psi$ is not invariant under gauge transformations, but using eqs.(19), (21) and the commutation relations (18) we obtain the change of $\mathbf{D}_\mu\psi$ under an infinitesimal gauge transformation to be

$$\delta\left(\mathbf{D}_\mu\psi\right) = i\,\omega^a \mathbf{T}^a \mathbf{D}_\mu\psi, \tag{24}$$

which, together with eq.(20), tells us that the new Lagrangian density (23) is invariant under local isospin transformations ('SU(2) gauge transformations').

We can express the transformation rule for $\mathbf{D}_\mu\psi$ in terms of a transformation rule for the matrix $\mathbf{D}_\mu$ as

$$\delta\mathbf{D}_\mu = i\,[\omega^a \mathbf{T}^a, \mathbf{D}_\mu]. \tag{25}$$

The kinetic term for the gauge bosons is again constructed from the field strengths $F^a_{\mu\nu}$ which are defined from the commutator of two covariant derivatives:

$$\mathbf{F}_{\mu\nu} = -\frac{i}{g} \, [\mathbf{D}_\mu, \mathbf{D}_\nu], \tag{26}$$

where the matrix $\mathbf{F}_{\mu\nu}$ is given by

$$\mathbf{F}_{\mu\nu} = \mathbf{T}^a F_{\mu\nu}^a.$$

This gives us

$$F_{\mu\nu}^a = \partial_\mu A_\nu^a - \partial_\nu A_\mu^a - g\,\epsilon_{abc}\,A_\mu^b A_\nu^c. \tag{27}$$

From (24) we obtain the change in $\mathbf{F}_{\mu\nu}$ under an infinitesimal gauge transformation as

$$\delta\mathbf{F}_{\mu\nu} = i\omega^a\left[\mathbf{T}^a, \mathbf{F}_{\mu\nu}\right], \tag{28}$$

which leads to

$$\delta F_{\mu\nu}^a = \epsilon_{abc}\,F_{\mu\nu}^b\,\omega^c. \tag{29}$$

The gauge invariant term which contains the kinetic term for the gauge bosons is therefore

$$-\frac{1}{4}F_{\mu\nu}^a F^{a\,\mu\nu},$$

where a summation over the isospin index a is implied.

In sharp contrast with the Abelian case, this term does not only contain the terms which are quadratic in the derivatives of the gauge boson fields, but also the terms

$$g\,\epsilon_{abc}(\partial_\mu A_\nu^a)A_\mu^b A_\nu^c - \frac{1}{4}g^2\epsilon_{abc}\epsilon_{ade}A_\mu^b A_\nu^c A_\mu^d A_\nu^e.$$

This means that there is a very important difference between Abelian and non-Abelian gauge theories. For non-Abelian gauge theories the gauge bosons interact with each other via both three-point and four-point interaction terms. The three-point interaction term contains a derivative, which means that the Feynman rule for the three-point vertex involves the momenta of the particles going into the vertex. We shall write down the Feynman rules in detail later.

Once again, a mass term for the gauge bosons is forbidden, since a term proportional to $A_\mu^a A^{a\,\mu}$ is *not* invariant under gauge transformations.

3.3 The Lagrangian for a general non-Abelian gauge theory

Consider a gauge group, $\mathcal{G}$, of 'dimension' N, whose N generators, $\mathbf{T}^a$, obey the commutation relations

$$\left[\mathbf{T}^a, \mathbf{T}^b\right] = if_{abc}\mathbf{T}^c, \tag{30}$$

where f_{abc} are called the 'structure constants' of the group (they are antisymmetric in the indices a, b, c).

The Lagrangian density for a gauge theory with this group, with a fermion multiplet ψ_i, is given by

$$\mathcal{L} = -\frac{1}{4}F_{\mu\nu}^a F^{a\,\mu\nu} + i\bar{\psi}\left(\gamma^\mu D_\mu - m\mathbf{I}\right)\psi, \tag{31}$$

where

$$F_{\mu\nu}^a = \partial_\mu A_\nu^a - \partial_\nu A_\mu^a - g\,f_{abc}A_\mu^b A_\nu^c, \tag{32}$$

$$D_\mu = \partial_\mu\mathbf{I} + i\,g\,\mathbf{T}^a A_\mu^a. \tag{33}$$

Under an infinitesimal gauge transformation, the N gauge bosons, A^a_μ, change by an amount that contains a term which is not linear in A^a_μ:

$$\delta A^a_\mu(x) \;=\; f_{abc} A^b_\mu(x) \omega^c(x) - \frac{1}{g} \partial_\mu \omega^a(x), \tag{34}$$

whereas the field strengths $F^a_{\mu\nu}$ transform by a change

$$\delta F^a_{\mu\nu}(x) \;=\; f_{abc} F^b_{\mu\nu}(x) \, \omega^c. \tag{35}$$

In other words they transform as the 'adjoint' representation of the group (which has as many components as there are generators). This means that the quantity $F^a_{\mu\nu} F^{a\,\mu\nu}$ (summation over a implied) is invariant under gauge transformations.

3.4   Gauge fixing

As in the case of QED, we need to add a gauge fixing term in order to be able to derive a propagator for the gauge bosons. Again we impose the condition

$$\partial \cdot A^a \;=\; 0, \tag{36}$$

and in the Feynman gauge this means adding the term $-\frac{1}{2}(\partial^\mu A^a_\mu)^2$ to the Lagrangian density. The propagator (in momentum space) then becomes

$$-i\,\delta_{ab}\,\frac{g_{\mu\nu}}{p^2}.$$

There is one unfortunate complication, which is included briefly here for the sake of completeness, although one only needs to know about it for the purpose of performing higher loop calculations with non-Abelian gauge theories.

Consider a higher-order loop consisting of internal gauge bosons, for example arising in the one-loop self-energy of the gauge-boson itself. Unitarity (the optical theorem) tells us that the imaginary part of this graph is proportional to the cross-section for the production of two real gauge-bosons:

Unfortunately, this is not going to work as expected when we choose the Feynman gauge. The point is that in Feynman gauge the gauge-bosons going around the loop can have four possible degrees of freedom, whereas the physical states of a massless vector particle can only have two possible polarizations (i.e. there is a quantum field associated with each Lorentz component, A^μ). This results in a mismatch between the number of internal degrees of freedom either side of the cut on the left hand side and the physical degrees of freedom used to calculate the cross-section on the right hand side.

This is solved by introducing loops of 'Faddeev-Popov ghosts' which behave as a scalar particle associated with each gauge boson and associating a cancelling negative sign with each such loop. These are *not* to be interpreted as physical scalar particles which could in principle be observed experimentally, but merely as part of the gauge-fixing programme. For this reason they are referred to as 'ghosts'. They have two peculiarities.

(a) They only occur inside loops. This is because they are not really particles and cannot occur in initial or final states, but are introduced to clean up a difficulty that arises in the gauge-fixing mechanism.

(b) They behave like fermions even though they are scalars (spin zero). This means that we need to count a minus sign for each loop of Faddeev-Popov (Faddeev and Popov 1967, Feynman 1963) ghosts in any Feynman diagram.

This means that when we calculate the self-energy of a gauge-boson we need to subtract off a loop consisting of these ghost scalars, which give a (negative) contribution to the imaginary part of the self-energy

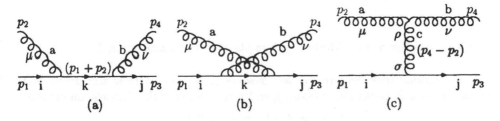

such that the unitarity relation is restored.

3.5 Feynman rules

The Feynman rules for a non-Abelian gauge theory (in Feynman gauge) are given below.

3.6 An example

As an example of the application of these Feynman rules, we consider the process of Compton scattering, but this time for the scattering of non-Abelian gauge-bosons and fermions, rather than photons. We need to calculate the amplitude for a gauge-boson of momentum p_2 and isospin a to scatter a fermion of momentum p_1 and isospin i producing a fermion of momentum p_3 and isospin j and a gauge-boson of momentum p_4. In addition to the two Feynman diagrams one gets in the QED case there is a third diagram involving the self-interaction of the gauge bosons.

Propagators:

Gluon

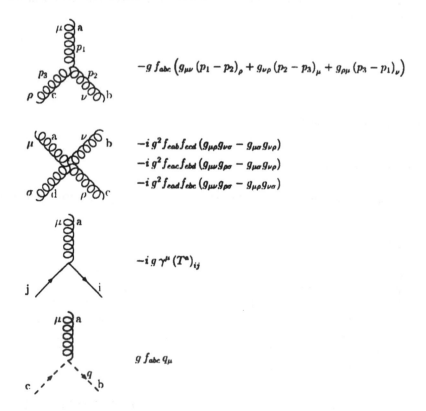

$$-i\,\delta_{ab}g_{\mu\nu}/p^2$$

Fermion

$$i\,\delta_{ij}(\gamma^\mu p_\mu + m)/(p^2 - m^2)$$

Faddeev-Popov ghost

$$i\,\delta_{ab}/p^2$$

Vertices:
(all momenta are flowing into the vertex).

$$-g\,f_{abc}\left(g_{\mu\nu}\,(p_1 - p_2)_\rho + g_{\nu\rho}\,(p_2 - p_3)_\mu + g_{\rho\mu}\,(p_3 - p_1)_\nu\right)$$

$$-i\,g^2 f_{eab}f_{ecd}\,(g_{\mu\rho}g_{\nu\sigma} - g_{\mu\sigma}g_{\nu\rho})$$
$$-i\,g^2 f_{eac}f_{ebd}\,(g_{\mu\nu}g_{\rho\sigma} - g_{\mu\sigma}g_{\nu\rho})$$
$$-i\,g^2 f_{ead}f_{ebc}\,(g_{\mu\nu}g_{\rho\sigma} - g_{\mu\rho}g_{\nu\sigma})$$

$$-i\,g\,\gamma^\mu\,(T^a)_{ij}$$

$$g\,f_{abc}\,q_\mu$$

Feynman rules for a non-Abelian gauge theory (Feynman gauge).

We will assume that the fermions are massless (i.e. that we are at sufficiently high energies that we may neglect their masses), and work in terms of the Mandelstam variables

$$s = (p_1 + p_2)^2 = (p_3 + p_4)^2,$$

$$t = (p_1 - p_3)^2 = (p_2 - p_4)^2,$$

$$u = (p_1 - p_4)^2 = (p_2 - p_3)^2.$$

The polarisations are accounted for by contracting the amplitude obtained for the above diagrams with the polarisation vectors $\epsilon^\mu(\lambda_2)$ and $\epsilon^\nu(\lambda_4)$. Each diagram consists of two vertices and a propagator and so their contributions can be read off from the Feynman rules.

For diagram (a) we get

$$\epsilon_\mu(\lambda_2)\epsilon_\nu(\lambda_4)\overline{u}^j(p_3)\left(-ig\,\gamma^\nu(T^b)^k_j\right)\left(i\frac{\gamma\cdot(p_2 + p_2)}{s}\right)\left(-ig\,\gamma^\mu(T^a)^i_k\right)u_i(p_1)$$

$$= -i\frac{g^2}{s}\epsilon_\mu(\lambda_2)\epsilon_\nu(\lambda_4)\overline{u}(p_3)\left(\gamma^\nu\gamma\cdot(p_1 + p_2)\gamma^\mu\right)\left(T^bT^a\right)u(p_1).$$

For diagram (b) we get

$$\epsilon_\mu(\lambda_2)\epsilon_\nu(\lambda_4)\overline{u}^j(p_3)\left(-ig\,\gamma^\mu(T^a)^k_j\right)\left(i\frac{\gamma\cdot(p_1 - p_4)}{u}\right)\left(-ig\,\gamma^\nu(T^b)^i_k\right)u_i(p_1)$$

$$= -i\frac{g^2}{u}\epsilon_\mu(\lambda_2)\epsilon_\nu(\lambda_4)\overline{u}(p_3)\left(\gamma^\nu\gamma\cdot(p_1 - p_4)\gamma^\mu\right)\left(T^aT^b\right)u(p_1).$$

Note here that the order of the T matrices is the other way around from diagram (a).

Diagram (c) involves the three-point gauge-boson self-coupling. Since the Feynman rule for this vertex is given with incoming momenta, it is useful to replace the outgoing gauge-boson momentum p_4 by $-p_4$ and understand this to be an incoming momentum. Note that the internal gauge-boson line carries momentum $p_4 - p_2$ coming into the vertex – the three incoming momenta that are to be substituted into the Feynman rule for the vertex are therefore p_2, $-p_4$, $p_4 - p_2$. The vertex thus becomes

$$-g\,f_{abc}\left(g_{\mu\nu}(p_2 + p_4)_\rho + g_{\rho\nu}(p_2 - 2p_4)_\mu + g_{\mu\rho}(p_4 - 2p_2)_\nu\right),$$

and the diagram gives

$$\epsilon^\mu(\lambda_2)\epsilon^\nu(\lambda_4)\overline{u}^j(p_3)\left(-ig\,\gamma_\sigma(T^c)^i_j\right)u_i(p_1)\left(-i\frac{g^{\rho\sigma}}{t}\right)$$

$$\times\;(-g\,f_{abc})\left(g_{\mu\nu}(p_2 + p_4)_\rho + g_{\rho\nu}(p_2 - 2p_4)_\mu + g_{\mu\rho}(p_4 - 2p_2)_\nu\right)$$

$$= -i\frac{g^3}{t}\epsilon^\mu(\lambda_2)\epsilon^\nu(\lambda_4)\overline{u}(p_3)\left[T^a, T^b\right]\gamma^\rho u(p_1)\left(g_{\mu\nu}(p_2 + p_4)_\rho - 2(p_4)_\mu g_{\nu\rho} - 2(p_2)_\nu g_{\mu\rho}\right),$$

where in the last step we have used that the polarisation vectors are transverse so that $p_2 \cdot \epsilon(\lambda_2) = 0$ and $p_4 \cdot \epsilon(\lambda_4) = 0$ and the commutation relations (30).

The contributions to the amplitude from all three of these graphs must be squared, summed or averaged over external state spin polarizations as appropriate and integrated over phase-space in order to yield the scattering cross-section.

4 Quantum chromodynamics (QCD)

4.1 Colour and the spin-statistics problem

When the quark model was first invented, it turned out that in order to reproduce the known spectrum of baryons, the quarks had to have wavefunctions which were symmetric under interchange of the spatial part of the wavefunction and spin. For example the resonance Δ^{++} consists of three u-type quarks and has spin $\frac{3}{2}$. This means that the spin part of the wavefunction must be symmetric (all quarks with the same helicity) and in the ground state the orbital angular momentum is zero.

This was unfortunate since quarks are fermions and should have a wavefunction which is *antisymmetric* under the interchange of the quantum numbers of any two fermions.

The problem was solved by Han and Nambu (Han and Nambu 1974) postulating that, as well as flavour (u-type, d-type etc.), quarks also carry one of three 'colours'. Thus as well as the flavour index, f, a quark field carries a colour index, i, so we write a quark field as q^i_f, $i = 1 \ldots 3$. It was further assumed that the strong interactions are invariant under colour SU(3) transformations. The quarks transform as a triplet representation of the group SU(3) which has eight generators. Furthermore it was assumed that all the observed hadrons are singlets of this new SU(3) group.

The spin statistics problem is now solved because a baryon consists of three quarks and if this is to be in a colour singlet state then the colour part of the wavefunction is required to be of the form

$$|B> = \epsilon_{ijk}|q^i_{f_1} q^j_{f_2} q^k_{f_3}>,$$

where f_1, f_2, f_3 are the flavours of the three quarks that make up the baryon B and i, j, k are the colours. The tensor ϵ_{ijk} is totally antisymmetric under the interchange of any two indices, so that if the part of the wavefunction of the baryon that does *not* depend on colour is symmetric, the total wavefunction (including the colour part) is antisymmetric, as required by the spin-statistics theorem.

4.2 QCD

QCD (Fritzsch *et al* 1973) is none other than the theory in which the invariance under colour SU(3) transformations is promoted to an invariance under local SU(3) (gauge) transformations. The eight gauge bosons which have to be introduced in order to preserve this invariance are the eight 'gluons'. These are taken to be the carriers which mediate the strong interactions in exactly the same way that photons are the carriers which mediate the electromagnetic interactions.

The Feynman rules for QCD are therefore simply the Feynman rules listed in the previous lecture, with the gauge coupling constant, g, taken to be the strong coupling, g_s, (more about this later), the generators T^a taken to be the eight generators of SU(3) in the triplet representation, and f_{abc}, a, b, c, $= 1 \ldots 8$ are the structure constants of SU(3).

Thus we now have a quantum field theory which can be used to describe the strong interactions.

4.3 Running coupling

The coupling for the strong interactions is the QCD gauge coupling, g_s. We usually work in terms of α_s defined as

$$\alpha_s = \frac{g_s^2}{4\pi}.$$

Since the interactions are strong, we would expect α_s to be too large to perform reliable calculations in perturbation theory. On the other hand the Feynman rules are only useful within the context of perturbation theory.

This difficulty is resolved when we understand that 'coupling constants' are not constant at all. The electromagnetic fine structure constant, α, only has the value $1/137$ at energies which are not large compared with the electron mass. At higher energies it is larger than this. For example, at LEP energies it takes a value closer to $1/128$. On the other hand, it turns out that for non-Abelian gauge theories the coupling *decreases* as the energy increases.

To see how this works within the context of QCD, we note that when we perform higher order perturbative calculations there are loop diagrams which have the effect of dressing the couplings. For example, the one-loop diagrams which dress the coupling between a quark and a gluon are:

where

are the diagrams needed to calculate the one-loop corrections to the gluon propagator.

These diagrams contain ultraviolet (UV) divergences and need to be renormalized by subtracting at some renormalization scale μ. This scale then appears inside a logarithm for the renormalized quantities. This means that if the square-momenta of all the external particles coming into the vertex are of order Q^2, where $Q \gg \mu$, then the above diagrams give rise to a correction which contains a logarithm of the ratio Q^2/μ^2:

$$-\alpha_s^2 \beta_0 \ln\left(Q^2/\mu^2\right).$$

This correction is interpreted as the correction to the effective QCD coupling, $\alpha_s(Q^2)$, at momentum scale Q. i.e.

$$\alpha_s(Q^2) = \alpha_s(\mu^2) - \alpha_s(\mu^2)^2\, \beta_0\, \ln\left(Q^2/\mu^2\right) + \cdots \tag{37}$$

β_0 is calculated to be

$$\beta_0 = \frac{11\, N_c - 2n_f}{12\,\pi}, \tag{38}$$

where N_c is the number of colours (=3), n_f is the number of active flavours, i.e. the number of flavours whose mass threshold is below the momentum scale, Q. Note that β_0 is *positive*, which means that the coefficient in front of the logarithm in (37) is *negative* (Politzer 1973, Politzer 1974, Gross and Wilczek 1973), so that the effective coupling *decreases* as the momentum scale is increased.

A more precise analysis shows that the effective coupling obeys the differential equation

$$\frac{\partial\,\alpha_s(Q^2)}{\partial\,\ln(Q^2)} = \beta\left(\alpha_s(Q^2)\right), \tag{39}$$

where β has a perturbative expansion

$$\beta(\alpha) = -\beta_0\,\alpha^2 + \mathcal{O}(\alpha^3) + \cdots \tag{40}$$

In order to solve this differential equation we need a boundary value. Nowadays this is taken to be the measured value of the coupling at the Z-boson mass (= 91 GeV), which is measured to be

$$\alpha_s(M_Z^2) = 0.118 \pm 0.002. \tag{41}$$

This is one of the free parameters of the standard model.

Previously the solution to equation (39) (to leading order) was written as

$$\alpha_s(Q^2) = 4\pi/\beta_0 \ln(Q^2/\Lambda_{QCD}^2),$$

and the scale Λ_{QCD} was used as the standard parameter which sets the scale for the magnitude of the strong coupling. This turns out to be rather inconvenient since it needs to be adjusted every time higher order corrections are taken into consideration and the number of active flavours has to be specified. This parametrisation is now hardly ever used.

The running of $\alpha_s(Q^2)$ is shown in the figure. We can see that for momentum scales above about 2 GeV the coupling is less than 0.3 so that one can hope to carry out reliable perturbative calculations for QCD processes with energy scales larger than this.

Gauge invariance requires that the gauge coupling for the interaction between gluons must be exactly the same as the gauge coupling for the interaction between quarks and gluons. The β-function could therefore have been calculated from the higher order corrections to the three-gluon (or four-gluon) vertex and must yield the same result, despite the fact that it is calculated from a completely different set of diagrams.

4.4   Quark (and gluon) confinement

This argument can be inverted to provide an answer to the question 'why have we never seen quarks or gluons in a laboratory?'

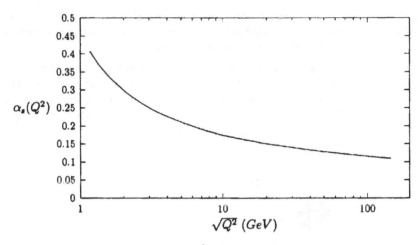

The running of $\alpha_s(Q^2)$ with β taken to two loops.

Asymptotic freedom, which tells us that the effective coupling between quarks becomes weaker as we go to short distances (this is equivalent to going to high energies) implies, conversely, that effective couplings grow as we go to large distances. Therefore, the complicated system of gluon exchanges, which leads to the binding of quarks (and antiquarks) inside hadrons, leads to a stronger and stronger binding as we attempt to pull the quarks apart. This means that we can never isolate a quark (or a gluon) at large distances since we require more and more energy to overcome the binding as the distance between the quarks grows.

The upshot of this is that the only free particles which can be observed at macroscopic distances from each other are colour singlets. This mechanism is known as 'quark confinement'. The details of how it works are not fully understood. Nevertheless the argument presented here is suggestive of such confinement and at the level of non-perturbative field theory, lattice calculations have confirmed that for non-Abelian gauge theories the binding energy does indeed grow as the distance between quarks increases.

Thus we have two different pictures of the world. At sufficiently short distances, which can be probed at sufficiently large energies, we can consider quarks and gluons interacting with each other. We can perform calculations of the scattering cross-sections between quarks and gluons (called the 'hard cross-section') at short distances because the running coupling is sufficiently small that we can rely on perturbation theory.

4.5 Infrared divergences

It would be very convenient if all this meant that when considering processes at sufficiently high energies we could calculate the relevant cross-sections using perturbative QCD in the quark-gluon picture.

Unfortunately, this is not the case because as well as ultraviolet divergences which occur in higher order perturbative orders that can be absorbed by a programme of renormalization, there are other divergences which occur when a gluon is very soft ($k \to 0$) or collinear to the massless particle that emits it. These are both called 'infrared divergences' although in the latter case the gluon is not necessarily soft.

We will assume that we are dealing with sufficiently high energies that the masses of the quarks can be neglected. This means that we expect collinear divergences for gluons emitted parallel to a quark. In practice, the divergences are regulated by the quark masses, m_q, but yield a large logarithm $\sim \ln(s/m_q^2)$. A study of the collinear divergences in the limit $m_q \to 0$ is a way of investigating these large logarithms.

Consider a loop graph of the form:

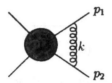

The internal quark propagators are proportional to

$$\frac{1}{(k^2 + 2k \cdot p_1)} \quad \text{and} \quad \frac{1}{(k^2 - 2k \cdot p_2)}.$$

These diverge as $k \to 0$ such that the loop integral over the gluon momentum, k, diverges.

This soft divergence, however, is cancelled exactly by soft gluon emission processes (Bloch and Nordsieck 1937, Yennie *et al* 1961) which can occur at the same order in perturbation theory, such as

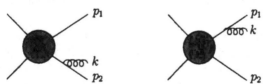

In these graphs the propagators of the 'parent' quarks that emit the gluon are proportional to

$$\frac{1}{2k \cdot p_1} \quad \text{and} \quad \frac{1}{2k \cdot p_2},$$

such that the integral over the phase-space of the emitted gluon is also divergent. The coefficients of the divergences from the virtual soft-gluon correction and the real gluon emission are equal and opposite. This means that provided we calculate the cross-section for a process which includes an emitted gluon that carries off some energy, ΔE, the total cross-section will be free of soft divergences. ΔE may be the energy resolution of the final state particle calorimeter, or one may be genuinely interested in inclusive processes.

Collinear divergences occur whenever the emitted gluon has a momentum

$$k = \lambda p, \quad \text{and} \quad p^2 = 0,$$

which happens when the gluon is emitted parallel to a massless particle (this could be another gluon which is strictly massless as well as a quark whose mass is being neglected).

In such a case the square-momentum of the parent particle, $(k+p)^2 \propto p^2 = 0$ so that its propagator diverges.

There is once again an interplay between collinear divergences associated with virtual gluon corrections and those associated with real emission. However, in this case, the cancellation only works (Kinoshita 1962, Lee and Nauenberg 1964) if we sum over all possible *initial* and *final* states (to a given order in perturbation theory). In practice, it is sufficient to sum over all states which can contribute to infrared divergences.

For example we need to sum over processes in which a massless outgoing particle is replaced by a jet of nearly parallel outgoing massless particles.

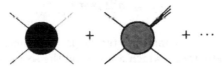

In addition, we need to sum over processes in which an incoming massless particle is replaced by an incoming jet of nearly parallel massless particles.

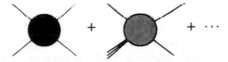

For the summation over final states, this is not really a problem, since one cannot distinguish experimentally between a single particle and a sufficiently narrow jet of particles any more than one can detect soft photon (or gluon) radiation which takes off energy less than the energy resolution of the experiment.

The requirement that one sums over incoming jets in order to cancel the collinear divergences is more problematic. What this means is that if we calculate in perturbation theory the QCD process of quark-quark (or quark-gluon, or gluon-gluon) scattering, we will not get a finite result even when summing over all possible final states. On the other hand, it is important to note that in practice one cannot prepare an initial state which consists of free quarks and/or gluons. The initial states are hadrons which contain quarks and gluons. The remaining divergence arising from the calculation of a process with initial quarks and/or gluons is absorbed into the (momentum scale dependence) of the parton 'distribution function' (Gribov and Lipatov 1972, Lipatov 1975, Dokshitzer 1977, Altarelli and Parisi 1977), i.e. the probability that a parent hadron contains a parton with a given flavour and momentum fraction.

Moreover, we do not usually want to sum over all possible final states – i.e. to calculate a total cross-section. Normally we are interested in a differential cross-section with respect to some variable T, which is a function of the kinematic configuration of the final state particles. For certain variables, T, known as 'infrared safe' variables, it is possible to restrict the integration over the phase-space of the final particles to a kinematic configuration which gives a particular value for T, without re-introducing soft or collinear divergences. For other variables this is not possible. For example, it is not possible to specify how many partons there are in the final state, since we have already seen that real gluon emission is required in order to cancel soft infrared divergences. On the other hand, it is possible to calculate the differential cross-section into a given number of jets, provided that a jet is defined in terms of some opening angle which allows configurations in which two or more almost parallel particles are included in a single jet.

4.6 θ-parameter

There is one more gauge invariant term that can be added to the Lagrangian density for QCD. This term is

$$\mathcal{L}_\theta = \theta \frac{\alpha_s}{8\pi} \epsilon^{\mu\nu\rho\sigma} F_{\mu\nu}^a F_{\rho\sigma}^a,$$

where $\epsilon^{\mu\nu\rho\sigma}$ is the totally antisymmetric tensor (in four dimensions). Such a term arises when one considers 'instantons' (which are beyond the scope of these lectures).

This term violates CP. In QED we would have

$$\epsilon^{\mu\nu\rho\sigma} F_{\mu\nu} F_{\rho\sigma} = \mathbf{E} \cdot \mathbf{B},$$

and for QCD we have a similar expression except that $\mathbf{E}^a$ and $\mathbf{B}^a$ carry a colour index – they are known as the chromoelectric and chromomagnetic fields. Under charge conjugation both the electric and magnetic field change sign, but under parity the electric field, which is a proper vector, changes sign, whereas the magnetic field, which is a polar vector, does not change sign. Thus we see that the term $\mathbf{E} \cdot \mathbf{B}$ is odd under CP.

For this reason, the parameter θ in front of this term must be exceedingly small in order not to give rise to strong interaction contributions to CP violating quantities such as the electric dipole moment of the neutron. The current experimental limits on this dipole moment tell us that $\theta < 10^{-9}$ and it is probably zero. Nevertheless, strictly speaking θ is a free parameter of QCD.

5   Spontaneous symmetry breaking

We have seen that in an unbroken gauge theory the gauge bosons must be massless. The only observed massless spin-1 particles are photons. In the case of QCD, the gluons are also massless, but they cannot be observed because of confinement.

If we wish to extend the ideas of describing interactions by a gauge theory to the weak interactions, the symmetry must somehow be broken since the carriers of weak interactions (W- and Z-bosons) are massive (weak interactions are very short range). We could simply break the symmetry by hand by adding a mass term for the gauge bosons, which we know violates the gauge symmetry. However, there is a far more elegant way of doing this which is called 'spontaneous symmetry breaking'. In this scenario, the Lagrangian maintains its symmetry under a set of local gauge transformations. On the other hand, the lowest energy state, which we interpret as the vacuum state, is *not* a singlet of the gauge symmetry. There is an infinite number of states each with the same ground-state energy and nature chooses one of these states as the 'true' vacuum.

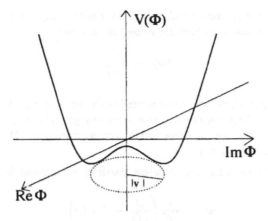

Figure 1. *A potential with spontaneous symmetry breaking*

5.1 Spontaneous symmetry breaking

We start by considering a complex scalar field theory with a mass term and a quartic self-interaction. The Lagrangian density for such a theory may be written

$$\mathcal{L} = \partial_\mu \Phi^* \, \partial^\mu \Phi - V(\Phi), \tag{42}$$

where the 'potential' $V(\Phi)$, is given by

$$V(\Phi) = \mu^2 \Phi^* \Phi + \lambda |\Phi^* \Phi|^2. \tag{43}$$

This Lagrangian is invariant under global U(1) transformations

$$\Phi \to e^{i\omega} \Phi.$$

Provided μ^2 is positive this potential has a minimum at $\Phi = 0$. We call the $\Phi = 0$ state the vacuum and expand Φ in terms of creation and annihilation operators that populate the higher energy states. In terms of a quantum field theory, where Φ is an operator, the precise statement is that the operator, Φ, has zero 'vacuum expectation value'.

Suppose now that we reverse the sign of μ^2, so that the potential becomes

$$V(\Phi) = -\mu^2 \Phi^* \Phi + \lambda |\Phi^* \Phi|^2. \tag{44}$$

We see that this potential no longer has a minimum at $\Phi = 0$, but a *maximum*. The minimum occurs at

$$\Phi = \frac{v}{\sqrt{2}} = e^{i\theta} \sqrt{\frac{\mu^2}{2\lambda}}, \tag{45}$$

where θ takes any value from 0 to 2π. There is an infinite number of states each with the same lowest energy – i.e. we have a degenerate vacuum. The symmetry breaking occurs in the choice made for the value of θ which represents the true vacuum. For convenience we shall choose $\theta = 0$ to be this vacuum. Such a choice constitutes a spontaneous breaking of the U(1) invariance, since a U(1) transformation takes us to a different lowest energy state. In other words the vacuum breaks U(1) invariance. In quantum field theory we say that the field, Φ, has a non-zero vacuum expectation value

$$\langle\Phi\rangle = \frac{v}{\sqrt{2}}.$$

But this means that there are 'excitations' with zero energy, that take us from the vacuum to one of the other states with the same energy. The only particles which can have zero energy are massless particles (with zero momentum). We therefore expect a massless particle in such a theory.

To see that we do indeed get a massless particle, we expand Φ around its vacuum expectation value as

$$\Phi = \frac{1}{\sqrt{2}}\left(\frac{\mu}{\sqrt{\lambda}} + H + i\phi\right). \tag{46}$$

The fields H and ϕ have zero vacuum expectation value and it is these fields that are expanded in terms of creation and annihilation operators of the particles that populate the excited states. If we now insert (46) into (44) we find

$$V = \mu^2 H^2 + \mu\sqrt{\lambda}\left(H^3 + \phi^2 H\right) + \frac{\lambda}{4}\left(H^4 + \phi^4 + 2H^2\,\phi^2\right) + \frac{\mu^4}{4\lambda}. \tag{47}$$

Note that in (47) there is a mass term for the field H, but *no* mass term for the field ϕ. Thus ϕ is a field for a massless particle called a 'Goldstone boson' (Goldstone 1961).

5.2   Goldstone bosons

Goldstone's theorem extends this to spontaneous breaking of a general symmetry.

Suppose we have a theory which is invariant under a symmetry group $\mathcal{G}$ with N generators, and that some operator (i.e. a function of the quantum fields – which might just be a component of one of these fields) has a non-zero vacuum expectation value, which breaks the symmetry down to a subgroup $\mathcal{H}$ of $\mathcal{G}$, with n generators (this means that the vacuum state is still invariant under transformations generated by the n generators of $\mathcal{H}$, but not the remaining $N - n$ generators of the original symmetry group $\mathcal{G}$). Goldstone's theorem states that there will be $N - n$ massless particles (one for each broken generator of the group).

5.3   The Higgs mechanism

Like all good general theorems, Goldstone's theorem has a loophole (Higgs 1964a, Higgs 1964b, Brout and Englert 1964, Guralnik *et al* 1964, Kibble 1967), which arises when one considers a gauge theory, i.e. when one allows the original symmetry transformations to be local.

In spontaneously broken gauge theory, the choice of which vacuum is the true vacuum is equivalent to choosing a gauge, which is necessary in order to be able to quantize the theory. What this means is that the Goldstone bosons, which can, in principle, transform the vacuum into one of the states degenerate with the vacuum, now affect transitions into states which are not consistent with the original gauge choice. This means that the Goldstone bosons are 'unphysical' and are often called 'Goldstone ghosts'.

On the other hand the quantum degrees of freedom associated with the Goldstone bosons are certainly there *ab initio* (before a choice of gauge is made). What happens to them?

A massless vector boson has only two degrees of freedom (the two directions of polarisation of a photon), whereas a massive vector (spin-one) particle has three possible values for the helicity of the particle. In a spontaneously broken gauge theory, the Goldstone boson associated with each broken generator provides the third degree of freedom to the gauge bosons. This means that the gauge bosons become massive. The Goldstone boson is said to be 'eaten' by the gauge boson.

To see how this works we return to the U(1) gauge theory, but now we promote the symmetry to a local symmetry and we must introduce a gauge boson, A_μ. The partial derivative of the field Φ is replaced by a covariant derivative

$$\partial_\mu \Phi \rightarrow D_\mu \Phi = (\partial_\mu + i e A_\mu) \Phi.$$

Including the kinetic term $-\frac{1}{4}F_{\mu\nu}F^{\mu\nu}$ for the gauge bosons, the Lagrangian density becomes

$$\mathcal{L} = -\frac{1}{4}F_{\mu\nu}F^{\mu\nu} + (D_\mu \Phi)^* D^\mu \Phi - V(\Phi). \tag{48}$$

Now note what happens if we insert the expansion (46) into the term $(D_\mu \Phi)^* D^\mu \Phi$. This generates the following terms

$$\frac{1}{2}\partial_\mu H \, \partial^\mu H + \frac{1}{2}\partial_\mu \phi \, \partial^\mu \phi + \frac{1}{2}e^2 v^2 A_\mu A^\mu + e v \, A^\mu \partial_\mu \phi - e \, A^\mu (\phi \partial_\mu H - H \partial_\mu \phi)$$

$$+ \frac{1}{2}e^2 A_\mu A^\mu \left(H^2 + \phi^2\right), \tag{49}$$

where $v = \mu/\sqrt{\lambda}$. The gauge boson has acquired a mass term, $M_A = e v$, even though the Lagrangian (48) is invariant under local U(1) transformations. (Note that for a real field ϕ representing a particle of mass m the mass term is $\frac{1}{2}m^2\phi^2$ whereas for a complex field the mass term is $m^2\phi^\dagger\phi$.)

5.4 Gauge fixing

There is also the bilinear term $e v \, A^\mu \partial_\mu \phi$ in eq.(49) which after integrating by parts (for the action S) may be written as $- M_A \phi \partial_\mu A^\mu$. This simply indicates that the Goldstone boson, ϕ, couples to the longitudinal component of the gauge boson with strength M_A.

Let us separate the gauge-boson field, A_μ, into its transverse and longitudinal components,

$$A_\mu = A_\mu^L + A_\mu^T,$$

where $\partial^\mu A_\mu^T = 0$. We can think of the longitudinal component of the gauge boson oscillating between the Goldstone boson with a mixing term given by $- M_A \phi \partial^\mu A_\mu^L$, so that the physical particle is described by a superposition of these fields.

We consider two special cases:

(a) The unitary gauge:
The physical field for the longitudinal component of the gauge boson is not simply A_μ^L, but the superposition

$$A_\mu^{ph} = A_\mu + \frac{1}{M_A}\partial_\mu\phi. \tag{50}$$

(Note that this only affects the longitudinal component.) The Goldstone boson field, ϕ, drops out when the Lagrangian is written in terms of the physical gauge-boson field, A_μ^{ph}. The terms involving ϕ in the original expression have been absorbed (or 'eaten') by the redefinition (50) of the gauge-boson field. The equation of motion for the free physical gauge boson is

$$\left(-g^{\mu\nu}\left(\Box + M_A^2\right) + \partial^\mu\partial^\nu\right)A_\nu^{ph} = 0,$$

leading to a (momentum-space) propagator

$$-i\left(g_{\mu\nu} - \frac{p_\mu p_\nu}{M_A^2}\right)\frac{1}{(p^2 - M_A^2)}, \tag{51}$$

which is the usual expression for the propagator of a massive spin-one particle.

The only other remaining particle is the scalar, H, with mass $m_H = \sqrt{2}\,\mu$ which is called the 'Higgs' particle (after its inventor). This is a physical particle, which interacts with the gauge boson and also has cubic and quartic self-interactions. The interaction terms involving the Higgs boson are

$$\mathcal{L}_I(H) = \frac{e^2}{2}A_\mu A^\mu H^2 + eM_A A_\mu A^\mu H - \frac{\lambda}{4}H^4 - m_H\sqrt{2\lambda}H^3, \tag{52}$$

which leads to the following vertices and Feynman rules:

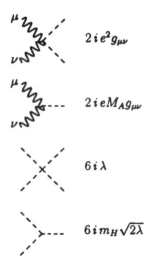

$$2\,i\,e^2 g_{\mu\nu}$$

$$2\,i\,eM_A g_{\mu\nu}$$

$$6\,i\,\lambda$$

$$6\,i\,m_H\sqrt{2\lambda}$$

(b) **Feynman gauge**:

We select the Feynman gauge by adding to the Lagrangian density the term

$$-\frac{1}{2}\left(\partial \cdot A + M_A \phi\right)^2.$$

After integration by parts, the cross-term $M_A \phi \partial \cdot A$ cancels the bilinear mixing term. The equation of motion for the free gauge boson is

$$-g^{\mu\nu}\left(\square + M_A^2\right) A_\nu = 0,$$

so that the gauge boson propagator simplifies to

$$-i\,\frac{g_{\mu\nu}}{(p^2 - M_A^2)}, \tag{53}$$

which is easy to handle. There is, however, a price to pay. The Goldstone boson is still present. It has acquired a mass, M_A, from the gauge fixing term, and it has interactions with the gauge boson, with the Higgs scalar and with itself. Furthermore, for the purposes of higher order corrections we again need to introduce Faddeev-Popov ghosts which in this case interact not only with the gauge boson, but also with the Higgs scalar and Goldstone boson.

5.5 Renormalizability

Generating a mass for the gauge bosons by spontaneous symmetry breaking solves the very important problem of renormalizability in the case of massive spin-1 particles.

Recall that for a theory to be renormalizable (i.e. that one can absorb all the UV divergences into the fields, masses and couplings of the theory), all the propagators have to decrease like $1/p^2$ as the momentum p tends to ∞.

If we look at the propagator for the gauge boson in unitary gauge (51) we see that it does *not* decrease as $p \to \infty$, because of the second term (proportional to $p_\mu p_\nu$). This would normally lead to a violation of renormalizability, thereby rendering the quantum field theory useless (at least as far as perturbative calculations were concerned).

However as was shown by 't Hooft ('t Hooft 1971), since any physical quantity that we measure should be gauge invariant, we are at liberty to choose a different gauge to do our calculations. If we look at the propagator (53), we see that in such gauges the propagator does decrease like $1/p^2$ as the momentum $p \to \infty$. This means that although it would appear from the unitary gauge that the theory was not renormalizable, any physical quantity can be calculated in a gauge where renormalizability is manifest. As mentioned above, the price we pay for this is that there are more particles and many more interactions, leading to a plethora of Feynman diagrams. We therefore only work in such gauges if we want to compute higher order corrections. For the rest of these lectures we shall confine ourselves to tree-level calculations and work solely in unitary gauge.

Nevertheless, one cannot over-stress the fact that only when the gauge bosons acquire masses through the Higgs mechanism do we have a renormalizable theory. It is this mechanism that makes it possible to write down a consistent quantum field theory which describes the weak interactions.

5.6   Spontaneous symmetry breaking in a non-Abelian gauge theory

It is a relatively straightforward matter to extend this to the case of non-Abelian gauge theories. We take as an example an SU(2) gauge theory and consider a complex doublet of scalar fields, Φ^i, $i = 1, 2$. The Lagrangian density is

$$\mathcal{L} = -\frac{1}{4}F^a_{\mu\nu}F^{a\,\mu\nu} + |D_\mu\Phi|^2 - V(\Phi), \tag{54}$$

where

$$D_\mu\Phi = \partial_\mu\Phi + ig\,W^a_\mu\,\mathbf{T}^a\,\Phi,$$

(we have changed notation for the gauge bosons from A^a_μ to W^a_μ), and

$$V(\Phi) = -\mu^2\Phi^\dagger_i\Phi^i + \lambda\left(\Phi^\dagger_i\Phi^i\right)^2. \tag{55}$$

This potential has a minimum at $\Phi^\dagger_i\Phi_i = \frac{1}{2}\mu^2/\lambda$. We choose the vacuum expectation value to be in the $T^3 = -\frac{1}{2}$ direction and to be real, i.e.

$$\langle\Phi\rangle = \frac{1}{\sqrt{2}}\begin{pmatrix} 0 \\ v \end{pmatrix}$$

($v = \mu/\sqrt{\lambda}$). This vacuum expectation value is not invariant under *any* SU(2) transformation. This means that there is *no* unbroken subgroup, so we expect three Goldstone bosons and all three of the gauge bosons will acquire a mass.

We expand Φ^i about its vacuum expectation value ('vev')

$$\Phi = \frac{1}{\sqrt{2}}\begin{pmatrix} \phi_1 - i\phi_2 \\ v + H + i\phi_0 \end{pmatrix}.$$

The ϕ_a, $a = 0\ldots2$ are the three Goldstone bosons and H is the physical Higgs scalar. All of these fields have zero vev. If we insert this expansion into the potential (55) then we find that we only get a mass term for the Higgs field, with value $m_H = \sqrt{2}\mu$.

For simplicity, we move directly into the unitary gauge by setting all the three ϕ_a to zero. In this gauge $D_\mu\Phi$ may be written

$$D_\mu\Phi = \frac{1}{\sqrt{2}}\left(\partial_\mu\begin{pmatrix} 0 \\ H \end{pmatrix} + i\frac{g}{2}\begin{pmatrix} W^0_\mu & \sqrt{2}W^-_\mu \\ \sqrt{2}W^+_\mu & -W^0_\mu \end{pmatrix}\begin{pmatrix} 0 \\ v + H \end{pmatrix}\right),$$

where we have introduced the notation $W^+_\mu = (W^1_\mu + iW^2_\mu)/\sqrt{2}$, and used the explicit form for the generators of SU(2) in the 2×2 representation given by (17). The term $|D_\mu\Phi|^2$ then becomes

$$|D_\mu\Phi|^2 = \frac{1}{2}\partial_\mu H\,\partial^\mu H + \frac{1}{4}g^2v^2\left(W^+_\mu W^{-\,\mu} + \frac{1}{2}W^0_\mu W^{0\,\mu}\right)$$
$$+\frac{1}{4}g^2H^2\left(W^+_\mu W^{-\,\mu} + \frac{1}{2}W^0_\mu W^{0\,\mu}\right) + \frac{1}{2}g^2vH\left(W^+_\mu W^{-\,\mu} + \frac{1}{2}W^0_\mu W^{0\,\mu}\right). \tag{56}$$

We see from the terms quadratic in W_μ that all three of the gauge bosons have acquired a mass

$$M_W = \frac{gv}{2}.$$

6   The electroweak model of leptons

Only one or two modifications are needed to the model described at the end of the last lecture to obtain the Glashow-Weinberg-Salam (GWS) (Glashow 1961, Weinberg 1967, Salam 1968) model of electroweak interactions. This was the first model that successfully unified different forces of Nature.

In this lecture we shall consider only leptons as matter fields, deferring the introduction of hadrons to the next lecture.

6.1   Left- and right- handed fermions

The weak interactions are known to violate parity – they are not symmetric under interchange of left-helicity and right-helicity fermions.

A Dirac field, ψ, representing a fermion, can be expressed as the sum of a left-handed part, ψ_L, and a right-handed part, ψ_R,

$$\psi = \psi_L + \psi_R, \tag{57}$$

where

$$\begin{aligned} \psi_L &= P_L \psi, \\ P_L &= \frac{(1 - \gamma_5)}{2}, \end{aligned} \tag{58}$$

$$\begin{aligned} \psi_R &= P_R \psi, \\ P_R &= \frac{(1 + \gamma_5)}{2}. \end{aligned} \tag{59}$$

P_L and P_R are projection operators in the sense that

$$P_L P_L = P_L, \quad P_R P_R = P_R, \quad \text{and } P_L P_R = 0. \tag{60}$$

They project out the left-handed (negative) and right-handed (positive) helicity states of the fermion respectively.

The kinetic term of the Dirac Lagrangian and the interaction term of a fermion with a vector field can also be written as a sum of two terms each involving only one helicity.

$$\overline{\psi} \gamma^\mu \partial_\mu \psi = \overline{\psi_L} \gamma^\mu \partial_\mu \psi_L + \overline{\psi_R} \gamma^\mu \partial_\mu \psi_R, \tag{61}$$

$$\overline{\psi} \gamma^\mu A_\mu \psi = \overline{\psi_L} \gamma^\mu A_\mu \psi_L + \overline{\psi_R} \gamma^\mu A_\mu \psi_R. \tag{62}$$

On the other hand, a mass term mixes the two helicities

$$m \overline{\psi} \psi = m \overline{\psi_L} \psi_R + m \overline{\psi_R} \psi_L.$$

Thus, if the fermions are massless, we can treat the left-handed and right-handed helicities as separate particles.

We can understand this physically from the fact that if a fermion is massive and is moving along the *positive* z-axis along which its spin has a *positive* component, so that the helicity is *positive* in this frame, one can always boost into a frame in which the fermion is moving along the *negative* z-axis, but the component of spin is unchanged. In the new frame the helicity will be *negative*. On the other hand if the particle is massless and travels with the speed of light no such boost is possible and in that case the helicity is a good quantum number.

As charged weak interactions are observed to occur only between left-helicity fermions, we consider a weak isospin, SU(2), under which the left-handed leptons transform as a doublet

$$l_L = \begin{pmatrix} \nu \\ e \end{pmatrix}_L,$$

under this SU(2), but the right-handed electron e_R is a singlet. (Until recently there was *no* evidence of right-handed neutrinos and the standard model was developed assuming that these particles did not exist. Clearly recent evidence for neutrino oscillations points to physics beyond the standard model.)

Since this separation of the electron into its left- and right-handed helicity only makes sense for a massless electron we also need to assume that the electron *is* massless in the exact SU(2) limit and that the mass for the electron arises as a result of spontaneous symmetry breaking in the same way that the masses for the gauge bosons arise.

6.2   SU(2)×U(1)

The GWS model describes both the weak and electromagnetic interactions. Thus in addition to a weak isospin gauge group, with gauge bosons W_μ^a and gauge coupling g_W, there is an Abelian U(1) gauge group with gauge boson B_μ and gauge coupling g'_W. We shall see shortly that the photon, A_μ, is not identified directly with B_μ but rather with a linear superposition of B_μ and W_μ^0.

We introduce immediately the weak mixing angle, θ_W, which measures the relative strength of the two gauge couplings, through the relation

$$g'_W = g_W \tan \theta_W. \tag{63}$$

All the weak-SU(2) multiplets (including singlets) will transform under this U(1) with a gauge coupling of

$$g_W \tan \theta_W Y,$$

where Y is known as the 'weak hypercharge' that relates the electric charge, Q, to the third component of isospin, T_3, by

$$Q = T_3 + Y. \tag{64}$$

Thus the doublet of left-handed leptons has $Y = -\frac{1}{2}$, since the neutrino ($T_3 = +\frac{1}{2}$) and electron ($T_3 = -\frac{1}{2}$) have electric charges, 0 and -1 respectively. The right-handed electron, which has zero weak-isospin has $Y = -1$.

For a weak isodoublet with weak hypercharge Y, the covariant derivative is

$$\mathbf{D}_\mu = \left(\partial_\mu + i\,g_W\,W^a_\mu \mathbf{T}^a + ig_W \tan\theta_W\,Y\,B_\mu\right), \tag{65}$$

whereas for a singlet (with non-zero electric charge) we have

$$D_\mu = (\partial_\mu + ig_W \tan\theta_W\,Y\,B_\mu). \tag{66}$$

6.3   W's, Z's and photons

The weak isodoublet of scalar particles, Φ, also transforms under the U(1) gauge group, with weak hypercharge $Y = +\frac{1}{2}$, so that it is the $T_3 = -\frac{1}{2}$ component that has zero electric charge and can therefore be endowed with a vev v. Thus, in unitary gauge where the Goldstone bosons are set to zero, the covariant derivative of Φ may be written

$$\mathbf{D}_\mu \Phi = \frac{1}{\sqrt{2}}\left(\partial_\mu + i\frac{g_W}{2}\left(\begin{array}{cc} W^0_\mu & \sqrt{2}\,W^-_\mu \\ \sqrt{2}\,W^+_\mu & -W^0_\mu \end{array}\right) + i\,\frac{g_W \tan\theta_W}{2}B_\mu\right)\left(\begin{array}{c} 0 \\ v+H \end{array}\right),$$

so that

$$|\mathbf{D}_\mu \Phi|^2 = \frac{1}{2}(\partial_\mu H)^2 + \frac{g_W^2 v^2}{4}W^{+\,\mu}W^-_\mu + \frac{v^2}{8}\left(g_W\,W^0_\mu - g_W \tan\theta_W B_\mu\right)^2$$
$$+ \text{ interaction terms.} \tag{67}$$

Here we see how the SU(2) and U(1) are unified (or at least 'entangled') in the sense that the neutral gauge boson that acquires a mass through the Higgs mechanism is the linear superposition

$$Z_\mu = \cos\theta_W\,W^0_\mu - \sin\theta_W B_\mu. \tag{68}$$

Identifying the mass M_W of the charged gauge bosons as $M_W = \frac{1}{2}g_W\,v$, the mass terms become

$$M_W^2\,W^{+\,\mu}W^-_\mu + \frac{M_W^2}{\cos^2\theta_W}\frac{1}{2}Z^\mu Z_\mu.$$

The Z-boson mediates the neutral current weak interactions. These were not observed until after the development of the model. From the magnitude of amplitudes involving weak neutral currents (exchange of a Z-boson), one can infer the magnitude of the weak mixing angle, θ_W. The ratio of the mass of the Z- and W-bosons is a prediction of the standard model. More precisely, we define a quantity known as the ρ-parameter by

$$M_W^2 = \rho\,M_Z^2 \cos^2\theta_W. \tag{69}$$

In the standard model $\rho = 1$ in leading order of perturbation theory. In higher orders there is some small correction to this and the actual standard model prediction is $\rho = 1.01$, which agrees extremely well with the experimental value. Note that the ρ-parameter would be very different from one if the symmetry breaking were due to a scalar multiplet which was not a doublet of weak isospin.

The orthogonal superposition to Z,

$$A_\mu = \cos\theta_W B_\mu + \sin\theta_W W^0_\mu, \tag{70}$$

remains massless and is identified as the photon. The spontaneous symmetry breaking mechanism breaks $SU(2) \times U(1)$ down to $U(1)$. It is this surviving $U(1)$ that is identified as the $U(1)$ of electromagnetism, but it is not the $U(1)$ of the original gauge group but a set of transformations generated by a particular linear combination of the original $U(1)$ and rotations about the third axis of weak isospin.

The coupling of any particle to the photon is always proportional to

$$g_W \sin \theta_W \, (Y + T_3) \; = \; g_W \sin \theta_W \, Q.$$

Thus we can identify $g_W \sin \theta_W$ with one unit of electric charge and we have a relationship between the weak coupling, g_W, and the electron charge, e,

$$e \; = \; g_W \sin \theta_W. \tag{71}$$

6.4   Weak interactions of leptons

If we now look at the fermionic terms (an isodoublet of left-handed fermions with hypercharge $Y = -\frac{1}{2}$, l_L and a right-handed electron with weak hypercharge $Y = -1$), the fermionic part of the Lagrangian is

$$\mathcal{L}_{Fermi} \; = \; i\overline{l_L}^T \gamma^\mu D_\mu \, l_L \; + \; i\overline{e_R} \gamma^\mu D_\mu \, e_R, \tag{72}$$

where the covariant derivatives in the two terms are replaced by (65) and (66) respectively. This gives the following interaction terms between the leptons and the gauge bosons:

$$- \frac{g_W}{2} \begin{pmatrix} \nu_L \\ e_L \end{pmatrix}^T \gamma^\mu \left(\begin{pmatrix} W^0_\mu & \sqrt{2} W^-_\mu \\ \sqrt{2} W^+_\mu & -W^0_\mu \end{pmatrix} - \tan \theta_W \, B_\mu \right) \begin{pmatrix} \nu_L \\ e_L \end{pmatrix}$$

$$- \, i \, g_W \tan \theta_W \, \overline{e_R} \gamma^\mu B_\mu e_R.$$

Expanding this out, using the physical particles Z_μ and A_μ in place of B_μ and W^0_μ and writing out the projection operators for left- and right-handed fermions, ((58) and (59)), we obtain the following interactions:

(a) A coupling of the charged vector bosons $W^\pm$, which mediate transitions between neutrinos and electrons with an interaction term

$$- \frac{g_W}{2\sqrt{2}} \overline{\nu} \gamma^\mu \left(1 - \gamma^5 \right) e \, W^-_\mu + \text{h.c.}$$

(h.c. means 'hermitian conjugate' and gives the interaction involving an emitted W^+_μ where the incoming particle is a neutrino and the outgoing particle is an electron.)

(b) The usual coupling of the electron with the photon (using the relation (71)),

$$g_W \sin \theta_W \, \overline{e} \gamma^\mu e \, A_\mu.$$

Note that the left- and right-handed electrons have exactly the same coupling to the photon so that the electromagnetic coupling turns out to be purely vector (i.e. no γ^5 term).

(c) The coupling of neutrinos to the neutral weak gauge boson, Z_μ,

$$-\frac{g_W}{4\cos\theta_W}\,\bar{\nu}\gamma^\mu\left(1-\gamma^5\right)\nu\,Z_\mu.$$

(d) The coupling of both the left- and right-handed electron to the Z

$$\frac{g_W}{4\cos\theta_W}\,\bar{e}\left(\gamma^\mu\left(1-\gamma^5\right)-4\sin^2\theta_W\gamma^\mu\right)e\,Z_\mu.$$

We can include other lepton families, the muon and its neutrino, and the tau-lepton with its neutrino, simply as copies of what we have for the electron and its neutrino. For each family we have a weak isodoublet of left-handed leptons and a right-handed isosinglet for the charged lepton.

Thus, the mechanism which determines the decay of the muon (μ) is one in which the muon converts into its neutrino and emits a charged W^-, which then decays into an electron and (anti-)neutrino. The Feynman diagram is

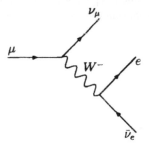

The amplitude for this process is given by the product of the vertex rules for the emission (or absorption) of a W^- with a propagator for the W-boson between them. Up to corrections of order m_μ^2/M_W^2, we may neglect the effect of the term $q^\mu q^\nu/M_W^2$ in the W-boson propagator, so that we have

$$\left(-i\frac{g_W}{2\sqrt{2}}\,\bar{\nu}_\mu\gamma^\rho(1-\gamma^5)\mu\right)\left(\frac{-ig_{\rho\sigma}}{q^2-M_W^2}\right)\left(-i\frac{g_W}{2\sqrt{2}}\,\bar{e}\gamma^\sigma(1-\gamma^5)\nu_e\right),$$

where q is the momentum transferred from the muon to its neutrino. Since this is negligible in comparison with M_W we may neglect it and the expression for the amplitude simplifies to

$$i\frac{g_W^2}{8M_W^2}\,\bar{\nu}_\mu\gamma^\rho(1-\gamma^5)\mu\,\bar{e}\gamma_\rho(1-\gamma^5)\nu_e.$$

Before the development of this model, weak interactions were described by the 'four-fermi model' with a weak interaction Hamiltonian given by

$$\mathcal{H}_{ijkl}\;=\;\frac{G_F}{\sqrt{2}}\,\bar{\psi}_i\gamma^\mu(1-\gamma^5)\psi_j\,\bar{\psi}_k\gamma_\mu(1-\gamma^5)\psi_l.$$

We now recognize this as an effective low-energy Hamiltonian which may be used when the energy scales involved in the weak process are negligible compared with the mass of the W-boson. The Fermi coupling constant, G_F, is related to the electric charge, e, the W-mass and the weak mixing angle by

$$G_F = \frac{e^2}{4\sqrt{2}\, M_W^2 \, \sin^2 \theta_W}. \tag{73}$$

This gives us a value for G_F,

$$G_F = 1.16 \times 10^{-5} \text{ GeV}^{-2},$$

which is very close to the value of 1.17×10^{-5} GeV^{-2}, measured from the lifetime of the muon.

We see that the weak interactions are 'weak', not because the coupling is particularly small (the SU(2) gauge coupling is about twice as large as the electromagnetic coupling), but because the exchanged boson is very massive, so that the Fermi coupling constant of the four-fermi theory is very small. The large mass of the W-boson is also responsible for the fact that the weak interactions are short range (of order 10^{-18} m).

In the standard model, however, we also have neutral weak currents. Thus, for example, we can have elastic scattering of muon-type neutrinos against electrons via the exchange of the Z-boson. The Feynman diagram for such a process is

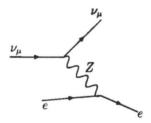

6.5   Fermion masses – Yukawa couplings

We cannot have an explicit mass term for the electrons since a mass term mixes left-handed and right-handed fermions and we have assigned these to different multiplets of weak isospin.

We can, however, have an interaction between the left-handed lepton doublet, the right-handed electron (or muon or tau-lepton) and the scalar doublet Φ. Such an interaction is called a 'Yukawa interaction' and is written

$$\mathcal{L}_{\text{Yukawa}} = -G_e \, \overline{L_L^i} \, \Phi_i e_R \; + \; \text{h.c.} \tag{74}$$

Note that this term has zero weak hypercharge. In the unitary gauge this is

$$-\frac{G_e}{\sqrt{2}} \begin{pmatrix} \overline{\nu_L} \\ \overline{e_L} \end{pmatrix}^T \begin{pmatrix} 0 \\ v + H \end{pmatrix} e_R \; + \; \text{h.c.}$$

The part proportional to the vev is simply

$$-\frac{G_e v}{\sqrt{2}} \left(\overline{e_L}\, e_R + \overline{e_R}\, e_L \right) \; = \; \frac{G_e v}{\sqrt{2}} \, \bar{e}e,$$

and we see that the electron has acquired a mass, which is proportional to the vev of the scalar field. This immediately gives us a relation for the Yukawa coupling in terms of the electron mass, m_e, and the W-mass, M_W,

$$G_e = g_W \frac{m_e}{\sqrt{2} M_W}.$$

There is, moreover, a Yukawa coupling between the electron and the scalar Higgs field

$$- g_W \frac{m_e}{2 M_W} \bar{e} H e.$$

Note that there is *no* coupling between the neutrino and the Higgs (and of course no neutrino mass term).

6.6 Classifying the free parameters

The free parameters in the GWS model for one generation of leptons are

- The two gauge couplings for the SU(2) and U(1) gauge groups, g_W and g_W'.

- The two parameters, μ and λ in the scalar potential $V(\Phi)$.

- The Yukawa coupling constant, G_e.

It is convenient to replace these five parameters by five parameters which are directly measurable experimentally, namely e, $\sin \theta_W$, and three masses m_H, M_W, m_e. The relation between these 'physical' parameters and the parameters that one writes down in the initial Lagrangian are

$$\tan \theta_W = \frac{g_W'}{g},$$

$$e = g_W \sin \theta_W,$$

$$m_H = \sqrt{2} \mu,$$

$$M_W = \frac{g_W \mu}{2 \sqrt{\lambda}},$$

$$m_e = G_e \frac{\mu}{\sqrt{\lambda}}.$$

Note that when we add more generations of leptons we acquire one more parameter for each added generation, namely the Yukawa coupling, which determines the mass of the charged lepton.

In terms of these measured quantities, the Z-mass, M_Z, and the Fermi-coupling, G_F are *predictions* (although historically G_F was known for many years before the discovery of the W-boson and its value was used to predict the W-mass).

For completeness, a full set of Feynman rules (in unitary gauge) for the case of a single family of leptons is given in the Appendix to this lecture.

7   Electroweak interactions of hadrons

7.1   One generation of quarks

We incorporate hadrons into the GWS model of electroweak interactions by adding weak isodoublets of left-handed quarks and weak isosinglets of right-handed quarks. A single generation consists of a u-quark and a d-quark. Remember that each of these is a triplet of colour SU(3) – we suppress this colour index for the moment but we must bear in mind that as far as the electroweak interactions are concerned we are really adding three copies of each of these quarks.

Thus we have an isodoublet

$$q_L = \begin{pmatrix} u \\ d \end{pmatrix}_L$$

and two isosinglets, u_R and d_R.

These two quarks have electric charges of $+\frac{2}{3}$ and $-\frac{1}{3}$ respectively, so we adjust the weak hypercharge, Y, accordingly. The isodoublet, q_L has $Y = \frac{1}{6}$, whereas u_R and d_R have $Y = +\frac{2}{3}$ and $Y = -\frac{1}{3}$ respectively (there is *no* reason why the right-handed components of the u- and d-quarks should have the same weak hypercharge, although the left-handed components *must* have the same weak hypercharge since they transform into each other under weak isospin transformations).

The covariant derivatives are therefore given by:

$$D_\mu q_L = \left(\partial_\mu + i g_W W_\mu^a T^a + i \frac{1}{6} g_W \tan \theta_W B_\mu \right) q_L, \tag{75}$$

$$D_\mu u_R = \left(\partial_\mu + i \frac{2}{3} g_W \tan \theta_W B_\mu \right) u_R, \tag{76}$$

$$D_\mu d_R = \left(\partial_\mu - i \frac{1}{3} g_W \tan \theta_W B_\mu \right) d_R. \tag{77}$$

These lead to the following interactions with the gauge bosons:

(a) The coupling of the charged vector bosons $W^\pm$, which mediate transitions between u- and d-quarks are analogous to the lepton case. The interaction term is

$$-\frac{g_W}{2\sqrt{2}} \bar{u} \gamma^\mu \left(1 - \gamma^5 \right) d W_\mu^- + \text{h.c.}$$

(b) The couplings of the quarks with the photon give us the required quark charges,

$$-\frac{2}{3} g_W \sin \theta_W \bar{u} \gamma^\mu u A_\mu + \frac{1}{3} g_W \sin \theta_W \bar{d} \gamma^\mu d A_\mu.$$

(c) The coupling of the quarks to the Z can be written in the general form

$$-\frac{g_W}{2 \cos \theta_W} \bar{q}_i \left(T_i^3 \gamma^\mu \left(1 - \gamma^5 \right) - 2 Q_i \sin^2 \theta_W \gamma^\mu \right) q_i Z_\mu,$$

where quark i has third component of weak isospin T_i^3 and electric charge Q_i.

Once again, we can use these vertices to calculate weak interactions at the quark level. This allows us, for example, to calculate the total decay width of the Z- or W-boson, by calculating the decay width into all possible quarks and leptons. However, for inclusive processes, in which we trigger on known initial or final state hadrons, information is needed about the probability to find a quark with given properties inside an initial hadron or the probability that a quark with given properties will decay ('fragment') into a final state hadron.

7.2 Quark masses

The quarks are also assumed to be massless in the symmetry limit and acquire a mass through the spontaneous symmetry breaking mechanism, via their Yukawa coupling with the scalars. The interaction term

$$- G_d \, \overline{q_L}^i \, \Phi_i \, d_R \; + \; \text{h.c.}$$

gives a mass, m_d, to the d-quark, when we replace Φ_i by its vev. This mass is given by

$$m_d \; = \; \frac{G_d}{\sqrt{2}} v \; = \; \sqrt{2} \frac{G_d \, M_W}{g_W}.$$

However, this does *not* generate a mass for the u-quark.

In the case of SU(2) there is a second way in which we can construct an invariant from such a Yukawa interaction. This is through the term

$$- G_u \, \epsilon_{ij} \overline{q_L}^i \, \Phi^{\dagger j} u_R + \; \text{h.c.}, \quad (i, j \, = \, 1, 2), \tag{78}$$

where ϵ_{ij} is the two-dimensional antisymmetric tensor. Note that this term also has zero weak hypercharge as required by the U(1) symmetry. This term does indeed give a mass m_u to the u-quark, where

$$m_u \; = \; \frac{G_u}{\sqrt{2}} v \; = \; \sqrt{2} \frac{G_u \, M_W}{g}.$$

The Higgs scalar couples to both the u-quark and the d-quark, with interaction terms

$$- g_W \frac{m_u}{2 \, M_W} \bar{u} \, H \, u \; - \; g_W \frac{m_d}{2 \, M_W} \bar{d} \, H \, d.$$

7.3 Adding another generation

The second generation of quarks consists of a c-quark, which has electric charge $+\frac{2}{3}$ and an s-quark, with electric charge $-\frac{1}{3}$. We can just add a copy of the left-handed isodoublet and copies of the right-handed singlets in order to include this generation.

The only difference would be in the Yukawa interaction terms where the coupling constants are chosen to reproduce the correct masses for the new quarks. But in this case there is a further complication. It is possible to write down Yukawa terms which mix quarks of different generations, e.g.

$$- G_{ds} \begin{pmatrix} \overline{u_L} \\ \overline{d_L} \end{pmatrix}^T \Phi \, s_R.$$

This term gives rise to a mass mixing between the d-quark and s-quark.

The 'physical' particles are those that diagonalize the mass matrix. It is therefore more convenient to assume that the mass matrix does *not* contain off-diagonal terms and that the quarks involved in the Yukawa terms are the physical quarks such that there is *no* Yukawa mixing between quarks of different generations.

This means that the quarks which couple to the gauge bosons *are*, in general, superposition of physical quarks. Thus we write the two isodoublets of left-handed quarks as

$$\begin{pmatrix} u \\ \tilde{d} \end{pmatrix}_L ,$$

and

$$\begin{pmatrix} c \\ \tilde{s} \end{pmatrix}_L ,$$

where $\tilde{d}$ and $\tilde{s}$ are related to the physical d-quark and s-quark by

$$\begin{pmatrix} \tilde{d} \\ \tilde{s} \end{pmatrix} = \mathbf{V_C} \begin{pmatrix} d \\ s \end{pmatrix}, \tag{79}$$

where $\mathbf{V_C}$ is a unitary 2×2 matrix.

Terms which are diagonal in the quarks are unaffected by this unitary transformation of the quarks. Thus the coupling to photons or Z-bosons is the same whether written in terms of $\tilde{d}$, $\tilde{s}$ or simply s, d. We return to this later.

On the other hand the coupling to the charged gauge bosons is

$$-\frac{g_W}{2\sqrt{2}} \bar{u}\gamma^\mu(1-\gamma^5)\tilde{d}\, W_\mu^- \; - \; \frac{g_W}{2\sqrt{2}} \bar{c}\gamma^\mu(1-\gamma^5)\tilde{s}\, W_\mu^- \; + \text{h.c.},$$

which we may write as

$$-\frac{g_W}{2\sqrt{2}} \begin{pmatrix} \bar{u} \\ \bar{c} \end{pmatrix}^T \gamma^\mu(1-\gamma^5)\mathbf{V_C} \begin{pmatrix} d \\ s \end{pmatrix} W_\mu^- \; + \text{h.c.}$$

We see now that there is no point in applying the transformation to *both* the $T^3 = +\frac{1}{2}$ and $T^3 = -\frac{1}{2}$ quarks, since we can absorb the product of two unitary matrices into what we call $\mathbf{V_C}$ here.

The most general 2×2 unitary matrix may be written as

$$\begin{pmatrix} e^{-i\gamma} & \\ & 1 \end{pmatrix} \begin{pmatrix} \cos\theta_C & \sin\theta_C \\ -\sin\theta_C & \cos\theta_C \end{pmatrix} \begin{pmatrix} e^{i\alpha} & \\ & e^{i\beta} \end{pmatrix}.$$

We have set one of the phases to unity since we can always absorb an overall phase by adjusting the remaining phases, α, β, and γ.

The phases, α, β, γ can be absorbed by performing a global phase transformation on the d-, s- and u-quarks respectively. This again has no effect on the neutral terms. Thus the most general observable unitary matrix is given by

$$\mathbf{V_C} = \begin{pmatrix} \cos\theta_C & \sin\theta_C \\ -\sin\theta_C & \cos\theta_C \end{pmatrix}, \tag{80}$$

where θ_C is the Cabibbo angle (Cabbibo 1983).

In terms of the physical quarks, the charged gauge boson interaction terms are

$$-\frac{g_W}{2\sqrt{2}} \Big(\cos\theta_C \,\bar{u}\,\gamma^\mu(1-\gamma^5)\,d \;+\; \sin\theta_C\,\bar{u}\,\gamma^\mu(1-\gamma^5)\,s$$
$$+\; \cos\theta_C\,\bar{c}\,\gamma^\mu(1-\gamma^5)\,s \;-\; \sin\theta_C\,\bar{c}\,\gamma^\mu(1-\gamma^5)\,d \Big)\,W_\mu^- \;+\; \text{h.c.} \tag{81}$$

This means that the u-quark can undergo weak interactions in which it is converted into an s-quark, with an amplitude that is proportional to $\sin\theta_C$. It is this that gives rise to strangeness violating weak interaction processes, such as the leptonic decay of K^- into a muon and antineutrino. The Feynman diagram for this process is

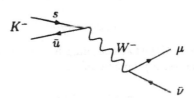

7.4 The GIM mechanism

Although there are charged weak interactions that violate strangeness conservation, there are no known neutral weak interactions which violate strangeness. For example, the K^0 does not decay into a muon pair or two neutrinos (branching ratio $< 10^{-5}$). This means that the the Z-boson only interacts with quarks of the same flavour. We can see this by noting that the Z-boson interaction terms are unaffected by a unitary transformation.

The Z-boson interactions with d- and s-quarks are proportional to

$$\bar{d}\tilde{d} + \bar{s}\tilde{s}.$$

(We have suppressed the γ-matrices which act between the fermion fields). Writing this out in terms of the physical quarks we get

$$\cos^2\theta_C\,\bar{d}d \;+\; \sin\theta_C\,\cos\theta_C\,\bar{s}d \;+\; \cos\theta_C\,\sin\theta_C\,\bar{d}s \;+\; \sin^2\theta_C\,\bar{s}s$$
$$+\; \cos^2\theta_C\,\bar{s}s \;-\; \sin\theta_C\,\cos\theta_C\,\bar{d}s \;-\; \cos\theta_C\,\sin\theta_C\,\bar{s}d \;+\; \sin^2\theta_C\,\bar{d}d.$$

We see that the cross-terms cancel out and we are left with simply

$$\bar{d}d \;+\; \bar{s}s.$$

This cancellation is known as the 'GIM' (Glashow-Iliopoulous-Maiani) mechanism (Glashow *et al* 1970). It was used to predict the existence of the c-quark.

There can be a small contribution to strangeness changing neutral processes from higher order corrections in which we do not exchange a Z-boson, but two charged W-bosons. The Feynman diagrams for such a contribution to the leptonic decay of a K^0 (which consists of a d-quark and an s-antiquark) are

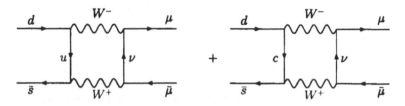

These diagrams differ in the flavour of the internal quark which is exchanged, being a u-quark in the first diagram and a c-quark in the second. Both of these diagrams are allowed because of the Cabibbo mixing. The first of these diagrams gives a contribution proportional to

$$+ \sin\theta_C \cos\theta_C,$$

which arises from the product of the two couplings involving the emission of the W-bosons. The second diagram gives a term proportional to

$$- \cos\theta_C \sin\theta_C.$$

If the c-quark and u-quark had identical mass then these two contributions would cancel precisely. However, because the c-quark is much more massive than the u-quark, there is some residual contribution. This was used to limit the mass of the c-quark to < 5 GeV, before it was discovered.

7.5   Adding a third generation

Adding a third generation is achieved in the same way. In this case the three weak isodoublets of left-handed fermions are

$$\begin{pmatrix} u \\ \tilde{d} \end{pmatrix}, \quad \begin{pmatrix} c \\ \tilde{s} \end{pmatrix}, \quad \begin{pmatrix} t \\ \tilde{b} \end{pmatrix},$$

where $\tilde{d}$, $\tilde{s}$ and $\tilde{b}$ are related to the physical d-, s- and b-quarks by

$$\begin{pmatrix} \tilde{d} \\ \tilde{s} \\ \tilde{b} \end{pmatrix} = \mathbf{V}_{\text{CKM}} \begin{pmatrix} d \\ s \\ b \end{pmatrix}. \tag{82}$$

The 3×3 unitary matrix $\mathbf{V}_{\text{CKM}}$ is the 'Cabibbo-Kobayashi-Maskawa (CKM) matrix' (Kobayashi and Maskawa 1972). Once again it only affects the charged weak processes in which a W-boson is exchanged. For this reason the elements are written as

$$\begin{pmatrix} V_{ud} & V_{us} & V_{ub} \\ V_{cd} & V_{cs} & V_{cb} \\ V_{td} & V_{ts} & V_{tb} \end{pmatrix}.$$

A 3×3 unitary matrix can have nine independent parameters (counting the real and imaginary parts of a complex element as two parameters). In this case there are six possible fermions involved in the charged weak processes and so we can have five relative phase transformations, thereby absorbing five of the nine parameters.

This means that whereas the Cabibbo matrix only has one parameter (the Cabibbo angle, θ_C) the CKM matrix has four independent parameters. If the CKM matrix were real it would only have three independent parameters. This means that in the case of the CKM matrix some of the elements may be complex. The four independent parameters can be thought of as three mixing angles between the three pairs of generations and a complex phase.

The requirement of unitarity puts various constraints on the elements of the CKM matrix. For example we have

$$V_{ud} V_{ub}^* + V_{cd} V_{cb}^* + V_{td} V_{tb}^* = 0. \tag{83}$$

This can be represented as a triangle in the complex plane known as the 'unitarity triangle'

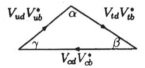

The angles of the triangle are related to ratios of elements of the CKM matrix

$$\alpha = -\arg\left\{\frac{V_{td} V_{tb}^*}{V_{ud} V_{ub}^*}\right\}, \tag{84}$$

$$\beta = -\arg\left\{\frac{V_{td} V_{tb}^*}{V_{cd} V_{cb}^*}\right\}, \tag{85}$$

$$\gamma = -\arg\left\{\frac{V_{ud} V_{ub}^*}{V_{cd} V_{cb}^*}\right\}. \tag{86}$$

A popular representation of the CKM matrix is due to Wolfenstein (Wolfenstein 1983) and uses parameters A, which is assumed to be of order unity, a complex number, $(\rho + i\eta)$, and a small number, λ, which is approximately equal to $\sin\theta_C$. In terms of these parameters the CKM matrix is written

$$\mathbf{V}_{\mathrm{CKM}} = \begin{pmatrix} 1 - \lambda^2/2 & \lambda & A\lambda^3(\rho - i\eta) \\ -\lambda & 1 - \lambda^2/2 & A\lambda^2 \\ A\lambda^3(1 - \rho - i\eta) & -A\lambda^2 & 1 \end{pmatrix} + \mathcal{O}(\lambda^4). \tag{87}$$

We see that whereas the W-bosons can mediate a transition between a u-quark and a b-quark (V_{ub}) or between a t-quark and a d-quark (V_{td}), the amplitude for such transitions are suppressed as the cube of the small quantity which determines the amplitude for transitions between the first and second generations, λ. The $\mathcal{O}(\lambda^4)$ corrections are needed to ensure the unitarity of the CKM matrix and these corrections have several matrix elements which are complex.

7.6 CP violation

The possibility that some of the elements of the CKM matrix may be complex provides a mechanism for the violation of CP conservation. Violation of CP conservation has been observed in the K^0, $\overline{K}^0$ system, and is currently being investigated for B-mesons.

Higher order corrections to the masses of B^0 and $\overline{B}^0$, give rise to mixing between the two states.

Since B^0 and $\overline{B}^0$ decay their 'masses' are considered to be complex and we write them in terms of a 2×2 matrix

$$\Lambda \;=\; M + i\Gamma,$$

where both M and Γ are Hermitian 2×2 matrices, whose off-diagonal elements correspond to the mixing.

A typical weak interaction contribution to the mass-mixing term, i.e. the off-diagonal elements of Λ, are given by the Feynman diagrams

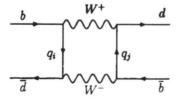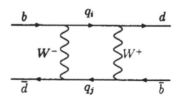

Note that on the left we have a B^0, consisting of a b-quark and an d-antiquark, whereas on the right we have a $\overline{B}^0$ consisting of an d-quark and a b-antiquark. The internal quarks q_i, q_j can be u-, c- or t-type. The contribution from these graphs is proportional to

$$\sum_{i,j=u,c,t} V_{ib}V_{id}^* V_{jb}V_{jd}^* \, a_{ij}.$$

If all the quark masses were equal then the coefficients a_{ij} would be equal and the contribution would vanish by virtue of the unitarity relation (83). The mass differences prevent this contribution from vanishing. Indeed the dominant contribution arises from the case where the internal fermion is a t-quark, as this is by far the most massive.

If the couplings at all the vertices were real, then the real part of these diagrams would contribute to the off-diagonal components of the mass-matrix, M, whereas the dispersive parts would contribute to the off-diagonal components of the decay matrix, Γ. However, the complex phases which arise in some of the couplings, as determined by the CKM matrix, means that a small part of the dispersive part of these diagrams contributes to M, whereas a small part of the real part contributes to Γ.

The upshot of this is that the mass eigenstates are given by

$$|B_1\rangle = p|B^0\rangle + q|\overline{B^0}\rangle,$$

$$|B_2\rangle = p|B^0\rangle - q|\overline{B^0}\rangle,$$

where the ratio $|q|/|p|$ is different from unity. These two states have a small mass difference, ΔM.

On the other hand, since the CP operator acts on the B^0 and $\overline{B^0}$ states as

$$CP|B^0\rangle = -|\overline{B^0}\rangle,$$

the CP eigenstates are given by

$$\frac{1}{\sqrt{2}}\left(|B^0\rangle \mp |\overline{B^0}\rangle\right),$$

(with eigenvalue ± 1 respectively). We see that the physical (mass) eigenstates are *not* eigenstates of CP, so that as they propagate they oscillate between $CP = +1$ and $CP = -1$ states, thus giving rise to observable CP violation.

Appendix: Feynman rules for GWS model with one generation of leptons (in the unitary gauge)

Propagators:
All propagators carry momentum p.

$$-i\left(g_{\mu\nu} - p_\mu p_\nu/M_W^2\right)/(p^2 - M_W^2)$$

$$-i\left(g_{\mu\nu} - p_\mu p_\nu/M_Z^2\right)/(p^2 - M_Z^2)$$

$$-i\,g_{\mu\nu}/p^2$$

$$i\left(\gamma\cdot p + m_e\right)/(p^2 - m_e^2)$$

$$i\,\gamma\cdot p/p^2$$

$$i/(p^2 - m_H^2)$$

Three-point gauge-boson couplings:
All momenta are incoming

$$i\,g_W\,\sin\theta_W\left((p_1 - p_2)_\rho\, g_{\mu\nu} + (p_2 - p_3)_\mu\, g_{\nu\rho} + (p_3 - p_1)_\nu\, g_{\rho\mu}\right)$$

$$i\,g_W\,\cos\theta_W\left((p_1 - p_2)_\rho\, g_{\mu\nu} + (p_2 - p_3)_\mu\, g_{\nu\rho} + (p_3 - p_1)_\nu\, g_{\rho\mu}\right)$$

Four-point gauge-boson couplings:

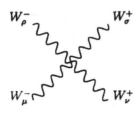

$i\, g_W^2 \left(2g_{\mu\rho}\, g_{\nu\sigma} - g_{\mu\nu}\, g_{\rho\sigma} - g_{\mu\sigma}\, g_{\nu\rho}\right)$

$i\, g_W^2 \cos^2\theta_W \left(2g_{\mu\nu}\, g_{\rho\sigma} - g_{\mu\rho}\, g_{\nu\sigma} - g_{\mu\sigma}\, g_{\nu\rho}\right)$

$i\, g_W^2 \sin^2\theta_W \left(2g_{\mu\nu}\, g_{\rho\sigma} - g_{\mu\rho}\, g_{\nu\sigma} - g_{\mu\sigma}\, g_{\nu\rho}\right)$

$i\, g_W^2 \cos\theta_W \sin\theta_W \left(2g_{\mu\nu}\, g_{\rho\sigma} - g_{\mu\rho}\, g_{\nu\sigma} - g_{\mu\sigma}\, g_{\nu\rho}\right)$

Three-point couplings with Higgs scalars:

$$-\tfrac{3}{2}\, i\, g_W\, m_H^2/M_W$$

$$-\tfrac{1}{2}\, i\, g_W\, m_e/M_W$$

$$i\, g_W\, M_W\, g_{\mu\nu}$$

$$i\, (g_W/\cos^2\theta_W)\, M_W\, g_{\mu\nu}$$

Four-point couplings with Higgs scalars:

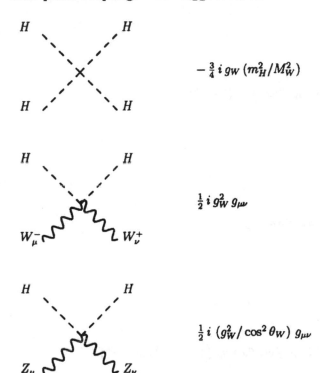

$$-\tfrac{3}{4}\, i\, g_W \left(m_H^2 / M_W^2\right)$$

$$\tfrac{1}{2}\, i\, g_W^2\, g_{\mu\nu}$$

$$\tfrac{1}{2}\, i\, \left(g_W^2 / \cos^2\theta_W\right) g_{\mu\nu}$$

Fermion interactions with gauge bosons:

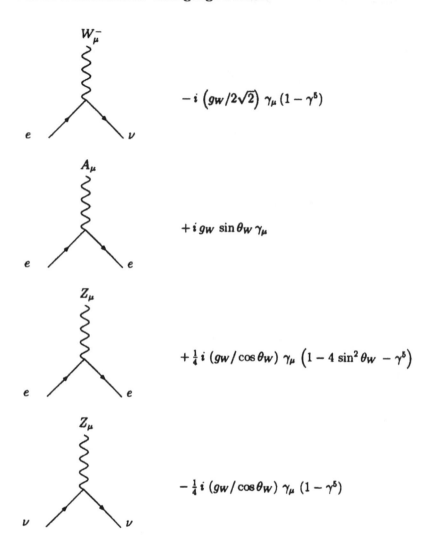

$$-i\left(g_W/2\sqrt{2}\right)\gamma_\mu\left(1-\gamma^5\right)$$

$$+i\,g_W\,\sin\theta_W\,\gamma_\mu$$

$$+\tfrac{1}{4}i\,(g_W/\cos\theta_W)\,\gamma_\mu\left(1-4\sin^2\theta_W-\gamma^5\right)$$

$$-\tfrac{1}{4}i\,(g_W/\cos\theta_W)\,\gamma_\mu\left(1-\gamma^5\right)$$

References

Altarelli G and Parisi G, 1977, *Nucl Phys* B126 298.

Bloch F and Nordsieck A, 1937, *Phys Rev* 52 54.

Brout R and Englert F, 1964, *Phys Rev Lett* 13 321.

Cabbibo N, 1963, *Phys Rev Lett* 10 531.

Dokshitzer, Yu L, 1977, *JETP* 47 641.

Faddeev L D and Popov V N, 1967, *Phys Lett* 25B 79.

Feynman R P, 1963, *Acta Phys Polonica* 24 297.

Fritzsch H et. al., 1973, *Phys Lett* 47B 365.

Glashow S L, 1961, *Nucl Phys* 72 579.

Glashow S L et. al., 1970, *Phys Rev* D2 185.

Goldstone J, 1961, *Nuov Cim* 19 154.

Gribov V N and Lipatov L N, 1972, *Sov J Nucl Phys* 15 438.

Gross D J and Wilczek F, 1973, *Phys Rev Lett* 30 1343.

Guralnik G S et. al., 1964 *Phys Rev Lett* 13 585.

Han N and Nambu Y, 1974 *Phys Rev* D10 674.

Higgs P W, 1964a, *Phys Lett* 12 132.

Higgs P W, 1964b, *Phys Rev* 145 1156.

Kibble T W B, 1967, *Phys Rev Lett* 18 507.

Kinoshita T, 1962, *J Math Phys* 3 650.

Kobayashi M and Maskawa K, 1972, *Prog Theor Phys* 49 782.

Lee T D and Nauenberg M, 1964, *Phys Rev* 133 B1549.

Lipatov L N, 1975, *Sov. J. Nucl Phys* 20 95.

Politzer H D, 1973, *Phys Rev Lett* 30 1346.

Politzer H D, 1974, *Phys Rep* 14C 129.

Salam A, 1968, Elementary Particle Physics, *ed. N. Svartholm, Almqvist and Wiksells, Stockholm.*

't Hooft G, 1971, *Nucl Phys* B35 167.

Yang C N and Mills R L, 1954, *Phys Rev* 99 191.

Yennie D R, Frautschi S C, and Suura H, 1961, *Ann Phys, NY* 13 379.

Weinberg S, 1967, *Phys Rev Lett* 19 1264.

Wolfenstein L, 1983, *Phys Rev Lett* 51 1945.

Standard model phenomenology

R. Keith Ellis

Fermilab, USA

These lectures address the phenomenological implications of the standard model and are a sequel to the introduction to the standard model given by Douglas Ross (Ross 2004). The standard model has proved to be a remarkably robust structure which has survived almost three decades of testing essentially unscathed. The shortcomings of the standard model and the search for physics beyond it will be covered by other lecturers. It is necessary to understand the standard model to look for physics which transcends it. This is particularly true at a hadron collider, such as the Tevatron or the Large Hadron Collider (LHC), because of the crucial role of the understanding of background in the identification of signals for new physics. But, in addition, the standard model provides a wonderful playground in which we can observe an interacting quantum field theory at work.

Much of the material in these lectures was originally formulated in collaboration with James Stirling and Bryan Webber and included in our book, 'QCD and collider physics', (Ellis et al. 1996). Thus, this writeup should not be construed as a complete rendering of the content of my lectures, but rather as a complement to the material in our book.

1 Lecture I

1.1 The Lagrangian of Quantum Chromodynamics

The first part of the standard model which we consider describes the strong interactions of quarks and gluons. Even though the Lagrangian of QCD is amenable to both perturbative and non-perturbative treatment, in these lectures we will only consider the perturbative sector of the theory. The Feynman rules for perturbative QCD follow from the Lagrangian

$$\mathcal{L} = -\frac{1}{4}F^A_{\alpha\beta}F^{\alpha\beta}_A + \sum_{\text{flavours}} \bar{q}_a(i\not{D} - m)_{ab}q_b + \mathcal{L}_{\text{gauge-fixing}} , \tag{1}$$

where $F^A_{\alpha\beta}$ is the field strength tensor for the spin-1 gluon field $\mathcal{A}^A_\alpha$,

$$F^A_{\alpha\beta} = \partial_\alpha \mathcal{A}^A_\beta - \partial_\beta \mathcal{A}^A_\alpha - g f^{ABC} \mathcal{A}^B_\alpha \mathcal{A}^C_\beta . \tag{2}$$

Capital indices A, B, C run over 8 colour degrees of freedom of the gluon field. It is the third 'non-Abelian' term which distinguishes QCD from QED, giving rise to triplet and quartic gluon self-interactions and ultimately to asymptotic freedom.

The QCD coupling strength, $\alpha_S \equiv g^2/4\pi$, is defined in analogy with the fine structure constant of QED. The numbers f^{ABC} $(A, B, C = 1, ..., 8)$ are the structure constants of the SU(3) colour group. The quark fields q_a $(a = 1, 2, 3)$ are in the triplet (or fundamental) colour representation. D is the covariant derivative, which acting on fields in the fundamental and adjoint representation can be written as

$$(D_\alpha)_{ab} = \partial_\alpha \delta_{ab} + ig\left(t^C A_\alpha^C\right)_{ab}, \tag{3}$$

$$(D_\alpha)_{AB} = \partial_\alpha \delta_{AB} + ig(T^C A_\alpha^C)_{AB}, \tag{4}$$

where t and T are matrices in the fundamental and adjoint representations of SU(3), respectively:

$$[t^A, t^B] = if^{ABC}t^C, \quad [T^A, T^B] = if^{ABC}T^C, \tag{5}$$

where $(T^A)_{BC} = -if^{ABC}$. We use the metric $g^{\alpha\beta} = \text{diag}(1, -1, -1, -1)$ and set $\hbar = c = 1$. D is a symbolic notation for $\gamma^\alpha D_\alpha$. The normalization of the t matrices is

$$\text{Tr } t^A t^B = T_R \delta^{AB}, \quad T_R = \frac{1}{2}. \tag{6}$$

With this normalization the colour matrices obey the relations:

$$\sum_A t_{ab}^A t_{bc}^A = C_F \delta_{ac}, \quad C_F = \frac{N^2 - 1}{2N}, \tag{7}$$

$$\text{Tr } T^C T^D = \sum_{A,B} f^{ABC} f^{ABD} = C_A \delta^{CD}, \quad C_A = N. \tag{8}$$

Thus $C_F = \frac{4}{3}$ and $C_A = 3$ for SU(3).

1.2   Feynman rules

The derivation of the Feynman rules follows in a simple way from the Lagrangian, Eq. (1). We use the free piece of the QCD Lagrangian to obtain the inverse quark and gluon propagators. The inverse quark propagator in momentum space is obtained by setting $\partial^\alpha = -ip^\alpha$ for an incoming field. The result for the propagator is given in Table 1. The $i\varepsilon$ prescription which fixes the positions of the poles in the propagator is determined by causality, just as in QED. The gluon propagator is impossible to define without a choice of gauge. The choice

$$\mathcal{L}_{\text{gauge-fixing}} = -\frac{1}{2\lambda}\left(\partial^\alpha A_\alpha^A\right)^2 \tag{9}$$

defines *covariant gauges* with gauge parameter λ. With this choice the inverse gluon propagator is

$$\Gamma_{\{AB, \alpha\beta\}}^{(2)}(p) = i\delta_{AB}\left[p^2 g_{\alpha\beta} - (1 - \frac{1}{\lambda})p_\alpha p_\beta\right]. \tag{10}$$

You can easily check that without the gauge-fixing term this function would have no inverse. The propagator is given by the inverse of Eq. (10) and is shown in Fig. 1. The choice $\lambda = 1$ (0) determines the *Feynman (Landau)* gauge.

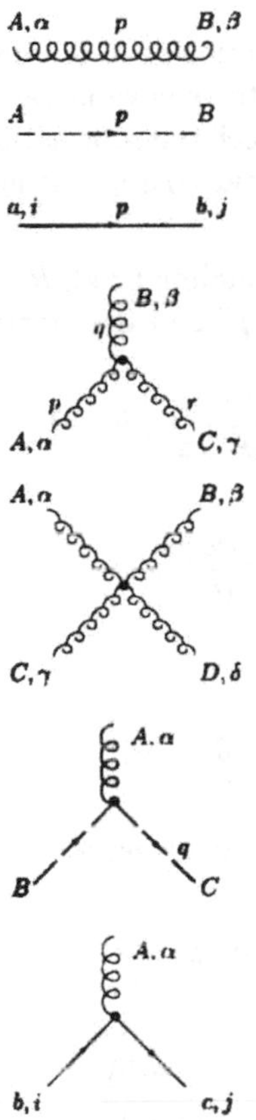

$$\delta^{AB}\left[-g^{\alpha\beta}+(1-\lambda)\frac{p^\alpha p^\beta}{p^2+i\varepsilon}\right]\frac{i}{p^2+i\varepsilon}$$

$$\delta^{AB}\frac{i}{p^2+i\varepsilon}$$

$$\delta^{ab}\frac{i}{(\not{p}-m+i\varepsilon)_{ji}}$$

$$-gf^{ABC}\left[g^{\alpha\beta}(p-q)^\gamma\right.$$
$$+g^{\beta\gamma}(q-r)^\alpha$$
$$\left.+g^{\gamma\alpha}(r-p)^\beta\right]$$
(all momenta incoming)

$$-ig^2 f^{XAC}f^{XBD}\left(g_{\alpha\beta}g_{\gamma\delta}-g_{\alpha\delta}g_{\beta\gamma}\right)$$
$$-ig^2 f^{XAD}f^{XBC}\left(g_{\alpha\beta}g_{\gamma\delta}-g_{\alpha\gamma}g_{\beta\delta}\right)$$
$$-ig^2 f^{XAB}f^{XCD}\left(g_{\alpha\gamma}g_{\beta\delta}-g_{\alpha\delta}g_{\beta\gamma}\right)$$

$$gf^{ABC}q^\alpha$$

$$-ig\left(t^A\right)_{cb}(\gamma^\alpha)_{ji}$$

Figure 1. Feynman rules for QCD in a covariant gauge.

The gauge fixing explicitly breaks gauge invariance. However, in the end physical results will be independent of the gauge. For convenience, we usually use the Feynman gauge because it minimizes the number of terms in the propagator and leads to less calculational burden.

In non-Abelian theories like QCD, the covariant gauge-fixing term must be supplemented by a *ghost term* which we do not discuss here. The ghost field, shown by dashed lines in the above table, cancels unphysical degrees of freedom of the gluon which would otherwise propagate in covariant gauges. Note that the propagators are determined from $-i\mathcal{L}$, whilst the interaction terms are determined from $i\mathcal{L}$.

1.3   Running coupling

Consider a dimensionless physical observable R which depends on a single large energy scale, Q, which is very much bigger than m, where m is any mass. Then we can set $m \to 0$ (assuming this limit exists), and dimensional analysis suggests that R should be independent of Q.

This is not true in quantum field theory. The calculation of R as a perturbation series in the coupling $\alpha_S = g^2/4\pi$ requires renormalization to remove ultraviolet divergences. This introduces a second mass scale μ – the point at which the subtractions which remove divergences are performed. Thus R depends on the ratio Q/μ and is not constant. The renormalized coupling α_S also depends on μ.

But μ is arbitrary! Therefore, if we hold the bare coupling fixed, R cannot depend on μ. Since R is dimensionless, it can only depend on Q^2/μ^2 and the renormalized coupling α_S. Hence

$$\mu^2 \frac{d}{d\mu^2} R\left(\frac{Q^2}{\mu^2}, \alpha_S\right) \equiv \left[\mu^2 \frac{\partial}{\partial \mu^2} + \mu^2 \frac{\partial \alpha_S}{\partial \mu^2} \frac{\partial}{\partial \alpha_S}\right] R = 0 . \tag{11}$$

Introducing

$$\tau = \ln\left(\frac{Q^2}{\mu^2}\right) , \quad \beta(\alpha_S) = \mu^2 \frac{\partial \alpha_S}{\partial \mu^2} , \tag{12}$$

Eq. (11) can be written as

$$\left[-\frac{\partial}{\partial \tau} + \beta(\alpha_S) \frac{\partial}{\partial \alpha_S}\right] R = 0. \tag{13}$$

This renormalization group equation is solved by defining a running coupling $\alpha_S(Q)$:

$$\tau = \int_{\alpha_S}^{\alpha_S(Q)} \frac{dx}{\beta(x)} , \quad \alpha_S(\mu) \equiv \alpha_S . \tag{14}$$

Then

$$\frac{\partial \alpha_S(Q)}{\partial \tau} = \beta(\alpha_S(Q)) , \quad \frac{\partial \alpha_S(Q)}{\partial \alpha_S} = \frac{\beta(\alpha_S(Q))}{\beta(\alpha_S)} . \tag{15}$$

and hence $R(Q^2/\mu^2, \alpha_S) = R(1, \alpha_S(Q))$. Thus all scale dependence in R comes from the running of $\alpha_S(Q)$.

We shall see that QCD is asymptotically free: $\alpha_S(Q) \to 0$ as $Q \to \infty$. Thus for large Q we can safely use perturbation theory. Then knowledge of $R(1, \alpha_S)$ to fixed order allows us to predict the variation of R with Q.

1.4   Beta function

The running of the QCD coupling α_S is determined by the β function, which has the expansion

$$\beta(\alpha_S) = -b\alpha_S^2(1 + b'\alpha_S) + \mathcal{O}(\alpha_S^4) ,$$
$$b = \frac{(11C_A - 2N_f)}{12\pi}, \quad b' = \frac{(17C_A^2 - 5C_A N_f - 3C_F N_f)}{2\pi(11C_A - 2N_f)} , \tag{16}$$

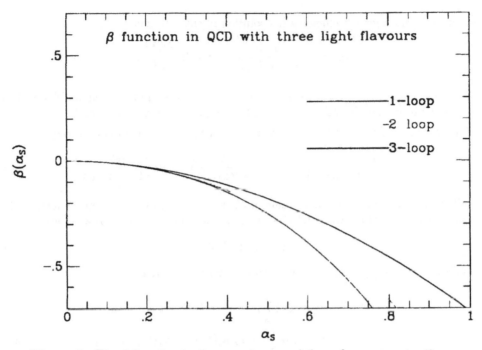

Figure 2. *The β function in the one-, two-, and three- loop approximations.*

Figure 3. *Graphs which contribute to the β function in the one loop approximation.*

where N_f is the number of 'active' light flavours. The terms in $\beta(\alpha_s)$ are known up to $\mathcal{O}(\alpha_S^5)$ (van Ritbergen *et al* 1997).

Roughly speaking, the quark loop diagram Fig. 3(a) contributes negative N_f terms in b, while the gluon loop Fig. 3(b) gives positive C_A contributions, which makes the β function negative overall. For comparison the QED β function is

$$\beta_{QED}(\alpha) = \frac{1}{3\pi}\alpha^2 + \dots \tag{17}$$

Thus the b coefficients in QED and QCD have opposite signs.

From our previous discussion, the running coupling, $\alpha_S(Q)$, satisfies Eq. (15)

$$\frac{\partial\alpha_S(Q)}{\partial\tau} = -b\alpha_S^2(Q)\left[1 + b'\alpha_S(Q)\right] + \mathcal{O}(\alpha_S^4). \tag{18}$$

Neglecting b' and higher coefficients the solution to the equation is

$$\alpha_S(Q) = \frac{\alpha_S(\mu)}{1 + \alpha_S(\mu)b\tau} \ , \quad \tau = \ln\left(\frac{Q^2}{\mu^2}\right). \tag{19}$$

As Q becomes large, $\alpha_S(Q)$ decreases to zero: this is asymptotic freedom. Notice that the sign of b is crucial. In QED, $b < 0$ and the coupling *increases* at large Q.

Including the next coefficient b' gives the following implicit equation for $\alpha_S(Q)$:

$$b\tau = \frac{1}{\alpha_S(Q)} - \frac{1}{\alpha_S(\mu)} + b'\ln\left(\frac{\alpha_S(Q)}{1 + b'\alpha_S(Q)}\right) - b'\ln\left(\frac{\alpha_S(\mu)}{1 + b'\alpha_S(\mu)}\right). \tag{20}$$

What type of terms does the solution of the renormalization group equation take into account for a physical quantity R? If we assume that R has the perturbative expansion

$$R = \alpha_S + \mathcal{O}(\alpha_S^2) \tag{21}$$

the solution $R(1, \alpha_S(Q))$ can be re-expressed in terms of $\alpha_S(\mu)$:

$$\begin{aligned}
R(1, \alpha_S(Q)) &= \alpha_S(\mu) \sum_{j=0}^{\infty} (-1)^j (\alpha_S(\mu) b\tau)^j \\
&= \alpha_S(\mu)\left[1 - \alpha_S(\mu)b\tau + \alpha_S^2(\mu)(b\tau)^2 + \ldots\right].
\end{aligned} \tag{22}$$

Thus there are logarithms of Q^2/μ^2 which are automatically resummed by using the running coupling with exactly one logarithm for every power of α_S. The higher order terms in the β-function give fewer logarithms per power of α_S.

1.5 Lambda parameter

Perturbative QCD tells us how $\alpha_S(Q)$ varies with Q, but its numerical value at a given scale has to be obtained from experiment. Nowadays we usually choose as the fundamental parameter the value of the coupling at $Q = M_Z$, which is simply a convenient reference scale large enough to be in the perturbative domain.

It is also useful to express $\alpha_S(Q)$ directly in terms of a dimensionful parameter (constant of integration) Λ:

$$\ln\frac{Q^2}{\Lambda^2} = -\int_{\alpha_S(Q)}^{\infty} \frac{dx}{\beta(x)} = \int_{\alpha_S(Q)}^{\infty} \frac{dx}{bx^2(1 + b'x + \ldots)}. \tag{23}$$

Then (if perturbation theory were the whole story) we would find that $\alpha_S(Q) \to \infty$ as $Q \to \Lambda$. More generally, Λ sets the scale at which $\alpha_S(Q)$ becomes large.

In leading order (LO) we keep only the first term in the β-function b and Eq. (23) can be solved exactly:

$$\alpha_S(Q) = \frac{1}{b\ln(Q^2/\Lambda^2)} \quad \text{(LO)}. \tag{24}$$

The most recent compilation of measurements of α_S from Bethke (Bethke 2002) is shown in Fig. 4. The evidence that $\alpha_S(Q)$ has a logarithmic fall-off with Q as predicted by Eq. (24) is persuasive.

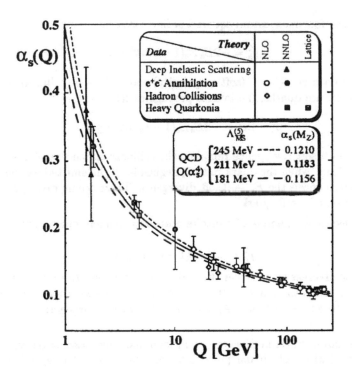

Figure 4. *The current status of α_s measurements (Bethke 2002).*

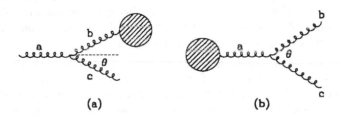

Figure 5. *Parton branching of (a) incoming and (b) outgoing partons.*

1.6 Infrared divergences

Even at high-energy when we are probing the short-distance regime, the long-distance aspects of QCD cannot be ignored. Soft or collinear gluon emission gives infrared divergences in perturbation theory. In addition, light quarks which have masses less than the QCD scale ($m_q \ll \Lambda$) also lead to divergences in the limit $m_q \to 0$. These latter are called mass singularities.

Fig. 5(a) shows an example of spacelike branching, namely gluon splitting on an incoming line. The four-momentum squared of line b is

$$p_b^2 = -2E_a E_c(1 - \cos \theta) \leq 0 \, . \tag{25}$$

which diverges both as $E_c \to 0$ (soft singularity) and as $\theta \to 0$ (collinear or mass singularity). If a and b are quarks, the inverse propagator factor is

$$p_b^2 - m_q^2 = -2E_a E_c(1 - v_a \cos \theta) \leq 0 \, . \tag{26}$$

Hence the $E_c \to 0$ soft divergence remains; the collinear enhancement becomes a divergence as $v_a \to 1$, i.e. when the quark mass is negligible. If the emitted parton c is a quark, the vertex factor cancels the $E_c \to 0$ soft divergence. There are no divergences associated with the emission of a soft quark.

Fig. 5(b) shows an example of timelike branching, namely gluon splitting on an outgoing line

$$p_a^2 = 2E_b E_c(1 - \cos \theta) \geq 0 \, . \tag{27}$$

This expression diverges when either the emitted gluon is soft (E_b or $E_c \to 0$) or when the opening angle θ tends to zero, $\theta \to 0$. If b and/or c are quarks, there is a collinear/mass singularity in the $m_q \to 0$ limit. Again, the soft quark divergences are cancelled by a vertex factor.

There are similar infrared divergences in loop diagrams, associated with soft and/or collinear configurations of virtual partons within the region of integration of loop momenta.

Infrared divergences indicate dependence on long-distance aspects of QCD not correctly described by perturbation theory. A parton close to its mass shell has little uncertainty in its energy. By the uncertainty principle, such a virtual state can exist for a long time. Thus divergent (or enhanced) propagators imply propagation of partons over long distances. When the distance travelled becomes comparable with the hadron size ~ 1 fm, the quasi-free partons of a perturbative calculation are confined/hadronized non-perturbatively, and apparent divergences disappear.

We can still use perturbation theory to perform calculations, provided we limit ourselves to two classes of observables: The first are *infrared safe quantities*, i.e. those insensitive to soft or collinear branching. Infrared divergences in a perturbation theory calculation either cancel between real and virtual contributions or are removed by kinematic factors. Such quantities are determined primarily by hard, short-distance physics; long-distance effects give power corrections, suppressed by inverse powers of a large momentum scale.

The second class are *factorizable quantities*, i.e. those in which infrared sensitivity can be absorbed into an overall non-perturbative factor, to be determined experimentally.

In either case, infrared divergences must be *regularized* during the perturbative calculation, even though they cancel or factorize in the end. One method is to introduce a finite gluon mass, set to zero at the end of the calculation. However, a gluon mass breaks gauge invariance. A better method is to use dimensional regularization to control the divergences. This is analogous to the method used for ultraviolet divergences. Dimensional regularization is discussed in Appendix A. The divergences appear as poles in ϵ, where $d = 4 - 2\epsilon$.

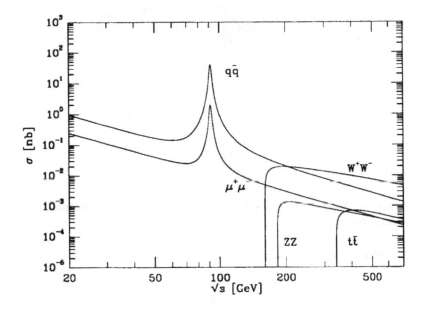

Figure 6. e^+e^- *cross sections vs. energy.*

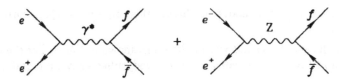

Figure 7. *Feynman diagrams for the process* $e^+e^- \to f\bar{f}$.

1.7 e^+e^- annihilation cross section

The process $e^+e^- \to \mu^+\mu^-$ is a fundamental electroweak reaction. A similar type of process, $e^+e^- \to q\bar{q}$, produces quarks which fragment into hadrons. The cross sections are roughly proportional as shown in Fig. 6. Since the formation of hadrons is non-perturbative, how can perturbation theory give a reliable estimate of a hadronic cross section? This can be understood by visualizing an event in space-time: The e^+ and e^- collide to form a γ or Z^0 with virtual mass $Q = \sqrt{s}$. This subsequently fluctuates into $q\bar{q}$, $q\bar{q}g$, . . . , which occupy a space-time volume $\sim 1/Q$. At large Q, the rate for this short-distance process is given by perturbation theory. Subsequently, at much later time $\sim 1/\Lambda$, the produced quarks and gluons form hadrons. This modifies the outgoing state, but occurs too late to change the original probability for the event to happen.

Well below the Z^0 pole, the process $e^+e^- \to f\bar{f}$ is purely electromagnetic, with a

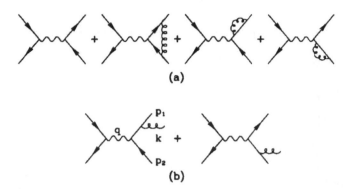

Figure 8. *Feynman diagrams for the $\mathcal{O}(\alpha_S)$ corrections to the total hadronic cross section in e^+e^- annihilation.*

lowest-order (Born) cross section (neglecting quark masses)

$$\sigma_0 = \frac{4\pi\alpha^2}{3s}\,Q_f^2\,. \tag{28}$$

Thus ($N = 3$ = number of $q\bar{q}$ colours)

$$R \equiv \frac{\sigma(e^+e^- \to \text{hadrons})}{\sigma(e^+e^- \to \mu^+\mu^-)} = \frac{\sum_q \sigma(e^+e^- \to q\bar{q})}{\sigma(e^+e^- \to \mu^+\mu^-)} = 3\sum_q Q_q^2\,. \tag{29}$$

The measured cross section is about 5% higher than σ_0, due to QCD corrections which we now consider.

To $\mathcal{O}(\alpha_S)$ we have the real and virtual diagrams shown in Fig. 8. For the real emission diagrams shown in Fig. 8(b) we can write the 3-body phase-space integration as

$$d\Phi_3 = [...]d\alpha\,d\beta\,d\gamma\,dx_1\,dx_2\,, \tag{30}$$

where α, β, γ are the Euler angles of the 3-parton plane, and $x_1 = 2p_1 \cdot q/q^2 = 2E_q/\sqrt{s}$, $x_2 = 2p_2 \cdot q/q^2 = 2E_{\bar{q}}/\sqrt{s}$. This is the result, familiar from the Dalitz plot, that three-particle phase space is proportional to $dE_1\,dE_2$ where E_i are the energies of two of the particles. Applying the Feynman rules and integrating over Euler angles we obtain

$$\sigma^{q\bar{q}g} = 3\sigma_0 C_F \frac{\alpha_S}{2\pi} \int dx_1\,dx_2 \frac{x_1^2 + x_2^2}{(1 - x_1)(1 - x_2)}\,. \tag{31}$$

The integration region is given by $0 \le x_1, x_2, x_3 \le 1$ where $x_3 = 2k \cdot q/q^2 = 2E_g/\sqrt{s} = 2 - x_1 - x_2$. The integral is divergent at $x_{1,2} = 1$:

$$1 - x_1 = \frac{1}{2}x_2 x_3(1 - \cos\theta_{qg})\,,$$

$$1 - x_2 = \frac{1}{2}x_1 x_3(1 - \cos\theta_{\bar{q}g})\,. \tag{32}$$

Thus we have collinear divergences when $\theta_{qg} \to 0$ or $\theta_{\bar{q}g} \to 0$. In addition, there are soft divergences when $E_g \to 0$, i.e. $x_3 \to 0$. These singularities are not physical, they simply indicate the breakdown of perturbation theory when energies and/or invariant masses approach the QCD scale Λ.

The collinear and/or soft regions do not in fact make important contributions to R. To see this, we regulate the integrals using dimensional regularization, $D = 4 - 2\epsilon$ with $\epsilon < 0$. Basic formulae for dimensional regularization are given in Appendix A. Then Eq. (31) becomes

$$\sigma^{q\bar{q}g} = 2\sigma_0 \frac{\alpha_S}{\pi} H(\epsilon)$$
$$\times \int \frac{dx_1 dx_2}{P(x_1, x_2)} \left[\frac{(1-\epsilon)(x_1^2 + x_2^2) + 2\epsilon(1 - x_3)}{[(1 - x_1)(1 - x_2)]} - 2\epsilon \right], \tag{33}$$

where

$$H(\epsilon) = \frac{3(1-\epsilon)(4\pi)^{2\epsilon}}{(3 - 2\epsilon)\Gamma(2 - 2\epsilon)} = 1 + \mathcal{O}(\epsilon), \tag{34}$$

and $P(x_1, x_2) = [(1 - x_1)(1 - x_2)(1 - x_3)]^\epsilon$. Performing the integration we obtain,

$$\sigma^{q\bar{q}g} = 2\sigma_0 \frac{\alpha_S}{\pi} H(\epsilon) \left[\frac{2}{\epsilon^2} + \frac{3}{\epsilon} + \frac{19}{2} - \pi^2 + \mathcal{O}(\epsilon) \right]. \tag{35}$$

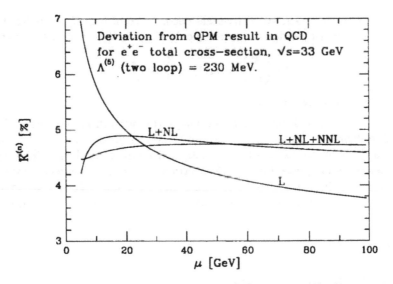

Figure 9. *The effect of higher order QCD corrections to R, as a function of the renormalization scale μ.*

The soft and collinear singularities are regulated, appearing instead as poles at $D = 4$. The virtual gluon contributions, Fig. 8(a), can be calculated using dimensional regularization. The calculation is performed in Appendix B and the answer is given in Eq. (B.20). For positive Q^2 we obtain

$$\sigma^{q\bar{q}} = 3\sigma_0 \left\{ 1 + \frac{2\alpha_S}{3\pi} H(\epsilon) \left[-\frac{2}{\epsilon^2} - \frac{3}{\epsilon} - 8 + \pi^2 + \mathcal{O}(\epsilon) \right] \right\}. \tag{36}$$

Adding real and virtual contributions, the poles cancel and the result is finite as $\epsilon \to 0$:

$$R = 3 \sum_q Q_q^2 \left\{ 1 + \frac{\alpha_S}{\pi} + \mathcal{O}(\alpha_S^2) \right\}. \tag{37}$$

Thus R is an infrared safe quantity. This is the mathematical expression of the argument given above that the total cross section should be calculable in perturbation theory.

The coupling α_S is evaluated at renormalization scale μ. The ultraviolet divergences in R cancel to $\mathcal{O}(\alpha_S)$, so the coefficient of α_S is independent of μ. At $\mathcal{O}(\alpha_S^2)$ and higher, ultraviolet divergences make the coefficients renormalization scheme dependent:

$$R = 3 K_{QCD} \sum_q Q_q^2 \,,$$

$$K_{QCD} = 1 + \frac{\alpha_S(\mu^2)}{\pi} + \sum_{n \geq 2} C_n \left(\frac{s}{\mu^2} \right) \left(\frac{\alpha_S(\mu^2)}{\pi} \right)^n . \tag{38}$$

In the $\overline{MS}$ scheme with scale $\mu = \sqrt{s}$, ($\zeta(3) = 1.2020569$)

$$C_2(1) = \frac{365}{24} - 11\zeta(3) - [11 - 8\zeta(3)]\frac{N_f}{12}$$
$$\simeq 1.986 - 0.115 N_f . \tag{39}$$

The coefficient C_3 is also known. The scale dependence of C_2, $C_3 \ldots$ is fixed by the requirement that, order-by-order, the series should be independent of μ. For example

$$C_2 \left(\frac{s}{\mu^2} \right) = C_2(1) - \frac{\beta_0}{4} \log \frac{s}{\mu^2} \,, \tag{40}$$

where $\beta_0 = 4\pi b = 11 - 2N_f/3$.

Scale and scheme dependence would cancel in perturbation theory only when the series is computed to all orders. A scale change at $\mathcal{O}(\alpha_S^n)$ induces changes at $\mathcal{O}(\alpha_S^{n+1})$. The more terms are added, the more stable is the prediction with respect to changes in μ. Residual scale dependence is an important source of uncertainty in QCD predictions. One can vary the scale over some 'physically reasonable' range, e.g. $\sqrt{s}/2 < \mu < 2\sqrt{s}$, to try to quantify this uncertainty, but there is no real substitute for a full higher-order calculation.

2   Lecture II

2.1   Deep inelastic scattering

Consider lepton-proton scattering via the exchange of a virtual photon. The amplitude shown in Fig. 10 is

$$\mathcal{A} = e\, \bar{u}(k')\gamma^\alpha u(k) \frac{1}{q^2} \left\langle X \middle| j_\alpha(0) \middle| P \right\rangle , \tag{41}$$

where j_α is the electromagnetic current and X represents the hadronic final state. After squaring the amplitude, the cross section factors into a leptonic and a hadronic piece,

$$\frac{d^2\sigma}{dx dy} \propto L_{\alpha\beta} W^{\alpha\beta} . \tag{42}$$

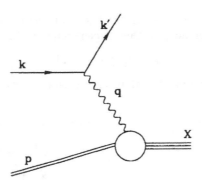

Figure 10. *Deep inelastic lepton scattering.*

The hadronic tensor, $W_{\alpha\beta}$, contains all the information about the structure of the target proton P:

$$W_{\alpha\beta}(p,q) = \frac{1}{4\pi}\sum_X \langle P|j_\beta^\dagger(0)|X\rangle \langle X|j_\alpha(0)|P\rangle$$
$$\times \ (2\pi)^4\delta^4(q+p-p_X)\,, \tag{43}$$

and because of current conservation it has the general expansion

$$W^{\alpha\beta}(p,q) = \left(-g^{\alpha\beta} + \frac{q^\alpha q^\beta}{q^2}\right)W_1(x,Q^2)$$
$$+ \left(p^\alpha + \frac{1}{2x}q^\alpha\right)\left(p^\beta + \frac{1}{2x}q^\beta\right)W_2(x,Q^2)\,. \tag{44}$$

The leptonic tensor is given by

$$L_{\alpha\beta} = 4\left[k_\alpha k_\beta' + k_\alpha k_\beta' - k\cdot k' g_{\alpha\beta}\right]\,. \tag{45}$$

The standard kinematic variables are:

$$x = \frac{-q^2}{2p\cdot q} = \frac{Q^2}{2M(E-E')},\quad \text{(Bjorken } x\text{)} \tag{46}$$

$$y = \frac{q\cdot p}{k\cdot p} = 1 - \frac{E'}{E}\,, \tag{47}$$

where $Q^2 = -q^2 > 0$, $M^2 = p^2$ and the energies refer to the target rest frame. Elastic scattering has $(p+q)^2 = M^2$, i.e. $x = 1$. Hence deep inelastic scattering (DIS) means $Q^2 \gg M^2$ and $x < 1$. We also define structure functions

$$F_1(x,Q^2) = W_1(x,Q^2)\,, \tag{48}$$
$$F_2(x,Q^2) = \nu W_2(x,Q^2). \tag{49}$$

which parametrize the target structure as 'seen' by the virtual photon. The cross section is given by

$$\frac{d^2\sigma}{dxdy} = \frac{8\pi\alpha^2 ME}{Q^4}\left[\left(\frac{1+(1-y)^2}{2}\right)2xF_1\right.$$
$$\left. +(1-y)(F_2 - 2xF_1) - (M/2E)xyF_2\right]\,. \tag{50}$$

The Bjorken limit is Q^2, $p \cdot q \to \infty$ with x fixed. In this limit, the structure functions obey an approximate scaling law, i.e. they depend only on the dimensionless variable x:

$$F_i(x, Q^2) \longrightarrow F_i(x). \tag{51}$$

Fig. 11 shows the F_2 structure function for a proton target. Although Q^2 varies by

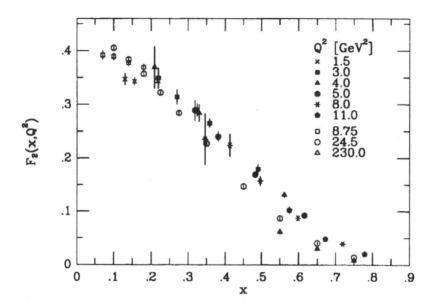

Figure 11. *The F_2 structure function from the SLAC-MIT (Miller* et al *1972) and BCDMS (Benvenuti* et al *1989) collaborations.*

two orders of magnitude, in first approximation the data points lie on a universal curve. Bjorken scaling implies that the virtual photon is scattered by *pointlike constituents* (partons) – otherwise the structure functions would depend on the ratio Q/Q_0, with $1/Q_0$ a length scale characterizing the size of constituents.

The parton model of DIS is formulated in a frame where the target proton is moving very fast – *the infinite momentum frame.* Suppose that, in this frame, the photon scatters from a pointlike quark with fraction ξ of the proton's momentum. Since $(\xi p + q)^2 = m_q^2 \ll Q^2$, we must have $\xi = Q^2/2p \cdot q = x$.

In terms of Mandelstam variables $\hat{s}, \hat{t}, \hat{u}$, the spin-averaged matrix element squared for massless $eq \to eq$ scattering is

$$\overline{\sum}|\mathcal{M}|^2 = 2e_q^2 e^4 \frac{\hat{s}^2 + \hat{u}^2}{\hat{t}^2}, \tag{52}$$

where $\overline{\sum}$ denotes the average (sum) over initial (final) colours and spins. Translating Eq. (52) in terms of DIS variables, $\hat{t} = -Q^2$, $\hat{u} = \hat{s}(y-1)$ and $\hat{s} = Q^2/xy$, the differential cross section becomes

$$\frac{d^2\hat{\sigma}}{dx dQ^2} = \frac{4\pi\alpha^2}{Q^4} \left[1 + (1-y)^2\right] \frac{1}{2} e_q^2 \delta(x - \xi). \tag{53}$$

From the structure function definition (neglecting M)

$$\frac{d^2\sigma}{dx dQ^2} = \frac{4\pi\alpha^2}{Q^4} \left\{ [1 + (1-y)^2] \, F_1 + \frac{(1-y)}{x} \, (F_2 - 2xF_1) \right\}. \tag{54}$$

Hence the structure function for scattering from a parton with momentum fraction ξ is

$$\hat{F}_2 = x e_q^2 \, \delta(x - \xi) = 2x\hat{F}_1 \, . \tag{55}$$

Suppose the probability that quark q carries momentum fraction between ξ and $\xi + d\xi$ is $q(\xi) \, d\xi$. Then

$$\begin{aligned}
F_2(x) &= \sum_q \int_0^1 d\xi \, q(\xi) \, x e_q^2 \, \delta(x - \xi) \\
&= \sum_q e_q^2 \, x q(x) = 2x F_1(x) \, .
\end{aligned} \tag{56}$$

The relationship $F_2 = 2xF_1$ (the Callan-Gross relation (Callan and Gross 1969)) follows from the spin-$\frac{1}{2}$ property of quarks. By contrast $F_1 = 0$ for spin-0 partons.

The proton consists of three valence quarks (uud), which carry its electric charge and baryon number, and an infinite sea of light $q\bar{q}$ pairs. Probed at a scale Q, the sea contains all quark flavours with $m_q \ll Q$. Thus at $Q \sim 1$ GeV we expect,

$$F_2^{em}(x) \simeq \frac{4}{9} x[u(x) + \bar{u}(x)] + \frac{1}{9} x[d(x) + \bar{d}(x) + s(x) + \bar{s}(x)] \, , \tag{57}$$

where

$$\begin{aligned}
u(x) &= u_V(x) + \bar{u}(x) \, , \\
d(x) &= d_V(x) + \bar{d}(x) \, , \\
s(x) &= \bar{s}(x) \, ,
\end{aligned} \tag{58}$$

with sum rules

$$\int_0^1 dx \, u_V(x) = 2 \, , \quad \int_0^1 dx \, d_V(x) = 1 \, . \tag{59}$$

Experimentally one finds

$$\sum_q \int_0^1 dx \, x[q(x) + \bar{q}(x)] \simeq 0.5 \, . \tag{60}$$

Thus quarks only carry about 50% of the proton's momentum. The remainder is carried by *gluons*. Although not directly measured in DIS, gluons participate in other hard scattering processes such as large-p_T jet and prompt photon production. Fig. 12 shows a typical set of parton distributions extracted from fits to DIS data, at $\mu^2 = 10$ GeV^2.

2.2   QCD corrections to the parton model

At large Q^2 the coupling constant is small, so we investigate perturbative corrections to the parton model. We introduce two light-like vectors p and n, with $n \cdot p = 1$,

$$k^\mu = \xi p^\mu + \frac{(k^2 + k_T^2)}{2\xi} n^\mu + k_T^\mu \, , \tag{61}$$

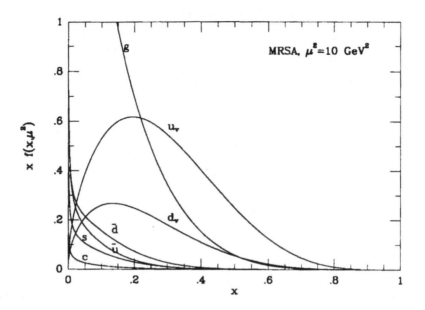

Figure 12. *A typical fit to parton distributions.*

where $p^2 = n^2 = n \cdot k_T = p \cdot k_T = 0$. k_T is a four-vector lying entirely in the transverse direction. We can write an explicit representation for the four-vectors using the notation (E, p_x, p_y, p_z):

$$
\begin{aligned}
p^\mu &= (P, 0, 0, P) , \\
n^\mu &= (\frac{1}{2P}, 0, 0, -\frac{1}{2P}) , \\
k_T^\mu &= (0, \mathbf{k_T}, 0) .
\end{aligned}
\tag{62}
$$

$\mathbf{k_T}$ is a Galileian two-vector in the transverse plane. The kinematic four-vectors are

$$
\begin{aligned}
P^\mu &= p^\mu + \frac{M^2}{2} n^\mu , \\
q^\mu &= \nu n^\mu + q_T^\mu ,
\end{aligned}
\tag{63}
$$

so that $q^2 = q_T^2 = -\mathbf{q}_T^2 = -Q^2$. The functions $W_{1,2}$ can be projected out of the hadronic tensor by

$$
\begin{aligned}
\nu n^\alpha n^\beta W_{\alpha\beta} &= \nu W_2 = F_2 , \\
\frac{4x^2}{\nu} p^\alpha p^\beta W_{\alpha\beta} &= \nu W_2 - 2x W_1 = F_L = F_2 - 2x F_1 .
\end{aligned}
\tag{64}
$$

2.3   Lowest order

To establish the normalization we first calculate the scattering of a virtual photon off a free quark with momentum p (*i.e.* $\xi = 1$), as shown in Fig. 13(a),

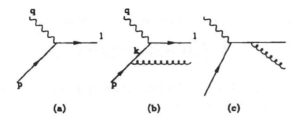

Figure 13. *Amplitudes for deep inelastic scattering off a quark.*

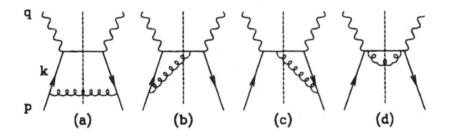

Figure 14. *Real gluon emission diagrams contributing to deep inelastic scattering.*

$$\gamma^*(q) + q(p) \to q(l) \, , \tag{65}$$

leading to a squared matrix element (summed and averaged over spins and colours and with the F_2 contribution projected out)

$$\nu n^\alpha n^\beta \overline{\sum} |\mathcal{M}|^2_{\alpha\beta} = 4\nu e_q^2 \, . \tag{66}$$

The one-dimensional phase space is

$$d\Phi_1 = 2\pi\delta((p+q)^2) \, . \tag{67}$$

Inserting a normalization factor of $1/4\pi$ the result for F_2 is

$$\hat{F}_2(x) = e_q^2\delta(1-x) \, . \tag{68}$$

2.4   Gluon emission

We shall now investigate the parton process in which the quark emits a gluon as shown in Fig. 13(b,c),

$$\gamma^*(q) + q(p) \to g(r) + q(l) \, . \tag{69}$$

The Lorentz invariant phase space for this contribution is

$$d\Phi_2 = \int \frac{d^4r}{(2\pi)^3} \frac{d^4l}{(2\pi)^3} \, \delta^+(r^2)\delta^+(l^2) \, (2\pi)^4\delta^4(p+q-r-l) \, . \tag{70}$$

Introducing k^μ to denote the momentum of the struck parton line we obtain

$$d\Phi_2 = \frac{1}{4\pi^2} \int d^4k\, \delta^+((p-k)^2)\, \delta^+((k+q)^2) \, . \tag{71}$$

k^μ can be written in terms of n, p and a four-vector k_T^μ lying entirely in the transverse plane, $k_T = (0, \mathbf{k_T}, 0)$ such that $k_T^2 = -\mathbf{k}_T^2$,

$$k^\mu = \xi p^\mu + \frac{\mathbf{k}_T^2 - |k^2|}{2\xi} n^\mu + k_T^\mu \, , \tag{72}$$

$$d^4k = \frac{d\xi}{2\xi} dk^2 d^2\mathbf{k_T} \, , \tag{73}$$

which gives

$$(p-k)^2 = (1-\xi)\frac{|k^2|}{\xi} - \frac{\mathbf{k}_T^2}{\xi} \, , \tag{74}$$

$$(k+q)^2 = 2\xi\nu - Q^2 - |k^2| - 2\mathbf{q_T} \cdot \mathbf{k_T} \, . \tag{75}$$

In terms of the components of k^μ

$$
\begin{aligned}
d\Phi_2 &= \frac{1}{16\nu\pi^2} \int d\xi\, d|k|^2\, d\mathbf{k}_T^2\, d\theta\, \delta\left(\mathbf{k}_T^2 - (1-\xi)|k^2|\right) \\
&\quad \times \delta\left(\xi - x - \frac{|k^2| + 2\mathbf{q_T} \cdot \mathbf{k_T}}{2\nu}\right) \\
&= \frac{1}{16\nu\pi^2} \int_0^{2\nu} d|k|^2 \int_{\xi_-}^{\xi_+} d\xi \frac{1}{\sqrt{(\xi_+ - \xi)(\xi - \xi_-)}} \, ,
\end{aligned}
\tag{76}
$$

where

$$\xi_\pm(z, x) = x + z - 2zx \pm \sqrt{4x(1-x)z(1-z)} \, . \tag{77}$$

In order to simplify the notation we have introduced the variable $z = |k^2|/(2\nu)$. Requiring real roots for the quadratic form gives the limits $0 < z < 1$.

The four diagrams obtained by squaring Figs. 13(b,c) are shown in Fig. 14. We shall consider only the diagram of Fig. 14(a), for which the matrix element is

$$\mathcal{M}^\alpha = -ige_q\, \bar{u}(l)\, \gamma^\alpha\, \frac{1}{\not{k}}\, \not{\epsilon}\, t^A\, u(p) \, , \tag{78}$$

where t^A is the SU(3) colour matrix. Squaring and averaging over colours and spins gives

$$\overline{\sum}|\mathcal{M}|^2_{\alpha\beta} = \frac{1}{2}e_q^2 g^2 \sum_{\text{pol}} C_F\, \text{Tr}[\gamma^\beta\, (\not{k} + \not{q})\, \gamma^\alpha\, \not{k}\, \not{\epsilon}\, \not{\epsilon}^*\, \not{k}]\, \frac{1}{k^4} \, , \tag{79}$$

$(\sum_A (t^A t^A)_{ij} = C_F\, \delta_{ij})$.

To sum over the polarization of the real gluon we use the projector

$$\sum_{\text{pol}} \varepsilon_\mu(r)\varepsilon_\nu^*(r) = -g_{\mu\nu} + \frac{n_\mu r_\nu + n_\nu r_\mu}{n \cdot r} \, . \tag{80}$$

In addition to the Lorentz condition $\varepsilon \cdot r = 0$, the gluon satisfies the (light-cone) gauge condition $\varepsilon \cdot n = 0$. This ensures that only two physical (*i.e.* transverse) polarizations propagate. The quark gluon amplitude vanishes in the forward direction. Because helicity is conserved in the vector interactions, a positive helicity quark cannot decay to a collinear quark and a transverse gluon and conserve both the helicity of the quark and the component of angular momentum along the direction of travel. The numerator factor of the amplitude vanishes like $A(q \to qg) \sim k_T$. This is true diagram-by-diagram in the physical gauge.

We project out $\hat{F}_2$ by using the vector n and using the on-shellness of the emitted gluon. After a simple calculation the result is

$$\frac{1}{4\pi} n^\alpha n^\beta \overline{\sum} |\mathcal{M}|^2_{\alpha\beta} = \frac{8 e_q^2 \alpha_S}{|k^2|} \xi P(\xi) \,, \tag{81}$$

where the function $P(\xi)$ is known as the *splitting function*:

$$P(\xi) = C_F \frac{1 + \xi^2}{1 - \xi} \,. \tag{82}$$

Its form is specific to the qqg vertex of QCD; it is independent of the process in which the quark participates. The $1/k^4$ factor has been reduced to $1/k^2$ by terms coming from the trace over the Dirac matrices. This is because of the square of the amplitude factor described above. Putting everything together we obtain

$$\hat{F}_2 = e_q^2 \frac{\alpha_S}{2\pi^2} \int_0^{2\nu} \frac{d|k^2|}{|k^2|} \int_{\xi_-}^{\xi_+} d\xi \, \frac{\xi P(\xi)}{\sqrt{(\xi_+ - \xi)(\xi - \xi_-)}} \,. \tag{83}$$

The $|k^2|$ integral is logarithmically divergent at small $|k^2|$. We introduce a small cut-off κ^2 in the integral and calculate the coefficient of the resulting logarithm. Noting that $\xi_\pm \to x$ as $z \to 0$, and using the result

$$\int_{\xi_-}^{\xi_+} d\xi \frac{1}{\sqrt{(\xi_+ - \xi)(\xi - \xi_-)}} = \pi \tag{84}$$

we obtain

$$\hat{F}_2 \Big|_{\text{div}} = e_q^2 \frac{\alpha_S}{2\pi} x P(x) \int_{\kappa^2}^{2\nu} \frac{d|k^2|}{|k^2|} = e_q^2 \frac{\alpha_S}{2\pi} x P(x) \ln\left(\frac{2\nu}{\kappa^2}\right) \,. \tag{85}$$

In the light-cone gauge only Fig. 14(a) gives a logarithmic divergence – the other diagrams give finite corrections to the structure function.

The full result for the four diagrams of Fig. 14 plus the leading-order diagram is

$$\hat{F}_2(x, Q^2) = e_q^2 x \left[\delta(1 - x) + \frac{\alpha_S}{2\pi} \left(P(x) \ln \frac{Q^2}{\kappa^2} + C(x) \right) \right] \,, \tag{86}$$

where P is given above and C is a calculable function.

2.5   Factorization

Our attempt to calculate the structure functions from perturbation theory has encountered an obstacle. We find that the result is sensitive to the low momentum region where the coupling is large and perturbation theory is not valid. However this sensitivity is universal; it is independent of the hard process and only depends on the type of incoming initial line. It can therefore be factored out:

$$1 + \frac{\alpha_S}{2\pi} P(\xi) \ln \frac{Q^2}{\kappa^2} \sim \left(1 + \frac{\alpha_S}{2\pi} P(\xi) \ln \frac{Q^2}{\mu^2}\right) \otimes \left(1 + \frac{\alpha_S}{2\pi} P(\xi) \ln \frac{\mu^2}{\kappa^2}\right). \tag{87}$$

The idea of separating the high and low frequencies is common in many branches of physics. The low frequency behaviour is absorbed into the parton distributions. The price is that the parton distributions depend on the auxiliary scale μ in a *calculable* way.

We now have parton distribution functions which depend on the scale. Thus for the structure function F_2 we can write

$$F_2 = \sum_i c_2(e_i^2, \xi) \otimes f^{(i)}(x, \mu^2). \tag{88}$$

The physical result cannot depend on the scale, which has been introduced only to separate high and low scales. So, in a schematic way we may write for every contributing process

$$\frac{d}{d\ln\mu^2} F_2 = 0, \tag{89}$$

$$\frac{d}{d\ln\mu^2} c_2(\xi) \otimes f^{(i)}(z, \mu^2) + c_2(\xi) \otimes \frac{d}{d\ln\mu^2} f^{(i)}(z, \mu^2) = 0. \tag{90}$$

In lowest order we get

$$\left(\frac{\partial}{\partial\ln\mu^2} + \beta(\alpha_S)\frac{\partial}{\partial\alpha_S}\right) f^{(i)}(x, \mu^2) = \frac{\alpha_S}{2\pi} P(\xi) \otimes f^{(i)}(z, \mu^2). \tag{91}$$

So the independence from the factorization scale leads to the DGLAP equation (Gribov and Lipatov 1972, Lipatov 1975, Altarelli and Parisi 1977, Dokshitzer 1977)

$$\frac{d}{\ln\mu^2} f^{(i)}(x, \mu^2) = \frac{\alpha_S(\mu^2)}{2\pi} P(\xi) \otimes f^{(i)}(z, \mu^2). \tag{92}$$

What does the symbol $\otimes$ represent? In the present context this symbol represents a convolution integral of the type

$$\begin{aligned} f(x) &= g(y) \otimes h(z) \\ &\equiv \int_0^1 dy \int_0^1 dz \, g(y) \, h(z) \, \delta(x - yz) \tag{93} \\ &\equiv \int_0^1 \frac{dy}{y} \, g(y) \, h(\frac{x}{y}). \tag{94} \end{aligned}$$

If we take moments $f^{(N)} = \int_0^1 dx x^{N-1} f(x)$ the convolution reduce to a product.

$$f^{(N)} = g^{(N)} h^{(N)}. \tag{95}$$

2.6 Splitting functions

We can calculate the splitting functions for all possible processes

$$\hat{P}_{qq}(z) = C_F \frac{1+z^2}{1-z}, \tag{96}$$

$$\hat{P}_{qg}(z) = T_R((1-z)^2 + z^2), \tag{97}$$

$$\hat{P}_{gq}(z) = C_F \frac{1+(1-z)^2}{z}, \tag{98}$$

$$\hat{P}_{gg}(z) = C_A\left[\frac{1}{z} + \frac{1}{(1-z)} - 2 + z(1-z)\right]. \tag{99}$$

The $\hat{P}$ are technically the unregularized splitting functions. The full functions have virtual contributions which contribute only at the endpoint $z = 1$. They can be fixed by the conservation of quark-number

$$\int_0^1 dz \, P_{qq}(z) = 0 \tag{100}$$

and by the conservation of parton momentum

$$\int_0^1 dz \, z\left[P_{qq}(z) + P_{gq}(z)\right] = 0, \tag{101}$$

$$\int_0^1 dz \, z\left[2N_f P_{qg}(z) + P_{gg}(z)\right] = 0. \tag{102}$$

We therefore introduce a plus-prescription with the definition

$$\int_0^1 dx \, f(x) \, g(x)_+ = \int_0^1 dx \, [f(x) - f(1)] \, g(x). \tag{103}$$

The distribution $g(x)_+$ is identical with the normal function $g(x)$ except at $x = 1$. Using this we can define the regularized splitting function

$$P(z) = \hat{P}(z)_+. \tag{104}$$

The plus-prescription applies only to the P_{qq} and P_{gg} parts giving

$$P_{qq}(z) = \hat{P}_{qq}(z)_+ = C_F \left(\frac{1+z^2}{1-z}\right)_+, \tag{105}$$

$$P_{qg}(z) = \hat{P}_{qg}(z) = T_R \left[z^2 + (1-z)^2\right]. \tag{106}$$

P_{qq} and P_{gg} can be written in the common forms

$$P_{qq}(z) = C_F \left[\frac{1+z^2}{(1-z)_+} + \frac{3}{2}\delta(1-z)\right], \tag{107}$$

$$P_{gg}(z) = 2C_A \left[\frac{z}{(1-z)_+} + \frac{1-z}{z} + z(1-z)\right]$$
$$+ \frac{1}{6}(11C_A - 4N_f T_R)\,\delta(1-z). \tag{108}$$

2.7   Scaling violation

Bjorken scaling is not exact. The structure functions decrease at large x and grow at small x with increasing Q^2. This is due to the Q^2 dependence of parton distributions, described in the last section. The quark distributions satisfy the DGLAP evolution equations of the form $(t = \mu^2)$

$$t\frac{\partial}{\partial t}q(x,t) = \frac{\alpha_S(t)}{2\pi}\int_x^1 \frac{dz}{z}P(z)q\left(\frac{x}{z},t\right) \equiv \frac{\alpha_S(t)}{2\pi}P\otimes q\,, \tag{109}$$

where P is $q \to qg$ splitting function. Taking into account other types of parton branching that can occur in addition to $q \to qg$, we obtain coupled evolution equations

$$\begin{aligned}
t\frac{\partial q_i}{\partial t} &= \frac{\alpha_S(t)}{2\pi}\left[P_{qq}\otimes q_i + P_{qg}\otimes g\right]\,, \\
t\frac{\partial \bar{q}_i}{\partial t} &= \frac{\alpha_S(t)}{2\pi}\left[P_{qq}\otimes \bar{q}_i + P_{qg}\otimes g\right]\,, \\
t\frac{\partial g}{\partial t} &= \frac{\alpha_S(t)}{2\pi}\left[P_{gq}\otimes \sum(q_i + \bar{q}_i) + P_{gg}\otimes g\right]\,.
\end{aligned} \tag{110}$$

We have illustrated the derivation of lowest order splitting functions, specificly looking at the case of qq splitting. More generally the splitting functions are power series in α_S. They are the same for jet fragmentation (timelike branching) and deep inelastic scattering (spacelike branching) in leading order, but differ in higher orders. For the present, we concentrate on larger x values ($x \gtrsim 0.01$), where the perturbative expansion converges better.

We first derive the solution of the evolution equations for flavour non-singlet combinations V, e.g. $q_i - \bar{q}_i$ or $q_i - q_j$. In this case the mixing with gluons drops out and the equation takes the simple form

$$t\frac{\partial}{\partial t}V(x,t) = \frac{\alpha_S(t)}{2\pi}P_{qq}\otimes V\,. \tag{111}$$

Taking moments and using Eq. (95) we find that

$$\begin{aligned}
\tilde{V}(N,t) &= \int_0^1 dx\, x^{N-1}\,V(x,t)\,, \tag{112} \\
t\frac{\partial}{\partial t}\tilde{V}(N,t) &= \frac{\alpha_S(t)}{2\pi}\gamma_{qq}^{(0)}(N)\,\tilde{V}(N,t)\,, \tag{113}
\end{aligned}$$

where $\gamma_{qq}^{(0)}(N)$ is Mellin transform of $P_{qq}^{(0)}$. The solution is

$$\tilde{V}(N,t) = \tilde{V}(N,0)\left(\frac{\alpha_S(0)}{\alpha_S(t)}\right)^{d_{qq}(N)}\,, \tag{114}$$

where $d_{qq}(N) = \gamma_{qq}^{(0)}(N)/2\pi b$.

Now $d_{qq}(1) = 0$ and $d_{qq}(N) < 0$ for $N \geq 2$. Thus as t increases V *decreases* at large x and *increases* at small x. Physically, this is due to increase in the phase space for gluon emission by quarks as t increases, leading to loss of momentum. This is clearly visible in the data shown in Fig. 15.

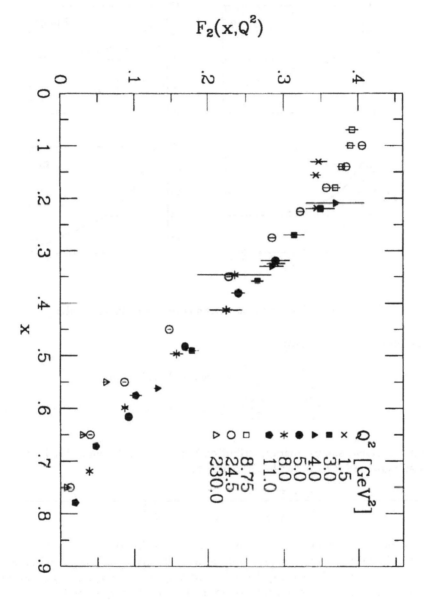

Figure 15. *Data on the structure function F_2 for deuterium.*

2.8 Singlet evolution

For the flavour-singlet combination, we define

$$\Sigma = \sum_i (q_i + \bar{q}_i) \, .$$

(115)

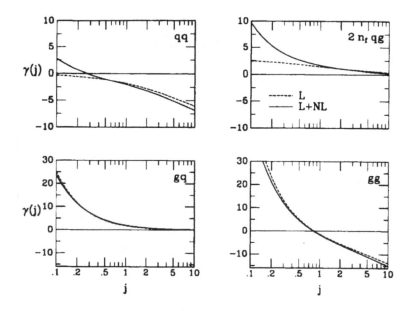

Figure 16. *Anomalous dimensions versus moment number j.*

Then we obtain from Eq. (110)

$$t\frac{\partial \Sigma}{\partial t} = \frac{\alpha_S(t)}{2\pi}[P_{qq} \otimes \Sigma + 2N_f P_{qg} \otimes g] , \tag{116}$$

$$t\frac{\partial g}{\partial t} = \frac{\alpha_S(t)}{2\pi}[P_{gq} \otimes \Sigma + P_{gg} \otimes g] . \tag{117}$$

Thus the flavour-singlet quark distribution Σ mixes with the gluon distribution g: The evolution equation for moments has a matrix form

$$t\frac{\partial}{\partial t}\begin{pmatrix} \tilde{\Sigma} \\ \tilde{g} \end{pmatrix} = \begin{pmatrix} \gamma_{qq} & 2N_f\gamma_{qg} \\ \gamma_{gq} & \gamma_{gg} \end{pmatrix}\begin{pmatrix} \tilde{\Sigma} \\ \tilde{g} \end{pmatrix} . \tag{118}$$

The form of the anomalous dimension matrix as a function of $j = N - 1$ is shown in Fig. 16. From Fig. 16 we see the rapid growth at small j in the gq and gg elements at lowest order. Note also the $\ln j$ behaviour at large j in the qq and gg elements. The singlet anomalous dimension matrix has two real eigenvalues $\gamma_\pm$ given by

$$\gamma_\pm = \frac{1}{2}[\gamma_{gg} + \gamma_{qq} \pm \sqrt{(\gamma_{gg} - \gamma_{qq})^2 + 8N_f\gamma_{gq}\gamma_{qg}}] . \tag{119}$$

Expressing $\tilde{\Sigma}$ and $\tilde{g}$ as linear combinations of eigenvectors $\tilde{\Sigma}_+$ and $\tilde{\Sigma}_-$, we find they evolve as superpositions of terms of above form with $\gamma_\pm$ in place of γ_{qq}.

The reduced DGLAP equation can be written as

$$\frac{d}{du}\begin{pmatrix} \tilde{\Sigma}(u) \\ \tilde{g}(u) \end{pmatrix} = \mathbf{P}\begin{pmatrix} \tilde{\Sigma}(u) \\ \tilde{g}(u) \end{pmatrix} , \tag{120}$$

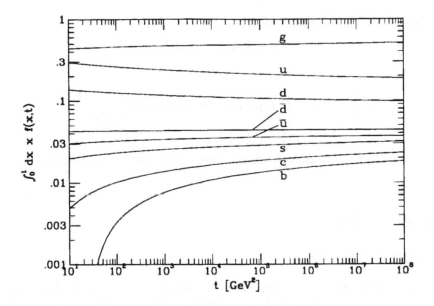

Figure 17. *Growth of the momentum fractions with $t = Q^2$.*

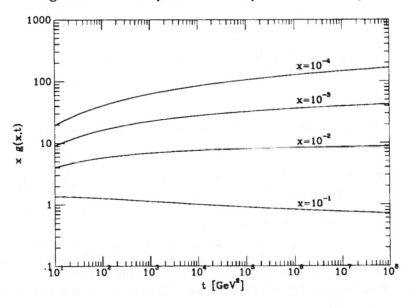

Figure 18. *Variation of the gluon distribution with $t = Q^2$.*

where $u = \frac{1}{2\pi b} \ln \frac{\alpha_S(\mu_0^2)}{\alpha_S(\mu^2)}$ and $\mathbf{P}$ is a matrix in the space of quarks and gluons. Define projection operators, $\mathbf{M}_\pm$,

$$\mathbf{M}_+ = \frac{1}{\gamma_+ - \gamma_-}\big[+\mathbf{P} - \gamma_- 1\big], \quad \mathbf{M}_- = \frac{1}{\gamma_+ - \gamma_-}\big[-\mathbf{P} + \gamma_+ 1\big], \tag{121}$$

where $M_\pm M_\pm = M_\pm, M_+ M_- = M_- M_+ = 0, M_+ + M_- = 1$ and

$$P = \gamma_+ M_+ + \gamma_- M_- . \tag{122}$$

The solution is

$$\begin{pmatrix} \tilde{\Sigma}(u) \\ \tilde{g}(u) \end{pmatrix} = \big[M_+ \exp(\gamma_+ u) + M_- \exp(\gamma_- u) \big] \begin{pmatrix} \tilde{\Sigma}(0) \\ \tilde{g}(0) \end{pmatrix} . \tag{123}$$

The scaling violation depends logarithmically on Q^2, so in Fig. 17 we see a large variation at low Q^2. Evolution also leads to a large number of gluons per unit rapidity as seen in Fig. 18.

3   Lecture III

3.1   DGLAP equation

We will now look in detail at the terms summed by the DGLAP equation. We consider the enhancement of higher-order contributions due to multiple small-angle parton emission, for example, in deep inelastic scattering (DIS) as shown in Fig. 19.

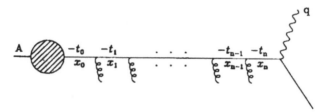

Figure 19. *Initial-state branching in deep inelastic scattering.*

The incoming quark from the target hadron, initially with a low virtual mass-squared $-t_0$ and carrying a fraction x_0 of the hadron's momentum, moves to more virtual masses and lower momentum fractions by successive small-angle emissions, and is finally struck by a photon of virtual mass-squared $q^2 = -Q^2$.

The cross section will depend on Q^2 and on the momentum fraction distribution of the partons seen by the virtual photon at this scale, $D(x, Q^2)$. To derive an evolution equation for the Q^2-dependence of $D(x, Q^2)$, we first introduce a pictorial representation of the evolution, shown in Fig. 20. This diagram will also be useful later when we discuss Monte Carlo simulation.

We represent the sequence of branchings by a path in (t, x)-space. Each branching is a step downwards in x, at a value of t equal to (minus) the virtual mass-squared after the branching. At $t = t_0$, the paths have a distribution of starting points $D(x_0, t_0)$ characteristic of the target hadron at that scale. Then the distribution $D(x, t)$ of partons at scale t is just the x-distribution of paths at that scale.

Consider the change in the parton distribution $D(x, t)$ when t is increased to $t + \delta t$. This is the number of paths arriving in element $(\delta t, \delta x)$ minus the number leaving that

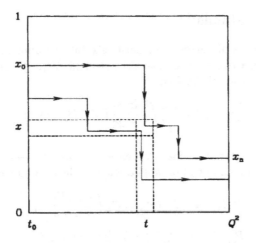

Figure 20. *Representation of parton branching by paths in (t, x)-space.*

element, divided by δx. The number arriving is the branching probability times the parton density integrated over all higher momenta $x' = x/z$,

$$\delta D_{\rm in}(x,t) = \frac{\delta t}{t} \int_x^1 dx' \, dz \frac{\alpha_S}{2\pi} \hat{P}(z) D(x',t) \delta(x - zx')$$

$$= \frac{\delta t}{t} \int_0^1 \frac{dz}{z} \frac{\alpha_S}{2\pi} \hat{P}(z) D(x/z,t) . \qquad (124)$$

For the number leaving the element, we must integrate over lower momenta $x' = zx$:

$$\delta D_{\rm out}(x,t) = \frac{\delta t}{t} D(x,t) \int_0^x dx' \, dz \frac{\alpha_S}{2\pi} \hat{P}(z) \delta(x' - zx)$$

$$= \frac{\delta t}{t} D(x,t) \int_0^1 dz \frac{\alpha_S}{2\pi} \hat{P}(z) . \qquad (125)$$

The change in population of the element is therefore the difference between those arriving and those leaving,

$$\delta D(x,t) = \delta D_{\rm in} - \delta D_{\rm out} \qquad (126)$$

$$= \frac{\delta t}{t} \int_0^1 dz \frac{\alpha_S}{2\pi} \hat{P}(z) \left[\frac{1}{z} D(x/z,t) - D(x,t)\right] . \qquad (127)$$

We now use the plus-prescription with the definition given in Eq. (103) to define a regularized splitting function,

$$P(z) = \hat{P}(z)_+ , \qquad (128)$$

and obtain the Dokshitzer-Gribov-Lipatov-Altarelli-Parisi (DGLAP) evolution equation:

$$t \frac{\partial}{\partial t} D(x,t) = \int_x^1 \frac{dz}{z} \frac{\alpha_S}{2\pi} P(z) D(x/z,t) . \qquad (129)$$

Here $D(x,t)$ represents the parton momentum fraction distribution inside an incoming hadron probed at scale t. In timelike branching, it represents instead the hadron momentum fraction distribution produced by an outgoing parton. The boundary conditions and direction of evolution are different, but the structure of the evolution equation remains the same.

3.2   Quarks and gluons

For several different types of partons, we must take into account different processes by which a parton of type i can enter or leave the element $(\delta t, \delta x)$. This leads to coupled DGLAP evolution equations of the form

$$t\frac{\partial}{\partial t}D_i(x,t) = \sum_j \int_x^1 \frac{dz}{z}\frac{\alpha_S}{2\pi}P_{ij}(z)D_j(x/z,t) . \tag{130}$$

A quark ($i = q$) can enter an element via either $q \to qg$ or $g \to q\bar{q}$, but can only leave via $q \to qg$. Thus the plus-prescription applies only to the $q \to qg$ part, giving

$$P_{qq}(z) = \hat{P}_{qq}(z)_+ = C_F\left(\frac{1+z^2}{1-z}\right)_+ , \tag{131}$$

$$P_{qg}(z) = \hat{P}_{qg}(z) = T_R\left[z^2 + (1-z)^2\right] . \tag{132}$$

A gluon can arrive either from $g \to gg$ (2 contributions) or from $q \to qg$ (or $\bar{q} \to \bar{q}g$). Thus the number arriving is

$$\begin{aligned}
\delta D_{g,\text{in}} &= \frac{\delta t}{t}\int_0^1 dz\frac{\alpha_S}{2\pi}\left\{\hat{P}_{gg}(z)\left[\frac{D_g(x/z,t)}{z} + \frac{D_g(x/(1-z),t)}{1-z}\right]\right. \\
&\quad \left. + \frac{\hat{P}_{qq}(z)}{1-z}\left[D_q\left(\frac{x}{1-z},t\right) + D_{\bar{q}}\left(\frac{x}{1-z},t\right)\right]\right\} \\
&= \frac{\delta t}{t}\int_0^1\frac{dz}{z}\frac{\alpha_S}{2\pi}\left\{2\hat{P}_{gg}(z)D_g\left(\frac{x}{z},t\right)\right. \\
&\quad \left. + \hat{P}_{qq}(1-z)\left[D_q\left(\frac{x}{z},t\right) + D_{\bar{q}}\left(\frac{x}{z},t\right)\right]\right\} .
\end{aligned} \tag{133}$$

A gluon can leave by splitting into either gg or $q\bar{q}$, so that

$$\delta D_{g,\text{out}} = \frac{\delta t}{t}D_g(x,t)\int_0^1 dz\frac{\alpha_S}{2\pi}\left[\hat{P}_{gg}(z) + N_f\hat{P}_{qg}(z)\,dz\right] . \tag{134}$$

After some manipulation we find

$$\begin{aligned}
P_{gg}(z) &= 2C_A\left[\left(\frac{z}{1-z} + \frac{1}{2}z(1-z)\right)_+ + \frac{1-z}{z}\right. \\
&\quad \left. + \frac{1}{2}z(1-z)\right] - \frac{2}{3}N_f T_R\,\delta(1-z) ,
\end{aligned} \tag{135}$$

$$P_{gq}(z) = P_{g\bar{q}}(z) = \hat{P}_{qq}(1-z) = C_F\frac{1+(1-z)^2}{z} . \tag{136}$$

Using the definition of the plus-prescription, Eq. (103), we can check that

$$\begin{aligned}
\left(\frac{z}{1-z} + \frac{1}{2}z(1-z)\right)_+ &= \frac{z}{(1-z)_+} + \frac{1}{2}z(1-z) \\
&\quad + \frac{11}{12}\delta(1-z) ,
\end{aligned} \tag{137}$$

$$\left(\frac{1+z^2}{1-z}\right)_+ = \frac{1+z^2}{(1-z)_+} + \frac{3}{2}\delta(1-z) , \tag{138}$$

so that P_{qq} and P_{gg} can be written in more common forms

$$P_{qq}(z) = C_F \left[\frac{1+z^2}{(1-z)_+} + \frac{3}{2}\delta(1-z) \right] , \tag{139}$$

$$P_{gg}(z) = 2C_A \left[\frac{z}{(1-z)_+} + \frac{1-z}{z} + z(1-z) \right] + \frac{1}{6}(11C_A - 4N_f T_R)\delta(1-z) . \tag{140}$$

3.3 Solution by moments

Given $D_i(x,t)$ at some scale $t = t_0$, the factorized structure of the DGLAP equation means we can compute its form at any other scale. One strategy for doing this is to take moments (Mellin transforms) with respect to x:

$$\tilde{D}_i(N,t) = \int_0^1 dx\, x^{N-1}\, D_i(x,t) . \tag{141}$$

The inverse Mellin transform is

$$D_i(x,t) = \frac{1}{2\pi i}\int_C dN\, x^{-N}\, \tilde{D}_i(N,t) , \tag{142}$$

where the contour C is parallel to the imaginary axis to the right of all singularities of the integrand.

After the Mellin transformation, the convolution in the DGLAP equation becomes simply a product:

$$t\frac{\partial}{\partial t}\tilde{D}_i(x,t) = \sum_j \gamma_{ij}(N,\alpha_S)\tilde{D}_j(N,t) , \tag{143}$$

where the moments of the splitting functions give the perturbative expansion of the anomalous dimensions γ_{ij}:

$$\gamma_{ij}(N,\alpha_S) = \sum_{n=0}^{\infty} \gamma_{ij}^{(n)}(N)\left(\frac{\alpha_S}{2\pi}\right)^{n+1} , \tag{144}$$

$$\gamma_{ij}^{(0)}(N) = \tilde{P}_{ij}(N) = \int_0^1 dz\, z^{N-1}\, P_{ij}(z) . \tag{145}$$

From the above expressions for $P_{ij}(z)$ we find

$$\gamma_{qq}^{(0)}(N) = C_F\left[-\frac{1}{2} + \frac{1}{N(N+1)} - 2\sum_{k=2}^{N}\frac{1}{k} \right] , \tag{146}$$

$$\gamma_{qg}^{(0)}(N) = T_R\left[\frac{(2+N+N^2)}{N(N+1)(N+2)} \right] , \tag{147}$$

$$\gamma_{gq}^{(0)}(N) = C_F\left[\frac{(2+N+N^2)}{N(N^2-1)} \right] , \tag{148}$$

$$\gamma_{gg}^{(0)}(N) = 2C_A\left[-\frac{1}{12} + \frac{1}{N(N-1)} + \frac{1}{(N+1)(N+2)} - \sum_{k=2}^{N}\frac{1}{k} \right] - \frac{2}{3}N_f T_R . \tag{149}$$

Consider the combination of parton distributions which is flavour non-singlet, e.g. $D_V = D_{q_i} - D_{\bar{q}_i}$ or $D_{q_i} - D_{q_j}$. Then mixing with the flavour-singlet gluons drops out and the solution for the case of fixed α_S is

$$\tilde{D}_V(N,t) = \tilde{D}_V(N,t_0) \left(\frac{t}{t_0}\right)^{\gamma_{qq}(N,\alpha_S)} . \qquad (150)$$

We see that the dimensionless function D_V, instead of being a scale-independent function of x as expected from dimensional analysis, has scaling violation: its moments vary like powers of the scale t (hence the name anomalous dimensions).

For the running coupling, $\alpha_S(t)$, scaling violation is power-behaved in $\ln t$ rather than t. Using leading-order formula $\alpha_S(t) = 1/b\ln(t/\Lambda^2)$, we find

$$\tilde{D}_V(N,t) = \tilde{D}_V(N,t_0) \left(\frac{\alpha_S(t_0)}{\alpha_S(t)}\right)^{d_{qq}(N)} , \qquad (151)$$

where $d_{qq}(N) = \gamma_{qq}^{(0)}(N)/2\pi b$.

3.4 Sudakov form factor

The DGLAP equations are convenient for the evolution of parton distributions. To study the structure of final states, a slightly different form is useful. Consider again the simplified treatment with only one type of branching. Introduce the Sudakov form factor:

$$\Delta(t) \equiv \exp\left[-\int_{t_0}^t \frac{dt'}{t'} \int dz \frac{\alpha_S}{2\pi} \hat{P}(z)\right] . \qquad (152)$$

Then

$$t\frac{\partial}{\partial t}D(x,t) = \int \frac{dz}{z} \frac{\alpha_S}{2\pi} \hat{P}(z) D(x/z,t) + \frac{D(x,t)}{\Delta(t)} t\frac{\partial}{\partial t}\Delta(t) , \qquad (153)$$

$$t\frac{\partial}{\partial t}\left(\frac{D}{\Delta}\right) = \frac{1}{\Delta} \int \frac{dz}{z} \frac{\alpha_S}{2\pi} \hat{P}(z) D(x/z,t) . \qquad (154)$$

This is similar to DGLAP, except D replaced by D/Δ and the regularized splitting function P replaced by the unregularized splitting function $\hat{P}$. Integrating,

$$D(x,t) = \Delta(t)D(x,t_0) + \int_{t_0}^t \frac{dt'}{t'} \frac{\Delta(t)}{\Delta(t')} \int \frac{dz}{z} \frac{\alpha_S}{2\pi} \hat{P}(z) D(x/z,t') . \qquad (155)$$

Eq. (155) has a simple interpretation. The first term is the contribution from paths that do not branch between scales t_0 and t. Thus the Sudakov form factor $\Delta(t)$ is the probability of evolving from t_0 to t without branching. The second term is the contribution from paths which have their last branching at scale t'. The factor of $\Delta(t)/\Delta(t')$ is the probability of evolving from t' to t without branching.

The generalization to several species of partons is straightforward. The species i has the Sudakov form factor

$$\Delta_i(t) \equiv \exp\left[-\sum_j \int_{t_0}^t \frac{dt'}{t'} \int dz \frac{\alpha_S}{2\pi} \hat{P}_{ji}(z)\right] , \qquad (156)$$

which is the probability of it evolving from t_0 to t without branching. The result for many species of partons is then

$$t\frac{\partial}{\partial t}\left(\frac{D_i}{\Delta_i}\right) = \frac{1}{\Delta_i}\sum_j \int \frac{dz}{z}\frac{\alpha_S}{2\pi}\hat{P}_{ij}(z)D_j(x/z,t) \; . \tag{157}$$

3.5 Infrared cutoff

In the DGLAP equation, infrared singularities of splitting functions at $z = 1$ are regularized by the plus-prescription. However, in the above form we must introduce an explicit infrared cutoff, $z < 1 - \epsilon(t)$. Branchings with z above this range are unresolvable: the emitted parton is too soft to detect. The Sudakov form factor with this cutoff is the probability of evolving from t_0 to t without any *resolvable* branching.

The Sudakov form factor sums enhanced virtual (parton loop) as well as real (parton emission) contributions. The no-branching probability is the sum of virtual and unresolvable real contributions: both are divergent but their sum is finite. The infrared cutoff $\epsilon(t)$ depends on what we classify as a resolvable emission. For timelike branching, the natural resolution limit is given by the cutoff on the parton virtual mass-squared, $t > t_0$. When the parton energies are much larger than the virtual masses, the transverse momentum in $a \rightarrow bc$ is

$$p_T^2 = z(1-z)p_a^2 - (1-z)p_b^2 - zp_c^2 > 0 \; . \tag{158}$$

Hence for $p_a^2 = t$ and $p_b^2, p_c^2 > t_0$ we require

$$z(1-z) > t_0/t \; , \tag{159}$$

that is,

$$z, \; 1-z > \epsilon(t) = \frac{1}{2} - \frac{1}{2}\sqrt{1 - 4t_0/t} \simeq t_0/t \; . \tag{160}$$

The quark Sudakov form factor is then

$$\Delta_q(t) \simeq \exp\left[-\int_{2t_0}^t \frac{dt'}{t'}\int_{t_0/t'}^{1-t_0/t'} dz\frac{\alpha_S}{2\pi}\hat{P}_{qq}(z)\right] \; . \tag{161}$$

Careful treatment of the running coupling suggests that its argument should be $p_T^2 \sim z(1-z)t'$. Then at large t

$$\Delta_q(t) \sim \left(\frac{\alpha_S(t)}{\alpha_S(t_0)}\right)^{p\ln t} \; , \tag{162}$$

(p = a constant), which tends to zero faster than any negative power of t.

The infrared cutoff discussed here follows from kinematics. We shall see later that QCD dynamics effectively reduces the phase space for parton branching, leading to a more restrictive effective cutoff.

3.6 Monte Carlo method

The formulation in terms of the Sudakov form factor is well suited to computer implementation, and is the basis of 'parton shower' Monte Carlo programs.

The Monte Carlo branching algorithm operates as follows: given a virtual mass scale and momentum fraction (t_1, x_1) either after some step of the evolution, or as an initial condition, one generates values (t_2, x_2) after the next step as shown in Fig. 21.

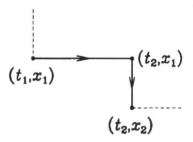

Figure 21. *Fundamental step in* (t, x)-*space.*

Since the probability of evolving from t_1 to t_2 without branching is $\Delta(t_2)/\Delta(t_1)$, t_2 can be generated with the correct distribution by solving

$$\frac{\Delta(t_2)}{\Delta(t_1)} = \mathcal{R} , \tag{163}$$

where $\mathcal{R}$ is a random number (uniform on $[0,1]$). If t_2 is higher than the hard process scale Q^2, this means the branching has finished. Otherwise, generate $z = x_2/x_1$ with a distribution proportional to $(\alpha_S/2\pi)P(z)$, where $P(z)$ is the appropriate splitting function, by solving

$$\int_\epsilon^{x_2/x_1} dz \frac{\alpha_S}{2\pi} P(z) = \mathcal{R}' \int_\epsilon^{1-\epsilon} dz \frac{\alpha_S}{2\pi} P(z) , \tag{164}$$

where $\mathcal{R}'$ is another random number and ϵ is the cutoff for resolvable branching. The normalization factor on the right hand side ensures that for a random number $\mathcal{R}'$ between zero and one, all resolvable branchings are generated.

In DIS, the (t_i, x_i) values generated define virtual masses and momentum fractions of the exchanged quark, from which momenta of emitted gluons can be computed. Azimuthal emission angles are then generated uniformly in the range $[0, 2\pi]$. More generally, e.g. when the exchanged parton is a gluon, azimuths must be generated with polarization angular correlations.

Each emitted (timelike) parton can itself branch. In that case t evolves downwards towards the cutoff value t_0, rather than upwards towards the hard process scale Q^2. The probability of evolving downwards without branching between t_1 and t_2 is now given by

$$\frac{\Delta(t_1)}{\Delta(t_2)} = \mathcal{R} . \tag{165}$$

Thus the branching stops when $\mathcal{R} < \Delta(t_1)$.

Due to successive branching, a parton cascade or shower develops. Each outgoing line is the source of a new cascade, until all outgoing lines have stopped branching. At this stage, which depends on the cutoff scale t_0, outgoing partons have to be converted into hadrons via a hadronization model.

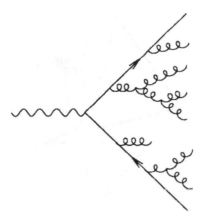

Figure 22. *Parton cascade in e^+e^- annihilation.*

3.7 Soft gluon emission

The parton branching formalism discussed so far takes account of collinear enhancements to all orders in perturbation theory. There are also soft enhancements: When an external line with momentum p and mass m (not necessarily small) emits a gluon with momentum q, the propagator factor is

$$\frac{1}{(p \pm q)^2 - m^2} = \frac{\pm 1}{2p \cdot q} = \frac{\pm 1}{2\omega E(1 - v\cos\theta)} , \tag{166}$$

where ω is the emitted gluon energy, E and v are energy and velocity of the parton emitting it, and θ is the angle of emission. This diverges as $\omega \to 0$, for any velocity and emission angle.

Including the numerator, soft gluon emission gives a colour factor times a universal, spin-independent factor in the amplitude

$$F_{\text{soft}} = \frac{p \cdot \epsilon}{p \cdot q} , \tag{167}$$

where ϵ is the polarization of the emitted gluon. For example, the emission from a quark gives a numerator factor $N \cdot \epsilon$, where

$$N^\mu = (\not{p} + \not{q} + m)\gamma^\mu u(p) \xrightarrow[\omega \to 0]{} (\gamma^\nu \gamma^\mu p_\nu + \gamma^\mu m)u(p) \tag{168}$$

$$= (2p^\mu - \gamma^\mu \not{p} + \gamma^\mu m)u(p) = 2p^\mu u(p) . \tag{169}$$

(using the Dirac equation for the on-mass-shell spinor $u(p)$). The universal factor F_{soft} coincides with the classical eikonal formula for radiation from a current p^μ, valid in the long-wavelength limit. There is no soft enhancement of radiation from off-mass-shell internal lines, since the associated denominator factor $(p + q)^2 - m^2$ tends to $p^2 - m^2 \neq 0$ as $\omega \to 0$.

The enhancement factor in the amplitude for each external line implies that the cross section enhancement is the sum over all pairs of external lines $\{i, j\}$:

$$d\sigma_{n+1} = d\sigma_n \frac{d\omega \, d\Omega}{\omega} \frac{\alpha_S}{2\pi} \sum_{i,j} C_{ij} W_{ij} , \tag{170}$$

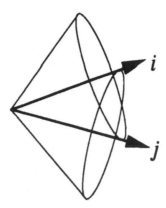

Figure 23. *Cones around partons i and j in which soft radiation is confined after azimuthal averaging.*

where $d\Omega$ is the element of the solid angle for the emitted gluon, C_{ij} is a colour factor, and the radiation function W_{ij} is given by

$$W_{ij} = \frac{\omega^2 p_i \cdot p_j}{p_i \cdot q \, p_j \cdot q} = \frac{1 - v_i v_j \cos\theta_{ij}}{(1 - v_i \cos\theta_{iq})(1 - v_j \cos\theta_{jq})} \,. \tag{171}$$

The colour-weighted sum of the radiation functions $C_{ij}W_{ij}$ is the antenna pattern of a hard process.

The radiation function can be separated into two parts containing collinear singularities along lines i and j. Consider for simplicity massless particles, $v_{i,j} = 1$. Then $W_{ij} = W_{ij}^i + W_{ij}^j$ where

$$W_{ij}^i = \frac{1}{2}\left(W_{ij} + \frac{1}{1 - \cos\theta_{iq}} - \frac{1}{1 - \cos\theta_{jq}}\right) \,. \tag{172}$$

This function has the remarkable property of angular ordering. Write the angular integration in polar coordinates with respect to the direction of i, $d\Omega = d\cos\theta_{iq}\, d\phi_{iq}$. Performing the azimuthal integration, we find

$$\int_0^{2\pi} \frac{d\phi_{iq}}{2\pi} W_{ij}^i = \frac{1}{1 - \cos\theta_{iq}} \quad \text{if } \theta_{iq} < \theta_{ij}, \text{ otherwise } 0. \tag{173}$$

Thus, after azimuthal averaging, the contribution from W_{ij}^i is confined to a cone, centred on the direction of i, extending in angle to the direction of j. Similarly, W_{ij}^j, averaged over ϕ_{jq}, is confined to a cone centred on line j extending to the direction of i.

3.8 Proof of angular ordering

To prove the angular ordering property, write

$$1 - \cos\theta_{jq} = a - b\cos\phi_{iq} \,, \tag{174}$$

where

$$a = 1 - \cos\theta_{ij}\cos\theta_{iq} , \quad b = \sin\theta_{ij}\sin\theta_{iq} . \tag{175}$$

Defining $z = \exp(i\phi_{iq})$, we have

$$I_{ij}^i \equiv \int_0^{2\pi} \frac{d\phi_{iq}}{2\pi} \frac{1}{1 - \cos\theta_{jq}} = \frac{1}{i\pi b} \oint \frac{dz}{(z_+ - z)(z - z_-)} , \tag{176}$$

where the z-integration contour is the unit circle and

$$z_\pm = \frac{a}{b} \pm \sqrt{\frac{a^2}{b^2} - 1} . \tag{177}$$

Now only the pole at $z = z_-$ can lie inside the unit circle, so by the calculus of residues we obtain

$$I_{ij}^i = \sqrt{\frac{1}{a^2 - b^2}} = \frac{1}{|\cos\theta_{iq} - \cos\theta_{ij}|} . \tag{178}$$

Hence

$$\int_0^{2\pi} \frac{d\phi_{iq}}{2\pi} W_{ij}^i = \frac{1}{2(1 - \cos\theta_{iq})}[1 + (\cos\theta_{iq} - \cos\theta_{ij})I_{ij}^i] \tag{179}$$

$$= \frac{1}{1 - \cos\theta_{iq}} \quad \text{if } \theta_{iq} < \theta_{ij}, \text{ otherwise } 0, \tag{180}$$

which proves the angular ordering property. Angular ordering is a coherence effect common to all gauge theories. In QED it causes the Chudakov effect – the suppression of soft bremsstrahlung from e^+e^- pairs.

3.9 Coherent branching

Angular ordering provides the basis for the coherent parton branching formalism, which includes leading soft gluon enhancements to all orders. In place of the virtual mass-squared variable t in the earlier treatment, we use the angular variable

$$\zeta = \frac{p_b \cdot p_c}{E_b E_c} \simeq 1 - \cos\theta \tag{181}$$

as the evolution variable for the branching $a \rightarrow bc$, and impose angular ordering $\zeta' < \zeta$ for successive branchings. The iterative formula for n-parton emission becomes

$$d\sigma_{n+1} = d\sigma_n \frac{d\zeta}{\zeta} dz \frac{\alpha_S}{2\pi} \hat{P}_{ba}(z) . \tag{182}$$

In place of the virtual mass-squared cutoff t_0, we must use the angular cutoff ζ_0 for coherent branching. This is to some extent arbitrary, depending on how we classify emission as unresolvable. The simplest choice is

$$\zeta_0 = t_0/E^2 \tag{183}$$

for the parton of energy E.

For radiation from particle i with finite mass-squared t_0, the radiation function becomes

$$\omega^2 \left(\frac{p_i \cdot p_j}{p_i \cdot q\, p_j \cdot q} - \frac{p_i^2}{(p_i \cdot q)^2} \right) \simeq \frac{1}{\zeta} \left(1 - \frac{t_0}{E^2 \zeta} \right), \tag{184}$$

so the angular distribution of radiation is cut off at $\zeta = t_0/E^2$. Thus t_0 can still be interpreted as the minimum virtual mass-squared.

With this cutoff, the most convenient definition of the evolution variable is not ζ itself but rather

$$\tilde{t} = E^2 \zeta \geq t_0 . \tag{185}$$

The angular ordering condition $\zeta_b, \zeta_c < \zeta_a$ for timelike branching $a \to bc$ (a outgoing) becomes

$$\tilde{t}_b < z^2 \tilde{t} , \quad \tilde{t}_c < (1-z)^2 \tilde{t} , \tag{186}$$

where $\tilde{t} = \tilde{t}_a$ and $z = E_b/E_a$. Thus the cutoff on z becomes

$$\sqrt{t_0/\tilde{t}} < z < 1 - \sqrt{t_0/\tilde{t}} . \tag{187}$$

Neglecting the masses of b and c, the virtual mass-squared of a and the transverse momentum of the branching are

$$t = z(1-z)\tilde{t} , \quad p_t^2 = z^2(1-z)^2 \tilde{t} . \tag{188}$$

Thus for coherent branching the Sudakov form factor of a quark becomes

$$\tilde{\Delta}_q(\tilde{t}) = \exp \left[-\int_{4t_0}^{\tilde{t}} \frac{dt'}{t'} \int_{\sqrt{t_0/t'}}^{1-\sqrt{t_0/t'}} \frac{dz}{2\pi} \alpha_S(z^2(1-z)^2 t') \hat{P}_{qq}(z) \right] . \tag{189}$$

At large $\tilde{t}$ this falls more slowly than the form factor without coherence, due to the suppression of soft gluon emission by angular ordering.

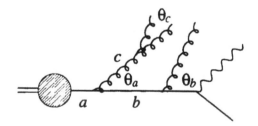

Figure 24. *Angular ordering in branching for DIS.*

Note that for spacelike branching $a \to bc$ (a incoming, b spacelike), the angular ordering condition is

$$\theta_b > \theta_a > \theta_c , \tag{190}$$

and so for $z = E_b/E_a$ we now have

$$\tilde{t}_b > z^2 \tilde{t}_a , \quad \tilde{t}_c < (1-z)^2 \tilde{t}_a . \tag{191}$$

Thus we can have either $\tilde{t}_b > \tilde{t}_a$ or $\tilde{t}_b < \tilde{t}_a$, especially at small z – spacelike branching becomes disordered at small x.

4 Lecture IV

4.1 Electroweak couplings

The basic structure of the electroweak model has been introduced by Douglas Ross in his lectures. In this lecture I want to develop the consequences of the model somewhat further and talk about what has been learned about the parameters of the model. The comparison of the standard model with the Fermi effective theory yields the relationship

$$\frac{G_\mu}{\sqrt{2}} = \frac{e^2}{8s_W^2 M_W^2} . \tag{192}$$

This equation, which can be easily derived using the Feynman rules of Fig. 25, was the basis for the prediction of the masses of the vector bosons, using the value of $s_W^2 = \sin^2 \theta_W$, the square of the sine of the Weinberg angle, measured in neutrino scattering. The values predicted are ($c_W^2 = 1 - s_W^2$)

$$M_W^2 = \frac{\pi\alpha}{\sqrt{2}G_\mu}\frac{1}{s_W^2} ,$$
$$M_Z^2 = \frac{\pi\alpha}{\sqrt{2}G_\mu}\frac{1}{s_W^2 c_W^2} . \tag{193}$$

Using the inputs from the particle data group (Hagiwara *et al* 2002), namely $G_\mu = 1.16637(1)10^{-5}$ GeV^2, $\alpha^{-1} = 137.035\,999\,6(50)$ and $s_W^2 = 0.2253 \pm 0.0021$ from neutrino deep inelastic scattering we obtain the numerical values

$$M_W = 78.54 \pm 0.4 \text{ GeV} ,$$
$$M_Z = 89.23 \pm 0.3 \text{ GeV} . \tag{194}$$

The experimentally measured values are (Hagiwara *et al* 2002)

$$M_W = 80.425 \pm 0.038 \text{ GeV} ,$$
$$M_Z = 91.1876 \pm 0.0021 \text{ GeV} . \tag{195}$$

The difference between prediction and experiment is an indication that the precision of the data is such that higher order effects have to be considered. As an another illustration of the precision of the data we can also consider the coupling of the Z-boson to fermions. Table 1 reports the expected values of the couplings. In Fig. 26 the measurements of the standard model couplings are shown in graphical form. Comparison with Table 1 provides a dramatic confirmation of the standard model.

The standard electroweak model contains many Yukawa couplings to fix the quark and charged lepton masses. But the primary parameters are the two couplings, g_W and g'_W, the vacuum expectation value of the Higgs field, v, and the mass of the Higgs boson, M_H. Unfortunately, the mass of the Higgs enters only peripherally into low energy phenomenology, so in most cases we have to provide three observables to make predictions within the standard model. It is normal to trade these three inputs, g_W, g'_W and v for the three most precisely measured parameters of the standard model. Their values

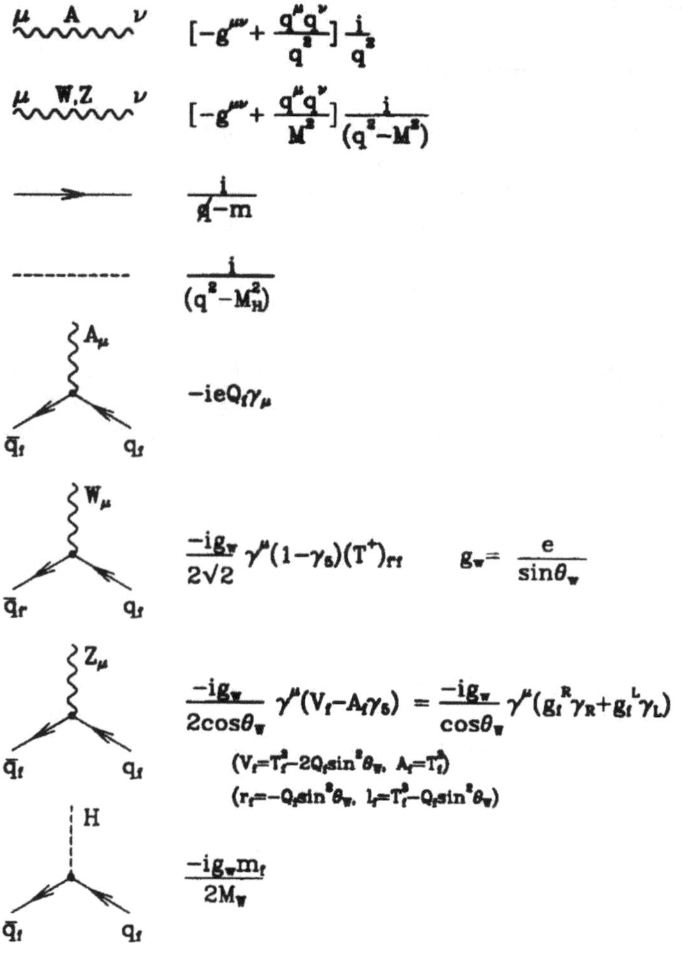
μ A ν
$[-g^{\mu\nu}+\frac{q^\mu q^\nu}{q^2}]\frac{i}{q^2}$
μ W,Z ν
$[-g^{\mu\nu}+\frac{q^\mu q^\nu}{M^2}]\frac{i}{(q^2-M^2)}$
$\frac{i}{\not{q}-m}$
$\frac{i}{(q^2-M_H^2)}$
A_μ
$-ieQ_f\gamma_\mu$
$\bar{q}_f$ q_f
W_μ
$\frac{-ig_w}{2\sqrt{2}}\gamma^\mu(1-\gamma_5)(T^+)_{rf}$
$g_w=\frac{e}{\sin\theta_w}$
$\bar{q}_r$ q_f
Z_μ
$\frac{-ig_w}{2\cos\theta_w}\gamma^\mu(V_f-A_f\gamma_5) = \frac{-ig_w}{\cos\theta_w}\gamma^\mu(g_f^R\gamma_R+g_f^L\gamma_L)$
$(V_f=T_f^3-2Q_f\sin^2\theta_w,\ A_f=T_f^3)$
$(r_f=-Q_f\sin^2\theta_w,\ l_f=T_f^3-Q_f\sin^2\theta_w)$
$\bar{q}_f$ q_f
H
$\frac{-ig_w m_f}{2M_W}$
$\bar{q}_f$ q_f

Figure 25. *Propagators and Feynman rules for fermion interactions.*

Fermions			Q_f	V_f	A_f
u	c	t	$+\frac{2}{3}$	$(+\frac{1}{2}-\frac{4}{3}\sin^2\theta_W)\sim+0.191$	$+\frac{1}{2}$
d	s	b	$-\frac{1}{3}$	$(-\frac{1}{2}+\frac{2}{3}\sin^2\theta_W)\sim-0.345$	$-\frac{1}{2}$
ν_e	ν_μ	ν_τ	0	$\frac{1}{2}$	$+\frac{1}{2}$
e	μ	τ	-1	$(-\frac{1}{2}+2\sin^2\theta_W)\sim-0.036$	$-\frac{1}{2}$

Table 1. *Couplings of fermions to the Z boson. The numerical values of V_f are for* $\sin^2\theta_W=0.232$.

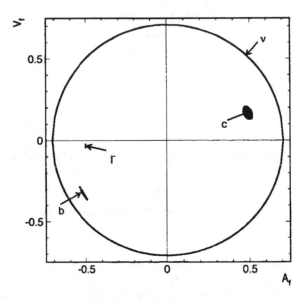

Figure 26. *Measured values of the fermion couplings to the Z boson.*

are (Hagiwara *et al* 2002)

$$
\begin{aligned}
\alpha^{-1} &= 137.035\,989\,6(50) \;, \\
G_F &= 1.166\,39(1) \times 10^{-5} \;\; \mathrm{GeV}^{-2} \;, \\
M_Z &= 91.1876(21) \;\; \mathrm{GeV} \;.
\end{aligned}
\tag{196}
$$

At tree level their relationships to the more fundamental parameters of the standard electroweak model are

$$
\begin{aligned}
\alpha &= \frac{g_W^2 \sin^2 \theta_W}{4\pi} \;, \\
\sin^2 \theta_W &= \frac{g_W'^2}{g_W'^2 + g_W^2} \;, \\
G_F &= \frac{1}{\sqrt{2}v^2} \;, \\
M_Z &= \frac{\frac{1}{2}g_W v}{\cos \theta_W} \;.
\end{aligned}
\tag{197}
$$

4.2 Precision electroweak phenomena

Although the electroweak couplings are small, the precision of the data requires the inclusion of electroweak radiative corrections[1]. In many circumstances the qualitative effect

---

[1]In preparing this section I have relied extensively on the work of other authors, especially, Peskin (Peskin 1989) and Hollik (Hollik 1996).

R. Keith Ellis

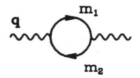

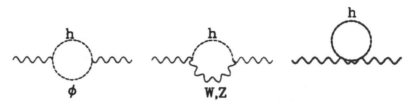

Figure 27. *Examples of oblique corrections.*

of these electroweak radiative corrections can be understood by considering the *oblique corrections*, corresponding to the self-energy corrections to the vector boson propagators. Examples of such oblique corrections at one loop are shown in Fig. 27. We write the one loop correction to the inverse propagators as

$$-i\Sigma_{\mu\nu}^{W,Z,\gamma} = -i\Sigma^{W,Z,\gamma} g_{\mu\nu} + \dots \tag{198}$$

The terms proportional to $q^\mu q^\nu$ are irrelevant for the electromagnetic case, and can generally be neglected for the weak case if we couple the vector bosons to fermions of small masses.

To lay the groundwork we first calculate the one loop corrections to vector boson propagators, due to a chiral fermion loop, because it turns out that fermion loops can have significant effects. The specific cases of W, Z or γ can be derived from this general case by inserting the left and right-handed couplings from Fig. 25. The results are

$$\begin{aligned}
\Sigma_{LL}(m_1^2, m_2^2, q^2) &= \Sigma_{RR}(m_1^2, m_2^2, q^2) \\
&= \frac{1}{4\pi^2} \int_0^1 dx \left(\Delta + \ln\left[\frac{\mu^2}{M^2 - x(1-x)q^2}\right] \right) \left(x(1-x)q^2 - \frac{1}{2}M^2 \right),
\end{aligned} \tag{199}$$

$$\begin{aligned}
\Sigma_{LR}(m_1^2, m_2^2, q^2) &= \Sigma_{RL}(m_1^2, m_2^2, q^2) \\
&= \frac{1}{4\pi^2} \int_0^1 dx \left(\Delta + \ln\left[\frac{\mu^2}{M^2 - x(1-x)q^2}\right] \right) \left(\frac{1}{2}m_1 m_2 \right),
\end{aligned} \tag{200}$$

where $M^2 = xm_1^2 + (1-x)m_2^2$ and the number of dimensions $D = 4 - 2\epsilon$ and $\Delta = \frac{1}{\epsilon} - \gamma_E + \ln(4\pi)$. In deriving these results we have taken γ_5 to anti-commute in D dimensions and included all factors except the overall charge. Notice that the chiral loops, unlike the vector loop, do not decouple in the $q \to 0$ limit.

However when we form the vector-vector combination the amplitude does vanish for $q^2 = 0$. Defining $\Pi^\gamma(m^2, q^2) = e^2 \Sigma_{VV}(m^2, m^2, q^2)/q^2$ to remove this overall factor of q^2

we obtain

$$\Pi^\gamma(m^2, m^2, q^2) = \frac{2\alpha}{\pi} \int_0^1 dx \left(\Delta + \ln \left[\frac{\mu^2}{m^2 - x(1-x)q^2} \right] \right) x(1-x) . \qquad (201)$$

To calculate the electromagnetic coupling at another scale it is helpful to define the subtracted quantity, $\hat{\Pi}^\gamma$,

$$\begin{aligned} \hat{\Pi}^\gamma(m^2, q^2) &= \mathrm{Re}\left[\Pi^\gamma(m^2, q^2) - \Pi^\gamma(m^2, 0) \right] \\ &\simeq \frac{2\alpha}{\pi} \int_0^1 dx \left(\ln \left[\frac{m^2}{x(1-x)q^2} \right] \right) x(1-x) , \end{aligned} \qquad (202)$$

where $\alpha = e^2/(4\pi)$. Note that in the limit $q^2 \gg m^2$ we have

$$\hat{\Pi}^\gamma(m^2, q^2) \to -\frac{\alpha}{3\pi} \left[\ln \left(\frac{q^2}{m^2} \right) - \frac{5}{3} \right] . \qquad (203)$$

After inclusion of a single fermion loop the photon propagator is modified

$$\frac{-ig^{\mu\nu}}{q^2} \to \frac{-ig^{\mu\nu}}{q^2} \left[1 - \Pi^\gamma(m^2, q^2) + \ldots \right] = \frac{-ig^{\mu\nu}}{q^2} \left[\frac{1}{1 + \Pi^\gamma(m^2, q^2)} \right] . \qquad (204)$$

The last term on right hand side of Eq. (204) actually corresponds to the bubble sum of multiple fermion loops. The fermion loops thus contribute to the renormalization of the charge. The renormalized charge at scale q^2, $\alpha(q^2)$, due to a single species of fermion is given in terms of the bare charge α_0 as

$$\alpha(q^2) = \frac{\alpha_0}{1 + \sum_f \Pi^\gamma(m_f^2, q^2)} . \qquad (205)$$

The Thompson charge, $\alpha(0)$ is the value of this quantity evaluated at $q^2 = 0$.

Therefore the difference between the renormalized charge at $q^2 = M_Z^2$ and at $q^2 = 0$ is given as

$$\frac{1}{\alpha(M_Z^2)} - \frac{1}{\alpha(0)} = \frac{\Pi^\gamma(m^2, M_Z^2)}{\alpha} . \qquad (206)$$

Therefore

$$\alpha(M_Z^2) = \alpha(0)(1 + \Delta\alpha) , \qquad (207)$$

where $\Delta\alpha = -\mathrm{Re}\,\hat{\Pi}^\gamma(m^2, M_Z^2)$.

Defining the charge at scale M_Z^2 in terms of the renormalized Thompson charge $\alpha(0)$ we have

$$\frac{1}{\alpha(M_Z^2)} - \frac{1}{\alpha(0)} = -\frac{1}{3\pi} \sum_f Q_f^2 N_C \left[\ln \left(\frac{M_Z^2}{m_f^2} \right) - \frac{5}{3} \right] , \qquad (208)$$

where the sum runs over all fermions with masses very much less than M_Z^2 and $N_C = 3(1)$ for quarks(leptons). We can estimate the shift in α by plugging in the current algebra masses for the quarks and the lepton masses, leading to a shift in the value of $1/\alpha$ of about 8 units. The result is given in Table 2. This estimate is not expected to be reliable at small q^2 because of strong interaction corrections. More accurate estimates using dispersion relations to estimate the hadronic contribution give (Erler 1999)

$$\alpha(M_Z)^{-1} = 127.934 \pm 0.027 . \qquad (209)$$

	e	μ	τ	u	d	s	c	b
m_f(MeV)	0.5	106	1784	5.5	8	150	1200	5000
$(\Delta\alpha^{-1})$	-2.4	-1.3	-0.7	-2.5	-0.6	-0.4	-0.1	-0.1

Table 2. *Contributions to the shift in the electromagnetic coupling from the charged fermions with mass below M_Z (Peskin 1989).*

After our success with the renormalization of the charge we can attempt to calculate genuine electroweak corrections. We first investigate the limits of the formula for chiral loops, Eqs. (199,200). In the limit in which we have massless fermions circulating in the loop we find

$$\Sigma_{LL}(0,0,q^2) = \Sigma_{RR}(0,0,q^2)$$
$$= \frac{1}{4\pi^2}\frac{q^2}{6}\Big[\Delta - \ln\Big(\frac{q^2}{\mu^2}\Big) + \frac{5}{3}\Big] , \qquad (210)$$
$$\Sigma_{LR}(0,0,q^2) = 0 . \qquad (211)$$

In the limit in which we have a heavy fermion of mass m circulating in the loop we find

$$\Sigma_{LL}(m^2,m^2,q^2) = \Sigma_{RR}(m,m^2,q^2)$$
$$= \frac{1}{4\pi^2}\Big(\frac{q^2}{6} - \frac{m^2}{2}\Big)\Big[\Delta - \ln(\frac{m^2}{\mu^2})\Big] , \qquad (212)$$
$$\Sigma_{LL}(0,m^2,q^2) = \Sigma_{RR}(0,m^2,q^2)$$
$$= \frac{1}{4\pi^2}\Big(\frac{q^2}{6}\Big[\Delta - \ln(\frac{m^2}{\mu^2})\Big] - \frac{m^2}{4}\Big[\Delta - \ln(\frac{m^2}{\mu^2}) + \frac{1}{2}\Big]\Big) , \qquad (213)$$
$$\Sigma_{LR}(m^2,m^2,q^2) = \frac{1}{4\pi^2}\frac{m^2}{2}\Big[\Delta - \ln(\frac{m^2}{\mu^2})\Big] . \qquad (214)$$

$\Sigma_{LR}(0,m^2,q^2)$ is not needed because in the standard model the flavour changing interactions are all left-handed. In the above formula we have only kept divergent terms or terms proportional to the large mass, m. Using these results and the couplings of Fig. 25, we can write down the expressions for Σ_W and Σ_Z.

The contribution of a doublet of light fermions is

$$\Sigma_W(0,0,q^2) = \frac{\alpha N_C}{3\pi}\Big(\frac{q^2}{4s_W^2}\Big)\Big[\Delta - \ln(\frac{q^2}{\mu^2}) + \frac{5}{3}\Big] , \qquad (215)$$
$$\Sigma_Z(0,0,q^2) = \frac{\alpha N_C}{3\pi}(l_-^2 + r_-^2 + l_+^2 + r_+^2)\Big(\frac{q^2}{8s_W^2 c_W^2}\Big)\Big[\Delta - \ln(\frac{q^2}{\mu^2}) + \frac{5}{3}\Big] . \qquad (216)$$

$l_\pm, r_\pm$ are the left- and right-handed couplings of the upper (+) and lower (-) members of the weak isospin doublet to the Z-boson. In the approximation in which we keep the effects of the top quark loop we have

$$\Sigma_W(m,0,q^2) = \frac{\alpha N_C}{3\pi}\Big[\Big(\frac{q^2}{4s_W^2}\Big)\Big[\Delta - \ln(\frac{m_+^2}{\mu^2})\Big] - \Big(\frac{3m^2}{8s_W^2}\Big)\Big[\Delta - \ln(\frac{m_+^2}{\mu^2}) + \frac{1}{2}\Big] , \qquad (217)$$

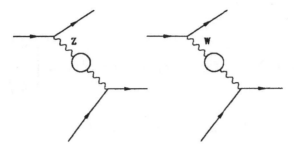

Figure 28. *Oblique corrections to low-energy neutrino scattering.*

$$\Sigma_Z(m, m, q^2) = \frac{\alpha N_C}{3\pi}\Big[(l_-^2 + r_-^2 + l_+^2 + r_+^2)\Big(\frac{q^2}{2s_W^2 c_W^2}\Big) - \frac{3m_+^2}{2s_W^2 c_W^2}(l_+ - r_+)^2\Big]$$
$$\times \Big[\Delta - \ln(\frac{m_+^2}{\mu^2})\Big] . \tag{218}$$

Let us consider, for example, radiative corrections to charged and neutral neutrino scattering as shown in Fig. 28. The insertion of the loop corrections modifies the propagator as follows:

$$\frac{-ig^{\mu\sigma}}{q^2 - M_V^2}(-i\Sigma_{\sigma\rho}^{W,Z})\frac{-ig^{\rho\nu}}{q^2 - M_V^2} = \frac{-ig^{\mu\sigma}}{q^2 - M_V^2}\Big(\frac{-\Sigma^{W,Z}}{q^2 - M_V^2}\Big) . \tag{219}$$

Performing many insertions of the fermionic loops we reproduce a geometrical series

$$\frac{-ig^{\mu\nu}}{q^2 - M_V^2 + \Sigma^V(q^2)} . \tag{220}$$

So the mass renormalization is quite clear. Writing the bare mass as a renormalized mass plus a shift we have that

$$M_W^{0\,2} = M_W^2 + \delta M_W^2, \quad M_Z^{0\,2} = M_Z^2 + \delta M_Z^2, \tag{221}$$

and the resummed propagator becomes

$$\frac{-ig^{\mu\nu}}{q^2 - M_V^2 - \delta M_V^2 + \Sigma^V(q^2)} . \tag{222}$$

We fix the physical masses by the definitions

$$\delta M_W^2 = \text{Re } \Sigma^W(M_W^2), \quad \delta M_Z^2 = \text{Re } \Sigma^Z(M_Z^2) . \tag{223}$$

Now we are in a position to write down the radiative corrections to the ρ parameter, defined as the ratio of the neutral to charged current strength in neutrino scattering. This constitutes a change in the relationship between G_F, coupling constants and M_Z in neutral current exchanges. At low energy we have the effective Lagrangian,

$$\mathcal{L} = \frac{4G_F}{\sqrt{2}}\Big[J_W^\mu J_{W\,\mu} + \rho J_Z^\mu J_{Z\,\mu}\Big] , \tag{224}$$

where the currents are

$$J_W^\mu = \sum_{l=e,\mu,\tau} \bar\nu_l \gamma^\mu \gamma_L \nu_l + (\bar u, \bar c, \bar t) \gamma^\mu (1 - \gamma_5) U_{CKM} \begin{pmatrix} d \\ s \\ b \end{pmatrix}, \qquad (225)$$

$$J_Z^\mu = \sum_f \bar\psi_f \, \gamma^\mu (g_L^{(f)} \gamma_L + g_R^{(f)} \gamma_R) \, \psi_f \,, \qquad (226)$$

where $g_L^{(f)} = T_3 - Q_f s_W^2, g_R^{(f)} = -Q_f s_W^2$. At zeroth order we have

$$\rho = \frac{M_W^2}{c_W^2 M_Z^2} \,. \qquad (227)$$

Retaining only the terms proportional to m_t at first order we obtain $\rho = \rho + \Delta\rho$ where $\Delta\rho$ is given by,

$$\Delta\rho = \frac{\Sigma_Z(m_t, m_t, 0)}{M_Z^2} - \frac{\Sigma_W(m_t, 0, 0)}{M_W^2} \,. \qquad (228)$$

We can evaluate $\Delta\rho$ using Eqs. (217,218). We hence obtain a finite 1% correction due to the large top mass:

$$\Delta\rho = \frac{\alpha N_c}{2\pi} \frac{m_t^2}{8 s_W^2 M_W^2} = \frac{N_C G_F m_t^2}{8\pi^2 \sqrt{2}} = 0.01 \left(\frac{m_t}{175 \text{ GeV}}\right)^2 . \qquad (229)$$

This term $\Delta\rho$ is also responsible for a large shift in the electroweak mixing angle. There are many different possible definitions for the weak mixing angle beyond leading order. We shall use the on-shell definition $s_W^2 = 1 - M_W^2/M_Z^2$ which is conceptually the simplest. The effect of higher order corrections on this definition is determined by the shift in the masses described above, ($x_W \equiv s_W^2$),

$$x_W^0 = 1 - \frac{M_W^2 + \delta M_W^2}{M_Z^2 + \delta M_Z^2} = x_W + (1 - x_W)\left(\frac{\delta M_Z^2}{M_Z^2} - \frac{\delta M_W^2}{M_W^2}\right) = x_W + (1 - x_W)\Delta\rho \,. \qquad (230)$$

We can now consider the radiative corrections to the muon decay process which fixes G_μ and ultimately determines the mass of the W. Including only the propagator terms we find

$$\frac{G_\mu}{\sqrt{2}} = \frac{e_0^2}{8 s_{W,0}^2 M_{W,0}^2}\left[1 + \frac{\Sigma_W W(q^2 = 0)}{M_W^2} + \ldots\right], \qquad (231)$$

where the ellipses denotes non-oblique terms and the suffix 0 denotes bare quantities. Expanding in terms of the renormalized parameters we have

$$\begin{aligned}
e_0^2 &= e^2(0)(1 + \Pi^\gamma(m^2, 0)) \simeq e^2(0)(1 + \Delta\alpha + \text{Re}\Pi^\gamma(0, M_Z^2)) \,, \\
M_{W,0}^2 &= M_W^2(1 - \hat\Pi^\gamma(m^2, M_Z^2) + \Pi^\gamma(m^2, M_Z^2) + \frac{\delta M_W^2}{M_W^2}) \,, \\
s_{W,0}^2 &= 1 - \frac{M_W^2 + \delta M_W^2}{M_Z^2 + \delta M_Z^2} = s_W^2 + c_W^2\left(\frac{\delta M_Z^2}{M_Z^2} - \frac{\delta M_W^2}{M_W^2}\right) .
\end{aligned} \qquad (232)$$

Keeping only the one loop terms we get,

$$
\begin{aligned}
\frac{G_\mu}{\sqrt{2}} &= \frac{\pi\alpha}{2s_W^2 M_W^2}\Big[1 + \Delta\alpha + \Pi^\gamma(0, M_Z^2) - \frac{c_W^2}{s_W^2}\Big(\frac{\delta M_Z^2}{M_Z^2} - \frac{\delta M_W^2}{M_W^2}\Big) + \frac{\Sigma^W(0) - \delta M_W^2}{M_W^2}\Big] \\
&= \frac{\pi\alpha}{2s_W^2 M_W^2}\Big[1 + \Delta r\Big].
\end{aligned}
\tag{233}
$$

The quantity Δr gives the finite correction to the tree level relation between the Fermi constant relevant for muon decay and the other parameters of the standard model. We first consider the contribution to Δr of a light doublet of fermions.

$$
\begin{aligned}
\frac{\Sigma^W(0)}{M_W^2} &= O\Big(\frac{m_\pm^2}{M_W^2}\Big) \simeq 0, \\
\frac{\delta M_Z^2}{M_Z^2} &= \frac{\alpha}{3\pi}\cdot\frac{l_+^2 + r_+^2 + l_-^2 + r_-^2}{2s_W^2 c_W^2}\Big[\Delta - \ln(\frac{M_Z^2}{\mu^2}) + \frac{5}{3}\Big], \\
\frac{\delta M_W^2}{M_W^2} &= \frac{\alpha}{3\pi}\cdot\frac{1}{4s_W^2}\Big[\Delta - \ln(\frac{M_Z^2}{\mu^2}) + \frac{5}{3}\Big] - \frac{\alpha}{12\pi s_W^2}\log c_W^2, \\
\mathrm{Re}\,\Pi^\gamma(0, M_Z^2) &= \frac{\alpha}{3\pi}(Q_+^2 + Q_-^2)\Big[\Delta - \ln(\frac{M_Z^2}{\mu^2}) + \frac{5}{3}\Big].
\end{aligned}
\tag{234}
$$

Adding everything together we obtain an expression for the contribution to Δr due to a light doublet of fermions.

$$
\Delta r = \Delta\alpha - \frac{\alpha}{3\pi}\frac{c_W^2 - s_W^2}{4s_W^4}\log c_W^2.
\tag{235}
$$

The singular piece cancels as a result of the identity

$$
(l_+^2 + r_+^2 + l_-^2 + r_-^2) = 2s_W^4(Q_+^2 + Q_-^2) - s_W^2 + \frac{1}{2}.
\tag{236}
$$

Thus, the main effect from the light fermions is to contribute to the running of the electric charge.

We now turn to the interesting case in which we have a heavy top quark. As we saw before this case gave rise to 1% correction to the ρ-parameter. To keep the treatment simple we will only keep only singular terms and terms quadratic in the top mass, m_t:

$$
\begin{aligned}
\mathrm{Re}\,\Pi^\gamma(0, M_Z^2) &= N_C\,\frac{\alpha}{3\pi}(Q_+^2 + Q_-^2)\Big(\Delta - \log\frac{m_t^2}{\mu^2}\Big) + \cdots, \\
\frac{\delta M_Z^2}{M_Z^2} &= N_C\,\frac{\alpha}{3\pi}\Big\{\frac{l_+^2 + r_+^2 + l_-^2 + r_-^2}{2s_W^2 c_W^2} - \frac{3m_t^2}{8s_W^2 c_W^2 M_Z^2}\Big\}\Big(\Delta - \log\frac{m_t^2}{\mu^2}\Big) + \cdots, \\
\frac{\delta M_W^2}{M_W^2} &= N_C\,\frac{\alpha}{3\pi}\Big\{\frac{1}{4s_W^2}\Big(\Delta - \log\frac{m_t^2}{\mu^2}\Big) - \frac{3m_t^2}{8s_W^2 M_W^2}\Big(\Delta - \log\frac{m_t^2}{\mu^2} + \frac{1}{2}\Big)\Big\} + \cdots, \\
\frac{\Sigma^W(0)}{M_W^2} &= -N_C\,\frac{\alpha}{3\pi}\cdot\frac{3m_t^2}{8s_W^2 M_W^2}\Big(\Delta - \log\frac{m_t^2}{\mu^2} + \frac{1}{2}\Big).
\end{aligned}
$$

$N_C = 3$ is the number of colours. Adding these contributions in the form mandated by Eq. (233) the singular parts cancel and in addition to the effect due to the running of α

from the b-quark loop, a finite term propotional to m_t^2 remains:

$$(\Delta r)_{b,t} = -\text{Re}\hat{\Pi}_b^{\gamma}(0, M_Z^2) - \frac{c_W^2}{s_W^2}\Delta\rho + \cdots \tag{237}$$

with $\Delta\rho$ from Eq. (229).

Thus the general structure for Δr is that it is dominated by the running of the electromagnetic charge and corrections from the heavy top quark. Following Hollik, (Hollik 1996) we may write an expression for Δr which is valid also after including the full non-fermionic contributions:

$$\Delta r = \Delta\alpha - \frac{c_W^2}{s_W^2}\Delta\rho + (\Delta r)_{\text{remainder}}. \tag{238}$$

$\Delta\alpha$ contains the large logarithmic corrections from the light fermions and $\Delta\rho$ the leading quadratic correction from a large top mass. All other terms are collected in $(\Delta r)_{\text{remainder}}$. Note that the remainder contains a term logarithmic in the top mass:

$$(\Delta r)_{\text{remainder}}^{\text{top}} = -\frac{\alpha}{4\pi s_W^2}\left(\frac{c_W^2}{s_W^2} - \frac{1}{3}\right)\log\frac{m_t}{M_Z} + \cdots \tag{239}$$

The Higgs boson also contributes to the remainder term in Δr. As discovered by Veltman (Veltman 1977) it depends only logarithmically on M_H:

$$(\Delta r)_{\text{remainder}}^{\text{Higgs}} \simeq \frac{\alpha}{16\pi s_W^2}\frac{11}{3}\left(\log\frac{M_H^2}{M_W^2} - \frac{5}{6}\right). \tag{240}$$

We have seen that the contributions are approximately resummed by including the running of the electromagnetic coupling and the effects of heavy top loops:

$$\frac{G_\mu}{\sqrt{2}} = \frac{\pi\alpha(M_Z)}{2s_W^2 M_W^2}\left[\frac{1}{1 + \frac{c_W^2}{s_W^2}\Delta\rho}\right]. \tag{241}$$

We may turn the argument around to estimate the top mass. Putting in the numbers from Eqs. (195,196,209) with $s_W^2 = 1 - M_W^2/M_Z^2$ we find that $\Delta\rho = 0.0103$, which, using Eq. (229), gives a value for the top quark mass of about 180 GeV.

4.3   Custodial symmetry

As described by Douglas Ross in his lectures the Higgs potential is a function of

$$\phi^\dagger\phi = \phi_1^2 + \phi_2^2 + \phi_3^2 + \phi_4^2. \tag{242}$$

It therefore has an SO(4) $\equiv$ SU(2) $\otimes$ SU(2) global symmetry, which is made manifest by rewriting the Higgs field as a matrix,

$$\Phi = \begin{pmatrix} \phi^{0\,*} & \phi^+ \\ -\phi^{+\,*} & \phi^0 \end{pmatrix} = \begin{pmatrix} \phi_2 - i\phi_4 & \phi_1 + i\phi_3 \\ -\phi_1 + i\phi_3 & \phi_2 + i\phi_4 \end{pmatrix}. \tag{243}$$

The potential $\mathcal{V}$ is a function of $\text{Tr}(\Phi^\dagger\Phi)$ and is therefore invariant under the global transformation

$$\Phi \to U_L \Phi U_R^\dagger , \tag{244}$$

where U_L and U_R are special unitary 2×2 matrices. The left one, U_L, is the familiar gauged SU(2) invariance. The potential has a SU(2)$_L$ × SU(2)$_R$ global symmetry which is larger than the local symmetry required by the SU(2)$_L$ × U(1)$_Y$ gauge invariance. The Higgs sector of the standard model can be written in this notation as

$$\mathcal{L} = \frac{1}{2}\left\{ \text{Tr}\left[(D_\mu\Phi)^\dagger D^\mu\Phi\right] + \mu^2\text{Tr}\left[\Phi^\dagger\Phi\right] - \lambda\text{Tr}\left[\Phi^\dagger\Phi\Phi^\dagger\Phi\right]\right\} , \tag{245}$$

where the covariant derivative in the present notation is

$$D_\mu\Phi = \partial_\mu\Phi + \frac{1}{2}ig_W W_\mu \cdot \tau\Phi - \frac{1}{2}ig'_W B_\mu\Phi\tau_3 . \tag{246}$$

The last term breaks the SU(2)$_R$ symmetry, so we first consider the limit $g'_W = 0$. In this limit the Lagrangian (including the gauge fields) has the full SU(2)$_L$ ⊗ SU(2)$_R$ symmetry if the W field transforms as a triplet under the gauged SU(2)$_L$ and as a singlet under the global SU(2)$_R$.

In this language the vacuum expectation value of the Higgs field is

$$\langle\Phi\rangle = \frac{v}{\sqrt{2}}\begin{pmatrix} 1 & 0 \\ 0 & 1 \end{pmatrix} . \tag{247}$$

After symmetry breaking, the SU(2)$_L$ ⊗ SU(2)$_R$ group is broken down to a single SU(2)$_V$ (or equivalently O(3)):

$$\langle\Phi\rangle \to U \langle\Phi\rangle U^\dagger = \langle\Phi\rangle , \tag{248}$$

where the left and right transformations are now the same. The structure of the electroweak gauge boson mass matrix is a consequence of the residual SU(2)$_V$ custodial symmetry plus the constraint that $m_\gamma = 0$. In the limit $g'_W \to 0$, the gauge bosons must be degenerate; they form an irreducible representation of the SU(2)$_V$ custodial group:

$$M^2 = \frac{v^2}{2}\begin{pmatrix} g_W^2 & & & \\ & g_W^2 & & \\ & & g_W^2 & -g_W g'_W \\ & & -g_W g'_W & g_W'^2 \end{pmatrix} , \tag{249}$$

leading to the relationship

$$\rho \equiv \frac{M_W^2}{M_Z^2 \cos^2\theta_W} = 1 , \tag{250}$$

which is thus a consequence of custodial symmetry.

4.3.1 Violations of custodial symmetry

Custodial SU(2)$_V$ is an accidental symmetry: it is a symmetry of all SU(2)$_L$× U(1)$_Y$ invariant terms of dimension 4 or less in the Higgs sector of the Lagrangian in the limit

$g'_W \to 0$. In dimension six we can have other operators which do not respect this accidental symmetry, such as

$$(\phi^\dagger D^\mu \phi)(\phi^\dagger D_\mu \phi) = \frac{1}{4}\left(\text{Tr}\,\sigma_3 \Phi^\dagger D^\mu \Phi\right)\left(\text{Tr}\,\sigma_3 \Phi^\dagger D_\mu \Phi\right). \tag{251}$$

Furthermore custodial $SU(2)_V$ is not a symmetry of the Higgs-fermion interactions. The Higgs boson must couple to the ordinary fermions in order to give rise to their observed masses. For example, for the third generation we have,

$$(\bar{t}_L \ \ \bar{b}_L)\,\Phi\begin{pmatrix} y_t & \\ & y_b \end{pmatrix}\begin{pmatrix} t_R \\ b_R \end{pmatrix}. \tag{252}$$

This violates custodial $SU(2)_V$ since

$$y_t \equiv \frac{\sqrt{2}m_t}{v} \gg y_b \equiv \frac{\sqrt{2}m_b}{v}. \tag{253}$$

As we have seen contributions to the gauge-boson self-energies violate Eq. (250) as a consequence of the violation of custodial symmetry and give rise to

$$\Delta\rho \approx 1\%\left(\frac{m_t}{175\ \text{GeV}}\right)^2. \tag{254}$$

Custodial symmetry is also violated by the presence of other Higgs representations such as Higgs triplets (see, for example, Forshaw et al 2001).

4.4 Vector boson scattering

The underlying gauge theory structure of the standard electroweak model leads to strong cancellations in the high-energy behaviour of amplitudes for the scattering of gauge bosons (Llewellyn Smith 1973). In this section we illustrate these cancellations by considering the scattering of W-bosons. The relevant diagrams, shown in Fig. 29, can be calculated using the Feynman rules of Fig. 30. The calculation will also allow us to display the role of the physical Higgs boson in WW scattering.

Consider the process

$$W^+(p_+) + W^-(p_-) \to W^+(q_+) + W^-(q_-) \tag{255}$$

in the centre-of-mass frame in which the incoming W bosons travel along the z axis. In this frame the kinematics of the reaction may be written as

$$\begin{aligned} p_\pm &= (E, 0, 0, \pm p), \\ q_\pm &= (E, 0, \pm p\sin\theta, \pm p\cos\theta), \end{aligned} \tag{256}$$

where $E^2 - p^2 = M_W^2$ and θ is the centre-of-mass scattering angle. The vector bosons can have longitudinal or transverse polarizations. Here we consider only the scattering of longitudinal bosons, which is responsible for the leading behaviour at high energy. The longitudinal polarization vectors for the W bosons are

$$\begin{aligned} \varepsilon_L(p_\pm) &= \left(\frac{p}{M_W}, 0, 0, \pm\frac{E}{M_W}\right), \\ \varepsilon_L(q_\pm) &= \left(\frac{p}{M_W}, 0, \pm\frac{E}{M_W}\sin\theta, \pm\frac{E}{M_W}\cos\theta\right). \end{aligned} \tag{257}$$

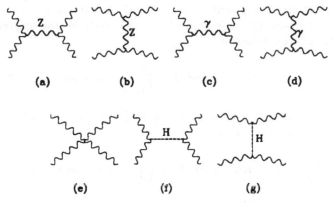

Figure 29. *Tree diagrams for $W^+ + W^- \rightarrow W^+ + W^-$ scattering.*

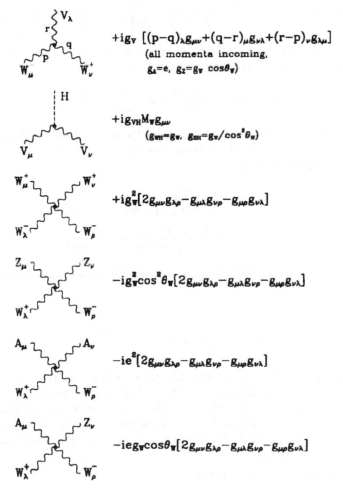

$+ig_V [(p-q)_\lambda g_{\mu\nu} + (q-r)_\mu g_{\nu\lambda} + (r-p)_\nu g_{\lambda\mu}]$
(all momenta incoming,
$g_A = e$, $g_Z = g_V \cos\theta_W$)

$+ig_{VH} M_W g_{\mu\nu}$
($g_{WH} = g_V$, $g_{ZH} = g_V/\cos^2\theta_W$)

$+ig_W^2 [2g_{\mu\nu}g_{\lambda\rho} - g_{\mu\lambda}g_{\nu\rho} - g_{\mu\rho}g_{\nu\lambda}]$

$-ig_W^2 \cos^2\theta_W [2g_{\mu\nu}g_{\lambda\rho} - g_{\mu\lambda}g_{\nu\rho} - g_{\mu\rho}g_{\nu\lambda}]$

$-ie^2 [2g_{\mu\nu}g_{\lambda\rho} - g_{\mu\lambda}g_{\nu\rho} - g_{\mu\rho}g_{\nu\lambda}]$

$-ieg_W \cos\theta_W [2g_{\mu\nu}g_{\lambda\rho} - g_{\mu\lambda}g_{\nu\rho} - g_{\mu\rho}g_{\nu\lambda}]$

Figure 30. *Propagators and Feynman rules for boson interactions.*

The polarizations satisfy the Lorentz condition $\varepsilon(k) \cdot k = 0$ and are normalized so that $\varepsilon^2 = -1$. In the high-energy limit we find that the result for the diagrams in Fig. 29 involving the three-boson vertex is

$$T^{a-d} = g_W^2 \left\{ \frac{p^4}{M_W^4} \left[3 - 6\cos\theta - \cos^2\theta \right] + \frac{p^2}{M_W^2} \left[\frac{9}{2} - \frac{11}{2}\cos\theta - 2\cos^2\theta \right] \right\}, \qquad (258)$$

neglecting terms not enhanced by factors of p^2/M_W^2. The corresponding result for the diagram with the four-boson vertex is

$$T^e = g_W^2 \left\{ \frac{p^4}{M_W^4} \left[-3 + 6\cos\theta + \cos^2\theta \right] + \frac{p^2}{M_W^2} \left[-4 + 6\cos\theta + 2\cos^2\theta \right] \right\}. \qquad (259)$$

Finally, the graphs involving Higgs exchange give

$$T^{f-g} = g_W^2 \left\{ \frac{p^2}{M_W^2} \left[-\frac{1}{2} - \frac{1}{2}\cos\theta \right] - \frac{M_H^2}{4M_W^2} \left[\frac{s}{s - M_H^2} + \frac{t}{t - M_H^2} \right] \right\}. \qquad (260)$$

Because of the longitudinal polarizations, the high-energy behaviour of individual graphs in Fig. 29 grows with the centre-of-mass momentum like

$$T(s,t) = A\left(\frac{p}{M_W}\right)^4 + B\left(\frac{p}{M_W}\right)^2 + C. \qquad (261)$$

On summing over graphs, the term A cancels without including the Higgs graphs, (f) and (g). The full cancellation of the B term involves the Higgs boson in an essential way. On summing all graphs the result is

$$T(s,t) = -g_W^2 \frac{M_H^2}{4M_W^2} \left[\frac{s}{s - M_H^2} + \frac{t}{t - M_H^2} \right] \qquad (262)$$

$$\equiv -4\left[\lambda + \frac{(\lambda v)^2}{s - 2\lambda v^2} + \frac{(\lambda v)^2}{t - 2\lambda v^2} \right]. \qquad (263)$$

Eq. (263) results from substituting for M_W and M_H using Eqs. (264) and (267):

$$M_W = \frac{1}{2} v g_W, \qquad (264)$$

$$M_Z = \frac{1}{2} v \sqrt{g_W^2 + g_W'^2} \equiv \frac{M_W}{\cos\theta_W}. \qquad (265)$$

To interpret this result we recall that the Higgs field ϕ acquires a non-zero vacuum expectation value at a particular point on the circle of minima away from $\phi = 0$ and the symmetry is spontaneously broken. However, this does not give rise to Goldstone bosons: instead, three of the four scalar field components get 'eaten' by the gauge particles to form massive vector bosons, $W^\pm, Z^0$, and there remains a single physical neutral scalar particle – the Higgs boson, H.

The self-couplings of the Higgs boson are

$$\mathcal{L}_{\text{Higgs}} = \frac{1}{2} \partial_\mu H \partial^\mu H - \mu^2 H^2 - \lambda v H^3 - \frac{1}{4}\lambda H^4, \qquad (266)$$

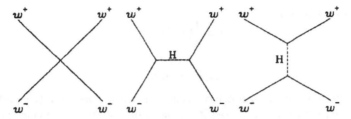

Figure 31. *Tree diagrams for $w^+ + w^- \to w^+ + w^-$ scattering.*

and its mass M_H is therefore

$$M_H = \sqrt{2}\mu \equiv \sqrt{2\lambda}v \, . \tag{267}$$

It is no accident that Eq. (263) looks like the amplitude for the scattering of charged scalar particles. The longitudinal components of the W-bosons are derived from the charged components of the Higgs doublet ϕ. The result of Eq. (263) can be simply calculated from the theory without gauge couplings in which there are three Goldstone bosons interacting with one massive scalar:

$$\mathcal{L}_{\text{Goldstone}} = \frac{1}{2}\partial_\mu H \partial^\mu H + \frac{1}{2}\partial_\mu z \partial^\mu z + \partial_\mu w^+ \partial^\mu w^- - \lambda v H^2$$

$$-\lambda v H (2w^+ w^- + z^2 + H^2) - \frac{1}{4}\lambda (2w^+ w^- + z^2 + H^2)^2 \, . \tag{268}$$

The two charged massless Goldstone bosons are denoted by $w^\pm$ and the diagrams are shown in Fig.. 31.

Although the leading-order WW scattering amplitude does not diverge at high energy, it might still be too large to be consistent with the fundamental physical limitation due to unitarity of the scattering matrix. To derive the unitarity limit on the high-energy behaviour it is convenient to perform a partial wave expansion,

$$T(s,t) = 16\pi \sum_J (2J+1)a_J(s)P_J(\cos\theta) \, , \tag{269}$$

where P_J are the Legendre polynomials, $P_0(x) = 1, P_1(x) = x$ and $P_2(x) = \frac{1}{2}(3x^2 - 1)$ etc., and $\cos\theta = 1 + 2t/s$ at high energy. The partial wave expansion decomposes the amplitudes into the contributions from various total angular momenta J. The differential cross section, neglecting particles masses, is given by

$$\frac{d\sigma}{d\Omega} = \frac{1}{64\pi^2 s}|T(s,t)|^2 \, . \tag{270}$$

Inserting Eq. (269) and using the orthogonality property of the Legendre polynomials,

$$\int_{-1}^{1} dx \, P_J(x)P_K(x) = \delta_{JK}\frac{2}{2J+1} \, , \tag{271}$$

we find that the total cross section is given by

$$\sigma = 16\pi \sum_J (2J+1)|a_J(s)|^2 \, . \tag{272}$$

Using the optical theorem

$$\sigma = \frac{1}{s} \text{Im } T(s,0) \, , \tag{273}$$

we find that

$$|a_J|^2 = \text{Im } a_J \, , \tag{274}$$

and hence the modulus of a_J cannot exceed one. Since J is conserved, the modulus of each partial wave amplitude cannot exceed its initial value, which for a plane-wave incoming state is one.

From Eq. (262), we find that the result for the $J = 0$ partial wave amplitude is

$$\begin{aligned} a_0(s) &= -\frac{G_F M_H^2}{8\pi\sqrt{2}} \left[2 + \frac{M_H^2}{s - M_H^2} - \frac{M_H^2}{s} \ln\left(1 + \frac{s}{M_H^2}\right) \right] \\ &\rightarrow -\frac{G_F M_H^2}{4\pi\sqrt{2}} \, . \end{aligned} \tag{275}$$

In the second line of this equation we have taken the limit $s \gg M_H^2$. Since unitarity will be violated if $|a_0| > 1$, this implies that $M_H^2 < 4\sqrt{2}\pi/G_F$. A slightly more sophisticated analysis (Lee et al 1977) involving consideration of all elastic boson-boson scattering channels, gives the limit

$$M_H < \left(\frac{8\sqrt{2}\pi}{3G_F}\right)^{\frac{1}{2}} \approx 1 \text{ TeV} \, . \tag{276}$$

We therefore find that if the limit in Eq. (276) is exceeded weak interactions will become strong, i.e. the high-energy scattering of weak bosons will not be adequately described by perturbation theory. It is important to stress the nature of the limit that we have derived. If Eq. (276) is violated the perturbative methods which we have used to derive the bound cannot be trusted. The Higgs mass can exceed the value given by the bound, but in that case its properties cannot be analyzed using perturbation theory. Therefore if we can investigate the scattering of longitudinal vector bosons at a high-energy hadron collider, we will either discover the Higgs boson or see strong interactions of the electroweak bosons, or both. This provides a compelling motivation for the construction of a machine capable of observing longitudinal vector boson scattering.

APPENDICES

A   Dimensional regularization

In the intermediate stages of calculations we must introduce some regularization procedure to control ultraviolet, soft and collinear divergences. The most effective regulator is the method of dimensional regularization which continues the dimension of space-time to $d = 4 - 2\epsilon$ dimensions ('t Hooft and Veltman 1972). This method of regularization has the advantage that the Ward identities of the theory are preserved at all stages of the calculation. Integrals over loop momenta are performed in d dimensions with the help of the following formula,

$$\int \frac{d^d k}{(2\pi)^d} \frac{(-k^2)^r}{\left[-k^2 + C - i\varepsilon\right]^m} =$$
$$\frac{i(4\pi)^\epsilon}{16\pi^2} [C - i\varepsilon]^{2+r-m-\epsilon} \frac{\Gamma(r + d/2)}{\Gamma(d/2)} \frac{\Gamma(m - r - 2 + \epsilon)}{\Gamma(m)}. \qquad (A.1)$$

To demonstrate Eq. (A.1), we first perform a Wick rotation of the k_0 contour anti-

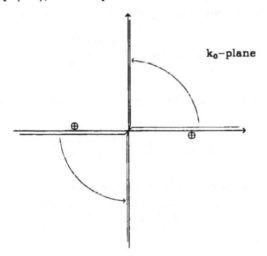

k_0-plane

Figure 32. *Wick rotation in the complex k_0 plane.*

clockwise. This is dictated by the $i\varepsilon$ prescription, since, for real C, the poles coming from the denominator of Eq. (A.1) lie in the second and fourth quadrant of the k_0 complex plane as shown in Fig. 32. Thus by anti-clockwise rotation of the contour of integration we encounter no poles. After rotation by an angle $\pi/2$, the k_0 integral runs along the imaginary axis in the k_0 plane, $(-i\infty < k_0 < i\infty)$. In order to deal only with real quantities we make the substitution $k_0 = i\kappa_d, k_j = \kappa_j$ for all $j \neq 0$ and introduce $|\kappa| = \sqrt{\kappa_1^2 + \kappa_2^2 \ldots + \kappa_d^2}$. We obtain a d-dimensional Euclidean integral which may be written as

$$\int d^d \kappa \, f(\kappa^2) = \int d|\kappa| \, f(\kappa^2) \, |\kappa|^{d-1} \sin^{d-2} \theta_{d-1} \sin^{d-3} \theta_{d-2} \ldots$$

$$\times \sin \theta_2 \; d\theta_{d-1} d\theta_{d-2} \ldots d\theta_2 d\theta_1 \; . \tag{A.2}$$

The range of the angular integrals is $0 \le \theta_i \le \pi$ except for $0 \le \theta_1 \le 2\pi$. Eq. (A.2) is best proved by induction. Assuming that it is true for an d-dimensional integral, in $(d+1)$ dimensions we can write

$$\int d^{d+1}\kappa \;=\; \int d\kappa_{d+1} \; d^d\kappa \tag{A.3}$$

$$=\; \int d\kappa_{d+1} \; d|\kappa| \; |\kappa|^{d-1} \sin^{d-2} \theta_{d-1} \sin^{d-3} \theta_{d-2} \ldots \sin \theta_2 \; d\theta_{d-1} d\theta_{d-2} \ldots d\theta_2 d\theta_1 \; . \tag{A.4}$$

The d-dimensional length, κ, can be written in terms of the $(d+1)$-dimensional length, ρ,

$$\kappa_{d+1} \;=\; \rho \cos \theta_d \; ,$$
$$|\kappa| \;=\; \rho \sin \theta_d \; . \tag{A.5}$$

Changing variables to ρ and θ_d we recover the $(d+1)$-dimensional version of Eq. (A.2).

The angular integrations, which only give an overall factor, can be performed using

$$\int_0^\pi d\theta \; \sin^d \theta = \sqrt{\pi} \frac{\Gamma\left(\frac{(d+1)}{2}\right)}{\Gamma\left(\frac{(d+2)}{2}\right)} \; . \tag{A.6}$$

We therefore find that the left hand side of Eq. (A.1) can be written as,

$$\frac{2i}{(4\pi)^{d/2}\Gamma\left(d/2\right)} \int_0^\infty d|\kappa| \; \frac{|\kappa|^{d+2r-1}}{\left[\kappa^2 + C\right]^m} \; . \tag{A.7}$$

This last integral can be reduced to a Beta function, (see Table 3),

$$\int_0^\infty dx \frac{x^s}{\left[x^2 + C\right]^m} = \frac{\Gamma\left(\frac{(s+1)}{2}\right)}{2} \frac{\Gamma(m - s/2 - 1/2)}{\Gamma(m)} C^{s/2+1/2-m} \; , \tag{A.8}$$

which demonstrates Eq. (A.1).

Feynman parameter identities are also useful for calculating virtual diagrams. The general form is

$$\frac{1}{A^\alpha \, B^\beta \ldots E^\epsilon} \;=\; \frac{\Gamma(\alpha + \beta + \cdots \epsilon)}{\Gamma(\alpha)\Gamma(\beta) \cdots \Gamma(\epsilon)}$$

$$\times \int_0^1 dx \, dy \cdots dz \, \delta(1 - x - y \cdots - z)$$

$$\times \frac{x^{\alpha-1} \, y^{\beta-1} \ldots z^{\epsilon-1}}{(Ax + By + \cdots + Ez)^{\alpha+\beta+\cdots+\epsilon}} \; . \tag{A.9}$$

$$\Gamma(z) = \int_0^\infty dt \ e^{-t} t^{z-1}$$

$$z\Gamma(z) = \Gamma(z+1)$$

$$\Gamma(2z) = \frac{2^{2z-1}}{\sqrt{\pi}}\Gamma(z)\Gamma(z+\tfrac{1}{2})$$

$$\Gamma(n+1) = n! \text{ for } n \text{ a positive integer}$$

$$\Gamma(1) = 1, \quad \Gamma(\tfrac{1}{2}) = \sqrt{\pi}$$

$$\Gamma'(1) = -\gamma_E, \quad \gamma_E \approx 0.57721566$$

$$\Gamma''(1) = \gamma_E^2 + \frac{\pi^2}{6}$$

$$B(a,b) = \int_0^1 dx \ x^a (1-x)^b$$

$$B(a,b) = \int_0^\infty dt \ \frac{t^{a-1}}{(1+t)^{a+b}} \quad \text{for Re } a, b > 0$$

$$B(a,b) = \frac{\Gamma(a)\Gamma(b)}{\Gamma(a+b)}$$

Table 3. *Useful properties of the Γ function and related functions.*

B   QCD corrections to the electromagnetic vertex

In this appendix we describe the calculation of the radiative corrections to the electromagnetic vertex. This calculation, which describes the modification of the interaction of a virtual photon with a quark due to strong interactions, is a good illustration of the use of dimensional regularization to control both ultraviolet and infrared singularities. The appropriate graphs are shown in Fig. 33.

At lowest order the expression for the vertex, Fig. 33(a), is

$$\Gamma_{(a)}^\mu(p',p) = -ie \ \bar{u}(p')\gamma^\mu u(p) \ . \tag{B.1}$$

In the dimensional regularization scheme, Figs. 33(c) and Figs. 33(d) give no contribution. For massless, on-shell quarks the dimensionally regularized integral vanishes,

$$\int \frac{d^d k}{k^4} = 0 \ , \tag{B.2}$$

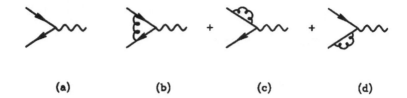

Figure 33. *Diagrams for QCD corrections to the electromagnetic vertex.*

because there is no dimensionful scale to which the integral could be proportional.

The calculation of the vertex diagram, Fig 33(b), proceeds as follows. Using the rules of Fig. 1 in the Feynman gauge ($\lambda = 1$) we have

$$\Gamma^\mu_{(b)}(p',p) = -i^3(-ig)^2(-ie) \int \frac{d^d l}{(2\pi)^d} \frac{\bar{u}(p')\gamma^\delta t^D (\slashed{l}+\slashed{p}')\gamma^\mu(\slashed{l}+\slashed{p})\gamma_\delta t^D u(p)}{((l+p')^2 + i\varepsilon)((l+p)^2 + i\varepsilon)(l^2 + i\varepsilon)} \cdot \quad (B.3)$$

Collecting terms and applying the color rule $t^D t^D = C_F I$ we get

$$\Gamma^\mu_{(b)}(p',p) = -eg^2 C_F \int \frac{d^d l}{(2\pi)^d} \frac{\bar{u}(p')\gamma^\delta (\slashed{l}+\slashed{p}')\gamma^\mu(\slashed{l}+\slashed{p})\gamma_\delta u(p)}{((l+p')^2 + i\varepsilon)((l+p)^2 + i\varepsilon)(l^2 + i\varepsilon)} \cdot \quad (B.4)$$

Notice the divergence structure of this expression. For large l the integral is logarithmically ultraviolet divergent:

$$\Gamma^\mu_{(b)}(p',p) \rightarrow -eg^2 C_F \int \frac{d^d l}{(2\pi)^d} \frac{\bar{u}(p')\gamma^\delta \slashed{l}\gamma^\mu \slashed{l}\gamma_\delta u(p)}{(l^2 + i\varepsilon)^3} \cdot \quad (B.5)$$

For small l (the soft region) the integral is also divergent:

$$\Gamma^\mu_{(b)}(p',p) \rightarrow -eg^2 C_F \int \frac{d^d l}{(2\pi)^d} \frac{4p \cdot p' \, \bar{u}(p')\gamma^\mu u(p)}{(2l \cdot p' + i\varepsilon)(2l \cdot p + i\varepsilon)(l^2 + i\varepsilon)} \cdot \quad (B.6)$$

In deriving this result we have used the commutation relation

$$\slashed{a}\gamma^\delta + \gamma^\delta \slashed{a} = 2a^\delta \quad (B.7)$$

to commute the momenta to positions where they can act on the spinors and use the equation of motion for massless free particles. The integral thus has ultraviolet and infrared divergences.

Returning to the full expression, Eq. (B.4), it proves useful to introduce Feynman parameters

$$\frac{1}{ABC} = 2 \int_0^1 d\alpha \int_0^1 d\beta \int_0^1 d\gamma \frac{\delta(1 - \alpha - \beta - \gamma)}{[\alpha A + \beta B + \gamma C]^3} \cdot \quad (B.8)$$

So the expression in Eq. (B.4) becomes

$$\Gamma^\mu_{(b)}(p',p) = -2eg^2 C_F \int \frac{d^d l}{(2\pi)^d} \int_0^1 d\alpha \int_0^1 d\beta \int_0^1 d\gamma \, \delta(1 - \alpha - \beta - \gamma)$$
$$\times \frac{\bar{u}(p')\gamma^\delta (\slashed{l}+\slashed{p}')\gamma^\mu(\slashed{l}+\slashed{p})\gamma_\delta u(p)}{\left[l^2 + 2\alpha p \cdot l + 2\beta p' \cdot l + i\varepsilon\right]^3} \cdot \quad (B.9)$$

We now perform the shift $l = l' + \alpha p + \beta p'$ so that the expression becomes

$$\Gamma^\mu_{(b)}(p',p) = -2eg^2 C_F \int \frac{d^d l}{(2\pi)^d} \int_0^1 d\alpha \int_0^1 d\beta \int_0^1 d\gamma \frac{\delta(1-\alpha-\beta-\gamma)\, N^\mu}{\left[l'^2 + \alpha\beta q^2 + i\varepsilon\right]^3} , \quad (B.10)$$

where the numerator is

$$N^\mu = \bar{u}(p')\gamma^\delta(\slashed{l}' + (1-\beta)\slashed{p}' - \alpha\slashed{p})\gamma^\mu(\slashed{l}' + (1-\alpha)\slashed{p} - \beta\slashed{p}')\gamma_\delta u(p) , \quad (B.11)$$

and $q = p' - p$. Terms odd in l' can be dropped so that the numerator becomes

$$N^\mu = \bar{u}(p')\gamma^\delta \slashed{l}'\gamma^\mu \slashed{l}'\gamma_\delta u(p) + \bar{u}(p')\gamma^\delta((1-\beta)\slashed{p}' - \alpha\slashed{p})\gamma^\mu((1-\alpha)\slashed{p} - \beta\slashed{p}')\gamma_\delta u(p) . \quad (B.12)$$

Using $l'_\mu l'_\nu = g_{\mu\nu} l'^2/n$ and $\gamma^\beta \gamma^\alpha \gamma_\beta = -2(1-\epsilon)\gamma^\alpha$, the first term in the numerator can be simplified to give

$$N^\mu_{(1)} = \frac{4(1-\epsilon)^2}{n} l'^2 \, \bar{u}(p')\gamma_\delta u(p) . \quad (B.13)$$

The second term in the numerator can also be simplified by using the equations of motion

$$\bar{u}(p')\slashed{p}' = 0, \quad \slashed{p} u(p) = 0 , \quad (B.14)$$

to give

$$N^\mu_{(2)} = 4\, p \cdot p'\, (1-\alpha)(1-\beta)\, \bar{u}(p')\gamma^\mu u(p) . \quad (B.15)$$

At first sight the term

$$+\alpha\beta\, \bar{u}(p')\gamma^\delta \slashed{p}'\gamma^\mu \slashed{p} \gamma_\delta u(p) \quad (B.16)$$

might appear to give a contribution, but since it is free from IR divergences it can be evaluated in four dimensions and hence vanishes using the equation of motion. The l' integral can easily be performed using Eq. (A.1)

$$\Gamma^\mu_{(b)}(p',p) = -ie\, \bar{u}(p')\gamma^\mu u(p) \frac{g^2 C_F}{16\pi^2}\left(\frac{4\pi\mu^2}{-q^2}\right)^\epsilon \Gamma(1+\epsilon) \int_0^1 d\alpha \int_0^1 d\beta \int_0^1 d\gamma$$

$$\times\; \delta(1-\alpha-\beta-\gamma)\left[2\frac{(1-\epsilon)^2}{\epsilon}(\alpha\beta)^{-\epsilon} - 2(1-\alpha)(1-\beta)(\alpha\beta)^{-1-\epsilon}\right] . \quad (B.17)$$

The integrals can be performed using the results in Table 3,

$$\int_0^1 d\alpha \int_0^1 d\beta \int_0^1 d\gamma\, \delta(1-\alpha-\beta-\gamma) \frac{(1-\alpha)(1-\beta)}{(\alpha\beta)^{1+\epsilon}} = \frac{\Gamma(1-\epsilon)^2}{\Gamma(2-2\epsilon)}\left[\frac{1}{\epsilon^2} + \frac{1}{2(1-\epsilon)}\right] ,$$

$$\int_0^1 d\alpha \int_0^1 d\beta \int_0^1 d\gamma\, \delta(1-\alpha-\beta-\gamma)\, (\alpha\beta)^{-\epsilon} = \frac{1}{2(1-\epsilon)} \frac{\Gamma(1-\epsilon)^2}{\Gamma(2-2\epsilon)} . \quad (B.18)$$

Using the identity

$$\frac{\Gamma(1+\epsilon)\Gamma(1-\epsilon)^2}{\Gamma(1-2\epsilon)} = \frac{1}{\Gamma(1-\epsilon)} + O(\epsilon^3) , \quad (B.19)$$

and collecting terms we may write the final answer for the lowest order term plus the radiative correction as

$$\Gamma^\mu(p',p) = -ie\bar{u}(p')\gamma^\mu u(p)\left\{1 + \frac{g^2 C_F}{16\pi^2 \Gamma(1-\epsilon)}\left(\frac{4\pi\mu^2}{-q^2 - i\varepsilon}\right)^\epsilon\left[-\frac{2}{\epsilon^2} - \frac{3}{\epsilon} - 8 + O(\epsilon)\right]\right\} . \quad (B.20)$$

Notice that for q^2 negative the result is real. For $q^2 > 0$ there is a branch cut extending from $q^2 = 0$ to $q^2 = \infty$. The path around the cut is indicated by the symbol ε.

References

Altarelli G and Parisi G, 1977, *Nucl Phys* B126 298.

Benvenuti A C *et al*, 1989, *Phys Lt B* 223 485.

Bhke S, 2002, arXiv:hep-ex/0211012.

Callan C G and Gross D J, 1969, *Phys Rev Lett* 22 156.

Dokshitzer Yu L, 1977, *Sov Phys JETP* 46 641.

Ellis R K *et al*, 1996, *Cambridge Monogr Part Phys Nucl Phys Cosmol* 8 1.

Erler J, 1999, *Phys Rev D* 59 054008.

Forshaw J R, White B E, and Ross D A, 2001, *JHEP* 0110 007.

Gribov V N and Lipatov L N, 1972, *Yad Fiz* 15 781 [*Sov J Nucl Phys* 15 438], *Yad Fiz* 15 1218 [*Sov J Nucl Phys* 15 675].

Hagiwara K *et al*, 2002, *Phys Rev D* 66 010001 and *www.pdg.lbl.gov*.

Hollik W, 1996, arXiv:hep-ph/9602380.

Lee B W, Quigg C, and Thacker H B, 1977, *Phys Rev D* 16 1519.

Lipatov L N, 1975, *Sov J Nucl Phys* 20 94.

Llewellyn Smith C H, 1973, *Phys Lett B* 46 233.

Miller G *et al*, 1972, *Phys Rev D* 5 528.

Ross D A, 2004, these proceedings.

Peskin M E, 1989, SLAC-PUB-5210.

't Hooft G and Veltman M J, 1972, *Nucl Phys B* 44 189.

van Ritbergen T, Vermaseren A M, and Larin S A , 1997, *Phys Lett B* 400 379.

Veltman M J, 1977, *Nucl Phys B* 123 89.

Looking beyond the standard model

John Ellis

CERN, Switzerland

1 Weighty problems

1.1 Roadmap to physics beyond the standard model

The standard model agrees with all confirmed experimental data from accelerators, but is theoretically very unsatisfactory [1, 2, 3, 4]. It does not explain the particle quantum numbers, such as the electric charge Q, weak isospin I, hypercharge Y and colour, and contains at least 19 arbitrary parameters. These include three independent vector-boson couplings and a possible CP-violating strong-interaction parameter, six quark and three charged-lepton masses, three generalized Cabibbo weak mixing angles and the CP-violating Kobayashi-Maskawa phase, as well as two independent masses for weak bosons. As seen in Fig. 1, the experimental data from LEP agree only too well with the theoretical curves, at all energies up to above 200 GeV [5]. This sounds great, but there are plenty of questions left open by the standard model.

The big issues in physics beyond the standard model are conveniently grouped into three categories [1, 2, 3]. These include the problem of mass: what is the origin of particle masses, are they due to a Higgs boson, and, if so, why are the masses so small; **unification**: is there a simple group framework for unifying all the particle interactions, a so-called grand unified theory (GUT); and **flavour**: why are there so many different types of quarks and leptons and why do their weak interactions mix in the peculiar way observed? Solutions to all these problems should eventually be incorporated in a theory of everything (TOE) that also includes gravity, reconciles it with quantum mechanics, explains the origin of space-time and why it has four dimensions, makes coffee, etc. String theory, perhaps in its current incarnation of M theory, is the best (only?) candidate we have for such a TOE [6], but we do not yet understand it well enough to make clear predictions at accessible energies.

As if the above 19 parameters were insufficient to appall you, at least nine more parameters must be introduced to accommodate the neutrino oscillations discussed later: 3 neutrino masses, 3 real mixing angles, and 3 CP-violating phases, of which one is in principle observable in neutrino-oscillation experiments and the other two in neutrinoless

John Ellis

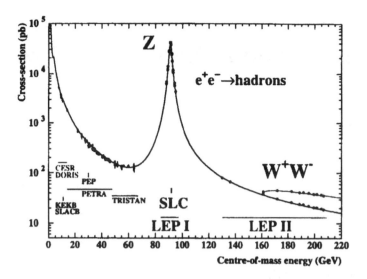

Figure 1. *Data from LEP and other e^+e^- experiments agree with the predictions of the standard model [5].*

double-beta decay experiments. In fact even the simplest models for neutrino masses involve 9 further parameters, as discussed later.

Moreover, there are many other cosmological parameters that we should also seek to explain. Gravity is characterized by at least two parameters, the Newton constant G_N and the cosmological vacuum energy. We may also want to construct a field-theoretical model for inflation, and we certainly need to explain the baryon asymmetry of the universe. So there is plenty of scope for physics beyond the standard model.

The mass problem is likely to require new physics, such as a Higgs boson or super-symmetry, to appear at some energy scale $\lesssim 1$ TeV, whereas GUTs may appear only at an energy $\sim 10^{16}$ GeV, and a TOE not until $\sim 10^{19}$ GeV. Measurements of the standard model gauge couplings also suggest that new physics such as supersymmetry may be needed at ~ 1 TeV, in order for a GUT to be possible. Also, one of the most natural candidates for astrophysical cold dark matter weighs $\lesssim 1$ TeV, and another hint for some new physics at the TeV scale may be provided by the on-again-off-again discrepancy between measurements of the anomalous magnetic moment of the muon and standard model calculation. However, there is no clear hint of the energy scale where the flavour problem might be solved, nor where extra dimensions might appear.

Foreseeable collider experiments will be able to probe directly physics at energy scales $\lesssim 1$ TeV, and the first clear evidence for physics beyond the standard model has been provided by neutrino physics, which may be offering us a window on the GUT scale. As we see later, neutrino physics might be the key to both inflation and the origin of the matter in the universe. Cosmology, via inflation, and high-energy astrophysics, via ultra-high-energy cosmic rays (UHECRs) may also offer windows on physics beyond the TeV scale. These lectures give emphasis to physics ideas that could be probed at forthcoming accelerators, but also include neutrino physics and cosmology.

1.2   Bosons of mass construction

Since the standard model is the rock on which our quest for new physics must be built, we first review its basic features [7] and examine whether its successes offer any hint of the direction in which to search for new physics. Let us first recall the structure of the charged-current weak interactions, which have the current-current form:

$$\frac{1}{4}L_{cc} = \frac{G_F}{\sqrt{2}} \ J_\mu^+ \ J^{-\mu},$$ (1)

where the charged currents violate parity maximally:

$$J_\mu^+ = \Sigma_{\ell=e,\mu,\tau}\bar{\ell}\gamma_\mu(1 - \gamma_5)\nu_\ell + \text{ similarly for quarks.}$$ (2)

The charged current (2) can be interpreted as a generator of a weak SU(2) isospin symmetry acting on the matter-particle doublets. The matter fermions with left-handed helicities are doublets of this weak SU(2), whereas the right-handed matter fermions are singlets. It was suggested already in the 1930's, and with more conviction in the 1960's, that the structure (2) could most naturally be obtained by exchanging massive $W^\pm$ vector bosons with coupling g and mass m_W:

$$\frac{G_F}{\sqrt{2}} = \frac{g^2}{8m_W^2}.$$ (3)

In 1973, neutral weak interactions with an analogous current-current structure were discovered at CERN:

$$\frac{1}{4}L_{NC} = \frac{G_F^{NC}}{\sqrt{2}} \ J_\mu^0 \ J^{\mu 0},$$ (4)

and it was natural to suggest that these might also be carried by massive neutral vector bosons Z^0.

The $W^\pm$ and Z^0 bosons were discovered at CERN in 1983, so let us now review the theory of them, as well as the Higgs mechanism of spontaneous symmetry breaking by which we believe they acquire masses [8]. The vector bosons are described by the Lagrangian

$$L = -\frac{1}{4} \ G_{\mu\nu}^i G^{i\mu\nu} - \frac{1}{4}F_{\mu\nu}F^{\mu\nu} ,$$ (5)

where $G_{\mu\nu}^i \equiv \partial_\mu W_\nu^i - \partial_\nu W_\mu^i + ig\epsilon_{ijk}W_\mu^j W_\nu^k$ is the field strength for the SU(2) vector boson W_μ^i, and $F_{\mu\nu} \equiv \partial_\mu W_\nu^i - \partial_\nu W_\mu^i$ is the field strength for a U(1) vector boson B_μ that is needed when we incorporate electromagnetism. The Lagrangian (5) contains bilinear terms that yield the boson propagators, and also trilinear and quartic vector-boson interactions.

The vector bosons couple to quarks and leptons via

$$L_F = -\sum_f i \ \left[\bar{f}_L\gamma^\mu D_\mu f_L + \bar{f}_R\gamma^\mu D_\mu f_R \right] ,$$ (6)

where the D_μ are covariant derivatives:

$$D_\mu \equiv \partial_\mu - i \, g \, \sigma_i \, W_\mu^i - i \, g' \, Y \, B_\mu .$$ (7)

The SU(2) piece appears only for the left-handed fermions f_L, whereas the U(1) vector boson B_μ couples to both left- and right-handed components, via their respective hypercharges Y.

The origin of all the masses in the standard model is postulated to be a weak doublet of scalar Higgs fields, whose kinetic term in the Lagrangian is

$$L_\phi = -|D_\mu \phi|^2 \tag{8}$$

and which has the magic potential:

$$L_V = -V(\phi) : V(\phi) = -\mu^2 \phi^\dagger \phi + \frac{\lambda}{2}(\phi^\dagger \phi)^2 . \tag{9}$$

Because of the negative sign for the quadratic term in (9), the symmetric solution $\langle 0|\phi|0\rangle = 0$ is unstable, and if $\lambda > 0$ the favoured solution has a non-zero vacuum expectation value which we may write in the form:

$$\langle 0|\phi|0\rangle = \langle 0|\phi^\dagger|0\rangle = v\left(0, \frac{1}{\sqrt{2}}\right) : v^2 = \frac{\mu^2}{2\lambda} , \tag{10}$$

corresponding to spontaneous breakdown of the electroweak symmetry.

Expanding around the vacuum: $\phi = \langle 0|\phi|0\rangle + \hat{\phi}$, the kinetic term (8) for the Higgs field yields mass terms for the vector bosons:

$$L_\phi \ni -\frac{g^2 v^2}{2} W_\mu^+ W^{\mu-} - g'^2 \frac{v^2}{2} B_\mu B^\mu + g g' v^2 B_\mu W^{\mu 3} - g^2 \frac{v^2}{2} W_\mu^3 W^{\mu 3} , \tag{11}$$

corresponding to masses

$$m_{W^\pm} = \frac{gv}{2} \tag{12}$$

for the charged vector bosons. The neutral vector bosons (W_μ^3, B_μ) have a 2×2 mass-squared matrix:

$$\begin{pmatrix} \frac{g^2}{2} & \frac{-gg'}{2} \\ \frac{-gg'}{2} & \frac{g'^2}{2} \end{pmatrix} v^2 . \tag{13}$$

This is easily diagonalized to yield the mass eigenstates:

$$Z_\mu = \frac{gW_\mu^3 - g'B_\mu}{\sqrt{g^2 + g'^2}} : m_Z = \frac{1}{2}\sqrt{g^2 + g'^2}\, v ; \quad A_\mu = \frac{g'W_\mu^3 + gB_\mu}{\sqrt{g^2 + g'^2}} : m_A = 0 , \tag{14}$$

that we identify with the massive Z^0 and massless γ, respectively. It is useful to introduce the electroweak mixing angle θ_W defined by

$$\sin\theta_W = \frac{g'}{\sqrt{g^2 + g'^2}} \tag{15}$$

in terms of the weak SU(2) coupling g and the weak U(1) coupling g'. Many other quantities can be expressed in terms of $\sin\theta_W$ (15): for example, $m_W^2/m_Z^2 = \cos^2\theta_W$.

With these boson masses, one indeed obtains charged-current interactions of the current-current form (2) shown above, and the neutral currents take the form:

$$J_\mu^0 \equiv J_\mu^3 - \sin^2\theta_W J_\mu^{em} , \quad G_F^{NC} \equiv \frac{g^2 + g'^2}{8m_Z^2} . \tag{16}$$

The ratio of neutral- and charged-current interaction strengths is often expressed as

$$\rho = \frac{G_F^{NC}}{G_F} = \frac{m_W^2}{m_Z^2 \cos^2 \theta_W} , \tag{17}$$

which takes the value unity in the standard model, apart from quantum corrections (loop effects).

The previous field-theoretical discussion of the Higgs mechanism can be rephrased in more physical language. It is well known that a massless vector boson such as the photon γ or gluon g has just two polarization states: $\lambda = \pm 1$. However, a massive vector boson such as the ρ has three polarization states: $\lambda = 0, \pm 1$. This third polarization state is provided by a spin-0 field. In order to make $m_{W^\pm, Z^0} \neq 0$, this should have non-zero electroweak isospin $I \neq 0$, and the simplest possibility is a complex isodoublet (ϕ^+, ϕ^0), as assumed above. This has four degrees of freedom, three of which are eaten by the $W^\pm$ and Z^0 as their third polarization states, leaving us with one physical Higgs boson H. Once the vacuum expectation value $|\langle 0|\phi|0\rangle| = v/\sqrt{2} : v = \mu/\sqrt{2\lambda}$ is fixed, the mass of the remaining physical Higgs boson is given by

$$m_H^2 = 2\mu^2 = 4\lambda v^2, \tag{18}$$

which is a free parameter in the standard model.

1.3   Testing the standard model

The standard model has been probed in many accelerator experiments, some of the most important being those carried out at LEP. The quantity that was measured most accurately there was the mass of the Z^0 boson [5]:

$$m_Z = 91,187.5 \pm 2.1 \text{ MeV}, \tag{19}$$

as seen in Fig. 2. Strikingly, m_Z is now known more accurately than the muon decay constant! Attaining this precision required understanding astrophysical effects – those of terrestrial tides on the LEP beam energy, which were $O(10)$ MeV, as well as meteorological – when it rained, the water expanded the rock in which LEP was buried, again changing the beam energy, and seasonal – variations in the level of water in Lake Geneva also caused the rock around LEP to expand and contract – as well as electrical – stray currents from the nearby electric train line affected the LEP magnets [9].

LEP experiments also made precision measurements of many properties of the Z^0 boson [5], such as the total cross section:

$$\sigma = \frac{12\pi}{m_Z^2} \frac{\Gamma_{ee}\Gamma_{had}}{\Gamma_Z^2}, \tag{20}$$

where $\Gamma_Z(\Gamma_{ee}, \Gamma_{had})$ is the total Z^0 decay rate (rate for decays into e^+e^-, hadrons). Eq. (20) is the classical (tree-level) expression, which is reduced by about 30 % by radiative corrections. The total decay rate is given by:

$$\Gamma_Z = \Gamma_{ee} + \Gamma_{\mu\mu} + \Gamma_{\tau\tau} + N_\nu \Gamma_{\nu\nu} + \Gamma_{had}, \tag{21}$$

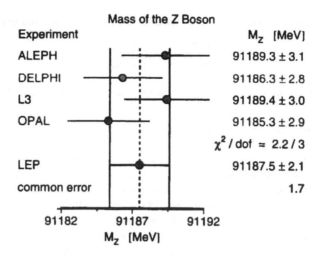

Figure 2. *The mass of the Z^0 vector boson is one of the parameters of the standard model that has been measured most accurately [5].*

where we expect $\Gamma_{ee} = \Gamma_{\mu\mu} = \Gamma_{\tau\tau}$ because of lepton universality, which has been verified experimentally, as seen in Fig. 3 [5]. Other partial decay rates have been measured via the branching ratios

$$R_{b,c} \equiv \frac{\Gamma_{b\bar{b},c\bar{c}}}{\Gamma_{had}}, \tag{22}$$

as seen in Fig. 4.

Also measured have been various forward-backward asymmetries $A_{\ell,q}$ in the production of leptons and quarks, as well as the polarization of τ leptons produced in Z^0 decay, as also seen in Fig. 4. Various other measurements are also shown there, including the mass and decay rate of the $W^{\pm}$, the mass of the top quark, and low-energy neutral-current measurements in ν-nucleon scattering and parity violation in atomic Cesium. The standard model is quite compatible with all these measurements, although some of them may differ by a couple of standard deviations: if they did not, we should be suspicious! Overall, the electroweak measurements tell us that [5]:

$$\sin^2\theta_W = 0.23143 \pm 0.00014, \tag{23}$$

providing us with a strong hint for grand unification, as we see later.

The precision electroweak measurements at LEP and elsewhere are sensitive to radiative corrections via quantum loop diagrams, in particular those involving particles such as the top quark and the Higgs boson that are too heavy to be observed directly at LEP [10, 11]. Many of the electroweak observables mentioned above exhibit quadratic sensitivity to the mass of the top quark:

$$\Delta \propto G_F m_t^2. \tag{24}$$

The measurements of these electroweak observables enabled the mass of the top quark to be predicted before it was discovered, and the measured value:

$$m_t = 174.3 \pm 5.1 \text{ GeV} \tag{25}$$

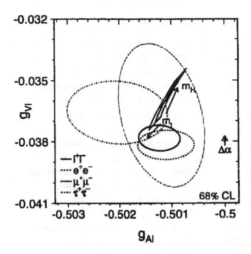

Figure 3. *Precision measurements of the properties of the charged leptons e, μ and τ indicate that they have universal couplings to the weak vector bosons [5], whose value favours a relatively light Higgs boson.*

agrees quite well with the prediction

$$m_t = 172^{+12}_{-9} \text{ GeV} \tag{26}$$

derived from precision electroweak data [5]. Electroweak observables are also sensitive logarithmically to the mass of the Higgs boson:

$$\Delta \propto \left(\frac{\alpha}{\pi}\right) \ln \left(\frac{m_H^2}{m_Z^2}\right), \tag{27}$$

so their measurements can also be used to predict the mass of the Higgs boson. This prediction can be made more definite by combining the precision electroweak data with the measurement (25) of the mass of the top quark. Making due allowance for theoretical uncertainties in the standard model calculations, as seen in Fig. 5, one may estimate that [5]:

$$m_H = 96^{+60}_{-38} \text{ GeV}, \tag{28}$$

whereas m_H is not known from first principles in the standard model.

1.4   The electroweak vacuums

Before considering the direct search for the Higgs boson, let us first cast around for any alternatives. Quite generally, the generation of particle masses requires the breaking of gauge symmetry in the vacuum:

$$m_{W,Z} \neq 0 \Leftrightarrow \langle 0|X_{I,I_3}|0\rangle \neq 0 \tag{29}$$

Summer 2003

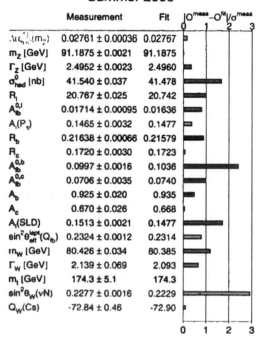

	Measurement	Fit	$\|O^{meas}-O^{fit}\|/\sigma^{meas}$ 0 1 2 3
$\Delta\alpha_{had}^{(5)}(m_z)$	0.02761 ± 0.00036	0.02767	
m_Z [GeV]	91.1875 ± 0.0021	91.1875	
Γ_Z [GeV]	2.4952 ± 0.0023	2.4960	
σ_{had}^0 [nb]	41.540 ± 0.037	41.478	
R_l	20.767 ± 0.025	20.742	
$A_{fb}^{0,l}$	0.01714 ± 0.00095	0.01636	
$A_l(P_\tau)$	0.1465 ± 0.0032	0.1477	
R_b	0.21638 ± 0.00066	0.21579	
R_c	0.1720 ± 0.0030	0.1723	
$A_{fb}^{0,b}$	0.0997 ± 0.0016	0.1036	
$A_{fb}^{0,c}$	0.0706 ± 0.0035	0.0740	
A_b	0.925 ± 0.020	0.935	
A_c	0.670 ± 0.026	0.668	
A_l(SLD)	0.1513 ± 0.0021	0.1477	
$\sin^2\theta_{eff}^{lept}(Q_{fb})$	0.2324 ± 0.0012	0.2314	
m_W [GeV]	80.426 ± 0.034	80.385	
Γ_W [GeV]	2.139 ± 0.069	2.093	
m_t [GeV]	174.3 ± 5.1	174.3	
$\sin^2\theta_W(\nu N)$	0.2277 ± 0.0016	0.2229	
Q_W(Cs)	-72.84 ± 0.46	-72.90	

Figure 4. *Precision electroweak measurements and the pulls they exert in a global fit [5].*

for some field X with isospin I and third component I_3. The measured ratio

$$\rho \equiv \frac{m_W^2}{m_Z^2 \cos^2\theta_W} \simeq 1 \tag{30}$$

tells us that X mainly has $I = 1/2$ [12], which is also what is needed to generate fermion masses. The key question is the nature of the field X: is it elementary or composite? A fermion-antifermion condensate $v \equiv \langle 0|X|0\rangle = \langle 0|\bar{F}F|0\rangle \neq 0$ would be analogous to what we know from QCD, where $\langle 0|\bar{q}q|0\rangle \neq 0$, and conventional superconductivity, where $\langle 0|e^-e^-|0\rangle \neq 0$.

In order to break the electroweak symmetry at a large enough scale, fermions with new interactions that become strong at a higher mass scale would be required. One suggestion was that the Yukawa interactions of the top quark might be strong enough to condense them: $\langle 0|\bar{t}t|0\rangle \neq 0$ [13], but this would have required the top quark to weigh more than 200 GeV, in simple models. Alternatively, theorists proposed the existence of new fermions held together by completely new interactions that became strong at a scale ~ 1 TeV, commonly called *Technicolour* models [14].

Specifically, the technicolour concept was to clone the QCD quark-antiquark condensate

$$\langle 0|\bar{q}_L q_R|0\rangle \sim \Lambda_{QCD}^3 : \quad \Lambda_{QCD} \sim 1\text{GeV}, \tag{31}$$

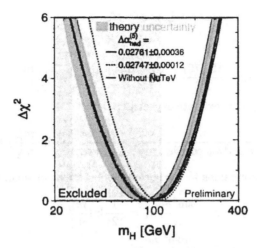

Figure 5. *Estimate of the mass of the Higgs boson obtained from precision electroweak measurements. The blue band indicates theoretical uncertainties, and the different curves demonstrate the effects of different plausible estimates of the renormalization of the fine-structure constant at the Z^0 peak [5].*

on a much larger scale, postulating a condensate of new massive fermions $\langle 0|\bar{Q}_L Q_R|0\rangle \sim \Lambda_{TC}^3$ where $\Lambda_{TC} \sim 1$ TeV. Assigning the techniquarks to the same weak representations as conventional quarks, $I_L = 1/2, I_R = 0$, the technicondensate breaks electroweak symmetry in just the way required to reproduce the relation (30). Just as QCD with two massless flavours would have three massless pions and a massive scalar meson, this simplest version of technicolour would predict three massless technipions that are eaten by the $W^\pm$ and Z^0 to provide their masses via the Higgs-Brout-Englert mechanism, leaving over a single physical scalar state weighing about 1 TeV, that would behave in some ways like a heavy Higgs boson.

Unfortunately, this simple technicolour picture must be complicated, in order to cancel triangle anomalies and to give masses to fermions [15], so the minimal default option has become a model with a single technigeneration:

$$\binom{\nu}{\ell}\binom{u}{d}_{1,2,3} \longrightarrow \binom{N}{L}_{1,...,N_{TC}} \binom{U}{D}_{1,...,N_{TC};1,2,3}. \tag{32}$$

One may then study models with different numbers N_{TC} of technicolours, and also different numbers of techniflavours N_{TF} if one wishes. The single-technigeneration model (32) above has $N_{TF} = 4$, corresponding to the technilepton doublet (N, L) and the three coloured techniquark doublets $(U, D)_{1,2,3}$.

The absence of any light technipions is already a problem for this scenario [16], as is the observed suppression of flavour-changing neutral interactions [17]. Another constraint is provided by precision electroweak data, which limit the possible magnitudes of one-loop radiative corrections due to virtual techniparticles. These may conveniently be parameterized in terms of three combinations of vacuum polarizations: for example [18]

$$T \equiv \frac{\epsilon_1}{\alpha} \equiv \frac{\Delta\rho}{\alpha}, \tag{33}$$

where

$$\Delta\rho = \frac{\Pi_{ZZ}(0)}{m_Z^2} - \frac{\Pi_{WW}(0)}{m_W^2} - 2\tan\theta_W \frac{\Pi_{\gamma Z}(0)}{m_Z^2}, \tag{34}$$

leading to the following approximate expression:

$$T = \frac{3}{16\pi \sin^2\theta_W \cos^2\theta_W}\left(\frac{m_t^2}{m_Z^2}\right) - \frac{3}{16\pi \cos^2\theta_W}\ln\left(\frac{m_H^2}{m_Z^2}\right) + \dots \tag{35}$$

There are analogous expressions for two other combinations of vacuum polarizations:

$$S \equiv \frac{4\sin^2\theta_W}{\alpha}\epsilon_3 = \frac{1}{12\pi}\ln\left(\frac{m_H^2}{m_Z^2}\right) + \dots \tag{36}$$

$$U \equiv -\frac{4\sin^2\theta_W}{\alpha}\epsilon_2. \tag{37}$$

The electroweak data may then be used to constrain $\epsilon_{1,2,3}$ (or, equivalently, S, T, U), and thereby extensions of the standard model with the same SU(2) × U(1) gauge group and extra matter particles that do not have important other interactions with ordinary matter. This approach does not include vertex corrections, so the most important one, that for $Z^0 \rightarrow b\bar{b}$, is treated by introducing another parameter ϵ_b.

This simple parameterization is perfectly sufficient to provide big headaches for the simple technicolour models described above. Fig. 6 compares the values of the parameters ϵ_i extracted from the final LEP data with the values calculated in the standard model for m_t within the range measured by CDF and D0, and for 113 GeV $< m_H <$ 1 TeV. We see that the agreement is quite good, which does not leave much room for new physics beyond the standard model to contribute to the ϵ_i. Fig. 7 compares these measured values also with the predictions of the simplest one-generation technicolour model, with $N_{TC} = 2$ and other assumptions described in [19].

We see that the data seem to disagree quite strongly with these technicolour predictions. Does this mean that technicolour is dead? Not quite [21], but it has motivated technicolour enthusiasts to pursue epicyclic variations on the original idea, such as walking technicolour [22], in which the technicolour dynamics is not scaled up from QCD in such a naive way. One cannot exclude the possibility that some calculable variant of technicolour might emerge that is consistent with the data, but for now we focus on elementary Higgs models.

1.5   Search for the Higgs boson

The Higgs production and decay rates are completely fixed as functions of the unknown mass m_H, enabling the search for the Higgs boson to be planned as a function of m_H [23]. This search was one of the main objectives of experiments at LEP, which established the lower limit:

$$m_H > 114.4 \text{ GeV}, \tag{38}$$

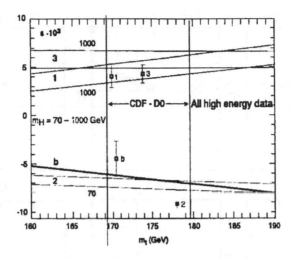

Figure 6. *The ranges of the vacuum-polarization parameters $\epsilon_{1,2,3,b}$ allowed by the precision electroweak data [20] are compared with the predictions of the standard model, as functions of the masses of the t quark and the Higgs boson.*

that is shown as the light yellow shaded region in Fig. 5 [24]. Combining this limit with the estimate (28), we see that there is good reason to expect that the Higgs boson may not be far away. Convoluting the likelihood function for the precision electroweak measurements with the lower limit established by the direct searches suggests that the Higgs mass is very likely to be below 125 GeV, as seen in Fig. 8 [25]. Indeed, in the closing weeks of the LEP experimental programme, there was a hint for the discovery of the Higgs boson at LEP with a mass $\sim$ 116 GeV, but this could not be confirmed [24].

If m_H is indeed as low as about 115 GeV, this would be *prima facie* evidence for physics beyond the standard model at a relatively low energy scale, as seen in Fig. 9, which might well be supersymmetry [27]. Consider first the case that m_H is larger than the central range marked in Fig. 9 [26]. The large Higgs self-coupling in the renormalization-group running of the effective Higgs potential would then cause it to blow up at some scale below the corresponding scale of Λ marked on the horizontal axis. Conversely, if m_H is below the central band, the larger Higgs-top Yukawa coupling overwhelms the relatively small Higgs self-coupling, driving the effective Higgs potential negative at some scale below the corresponding value of Λ. As a result, our present electroweak vacuum would be unstable, or at least metastable with a lifetime that might be longer than the age of the universe [28]. In the special case $m_H \sim 115$ GeV, this potential disaster could be averted only by postulating new physics at some scale $\Lambda \lesssim 10^6$ GeV.

This new physics should be bosonic [27], in order to counteract the negative effect of the fermionic top quark. Let us consider introducing N_I isomultiplets of bosons ϕ with isospin I, coupled to the conventional Higgs boson by

$$\lambda_{22}|H|^2|\phi|^2. \tag{39}$$

It turns out [27] that the coupled renormalization-group equations for the H, ϕ system

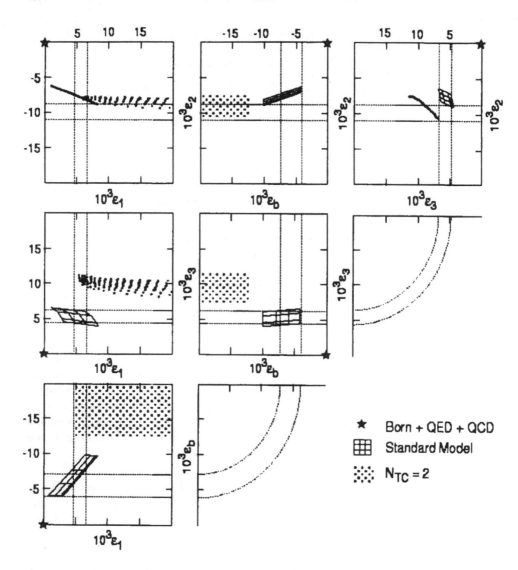

Figure 7. *Two-dimensional projections comparing the allowed ranges of the ϵ_i shown in Fig. 6 with the predictions of the standard model (hatched regions) and a minimal one-generation technicolour model (chicken-pox regions) [19].*

are very sensitive to the chosen value of λ_{22} in (39). As seen in Fig. 10, if the coupling

$$M_0^2 \equiv \lambda_{22}\langle 0|H|0\rangle^2 \tag{40}$$

is too large, the effective Higgs potential blows up, but it collapses if M_0^2 is too small, and the typical amount of fine-tuning required is 1 in 10^3! Radiative corrections may easily upset this fine-tuning, as seen in Fig. 11. The fine-tuning is maintained naturally in a supersymmetric theory, but is destroyed if one has top quarks and their supersymmetric

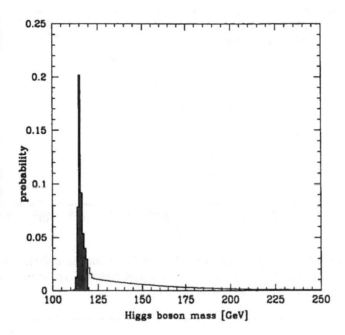

Figure 8. *An estimated probability distribution for the Higgs mass [25], obtained by convoluting the blue-band plot in Fig. 5 [5] with the experimental exclusion [24].*

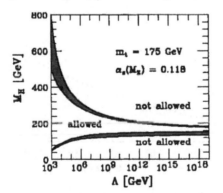

Figure 9. *The effective Higgs potential is well-behaved up the the Planck scale $m_P \simeq 10^{19}$ GeV only for a narrow range of Higgs masses ~ 180 GeV. A larger Higgs mass would cause the coupling to blow up at lower energies, and a smaller Higgs mass would cause the potential to turn negative at some scale $\Lambda \ll m_P$ [26].*

partners $\tilde{t}$, but not the supersymmetric partners $\tilde{H}$ of the Higgs bosons.

If the new physics below 10^6 GeV is not supersymmetry, it must quack very much like it!

Let us now turn, finally, to future direct searches for the Higgs boson. First at bat is the Fermilab Tevatron collider, the highest-energy accelerator currently operating. This

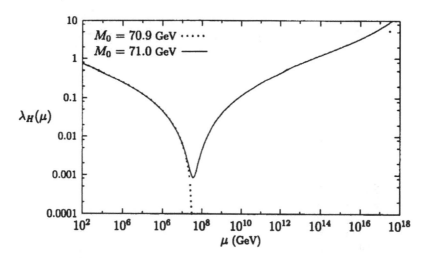

Figure 10. *Renormalization of the effective Higgs self-coupling for different values of the coupling M_0 to new bosons ϕ. It is seen that the coupled system must be tuned very finely in order for the potential not to collapse or blow up [27].*

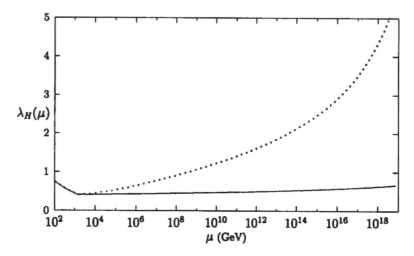

Figure 11. *An example of the role played by fermionic superpartners in the running of the Higgs self-coupling. The solid line corresponds to a supersymmetric model, whereas the dotted line gives the running of the quartic Higgs coupling when the contributions from fermionic Higgsino and gaugino superpartners have been removed [27].*

has a chance to detect the Higgs boson if it is light enough, and if the Tevatron collider can accumulate sufficient luminosity [29], as seen in Fig. 12. Unfortunately, the latest estimates suggest that the Tevatron collider will not achieve the required luminosity before the CERN LHC comes into operation.

The LHC will be able to discover the Higgs boson, whatever it mass below about 1 TeV, as seen in Fig. 13 [30]. For any value of the Higgs mass, the LHC experiments

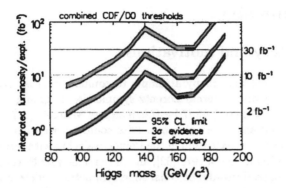

Figure 12. *The Tevatron collider experiments will have a chance to exclude a light Higgs boson, or perhaps find 3-σ evidence, or even discover it at the 5-σ level, if the collider can accumulate sufficient luminosity [29].*

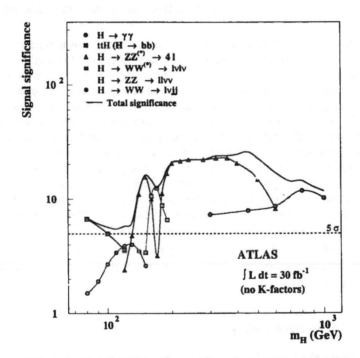

Figure 13. *The LHC experiments will be able to discover the Higgs boson with high significance, whatever its mass, and may observe several of its decay modes [30].*

ATLAS and CMS should be able to observe two or three of its decay modes, including $H \rightarrow \gamma\gamma, \bar{b}b$ and $\tau^+\tau^-$ at low masses, $H \rightarrow 4$ charged leptons at intermediate masses and $H \rightarrow W^+W^-$ and ZZ at high masses. Depending on the Higgs mass, they should also be able to measure it to 1% or better as well as measure at least the ratios of some of its couplings. The days of the Higgs boson are numbered!

2   Supersymmetry

2.1   Early history and motivations

Back in the 1960's, there were many attempts to combine internal symmetries such as flavour SU(2) or SU(3) with external Lorentz symmetries, in groups such as SU(6) and $\bar{U}(12)$. However, it was shown in 1967 by Coleman and Mandula [31] that no non-trivial combination of internal and external symmetries could be achieved using just bosonic charges. The first non-trivial extension of the Poincaré algebra with fermionic charges was made by Golfand and Likhtman in 1971 [32], and in the same year Neveu and Schwarz [33], and Ramond [34], proposed two-dimensional supersymmetric models in attempts to make fermionic string theories that could accommodate baryons. Two years later, the first interesting four-dimensional supersymmetric field theories were written down. Volkov and Akulov [35] wrote down a non-linear realization of supersymmetry with a massless fermion, that they hoped to identify with the neutrino, but this identification was soon found not to work, because the low-energy interactions of neutrinos differed from those in the non-linear supersymmetric model.

Also in 1973, Wess and Zumino [36, 37] started writing down renormalizable four-dimensional supersymmetric field theories, with the objective of describing mesons and baryons. Soon afterwards, together with Iliopoulos and Ferrara, they were able to show that supersymmetric field theories lacked many of the quadratic and other divergences found in conventional field theories [38], and some thought this was an attractive feature, although the physical application remained obscure for several years. Instead, for some time, phenomenological interest in supersymmetry was focused on the possibility of unifying fermions and bosons, for example matter particles (with spin 1/2) and force particles (with spin 1), or alternatively matter and Higgs particles, in the same supermultiplets [39]. With the discovery of local supersymmetry, or supergravity, in 1976 [40], this hope was extended to the unification of the graviton with lower-spin particles. Indeed, for a short while, the largest supergravity theory was touted as the TOE: in the words of Hawking [41], 'Is the end in sight for theoretical physics?'.

These are all attractive ideas, and many play rôles in current theories, but the only real motivation for expecting supersymmetry at accessible energies $\lesssim 1$ TeV is the naturalness of the mass hierarchy [42].

2.2   Why supersymmetry?

One may formulate the hierarchy problem [42] by posing the question why is $m_W \ll m_P$, or, equivalently, why is $G_F \sim 1/m_W^2 \gg G_N = 1/m_P^2$? Another equivalent question is why the Coulomb potential in an atom is so much greater than the Newton potential: $e^2 \gg G_N m^2 = m^2/m_P^2$, where m is a typical particle mass?

Your first thought might simply be to set $m_P \gg m_W$ by hand, and forget about the problem. Life is not so simple, because quantum corrections to m_H and hence m_W, such as those shown in Fig. 14(a), are quadratically divergent in the standard model:

$$\delta m_{H,W}^2 \simeq O(\frac{\alpha}{\pi})\Lambda^2, \tag{41}$$

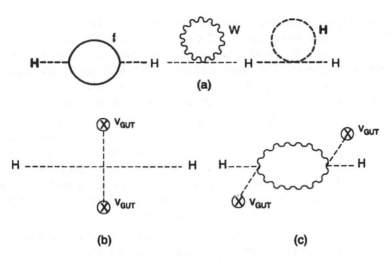

Figure 14. *(a) One-loop quantum corrections to m_H^2 in the standard model. (b) Tree-level and (c) one-loop corrections to m_H^2 in a GUT.*

which is $\gg m_W^2$ if the cutoff Λ, which represents the scale where new physics beyond the standard model appears, is comparable to the GUT or Planck scale. For example, if the standard model were to hold unscathed all the way up the Planck mass $m_P \sim 10^{19}$ GeV, the radiative correction (41) would be 36 orders of magnitude greater than the physical values of $m_{H,W}^2$!

In principle, this is not a problem from the mathematical point of view of renormalization theory. All one has to do is postulate a tree-level value of m_H^2 that is (very nearly) equal and opposite to the 'correction' (41), and the correct physical value may be obtained. However, this strikes many physicists as rather unnatural: they would prefer a mechanism that keeps the 'correction' (41) comparable at most to the physical value.

This was one of the motivations for technicolour theories. Above the characteristic technicolour energy scale, the Higgs boson 'dissolves' into its constituents, and the quadratic divergences are softened. However, as we saw earlier, extant technicolour theories disagree with the precision electroweak data from LEP and elsewhere. The alternative is to cancel the quadratic divergences, as occurs in a supersymmetric theory, in which there are equal numbers of bosons and fermions with identical couplings. Since bosonic and fermionic loops have opposite signs, the residual one-loop correction is of the form

$$\delta m_{H,W}^2 \simeq O(\frac{\alpha}{\pi})(m_B^2 - m_F^2), \tag{42}$$

which is $\lesssim m_{H,W}^2$ and hence naturally small if the supersymmetric partner bosons B and fermions F have similar masses:

$$|m_B^2 - m_F^2| \lesssim 1 \text{ TeV}^2. \tag{43}$$

This is the best motivation we have for finding supersymmetry at relatively low energies [42].

In addition to this first supersymmetric miracle of removing (42) the quadratic divergence (41), many logarithmic divergences are also absent in a supersymmetric theory.

This is the underlying reason why supersymmetry solves the fine-tuning problem of the effective Higgs potential when $m_H \sim 115$ GeV, as advertised in the previous Lecture. Note that this argument is logically distinct from the absence of quadratic divergences in a supersymmetric theory. The absence of some logarithmic divergences also prevents the high mass scales in a supersymmetric GUT theory from 'leaking' into the electroweak sector, as shown in Fig. 14(b), and spoiling the hierarchy, as discussed in Lecture 6.

By now, you may be wondering whether it makes sense to introduce so many new particles just to deal with a paltry little hierarchy or naturalness problem. But, as they used to say during the First World War, 'if you know a better hole, go to it.' As we learnt above, technicolour no longer seems to be a viable hole, and I am not convinced that theories with large extra dimensions really solve the hierarchy problem, rather than just rewrite it. Fortunately, there are two hints from the high-precision electroweak data that supersymmetry may not be such a bad hole, after all.

One is the fact that there probably exists a Higgs boson weighing less than about 200 GeV [5]. This is perfectly consistent with calculations in the minimal supersymmetric extension of the standard model (MSSM) [44], in which the lightest Higgs boson weighs less than about 130 GeV [43], as we discuss later in more detail.

The other hint is provided by the strengths of the different gauge interactions, as measured at LEP [45]. These may be run up to high energy scales using the renormalization-group equations [46, 47], to see whether they unify as predicted in a GUT. The answer is no, if supersymmetry is not included in the calculations, as seen in Fig. 15. In that case, GUTs would require

$$\sin^2 \theta_W = 0.214 \pm 0.004, \qquad (44)$$

whereas the experimental value of the effective neutral weak mixing parameter at the Z^0 peak is $\sin^2 \theta = 0.23149 \pm 0.00017$ [5]. On the other hand, as also seen in Fig. 15, unification is possible in minimal supersymmetric GUTs, which predict

$$\sin^2 \theta_W \sim 0.232, \qquad (45)$$

where the error depends on the assumed sparticle masses, the preferred value being around 1 TeV, as suggested completely independently by the naturalness of the electroweak mass hierarchy. This argument is also discussed later in more detail.

Another hint for supersymmetry may be provided by the astrophysical need for cold dark matter, discussed in Lecture 4, which could be met by the lightest supersymmetric particle (LSP), if it weighs less than about a TeV. From time to time, the measured value of the muon anomalous magnetic moment $g_\mu - 2$ also suggests there may be new physics beyond the standard model at accessible energies below about 1 TeV, which could well be supersymmetry. However, there are still significant uncertainties in the standard model contributions to $g_\mu - 2$, so this should not be relied upon too heavily.

2.3   What is supersymmetry?

The basic idea of supersymmetry is the existence of fermionic charges Q_α that relate bosons to fermions. Recall that all previous symmetries, such as flavour SU(3) or electromagnetic U(1), have involved scalar charges Q that link particles with the same spin into

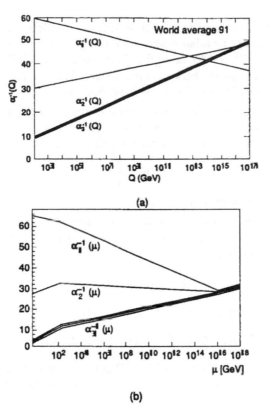

Figure 15. *The measurements of the gauge coupling strengths at LEP (a) do not evolve to a unified value if there is no supersymmetry but do (b) if supersymmetry is included [45].*

multiplets:

$$Q\,|\text{Spin}J> = \;|\text{Spin}J>. \qquad (46)$$

Indeed, Coleman and Mandula [31] proved that it was 'impossible' to mix internal and Lorentz symmetries: $J_1 \leftrightarrow J_2$. However, their 'no-go' theorem assumed implicitly that the prospective charges should have integer spins.

The basic element in their 'proof' was the observation that the only possible conserved tensor charges were those with no Lorentz indices, i.e., scalar charges, and the energy-momentum vector P_μ. To see how their 'proof' worked, consider two-to-two elastic scattering, $1 + 2 \rightarrow 3 + 4$, and imagine that there exists a conserved two-index tensor charge, $\Sigma_{\mu\nu}$. By Lorentz invariance, its diagonal matrix elements between single-particle states $|a>$ must take the general form:

$$\langle a|\Sigma_{\mu\nu}|a\rangle = \alpha P_\mu^{(a)} P_\nu^{(a)} + \beta g_{\mu\nu}, \qquad (47)$$

where α, β are arbitrary reduced matrix elements, and $g_{\mu\nu}$ is the metric tensor. For $\Sigma_{\mu\nu}$ to be conserved in a two-to-two scattering process, one must have

$$P_\mu^{(1)} P_\nu^{(1)} + P_\mu^{(2)} P_\nu^{(2)} = P_\mu^{(3)} P_\nu^{(3)} + P_\mu^{(4)} P_\nu^{(4)}, \qquad (48)$$

where we assume that the symmetry is local, so that two-particle matrix elements of $\Sigma_{\mu\nu}$ play no rôle. Since Lorentz invariance also requires $P_\mu^{(1)} + P_\mu^{(2)} = P_\mu^{(3)} + P_\mu^{(4)}$, the only possible outcomes are $P_\mu^{(1)} = P_\mu^{(3)}$ or $P_\mu^{(4)}$. Thus the only possibilities are completely forward scattering or completely backward scattering. This disagrees with observation, and is in fact theoretically impossible in any local field theory.

This rules out any non-trivial two-index tensor charge, and the argument can clearly be extended to any higher-rank tensor with more Lorentz indices. But what about a spinorial charge Q_α? This can have no diagonal matrix element:

$$\langle a|Q_\alpha|a\rangle \neq 0, \tag{49}$$

and hence the Coleman-Mandula argument fails.

So what is the possible form of a 'supersymmetry' algebra that includes such spinorial charges Q_α^i [1]? Since the different Q^i are supposed to generate symmetries, they must commute with the Hamiltonian:

$$[Q^i, H] = 0 : i = 1, 2, , N. \tag{50}$$

So also must the anticommutator of two spinorial charges:

$$[\{Q^i, Q^j\}, H] = 0 : i, j = 1, 2, , N. \tag{51}$$

However, the part of the anticommutator $\{Q^i, Q^j\}$ that is symmetric in the internal indices i, j cannot have spin 0. Instead, as we discussed just above, the only possible non-zero spin choice is $J = 1$, so that

$$\{Q^i, Q^j\} \propto \delta^{ij} P_\mu + \ldots : i, j = 1, 2, \ldots, N. \tag{52}$$

In fact, as was proved by Haag, Lopuszanski and Sohnius [48], the only allowed possibility is

$$\{Q^i, Q^j\} = 2\delta^{ij} \gamma^\mu P_\mu C + \ldots : i, j = 1, 2, \ldots, N, \tag{53}$$

where C is the charge-conjugation matrix discussed in more detail in Lecture 2, and the dots denote a possible 'central charge' that is antisymmetric in the indices i, j, and hence can only appear when $N > 1$.

According to a basic principle of Swiss law, anything not illegal is compulsory, so there MUST exist physical realizations of the supersymmetry algebra (53). Indeed, non-trivial realizations of the non-relativistic analogue of (53) are known from nuclear physics [49], atomic physics and condensed-matter physics. However, none of these is thought to be fundamental.

At the fundamental level, in the relativistic limit supermultiplets consist of massless particles with spins differing by half a unit. In the case of simple $N = 1$ supersymmetry, the basic building blocks are *chiral supermultiplets*:

$$\left(\frac{1}{2}, 0\right) \quad e.g., \quad \left(\ell(\text{lepton}), \tilde{\ell}(\text{slepton})\right) \quad \text{or} \quad \left(q(\text{quark}), \tilde{q}(\text{squark})\right), \tag{54}$$

---

[1] In what follows, I shall suppress the spinorial subscript α whenever it is not essential. The superscripts $i, j, \ldots, N$ denote different supersymmetry charges.

gauge supermultiplets:

$$\left(1, \frac{1}{2}\right) \; e.g., \; (\gamma \, (\text{photon}), \tilde{\gamma} \, (\text{photino})) \; or \; (g \, (\text{gluon}), \tilde{g} \, (\text{gluino})), \qquad (55)$$

and the *graviton supermultiplet* consisting of the spin-2 graviton and the spin-3/2 gravitino.

2.4 Deconstructing Dirac

In this Section, we tackle some unavoidable spinorology. The most familiar spinors used in four-dimensional field theories are four-component Dirac spinors ψ. You may recall that it is possible to introduce projection operators

$$P_{L,R} \equiv \frac{1}{2}(1 \mp \gamma_5), \qquad (56)$$

where $\gamma_5 \equiv i\gamma^0 \gamma^1 \gamma^2 \gamma^3$, and the γ_μ can be written in the forms

$$\gamma_\mu = \begin{pmatrix} 0 & \sigma_\mu \\ \bar{\sigma}_\mu & 0 \end{pmatrix}, \qquad (57)$$

where $\sigma_\mu \equiv (1, \sigma_i)$, $\bar{\sigma}_\mu \equiv (1, -\sigma_i)$. Then γ_5 can be written in the form diag$(-1, 1)$, where $-1, 1$ denote 2×2 matrices. Next, we introduce the corresponding left- and right-handed spinors

$$\psi_{L,R} \equiv P_{L,R}\psi, \qquad (58)$$

in terms of which one may decompose the four-component spinor into a pair of two-component spinors:

$$\psi = \begin{pmatrix} \psi_L \\ \psi_R \end{pmatrix}. \qquad (59)$$

These will serve as our basic fermionic building blocks.

Antifermions can be represented by adjoint spinors

$$\bar{\psi} \equiv \psi^\dagger \gamma^0 = (\bar{\psi}_R, \bar{\psi}_L) \qquad (60)$$

where the γ^0 factor has interchanged the left- and right-handed components $\psi_{L,R}$. We can now decompose in terms of these the conventional fermion kinetic term

$$\bar{\psi}\gamma_\mu \partial^\mu \psi = \bar{\psi}_L \gamma_\mu \partial^\mu \psi_L + \bar{\psi}_R \gamma_\mu \partial^\mu \psi_R \qquad (61)$$

and the conventional mass term

$$\bar{\psi}\psi = \bar{\psi}_R \psi_L + \bar{\psi}_L \psi_R. \qquad (62)$$

We see that the kinetic term keeps separate the left- and right-handed spinors, whereas the mass term mixes them.

The next step is to introduce the charge-conjugation operator C, which changes the overall sign of the vector current $\bar{\psi}\gamma^\mu\psi$. It transforms spinors into their conjugates:

$$\psi^c \equiv C \, \bar{\psi}^T = C(\psi^\dagger \gamma^0)^T = \begin{pmatrix} \bar{\psi}_R \\ \bar{\psi}_L \end{pmatrix}, \qquad (63)$$

and operates as follows on the γ matrices:

$$C^{-1}\gamma^\mu C = -\gamma^{\mu T}. \tag{64}$$

A convenient representation of C is:

$$C = i\gamma^0\gamma^2. \tag{65}$$

It is apparent from the above that the conjugate of a left-handed spinor is right-handed:

$$(\psi_L)^c = \begin{pmatrix} 0 \\ \bar{\psi}_L \end{pmatrix}, \tag{66}$$

so that the combination

$$\bar{\psi}_L^c \psi_L = \psi_L \sigma_2 \psi_L \tag{67}$$

mixes left- and right-handed spinors, and has the same form as a mass term (62).

It is apparent from (66) that we can construct four-component Dirac spinors entirely out of two-component left-handed spinors and their conjugates:

$$\psi = \begin{pmatrix} \psi_i \\ \psi_j^c \end{pmatrix}, \tag{68}$$

a trick that will be useful later in our supersymmetric model-building. As examples, instead of working with left- and right-handed quark fields q_L and q_R, or left- and right-handed lepton fields ℓ_L and ℓ_R, we can write the theory in terms of left-handed antiquarks and antileptons: $q_R \to q_L^c$ and $\ell_R \to \ell_L^c$.

2.5   Simplest supersymmetric field theories

Let us now consider a field theory [50] containing just a single left-handed fermion ψ_L and a complex boson ϕ, without any interactions, as described by the Lagrangian

$$L_0 = i\bar{\psi}_L\gamma^\mu\partial_\mu\psi_L + |\partial\phi|^2. \tag{69}$$

We consider the simplest possible non-trivial transformation law for the free theory (69):

$$\phi \to \phi + \delta\phi, \text{ where } \delta\phi = \sqrt{2}\bar{E}\psi_L, \tag{70}$$

where E is some constant right-handed spinor. In parallel with (70), we also consider the most general possible transformation law for the fermion ψ:

$$\psi_L \to \psi_L + \delta\psi_L, \text{ where } \delta\psi_L = -a\,i\,\sqrt{2}(\gamma_\mu\partial^\mu\phi)E - FE^c, \tag{71}$$

where a and F are constants to be fixed later, and we recall that E^c is a left-handed spinor. We can now consider the resulting transformation of the full Lagrangian (69), which can easily be checked to take the form

$$\delta L_0 = \sqrt{2}\partial_\mu[\bar{\psi}E\partial^\mu\phi + \bar{E}\gamma^\mu\phi^*\gamma_\nu\partial^\nu\psi], \tag{72}$$

if and only if we choose

$$a = 1 \text{ and } F = 0 \tag{73}$$

in this free-field model. With these choices, and the resulting total-derivative trans-formation law (72) for the free Lagrangian, the free action A_0 is invariant under the transformations (70,71), since

$$\delta A_0 = \delta \int d^4 x L_0 = 0. \tag{74}$$

Fine, you may say, but is this symmetry actually supersymmetry? To convince yourself that it is, consider the sequences of pairs (70,71) of transformations starting from either the boson ϕ or the fermion ψ:

$$\phi \to \psi \to \partial\phi, \ \psi \to \partial\phi \to \partial\psi. \tag{75}$$

In both cases, the action of two symmetry transformations is equivalent to a derivative, i.e., the momentum operator, corresponding exactly to the supersymmetry algebra. A free boson and a free fermion together realize supersymmetry: like the character in Molière, we have been talking prose all our lives without realizing it!

Now we look at interactions in a supersymmetric field theory [36]. The most general interactions between spin-0 fields ϕ^i and spin-1/2 fields ψ^i that are at most bilinear in the latter, and hence have a chance of being renormalizable in four dimensions, can be written in the form

$$L = L_0 - V(\phi^i, \phi_j^*) - \frac{1}{2} M_{ij}(\phi, \phi^*) \bar{\psi}^{ci} \psi^j , \tag{76}$$

where V is a general effective potential, and M_{ij} includes both mass terms and Yukawa interactions for the fermions. Supersymmetry imposes strong constraints on the allowed forms of V and M, as we now see. Suppose that M depended non-trivially on the conjugate fields ϕ^*: then the supersymmetric variation $\delta(M\bar{\psi}^c\psi)$ would contain a term

$$\frac{\partial M}{\partial \phi^*} \psi^* \bar{\psi}^c \psi \tag{77}$$

that could not be compensated by the variation of any other term. We conclude that M must be independent of ϕ^*, and hence $M = M(\phi)$ alone.

Another term in the variation of the last term in (76) is

$$\frac{\partial M_{ij}}{\partial \phi^k} \bar{E} \psi^k \bar{\psi}^{ci} \psi^j. \tag{78}$$

This term cannot be cancelled by the variation of any other term, but can vanish by itself if $\partial M_{ij}/\partial \phi^k$ is completely symmetric in the indices i, j, k. This is possible only if

$$M_{ij} = \frac{\partial W}{\partial \phi^i \partial \phi^j} \tag{79}$$

for some function $W(\phi)$ called the *superpotential*. If the theory is to be renormalizable, W can only be cubic. The trilinear term of W determines the Yukawa couplings, and the bilinear part the mass terms.

We now re-examine the form of the supersymmetric transformation law (71) itself. Yet another term in the variation of the second term in (76) has the form

$$i M_{jk} \bar{\psi}^{cj} \gamma_\mu \partial^\mu \phi^k E + (\text{Herm.Conj.}). \tag{80}$$

This can cancel against an F-dependent term in the variation of the fermion kinetic term

$$-i\bar{\psi}_i\gamma_\mu\partial^\mu F^i E^c + (\text{Herm.Conj.}),\tag{81}$$

if the following relation between F and M holds: $\frac{\partial F_i^*}{\partial\phi^j} = M_{ij}$, which is possible if and only if

$$F_i^* = \frac{\partial W}{\partial\phi^i}.\tag{82}$$

Thus the form of W also determines the required form of the supersymmetry transformation law.

The form of W also determines the effective potential V, as we now see. One of the terms in the variation of V is

$$\frac{\partial V}{\partial\phi^i}\bar{E}\psi^I + (\text{Herm.Conj.}),\tag{83}$$

which can only be cancelled by a term in the variation of $M_{ij}\bar{\psi}^{ci}\psi^j$, which can take the form $M_{ij}\bar{\psi}^{ci}F^j E^c$ if

$$\frac{\partial V}{\partial\phi^i} = M_{ij}F^i,\tag{84}$$

which is in turn possible only if

$$V = |\frac{\partial W}{\partial\phi^i}|^2 = |F^i|^2.\tag{85}$$

We now have the complete supersymmetric field theory for interacting chiral (matter) supermultiplets [36]:

$$L = i\bar{\psi}_i\gamma_\mu\partial^\mu\psi^i + |\partial_\mu\phi^i|^2 - |\frac{\partial W}{\partial\phi^i}|^2 - \frac{1}{2}\frac{\partial^2 W}{\partial\phi^i\partial^j}\bar{\psi}^{ci}\partial\psi^j + (\text{Herm.Conj.}).\tag{86}$$

This Lagrangian is invariant (up to a total derivative) under the supersymmetry transformations

$$\delta\phi^i = \sqrt{2}\bar{E}\psi^i,\ \delta\psi^i = -i\sqrt{2}\gamma_\mu\partial^\mu\phi^i E - F^i E^c:\ F^i = (\frac{\partial W}{\partial\phi^i})^*.\tag{87}$$

The simplest non-trivial superpotential involving a single superfield ϕ is

$$W = \frac{\lambda}{3}\phi^3 + \frac{m}{2}\phi^2.\tag{88}$$

It is a simple exercise for you to verify using the rules given above that the corresponding Lagrangian is

$$L = i\bar{\psi}\gamma_\mu\partial^\mu\psi + |\partial_\mu\phi|^2 - |m\phi + \lambda\phi^2|^2 - m\bar{\psi}^c\psi - \lambda\phi\bar{\psi}^c\psi.\tag{89}$$

We see explicitly that the bosonic component ϕ of the supermultiplet has the same mass as the fermionic component ψ, and that the Yukawa coupling λ fixes the effective potential.

We now turn to the possible form of a supersymmetric gauge theory [37]. Clearly, it must contain vector fields A_μ^a and fermions χ^a in the same adjoint representation of the gauge group. Once one knows the gauge group and the fermionic matter content, the form of the Lagrangian is completely determined by gauge invariance:

$$L = \frac{i}{2}\bar{\chi}^a\gamma^\mu D^\mu_{ab}\chi^b - \frac{1}{4}F^a_{\mu\nu}F^{a,\mu\nu} \left[-\frac{1}{2}(D^a)^2\right]. \tag{90}$$

Here, the gauge-covariant derivative

$$D^\mu_{ab} \equiv \delta_{ab}\partial^\mu - gf_{abc}A^\mu_c, \tag{91}$$

and the gauge field strength is

$$F^a_{\mu\nu} \equiv \partial_\mu A^a_\nu - \partial_\nu A^a_\mu + gf^{abc}A^b_\mu A^c_\nu, \tag{92}$$

as usual. We return later to the D term at the end of (90). Yet another of the miracles of supersymmetry is that the Lagrangian (90) is automatically supersymmetric, without any further monkeying around. The corresponding supersymmetry transformations may be written as

$$\delta A^a_\mu = -\bar{E}\gamma_\mu\chi^a, \tag{93}$$

$$\delta\chi^a = -\frac{i}{2}F^a_{\mu\nu}\gamma^\mu\gamma^\nu E + D^a E, \tag{94}$$

$$\delta D^a = -i\bar{E}\gamma_5\gamma^\mu D^{ab}_\mu\chi^b. \tag{95}$$

What about the D term in (90)? It is a trivial consequence of equations of motion derived from (90) that $D^a = 0$. However, this is no longer the case if matter is included. Then, it turns out, one must add to (90) the following:

$$\Delta L = -\sqrt{2}g\chi^a\phi^*_i(T^a)^i_j\psi^j + (\text{Herm.Conj.}) + g(\phi^*_i(T^a)^i_j\phi^j)D^a, \tag{96}$$

where T^a is the group representation matrix for the matter fields ϕ^i. With this addition, the equation of motion for D^a tells us that

$$D^a = g\phi^*_i(T^a)^i_j\phi^j, \tag{97}$$

and we find a D term in the full effective potential:

$$V = \Sigma_i|F_i|^2 + \Sigma_a\frac{1}{2}(D^a)^2, \tag{98}$$

where the form of D^a is given in (97).

2.6 Building supersymmetric models

Could any of the known particles in the standard model be linked together in supermultiplets? Unfortunately, none of the known fermions q, ℓ can be paired with any of the known bosons $\gamma, W^\pm Z^0, g, H$, because their internal quantum numbers do not match [39]. For example, quarks q sit in triplet representations of colour, whereas the known bosons are either singlets or octets of colour. Then again, leptons ℓ have non-zero lepton number $L = 1$, whereas the known bosons have $L = 0$. Thus, the only possibility seems to be to introduce new supersymmetric partners (spartners) for all the known particles: quark →

squark, lepton → slepton, photon → photino, Z → Zino, W → Wino, gluon → gluino, Higgs → Higgsino. The best that one can say for supersymmetry is that it economizes on principle, not on particles!

Any supersymmetric model is based on a Lagrangian that contains a supersymmetric part and a supersymmetry-breaking part:

$$\mathcal{L} = \mathcal{L}_{susy} + \mathcal{L}_{susy\times}. \tag{99}$$

We discuss later the supersymmetry-breaking part $\mathcal{L}_{susy\times}$: for now, we concentrate on the supersymmetric part $\mathcal{L}_{susy}$. The minimal supersymmetric extension of the standard model (MSSM) has the same gauge interactions as the standard model, and Yukawa interactions that are closely related. They are based on a superpotential W that is a cubic function of complex superfields corresponding to left-handed fermion fields. Conventional left-handed lepton and quark doublets are denoted L, Q, and right-handed fermions are introduced via their conjugate fields, which are left-handed, $e_R \to E^c, u_R \to U^c, d_R \to D^c$. In terms of these,

$$W = \Sigma_{L,E^c}\lambda_L LE^c H_1 + \Sigma_{Q,U^c}\lambda_U QU^c H_2 + \Sigma_{Q,D^c}\lambda_D QD^c H_1 + \mu H_1 H_2. \tag{100}$$

A few words of explanation are warranted. The first three terms in (100) yield masses for the charged leptons, charge-$(+2/3)$ quarks and charge-$(-1/3)$ quarks respectively. All of the Yukawa couplings $\lambda_{L,U,D}$ are 3×3 matrices in flavour space, whose diagonalizations yield the mass eigenstates and Cabibbo-Kobayashi-Maskawa mixing angles.

Note that two distinct Higgs doublets $H_{1,2}$ have been introduced, for two important reasons. One reason is that the superpotential must be an analytic polynomial: as we saw in (100), it cannot contain both H and H^*, whereas the standard model uses both of these to give masses to all the quarks and leptons with just a single Higgs doublet. The other reason is to cancel the triangle anomalies that destroy the renormalizability of a gauge theory. Ordinary Higgs boson doublets do not contribute to these anomalies, but the fermions in Higgs supermultiplets do, and two doublets are required to cancel each others' contributions. Once two Higgs supermultiplets have been introduced, there is the possibility, even the necessity, of a bilinear term $\mu H_1 H_2$ coupling them together.

Once the MSSM superpotential (100) has been specified, the effective potential is also fixed:

$$V = \Sigma_i |F^i|^2 + \frac{1}{2}\Sigma_a (D^a)^2 : \quad F_i^* \equiv \frac{\partial W}{\partial \phi^i}, \quad D^a \equiv g_a \phi_i^* (T^a)^i_j \phi^j, \tag{101}$$

according to the rules explained earlier in this Lecture, where the sums run over the different chiral fields i and the SU(3), SU(2) and U(1) gauge-group factors a.

There are important possible variations on the MSSM superpotential (100), which are impossible in the standard model, but are allowed by the gauge symmetries of the MSSM supermultiplets. These are additional superpotential terms that violate the quantity known as R parity:

$$R \equiv (-1)^{3B+L+2S}, \tag{102}$$

where B is baryon number, L is lepton number, and S is spin. It is easy to check that $R = +1$ for all the particles in the standard model, and $R = -1$ for all their spartners, which have identical values of B and L, but differ in spin by half a unit. Clearly, R would be conserved if both B and L were conserved, but this is not automatic. Consider the following superpotential terms:

$$\lambda_{ijk}L_iL_jE_k^c + \lambda'_{ijk}L_iQ_jD_k^c + \lambda''_{ijk}U_i^cD_j^cD_k^c + \epsilon_iHL_i, \tag{103}$$

which are visibly $SU(3) \times SU(2) \times U(1)$ symmetric. The first term in (103) would violate L, causing for example $\tilde{\ell} \to \ell + \ell$, the second would violate both B and L, causing for example $\tilde{q} \to q + \ell$, the third would violate B, causing for example $\tilde{q} \to \bar{q} + \bar{q}$, and the last would violate L by causing $H \leftrightarrow L_i$ mixing. These interactions would provide many exciting signatures for supersymmetry, such as dilepton events, jets plus leptons and multijet events. Such interactions are constrained by direct searches, by the experimental limits on flavour-changing interactions and other rare processes, and by cosmology: they would tend to wipe out the baryon asymmetry of the universe if they are too strong [51]. They would also cause the lightest supersymmetric particle to be unstable, not necessarily a disaster in itself, but it would remove an excellent candidate for the cold dark matter that apparently abounds throughout the universe. For simplicity, the conservation of R parity will be assumed in the rest of these Lectures.

Many remarkable no-renormalization theorems can be proved in supersymmetric field theories [38]. First and foremost, they have no quadratic divergences. One way to understand this is to compare the renormalizations of bosonic and fermionic mass terms:

$$m_B^2|\phi|^2 \leftrightarrow m_F\bar{\psi}\psi. \tag{104}$$

We know well that fermion masses m_F can only be renormalized logarithmically. Since supersymmetry guarantees that $m_B = m_F$, it follows that there can be no quadratic divergence in m_B. Going further, chiral symmetry guarantees that the one-loop renormalization of a fermion mass has the general multiplicative form:

$$\delta m_F = \mathcal{O}(\frac{\alpha}{\pi})\, m_F \ln(\frac{\mu_1}{\mu_2}), \tag{105}$$

where $\mu_{1,2}$ are different renormalization scales. This means that if m_F (and hence also m_B) vanish at the tree level in a supersymmetric theory, then both m_F and m_B remain zero after renormalization. This is one example of the reduction in the number of logarithmic divergences in a supersymetric theory.

In general, there is no intrinsic renormalization of any superpotential parameters, including the Yukawa couplings λ, apart from overall multiplicative factors due to wave-function renormalizations:

$$\Phi \to Z\Phi, \tag{106}$$

which are universal for both the bosonic and fermionic components ϕ, ψ in a given super-field Φ. However, gauge couplings *are* renormalized, though the β-function is changed:

$$\beta(g) \neq 0: \quad -11N_c \to -9N_c, \tag{107}$$

at one-loop order in an $SU(N_c)$ supersymmetric gauge theory with no matter, as a result of the extra gaugino contributions.

There are even fewer divergences in theories with more supersymmetries. For example, there is only a finite number of divergent diagrams in a theory with $N = 2$ supersymmetries, which may be cancelled by imposing a few simple relations on the spectrum of supermultiplets. Finally, there are no divergences at all in theories with $N = 4$ super-symmetries, which obey automatically the necessary finiteness conditions.

Many theorists from Dirac onwards have found the idea of a completely finite theory attractive, so it is natural to ask whether theories with $N \geq 2$ supersymmetries could be interesting as realistic field theories. Unfortunately, the answer is 'not immediately', because they do not allow the violation of parity, which is basic to the standard model, as we saw in (2) at the beginning of Lecture 1. To understand why theories with $N \geq 2$ cannot violate parity, consider the simplest possible extended supersymmetric theory con-taining an $N = 2$ matter multiplet, which contains both left- and right-handed fermions with helicities $\pm 1/2$. Suppose that the left-handed fermion with helicity $+1/2$ sits in a representation R of the gauge group. Now act on it with either of the two supersymmetry charges $Q_{1,2}$: they each yield bosons, that each sit in the same representation R. Now act on either of these with the other supercharge, to obtain a right-handed fermion with helicity $-1/2$: this must also sit in the same representation R of the gauge group. Hence, left- and right-handed fermions have the same interactions, and parity is conserved. There is no way out using gauginos, because they are forced to sit in adjoint representations of the gauge group, and hence also cannot distinguish between right and left.

Thus, if we want to make a supersymmetric extension of the standard model, it had better be with just $N = 1$ supersymmetry.

2.7   Supersymmetry breaking

This is clearly necessary: $m_e \neq m_{\tilde{e}}, m_\gamma \neq m_{\tilde{\gamma}}$, etc. The big issue is whether the breaking of supersymmetry is explicit, i.e., present already in the underlying Lagrangian of the theory, or whether it is spontaneous, i.e., induced by a non-supersymmetric vacuum state. There are in fact several reasons to disfavour explicit supersymmetry breaking. It is ugly, it would be unlike the way in which gauge symmetry is broken, and it would lead to inconsistencies in supergravity theory. For these reasons, theorists have focused on spontaneous supersymmetry breaking.

If the vacuum is not to be supersymmetric, there must be some fermionic state χ that is coupled to the vacuum by the supersymmetry charge Q:

$$\langle 0|Q|\chi\rangle \equiv f_\chi^2 \neq 0. \tag{108}$$

The fermion χ corresponds to a Goldstone boson in a spontaneously broken bosonic symmetry, and therefore is often termed a Goldstone fermion or a Goldstino.

There is just one small problem in globally supersymmetric models, i.e., those without gravity: spontaneous supersymmetry breaking necessarily entails a positive vacuum en-ergy E_0. To see this, consider the vacuum expectation value of the basic supersymmetry anticommutator:

$$\{Q,Q\} \propto \gamma_\mu P^\mu. \tag{109}$$

According to (108), there is an intermediate state χ, so that

$$\langle 0|\{Q,Q\}|0\rangle = |\langle 0|Q|\chi\rangle|^2 = f_\chi^4 \propto \langle 0|P_0|0\rangle = E_0, \tag{110}$$

where we have used Lorentz invariance to set the spatial components $\langle 0|P_i|0\rangle = 0$. Spon-taneous breaking of global supersymmetry (108) requires

$$E_0 = f_\chi^4 \neq 0. \tag{111}$$

The next question is how to generate non-zero vacuum energy. Hints are provided by the effective potential in a globally supersymmetric theory:

$$V = \Sigma_i \left| \frac{\partial W}{\partial \phi^i} \right|^2 + \frac{1}{2} \Sigma_\alpha g_\alpha^2 |\phi^* T^\alpha \phi|^2. \tag{112}$$

It is apparent from this expression that either the first 'F term' or the second 'D term' must be positive definite.

The option $D > 0$ requires constructing a model with a U(1) gauge symmetry [52]. The simplest example contains just one chiral (matter) supermultiplet with unit charge, for which the effective potential is:

$$V_D = \frac{1}{2}(\xi + g\phi^*\phi)^2. \tag{113}$$

the extra constant term ξ is not allowed in a non-Abelian theory, which is why one must use a U(1) theory. We see immediately that the minimum of the effective potential (113) is reached when $\langle 0|\phi|0 \rangle = 0$, in which case $V_F = 1/2\xi^2 > 0$ and supersymmetry is broken spontaneously. Indeed, it is easy to check that, in this vacuum:

$$m_\phi = g\xi, \; m_\psi = 0, \; m_V = m_{\tilde{V}} = 0, \tag{114}$$

exhibiting explicitly the boson-fermion mass splitting in the (ϕ, ψ) supermultiplet. Unfortunately, this example cannot be implemented with the U(1) of electromagnetism in the standard model, because there are fields with both signs of the hypercharge Y, enabling V_D to vanish. So, one needs a new U(1) gauge group factor, and many new fields in order to cancel triangle anomalies. For these reasons, D-breaking models did not attract much attention for quite some time, though they have had a revival in the context of string theory [53].

The option $F > 0$ also requires additional chiral (matter) fields with somewhat 'artificial' couplings [54]: again, those of the standard model do not suffice. The simplest example uses three chiral supermultiplets A, B, C with the superpotential

$$W = \alpha AB^2 + \beta C(B^2 - m^2). \tag{115}$$

Using the rules given in the previous Lecture, it is easy to calculate the corresponding F terms:

$$F_A = \alpha B^2, \; F_B = 2B(\alpha A + \beta C), \; F_C = \beta(B^2 - m^2), \tag{116}$$

and hence the effective potential

$$V_F = \Sigma_i |F_i|^2 = 4|B(\alpha A + \beta C)|^2 + |\alpha B^2|^2 + |\beta(B^2 - m^2)|^2. \tag{117}$$

Likewise, it is not difficult to check that the three different positive-semidefinite terms in (117) cannot all vanish simultaneously. Hence, necessarily $V_F > 0$, and hence supersymmetry *must* be broken.

The principal outcome of this brief discussion is that there are no satisfactory models of global supersymmetry breaking, which provided some of the motivation for studying local supersymmetry, i.e., supergravity theory.

2.8   Supergravity and local supersymmetry breaking

So far, we have considered global supersymmetry transformations, in which the infinitesimal transformation spinor E is constant throughout space. Now we consider the possibility of a space-time-dependent field $E(x)$. Why?

This step of making symmetries local has become familiar with bosonic symmetries, where it leads to gauge theories, so it is natural to try the analogous step with fermionic symmetries. Moreover, as we see shortly, it leads to an elegant mechanism for spontaneous supersymmetry breaking, again by analogy with gauge theories, the super-Higgs mechanism. Further, as we also see shortly, making supersymmetry local necessarily involves gravity, and even opens the prospect of unifying all the particle interactions and matter fields with extended supersymmetry transformations:

$$G(J=2) \;\rightarrow\; \tilde{G}(J=3/2) \;\rightarrow\; V(J=1) \;\rightarrow\; q, \ell(J=1/2) \;\rightarrow\; H(J=0) \qquad (118)$$

in supergravity with $N > 1$ supercharges. In (118), G denotes the graviton, and $\tilde{G}$ the spin-3/2 gravitino, which accompanies it in the graviton supermultiplet:

$$\begin{pmatrix} G \\ \tilde{G} \end{pmatrix} = \begin{pmatrix} 2 \\ \frac{3}{2} \end{pmatrix}. \qquad (119)$$

Supergravity is in any case an essential ingredient in the discussion of gravitational interactions of supersymmetric particles, needed, for example, for any meaningful discussion of the cosmological constant.

The mechanism for the spontaneous breaking of local supersymmetry is known as the super-Higgs effect [55, 56]. You recall that, in the conventional Higgs effect in spontaneously broken gauge theories, a massless Goldstone boson is 'eaten' by a gauge boson to provide it with the third polarization state it needs to become massive:

$$(2 \times V_{m=0}) + (1 \times GB) = (3 \times V_{m \neq 0}). \qquad (120)$$

In a locally supersymmetric theory, the two polarization states of the massless Goldstone fermion (Goldstino) are 'eaten' by a massless gravitino, giving it the total of four polarization states it needs to become massive:

$$(2 \times \psi^\mu_{m=0}) + (2 \times GF) = (4 \times \psi^\mu_{m \neq 0}). \qquad (121)$$

This process clearly involves the breakdown of local supersymmetry, since the end result is to give the gravitino a different mass from the graviton: $m_G = 0 \neq m_{\tilde{G}} \neq 0$. It is indeed the only known consistent way of breaking local supersymmetry, just as the Higgs mechanism is the only consistent way of breaking gauge symmetry. We shall not go here through all the details of the super-Higgs effect, but there is one noteworthy feature: this local breaking of supersymmetry can be achieved with zero vacuum energy:

$$\langle 0|V|0 \rangle = 0 \;\leftrightarrow\; \Lambda = 0. \qquad (122)$$

There is no inconsistency between local supersymmetry breaking and a vanishing cosmological constant Λ, unlike the case of global supersymmetry breaking that we discussed earlier.

2.9   Effective low-energy theory

The coupling of matter particles to supergravity is more complicated than the globally supersymmetric case discussed in the previous lecture. Therefore, it is not developed here in detail. Instead, a few key results are presented without proof. The form of the effective low-energy theory suggested by spontaneous supersymmetry breaking in supergravity is:

$$-\frac{1}{2}\Sigma_a m_{1/2_a} \bar{V}_\alpha \bar{V}_\alpha - \Sigma_i m_{0_i}^2 |\phi^i|^2 - (\Sigma_\lambda A_\lambda \lambda \phi^3 + \Sigma_\mu B_\mu \mu \phi^2 + \text{Herm.Conj.}), \qquad (123)$$

which contains many free parameters and phases. The breaking of supersymmetry in the effective low-energy theory (123) is explicit but 'soft', in the sense that the renormalization of the parameters $m_{1/2_a}, m_{0_i}, A_\lambda$ and B_μ is logarithmic. Of course, these parameters are not considered to be fundamental, and the underlying mechanism of supersymmetry breaking is thought to be spontaneous, for the reasons described at the beginning of this lecture.

The logarithmic renormalization of the parameters means that one can calculate their low-energy values in terms of high-energy inputs from a supergravity or superstring theory, using standard renormalization-group equations [57]. In the case of the low-energy gaugino masses M_a, the renormalization is multiplicative and identical with that of the corresponding gauge coupling α_a at the one-loop level:

$$\frac{M_a}{m_{1/2_a}} = \frac{\alpha_a}{\alpha_{\text{GUT}}}, \qquad (124)$$

where we assume GUT unification of the gauge couplings at the input supergravity scale. In the case of the scalar masses, there is both multiplicative renormalization and renormalization related to the gaugino masses:

$$\frac{\partial m_{0_i}^2}{\partial t} = \frac{1}{16\pi^2}[\lambda^2(m_0^2 + A_\lambda^2) - g_a^2 M_a^2], \qquad (125)$$

at the one-loop level, where $t \equiv \ln(Q^2/m_{\text{GUT}}^2)$, and the $\mathcal{O}(1)$ group-theoretical coefficients have been omitted. In the case of the first two generations, the first terms in (125) are negligible, and one may integrate (125) trivially to obtain effective low-energy parameters

$$m_{0_i}^2 = m_0^2 + C_i m_{1/2}^2, \qquad (126)$$

where universal inputs are assumed, and the coefficients C_i are calculable in any given model. The first terms in (125) are, however, important for the third generation and for the Higgs bosons of the MSSM, as we now see.

Notice that the signs of the first terms in (125) are positive, and that of the last term negative. This means that the last term tends to *increase* $m_{0_i}^2$ as the renormalization scale Q *decreases*, an effect seen in Fig. 16. The positive signs of the first terms mean that they tend to *decrease* $m_{0_i}^2$ as Q *decreases*, an effect seen for a Higgs squared-mass in Fig. 16. Specifically, the negative effect on H_u seen in Fig. 16 is due to its large Yukawa coupling to the t quark: $\lambda_t \sim g_{2,3}$. The exciting aspect of this observation is that spontaneous electroweak symmetry breaking is possible [57] when $m_H^2(Q) < 0$, as occurs in Fig. 16. Thus the spontaneous breaking of supersymmetry, which normally provides $m_0^2 > 0$, and renormalization, which then drive $m_H^2(Q) < 0$, conspire to make spontaneous electroweak

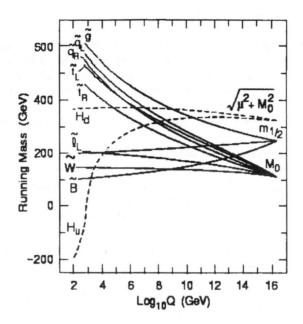

Figure 16. *The renormalization-group evolution of the soft supersymmetry-breaking parameters in the MSSM, showing the increase in the squark and slepton masses as the renormalization scale decreases, whilst the Higgs squared-mass may become negative, triggering electroweak symmetry breaking.*

symmetry breaking possible. Typically, this occurs at a renormalization scale that is exponentially smaller than the input supergravity scale:

$$\frac{m_W}{m_P} = exp(\frac{-\mathcal{O}(1)}{\alpha_t}): \quad \alpha_t \equiv \frac{\lambda_t^2}{4\pi}. \tag{127}$$

Typical dynamical calculations find that $m_W \sim 100$ GeV emerges naturally if $m_t \sim 60$ to 200 GeV, and this was in fact one of the first suggestions that m_t might be as high as was subsequently observed.

To conclude this section, let us briefly review the reasons why soft supersymmetry breaking might be universal, at least in some respects. There are important constraints on the mass differences of squarks and sleptons with the same internal quantum numbers, coming from flavour-changing neutral interactions [58]. These are suppressed in the standard model by the Glashow-Iliopoulos-Maiani mechanism [59], which limits them to magnitudes $\propto \Delta m_q^2/m_W^2$ for small squared-mass differences Δm_q^2. Depending on the process considered, it is either necessary or desirable that sparticle exchange contributions, which would have expected magnitudes $\sim \Delta m_{\tilde{q}}^2/m_{\tilde{q}}^2$, be suppressed by a comparable factor. In particular, one would like

$$m_0^2(\text{first generation}) - m_0^2(\text{second generation}) \sim \delta m_q^2 \times \frac{m_{\tilde{q}}^2}{m_W^2}. \tag{128}$$

The limits on third-generation sparticle masses from flavour-changing neutral interactions are less severe, and the first/second-generation degeneracy could be relaxed if $m_{\tilde{q}}^2 \gg$

m_W^2, but models with physical values of m_0^2 degenerate to $\mathcal{O}(m_q^2)$ are certainly preferred. However, this restriction is not respected in many low-energy effective theories derived from string models.

The desirability of degeneracy between sparticles of different generations help encourage some people to study models in which this property would emerge naturally, such as models of gauge-mediated supersymmetry breaking or extra dimensions [60]. However, for the rest of these lectures we shall mainly stick to familiar old supergravity.

2.10 Sparticle masses and mixing

We now progress to a more complete discussion of sparticle masses and mixing:

Sfermions: Each flavour of charged lepton or quark has both left- and right-handed components $f_{L,R}$, and these have separate spin-0 boson superpartners $\tilde{f}_{L,R}$. These have different isospins $I = \frac{1}{2}$, 0, but may mix as soon as the electroweak gauge symmetry is broken. Thus, for each flavour we should consider a 2×2 mixing matrix for the $\tilde{f}_{L,R}$, which takes the following general form:

$$M_{\tilde{f}}^2 \equiv \begin{pmatrix} m_{\tilde{f}_{LL}}^2 & m_{\tilde{f}_{LR}}^2 \\ m_{\tilde{f}_{LR}}^2 & m_{\tilde{f}_{RR}}^2 \end{pmatrix} . \tag{129}$$

The diagonal terms may be written in the form

$$m_{\tilde{f}_{LL,RR}}^2 = \tilde{m}_{\tilde{f}_{L,R}}^2 + m_{\tilde{f}_{L,R}}^{D^2} + m_f^2 , \tag{130}$$

where m_f is the mass of the corresponding fermion, $\tilde{m}_{\tilde{f}_{L,R}}^2$ is the soft supersymmetry-breaking mass discussed in the previous section, and $m_{\tilde{f}_{L,R}}^{D^2}$ is a contribution due to the quartic D terms in the effective potential:

$$m_{\tilde{f}_{L,R}}^{D^2} = m_Z^2 \cos 2\beta \ (I_3 + \sin^2 \theta_W Q_{em}) , \tag{131}$$

where the term $\propto I_3$ is non-zero only for the $\tilde{f}_L$. Finally, the off-diagonal mixing term takes the general form

$$m_{\tilde{f}_{L,R}}^2 = m_f \left(A_f + \mu_{\cot \beta}^{\tan \beta} \right) \quad \text{for} \quad f = {}_{u,c,t}^{e,\mu,\tau,d,s,b} . \tag{132}$$

It is clear that $\tilde{f}_{L,R}$ mixing is likely to be important for the $\tilde{t}$, and it may also be important for the $\tilde{b}_{L,R}$ and $\tilde{\tau}_{L,R}$ if $\tan \beta$ is large.

We also see from (130) that the diagonal entries for the $\tilde{t}_{L,R}$ would be different from those of the $\tilde{u}_{L,R}$ and $\tilde{c}_{L,R}$, even if their soft supersymmetry-breaking masses were universal, because of the m_f^2 contribution. In fact, we also expect non-universal renormalization of $m_{\tilde{t}_{LL,RR}}^2$ (and also $m_{\tilde{b}_{LL,RR}}^2$ and $m_{\tilde{\tau}_{LL,RR}}^2$ if $\tan \beta$ is large), because of Yukawa effects analogous to those discussed in the previous section for the renormalization of the soft Higgs masses. For these reasons, the $\tilde{t}_{L,R}$ are not usually assumed to be degenerate with the other squark flavours. Indeed, one of the $\tilde{t}$ could well be the lightest squark, perhaps even lighter than the t quark itself [61].

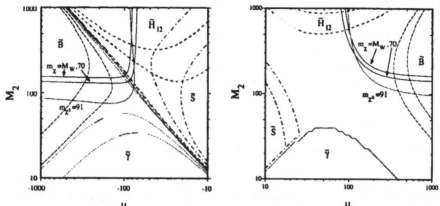

Figure 17. *The (μ, M_2) plane characterizing charginos and neutralinos, for (a) $\mu < 0$ and (b) $\mu > 0$, including contours of m_χ and $m_{\chi^\pm}$, and of neutralino purity [62].*

Charginos: These are the supersymmetric partners of the $W^\pm$ and $H^\pm$, which mix through a 2×2 matrix

$$-\frac{1}{2} \, (\tilde{W}^-, \tilde{H}^-) \; M_C \; \begin{pmatrix} \tilde{W}^+ \\ \tilde{H}^+ \end{pmatrix} \quad + \quad \text{herm.conj.} \tag{133}$$

where

$$M_C \equiv \begin{pmatrix} M_2 & \sqrt{2} m_W \sin\beta \\ \sqrt{2} m_W \cos\beta & \mu \end{pmatrix} . \tag{134}$$

Here M_2 is the unmixed SU(2) gaugino mass and μ is the Higgs mixing parameter introduced in (100). Fig. 17 displays (among other lines to be discussed later) the contour $m_{\chi^\pm} = 91$ GeV for the lighter of the two chargino mass eigenstates [62].

Neutralinos: These are characterized by a 4×4 mass mixing matrix [63], which takes the following form in the $(\tilde{W}^3, \tilde{B}, \tilde{H}_2^0, \tilde{H}_1^0)$ basis :

$$m_N = \begin{pmatrix} M_2 & 0 & \frac{-g_2 v_2}{\sqrt{2}} & \frac{g_2 v_1}{\sqrt{2}} \\ 0 & M_1 & \frac{g' v_2}{\sqrt{2}} & \frac{-g' v_1}{\sqrt{2}} \\ \frac{-g_2 v_2}{\sqrt{2}} & \frac{g' v_2}{\sqrt{2}} & 0 & \mu \\ \frac{g_2 v_1}{\sqrt{2}} & \frac{-g' v_1}{\sqrt{2}} & \mu & 0 \end{pmatrix} . \tag{135}$$

Note that this has a structure similar to M_C (134), but with its entries replaced by 2×2 submatrices. As has already been mentioned, one conventionally assumes that the SU(2) and U(1) gaugino masses $M_{1,2}$ are universal at the GUT or supergravity scale, so that

$$M_1 \simeq M_2 \, \frac{\alpha_1}{\alpha_2} , \tag{136}$$

so the relevant parameters of (135) are generally taken to be $M_2 = (\alpha_2/\alpha_{\rm GUT})m_{1/2}$, μ and $\tan\beta$.

Fig. 17 also displays contours of the mass of the lightest neutralino χ, as well as contours of its gaugino and Higgsino contents [62]. In the limit $M_2 \to 0$, χ would be approximately a photino and it would be approximately a Higgsino in the limit $\mu \to 0$. Unfortunately, these idealized limits are excluded by unsuccessful LEP and other searches for neutralinos and charginos, as discussed in more detail below.

Supersymmetric Higgs Bosons: As was discussed earlier, one expects two complex Higgs doublets $H_2 \equiv (H_2^+, H_2^0)$, $H_1 \equiv (H_1^+, H_1^0)$ in the MSSM, with a total of 8 real degrees of freedom. Of these, 3 are eaten via the Higgs mechanism to become the longitudinal polarization states of the $W^\pm$ and Z^0, leaving 5 physical Higgs bosons to be discovered by experiment. Three of these are neutral: the lighter CP-even neutral h, the heavier CP-even neutral H, the CP-odd neutral A, and charged bosons $H^\pm$. The quartic potential is completely determined by the D terms in the effective potential:

$$V_4 = \frac{g^2 + g'^2}{8} \left(|H_1^0|^2 - |H_2^0|^2 \right) , \tag{137}$$

for the neutral components, whilst the quadratic terms may be parametrized at the tree level by

$$\frac{1}{2} = m_{H_1}^2 \, |H_1|^2 + m_{H_2}^2 \, |H_2|^2 + (m_3^2 \, H_1 H_2 + \text{herm.conj.}) , \tag{138}$$

where $m_3^2 = B_\mu \mu$. One combination of the three parameters $(m_{H_1}^2, m_{H_2}^2, m_3^2)$ is fixed by the Higgs vacuum expectation $v = \sqrt{v_1^2 + v_2^2} = 246$ GeV, and the other two combinations may be rephrased as $(m_A, \tan\beta)$. These characterize all Higgs masses and couplings in the MSSM at the tree level. Looking back at (18), we see that the gauge coupling strength of the quartic interactions (137) suggests a relatively low mass for at least the lightest MSSM Higgs boson h, and this is indeed the case, with $m_h \le m_Z$ at the tree level:

$$m_h^2 = m_Z^2 \, \cos^2 2\beta . \tag{139}$$

This raised considerable hope that the lightest MSSM Higgs boson could be discovered at LEP, with its prospective reach to $m_H \sim 100$ GeV.

However, radiative corrections to the Higgs masses are calculable in a supersymmetric model (this was, in some sense, the whole point of introducing supersymmetry!), and they turn out to be non-negligible for $m_t \sim 175$ GeV [43]. Indeed, the leading one-loop corrections to m_h^2 depend quartically on m_t:

$$\Delta m_h^2 = \frac{3m_t^4}{4\pi^2 v^2} \, \ln \left(\frac{m_{\tilde{t}_1} m_{\tilde{t}_2}}{m_t^2} \right) + \frac{3m_t^4 \hat{A}_t^2}{8\pi^2 v^2} \, \left[2h(m_{\tilde{t}_1}^2, m_{\tilde{t}_2}^2) + \hat{A}_t^2 \; f(m_{\tilde{t}_1}^2, m_{\tilde{t}_2}^2) \right] + \dots , \tag{140}$$

where $m_{\tilde{t}_{1,2}}$ are the physical masses of the two stop squarks $\tilde{t}_{1,2}$, $\hat{A}_t \equiv A_t - \mu \cot\beta$, and

$$h(a,b) \equiv \frac{1}{a-b} \, \ln \left(\frac{a}{b} \right) , \quad f(a,b) = \frac{1}{(a-b)^2} \left[2 - \frac{a+b}{a-b} \, \ln \left(\frac{a}{b} \right) \right] . \tag{141}$$

Non-leading one-loop corrections to the MSSM Higgs masses are also known, as are corrections to coupling vertices, two-loop corrections and renormalization-group resummations. For $m_{\tilde{t}_{1m2}} \lesssim 1$ TeV and a plausible range of A_t, one finds

$$m_h \lesssim 130 \text{ GeV} , \tag{142}$$

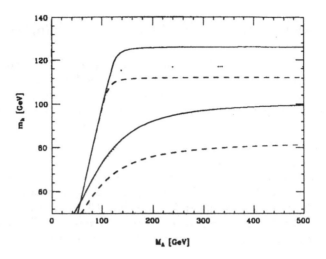

Figure 18. *The lightest Higgs boson mass in the MSSM, for different values of* tan β *and the CP-odd Higgs boson mass* M_A *[64].*

as seen in Fig. 14. There we see the sensitivity of m_h to $(m_A, \tan\beta)$, and we also see how m_A, m_H and $m_{H^\pm}$ approach each other for large m_A.

3   Search for supersymmetry

3.1   Constraints on the MSSM

Important experimental constraints on supersymmetric models have been provided by the unsuccessful direct searches at LEP and the Tevatron collider. When compiling these, the supersymmetry-breaking masses of the different unseen scalar particles are often assumed to have a universal value m_0 at some GUT input scale, and likewise the fermionic partners of the vector bosons are also commonly assumed to have universal fermionic masses $m_{1/2}$ at the GUT scale, as are the trilinear soft supersymmetry-breaking parameters A_0 – the so-called constrained MSSM (CMSSM) that might (but not necessarily) arise from a minimal supergravity theory. These input values are then renormalized by (supersymmetric) standard model interactions between the GUT and electroweak scales.

The allowed domains in some of the $(m_{1/2}, m_0)$ planes for different values of tan β and the sign of μ are shown in Fig. 19. Panel (a) of this figure features the limit $m_{\chi^\pm} \gtrsim$ 104 GeV provided by chargino searches at LEP [65]. The LEP neutrino counting and other measurements have also constrained the possibilities for light neutralinos, and LEP has also provided lower limits on slepton masses, of which the strongest is $m_{\tilde{e}} \gtrsim 99$ GeV [66], as also illustrated in panel (a) of Fig. 19. The most important constraints on the supersymmetric partners of the u, d, s, c, b squarks and on the gluinos are provided by the FNAL Tevatron collider: for equal masses $m_{\tilde{q}} = m_{\tilde{g}} \gtrsim 300$ GeV. In the case of the $\tilde{t}$, LEP provides the most stringent limit when $m_{\tilde{t}} - m_\chi$ is small, and the Tevatron for larger $m_{\tilde{t}} - m_\chi$ [65].

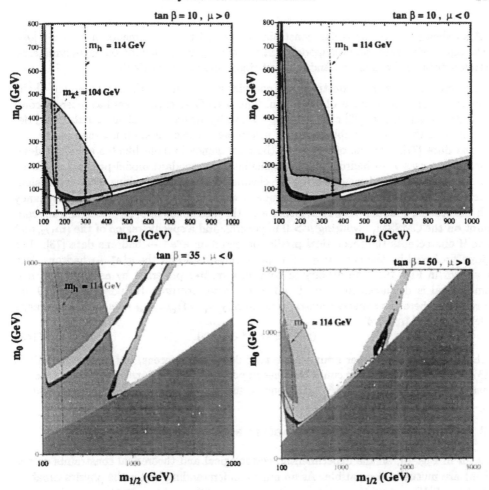

Figure 19. *Compilations of phenomenological constraints on the CMSSM for (a)* $\tan\beta = 10, \mu > 0$, *(b)* $\tan\beta = 10, \mu < 0$, *(c)* $\tan\beta = 35, \mu < 0$ *and (d)* $\tan\beta = 50, \mu > 0$ *[67]. The near-vertical lines are the LEP limits* $m_{\chi^\pm} = 104$ *GeV (dashed and black) [65], shown in (a) only, and* $m_h = 114$ *GeV (dotted and red) [24]. Also, in the lower left corner of (a), we show the* $m_{\tilde{e}} = 99$ *GeV contour [66]. The dark (brick red) shaded regions are excluded because the LSP is charged. The light (turquoise) shaded areas have* $0.1 \leq \Omega_\chi h^2 \leq 0.3$, *and the smaller dark (blue) shaded regions have* $0.094 \leq \Omega_\chi h^2 \leq 0.129$, *as favoured by WMAP [67]. The medium (dark green) shaded regions that are most prominent in panels (b) and (c) are excluded by* $b \to s\gamma$ *[68]. The shaded (pink) regions in panels (a) and (d) show the* $\pm 2\sigma$ *ranges of* $g_\mu - 2$ *[70].*

Another important constraint in Fig. 19 is provided by the LEP lower limit on the Higgs mass: $m_H > 114.4$ GeV [24]. Since m_h is sensitive to sparticle masses, particularly $m_{\tilde{t}}$, via the loop corrections (140), the Higgs limit also imposes important constraints on the soft supersymmetry-breaking CMSSM parameters, principally $m_{1/2}$ [69] as displayed in Fig. 19.

Also shown in Fig. 19 is the constraint imposed by measurements of $b \rightarrow s\gamma$ [68]. These agree with the standard model, and therefore provide bounds on supersymmetric particles, such as the chargino and charged Higgs masses, in particular.

The final experimental constraint we consider is that due to the measurement of the anomalous magnetic moment of the muon. The BNL E821 experiment has recently completed its measurements [70] of $a_\mu \equiv \frac{1}{2}(g_\mu - 2)$, which deviates by about 2.7 standard deviations from the best available standard model predictions based on low-energy $e^+e^- \rightarrow$ hadrons data [71]. On the other hand, the discrepancy is more like 0.8 standard deviations if one uses $\tau \rightarrow$ hadrons data to calculate the standard model prediction. Faced with this confusion, and remembering the chequered history of previous theoretical calculations [72], it is reasonable to defer judgement whether there is a significant discrepancy with the standard model. However, either way, the measurement of a_μ is a significant constraint on the CMSSM, favouring $\mu > 0$ in general, and a specific region of the $(m_{1/2}, m_0)$ plane if one accepts the theoretical prediction based on $e^+e^- \rightarrow$ hadrons data [73]. The regions preferred by the current $g - 2$ experimental data and the $e^+e^- \rightarrow$ hadrons data are shown in Fig. 19. The density of cold dark matter preferred by astrophysical and cosmological is discussed in more detail in the next Lecture. Included in Fig. 19 are the regions where the supersymmetric relic density $\rho_\chi = \Omega_\chi \rho_{critical}$ falls within the range preferred by WMAP [74]:

$$0.094 < \Omega_\chi h^2 < 0.129 \tag{143}$$

at the 2-σ level. The upper limit on the relic density is rigorous, but the lower limit in (143) is optional, since there could be other important contributions to the overall matter density. Smaller values of $\Omega_\chi h^2$ correspond to smaller values of $(m_{1/2}, m_0)$, in general.

3.2   Benchmark supersymmetric scenarios

As seen in Fig. 19, all the experimental, cosmological and theoretical constraints on the MSSM are mutually compatible. As an aid to understanding better the physics capabilities of the LHC and various other accelerators, as well as non-accelerator experiments, a set of benchmark supersymmetric scenarios have been proposed [75]. Their distribution in the $(m_{1/2}, m_0)$ plane is displayed in Fig. 20. These benchmark scenarios are compatible with all the accelerator constraints mentioned above, including the LEP searches and $b \rightarrow s\gamma$, and yield relic densities of LSPs in the range suggested by cosmology and astrophysics. The benchmarks are not intended to sample 'fairly' the allowed parameter space, but rather to illustrate the range of possibilities currently allowed.

In addition to a number of benchmark points falling in the 'bulk' region of parameter space at relatively low values of the supersymmetric particle masses, as seen along the 'WMAP lines' in Fig. 20, we also proposed [75, 76] some points out along the 'tails' of parameter space extending out to larger masses, including the coannihilation strips extending to large $m_{1/2}$ [77, 78], the rapid-annihilation 'funnels' extending to larger m_0 [79] and the focus-point regions above the left frame [80], which are discussed in more detail in the following Lecture. As discussed there, these clearly require some degree of fine-tuning to obtain the required relic density [81] and/or the correct $W^\pm$ mass [82], and some are also disfavoured by the supersymmetric interpretation of the $g_\mu - 2$ anomaly, but all are logically consistent possibilities. Fig. 21 displays estimates of the numbers of MSSM particles that could be detected at different accelerators discussed in subsequent sections.

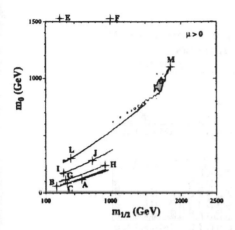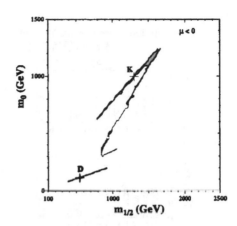

Figure 20. *The 'WMAP lines' display the regions of the* $(m_{1/2}, m_0)$ *plane that are compatible with* $0.094 < \Omega_\chi h^2 < 0.129$ *in the 'bulk', coannihilation 'tail', and rapid-annihilation 'funnel' regions, as well as the laboratory constraints, for (a)* $\mu > 0$ *and* $\tan\beta = 5, 10, 20, 35$ *and 50, and (b) for* $\mu < 0$ *and* $\tan\beta = 10$ *and 35. The parts of the* $\mu > 0$ *strips compatible with* $g_\mu - 2$ *at the 2-σ level have darker shading. The updated post-WMAP benchmark scenarios are marked in red. Points (E,F) in the focus-point region are at larger values of* m_0 *[75, 76].*

The Fermilab Tevatron collider has already established the best limits on squarks and gluinos, and will have the next shot at discovering sparticles. In the CMSSM, the regions of parameter space it can reach are disfavoured indirectly by the LEP limits on weakly-interacting sparticles, the absence of a light Higgs boson, and the agreement of $b \to s\gamma$ with the standard model [75]. However, the prospects may be improved in variants of the MSSM that abandon some of the CMSSM constraints [87].

Fig. 22 shows the physics reach for observing pairs of supersymmetric particles at the LHC. The prime signature for supersymmetry – multiple jets (and/or leptons) with a large amount of missing energy – is quite distinctive, as seen in Fig. 23 [88, 89]. Therefore, the detection of the supersymmetric partners of quarks and gluons at the LHC is expected to be quite easy if they weigh less than about 2.5 TeV [90]. Moreover, in many scenarios one should be able to observe their cascade decays into lighter supersymmetric particles, as seen in Fig. 24 [91], or into the lightest MSSM Higgs boson h.

The LHC collaborations have analyzed their reach for sparticle detection in both generic studies and specific benchmark scenarios proposed previously [92]. Based on these studies, Fig. 21 displays estimates how many different sparticles may be seen at the LHC in each of the newly-proposed benchmark scenarios [75]. The lightest Higgs boson is always found, and squarks and gluinos are usually found, though there are some scenarios where no sparticles are found at the LHC. This feature is also seen in Fig. 25, where the numbers of different observable sparticle species are indicated along the WMAP lines for $\tan\beta = 10$ and 50 for $\mu > 0$. Only at large $\tan\beta$ and $m_{1/2}$ does the LHC fail to see any sparticles, though it always sees the lightest Higgs boson. However, the LHC often misses heavier weakly-interacting sparticles such as charginos, neutralinos, sleptons and the other Higgs bosons, as seen in Fig. 25, leaving a physics opportunity for a linear e^+e^-

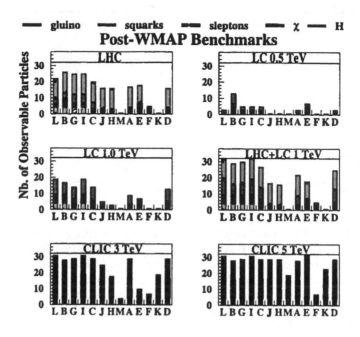

Figure 21. *Estimates of the numbers of different types of CMSSM particles that may be detectable [75] at (a) the LHC [30], (b) a 0.5-TeV and (c) a 1-TeV linear e^+e^- collider [83], (d) the combination of the LHC and a 1-TeV linear e^+e^- collider, and (e,f) a 3(5)-TeV e^+e^- [84] or $\mu^+\mu^-$ collider [85, 86]. Note the complementarity between the sparticles detectable at the LHC and at a 1-TeV linear e^+e^- collider.*

linear collider.

Many possible signatures of MSSM Higgs bosons at the LHC have been studied, and one or more of them can be detected in all the scenarios explored. As seen in Fig. 26, at large m_A, the lightest MSMM Higgs boson h may be detected via its $\gamma\gamma$ and/or $\bar{b}b$ decay modes, and many other channels are accessible at low m_A. At large $\tan\beta$, the heavier H, A and $H^\pm$ bosons may be detected. but there is a 'wedge' of parameter space extending out to large m_A at moderate $\tan\beta$ where only the lightest MSMM Higgs boson may be detectable at the LHC.

The question then arises whether, in this region, detailed LHC measurements of the lightest MSMM Higgs boson might be able to distinguish it from a standard model Higgs boson with the same mass. As seen in Fig. 27, the LHC $h \to \gamma\gamma$ and $\bar{b}b$ decay signatures are unlikely to be greatly suppressed (or enhanced) compared to those of a standard model Higgs boson [94], but the accuracy with which they can be measured may not be sufficient to distinguish the MSSM from the standard model. This may therefore be a task for the other accelerators discussed later [95].

Summarizing, a likely supersymmetric post-LHC physics scenario is that:

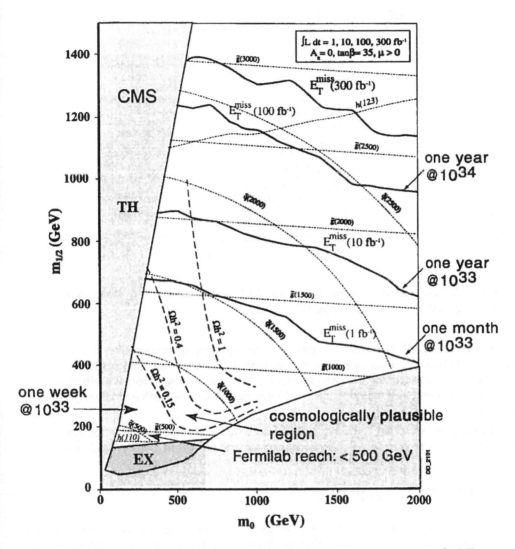

Figure 22. *The regions of the $(m_0, m_{1/2})$ plane that can be explored by the LHC with various integrated luminosities [90], using the missing energy + jets signature [89].*

• The lightest Higgs boson will have been discovered and some of its decay modes and other properties will have been measured, but its role in the generation of particle masses will not have been established, and the LHC will probably not be able to distinguish between a standard model Higgs boson and the lightest MSSM Higgs boson.

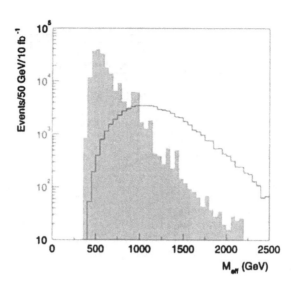

Figure 23. *The distribution expected at the LHC in the variable M_{eff} that combines the jet energies with the missing energy [92, 88, 89].*

• The LHC is likely to have discovered some supersymmetric particles, but not all of them, and there will in particular be gaps among the electroweakly-interacting sparticles. Furthermore, the accuracy with which sparticle masses and decay properties will have been measured will probably not be sufficient to distinguish between different supersymmetric models.

Thus, there are many supersymmetric issues that will require exploration elsewhere, and the same is true in many other scenarios for new physics at the TeV scale [96].

3.3   Linear e^+e^- colliders

Electron-positron colliders provide very clean experimental environments, with egalitarian production of all the new particles that are kinematically accessible, including those that have only weak interactions. Moreover, polarized beams provide a useful analysis tool, and $e\gamma$, $\gamma\gamma$ and e^-e^- colliders are readily available at relatively low marginal costs [83].

For these reasons, linear e^-e^- colliders are complementary to the LHC. Just as LEP built on the discoveries of the $W^\pm$ and Z^0 to establish the standard model and give us hints what might lie beyond it, a linear e^-e^- colliders in the TeV energy range will be essential to follow up on the discoveries made by the LHC, as well as make its own. The only question concerns the energy range that it should be able to cover.

At the low end, there is considerable interest in producing a large sample of $\sim 10^9$ Z^0 bosons with polarized beams, enabling electroweak measurements to be taken to the next level of precision [97]. A large sample of $e^+e^- \rightarrow W^+W^-$ events close to threshold would also be interesting for the same reason.

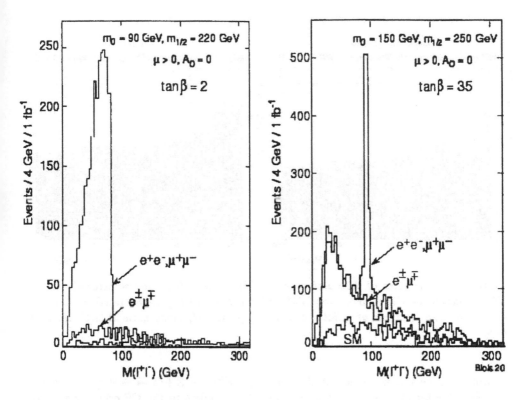

Figure 24. *The dilepton mass distributions expected at the LHC due to sparticle decays in two different supersymmetric scenarios [92, 90, 89].*

Looking to higher energies, we do know of one threshold that occurs around 350 GeV, namely that for $e^-e^- \to \bar{t}t$. As discussed earlier, we are also quite confident that the Higgs boson weighs $\lesssim 200$ GeV. However, we do not know where (if anywhere!) the thresholds for sparticle-pair production may appear. The first might appear just above the reach of LEP, but equally it might appear beyond 1 TeV in the centre of mass. We can hope that the LHC will provide crucial guidance, but for now we must envisage flexibility in the attainable energy range.

3.3.1 TeV-scale linear colliders

The physics capabilities of linear e^+e^- colliders are amply documented in various design studies [83]. If the Higgs boson indeed does weigh less than 200 GeV, its production and study would be easy at an e^+e^- collider with $E_{CM} \sim 500$ GeV. With a luminosity of 10^{34} cm^{-2}s^{-1} or more, many decay modes of the Higgs boson could be measured very accurately [83], as seen in Fig. 28.

One might be able to find a hint whether its properties were modified by supersymmetry, as seen in Fig. 29 [95]. The top panels show typical examples of the potential sensitivity of the reaction $e^+e^- \to Z + (h \to b\bar{b})$ to modifications expected within the CMSSM, and

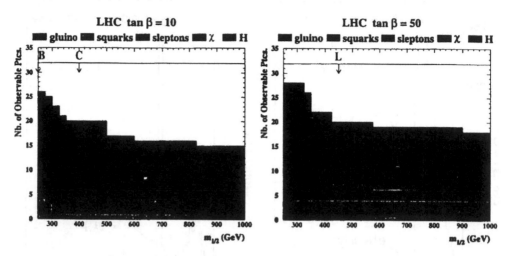

Figure 25. *Estimates of the numbers of MSSM particles that may be detectable at the LHC as functions of $m_{1/2}$ along the WMAP lines shown in Fig. 20 for $\tan\beta = 10$ and 50 for $\mu > 0$. The locations of updated benchmark points [76] along these WMAP lines are indicated.*

the bottom panels show the potential sensitivity in $e^+e^- \to Z + (h \to WW^*)$. Fig. 30 [95] compares the sensitivities of e^+e^-, $\gamma\gamma$ and $\mu^+\mu^-$ colliders to the CP-odd Higgs boson mass m_A and the Higgs mixing parameter μ, including the CMSSM as a special case.

Moreover, if sparticles are light enough to be produced directly, their masses and other properties can be measured very precisely, typical estimated precisions being [83]

$$\delta m_{\tilde\mu} \simeq 0.3 \text{ GeV}, \ \delta m_{\tilde\nu} \simeq 5 \text{ GeV}, \ \delta m_{\chi^\pm} \simeq 0.04 \text{ GeV},$$
$$\delta m_\chi \simeq 0.2 \text{ GeV}, \ \delta m_{\tilde\tau} \simeq 4 \text{ GeV}. \tag{144}$$

Moreover, the spin-parities and couplings of sparticles can be measured accurately. The mass measurements can be used to test models of supersymmetry breaking, as seen in Fig. 31 [98].

As seen in Fig. 21, the sparticles visible at an e^+e^- collider largely complement those visible at the LHC [75]. In most of the benchmark scenarios proposed, a 1-TeV linear collider would be able to discover and measure precisely several weakly-interacting sparticles that are invisible or difficult to detect at the LHC. However, there are some benchmark scenarios where the linear collider (as well as the LHC) fails to discover supersymmetry. Independently from the particular benchmark scenarios proposed, a linear e^+e^- collider with $E_{CM} < 0.5$ TeV would not cover all the supersymmetric parameter space allowed by cosmology, as seen in Fig. 32, whereas a combination of the LHC with a $E_{CM} = 1$ TeV linear e^+e^- collider would together discover a large fraction of the MSSM spectrum, as seen in Fig. 33.

There are compelling physics arguments for such a linear e^+e^- collider, which would be very complementary to the LHC in terms of its exploratory power and precision. It is

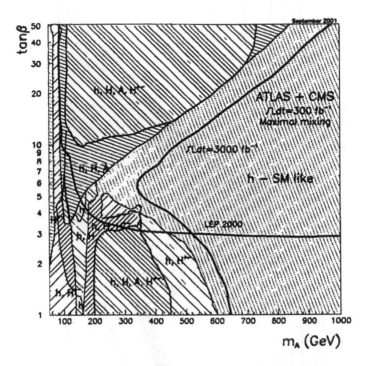

Figure 26. *Regions of the MSSM parameter space where the various Higgs bosons can be discovered at the LHC via their decays into standard model particles, combining ATLAS and CMS and assuming 300 /fb per experiment. In the dashed regions, at least two Higgs bosons can be discovered, whereas in the dotted region only the lightest MSSM Higgs boson h can be discovered. In the region to the left of the rightmost contour, at least two Higgs bosons could be discovered with an upgraded LHC delivering 3000/fb per experiment [93].*

to be hoped that the world community will converge on a single project with the widest possible energy range.

3.3.2 CLIC

Only a linear collider with a higher centre-of-mass energy appears sure to cover all the allowed CMSSM parameter space, as seen in the lower panels of Fig. 21, which illustrate the physics reach of a higher-energy lepton collider, such as CLIC [84] or a multi-TeV muon collider [85, 86].

CERN and its collaborating institutes are studying CLIC as a possible second step in linear e^+e^- colliders [84]. This would use a double-beam technique to attain accelerating gradients as high as 150 MV/m, and the viability of accelerating structures capable of achieving this field has been demonstrated in the CLIC test facility [99]. Parameter sets have been calculated for CLIC designs with $E_{CM} = 3$ and 5 TeV, and luminosities of 10^{35} cm^{-2}s^{-1} or more. The prospective layout of CLIC is shown in Fig. 34, illustrating

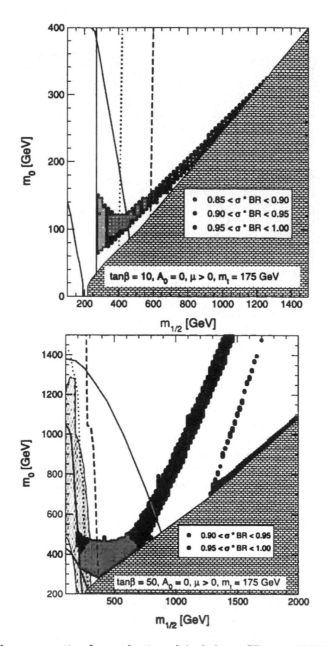

Figure 27. *The cross section for production of the lightest CP-even CMSSM Higgs boson in gluon fusion and its decay into a photon pair at the LHC, $\sigma(gg \to h) \times B(h \to \gamma\gamma)$, normalized to the standard model value for the same Higgs mass, is given in the regions of the $(m_{1/2}, m_0)$ planes allowed before the WMAP data for $\mu > 0$, $\tan\beta = 10, 50$, assuming $A_0 = 0$ and $m_t = 175$ GeV. The diagonal (red) solid lines are the $\pm 2 - \sigma$ contours for $g_\mu - 2$ [70, 73]. The near-vertical solid, dotted and dashed (black) lines are the $m_h = 113, 115, 117$ GeV contours. The light shaded (pink) regions are excluded by $b \to s\gamma$ [68]. The (brown) bricked regions are excluded since in these regions the LSP is the charged $\tilde{\tau}_1$.*

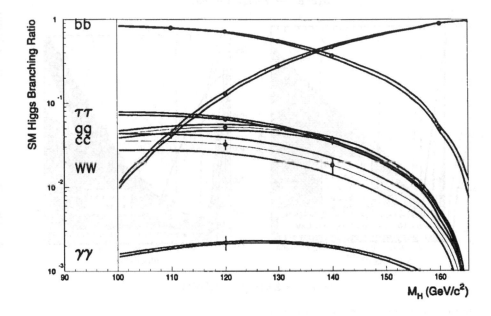

Figure 28. *Analysis of the accuracy with which Higgs decay branching ratios may be measured with a linear e^+e^- collider [83].*

how RF power from the high-intensity, low-energy drive beam is fed to the low-intensity, high-energy colliding beams.

Various topics in Higgs physics at CLIC have been studied [100]. For example, it may be possible to measure for the first time $H \to \mu^+\mu^-$ decay. Also, if the Higgs mass is light enough, $m_H \sim 120$ GeV, it will be possible to measure the triple-Higgs coupling λ_{HHH} more accurately than would be possible at a lower-energy machine, as seen in Fig. 35. CLIC would also have interesting capabilities for exploring the heavier MSSM Higgs bosons in the 'wedge' region left uncovered by direct searches at the LHC and a lower-energy linear e^+e^- collider.

In many of the proposed benchmark supersymmetric scenarios, CLIC would be able to complete the supersymmetric spectrum and/or measure in much more detail heavy sparticles found previously at the LHC, as seen in Fig. 36. CLIC produces more beam-strahlung than lower-energy linear e^+e^- colliders, but the supersymmetric missing-energy signature would still be easy to distinguish, and accurate measurements of masses and decay modes could still be made, as seen in Fig. 37 [102] for the example of $e^+e^- \to \tilde{\mu}^+\tilde{\mu}^-$ followed by $\tilde{\mu}^\pm \to \mu^\pm\chi$ decay. CLIC also has the potential to study heavier neutralinos and charginos well beyond the reach of the LHC and a lower-energy linear e^+e^- collider.

$$\sigma(e^+e^- \to Zh)B(h \to \bar{b}b)$$

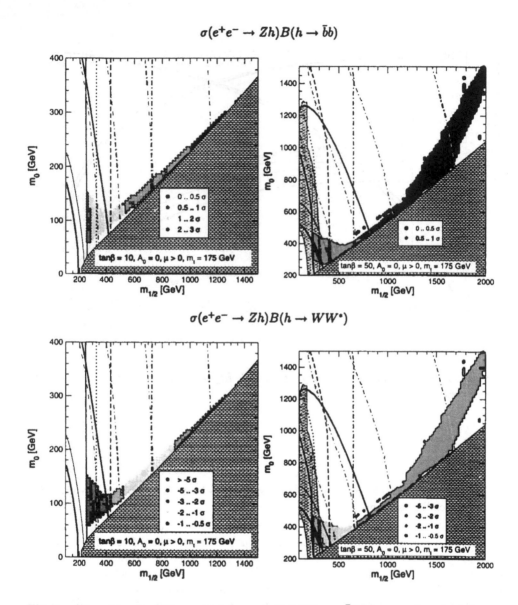

$$\sigma(e^+e^- \to Zh)B(h \to WW^*)$$

Figure 29. *The deviations of $\sigma(e^+e^- \to Zh)B(h \to \bar{b}b)$ (top row) and $\sigma(e^+e^- \to Zh)B(h \to WW^*)$ (bottom row) for the lightest CP-even CMSSM Higgs boson, normalized to the values in the standard model with the same Higgs mass, are given in the regions of the $(m_{1/2}, m_0)$ planes allowed before the WMAP data for $\mu > 0$, $\tan\beta = 10, 50$ and $A_0 = 0$ [95]. The diagonal red thick (thin) lines are the $\pm 2 - \sigma$ contours for $g_\mu - 2$: +56.3, +11.5 (+38.1, -4.7). The near-vertical solid, dotted short-dashed, dash-dotted and long-dashed (black) lines are the $m_h = 113, 115, 117, 120, 125$ GeV contours. The lighter dot-dashed (orange) lines correspond to $m_A = 500, 700, 1000, 1500$ GeV. The light shaded (pink) regions are excluded by $b \to s\gamma$. The (brown) bricked regions are excluded because the LSP is the charged $\tilde{\tau}_1$ in these regions.*

Variation of the σB with m_A

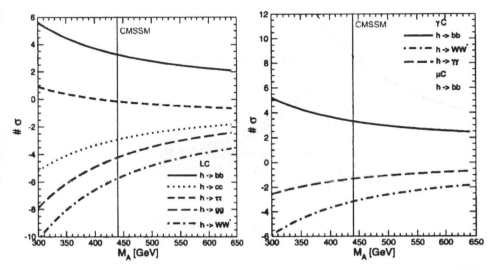

Figure 30. *The numbers of standard deviations of the predictions in the MSSM as compared to the standard model are shown in the different σB channels for e^+e^- (left column) and $\gamma\gamma$ and $\mu^+\mu^-$ colliders (right column), as functions of the CP-odd neutral Higgs boson mass m_A [95]. The corresponding CMSSM value of m_A is indicated by light vertical (orange) lines. The other parameters have been chosen as $m_{1/2} = 300\,\text{GeV}$, $m_0 = 100\,\text{GeV}$, $\tan\beta = 10$ and $A_0 = 0$.*

4   Constraints from cosmology

4.1   The Big Bang and particle physics

The universe is currently expanding almost homogeneously and isotropically, as discovered by Hubble, and the radiation it contains is cooling as it expands adiabatically:

$$a \times T \simeq \text{constant}, \tag{145}$$

where a is the scale factor of the universe and T is the temperature. There are two important pieces of evidence that the scale factor of the universe was once much smaller than it is today, and correspondingly that its temperature was much higher. One is the *Cosmic Microwave Background*, which bathes us in photons with a density

$$n_\gamma \simeq 400\ \text{cm}^{-3}, \tag{146}$$

with an effective temperature $T \simeq 2.7$ K. These photons were released when electrons and nuclei combined to form atoms, when the universe was some 3000 times hotter and the scale factor correspondingly 3000 times smaller than it is today. The second is the agreement of the *Abundances of Light Elements* [103], in particular those of ^{4}He, Deuterium and ^{6}Li, with calculations of cosmological nucelosynthesis. For these elements to have been produced by nuclear fusion, the universe must once have been some 10^9 times hotter and smaller than it is today.

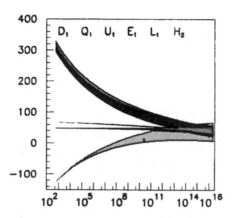

Figure 31. *Analogously to the unification of the gauge couplings shown in Fig. 15, measurements of the sparticle masses at future colliders (vertical axis, in units of GeV) can be evolved up to high scales (horizontal axis, in units of GeV) to test models of supersymmetry breaking, in particular whether squark and slepton masses are universal at some input GUT scale [98].*

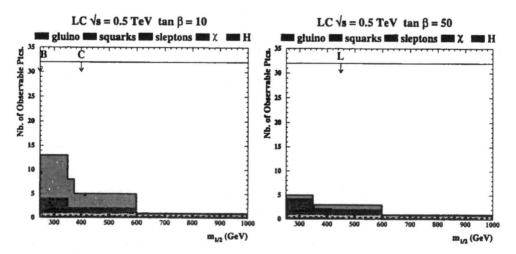

Figure 32. *Estimates of the numbers of MSSM particles that may be detectable at a 0.5-TeV linear e^+e^- collider as functions of $m_{1/2}$ along the WMAP lines for $\tan\beta = 10$ and 50 for $\mu > 0$. The locations of updated benchmark points [76] along these WMAP lines are indicated.*

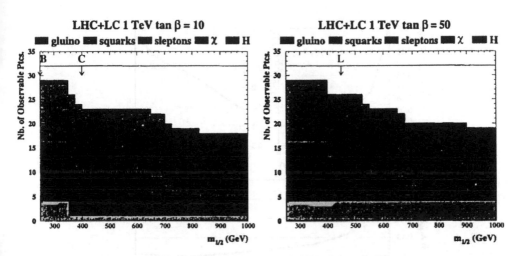

Figure 33. *Estimates of the combined numbers of MSSM particles that may be detectable at the LHC and a 1-TeV linear e^+e^- collider as functions of $m_{1/2}$ along the WMAP lines for $\tan\beta = 10$ and 50 for $\mu > 0$. The locations of updated benchmark points [76] along these WMAP lines are indicated.*

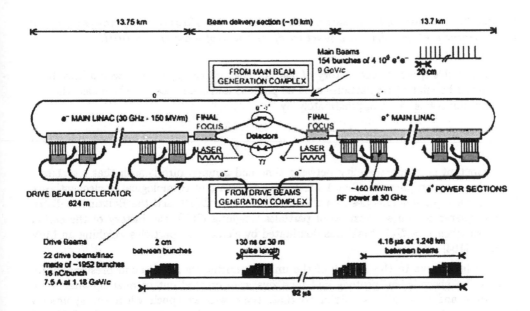

Figure 34. *Conceptual layout for the CLIC linear e^+e^- collider, which is designed to be capable of reaching a centre-of-mass energy of 3 TeV or more. CLIC uses high-power but low-energy drive beams to accelerate less-intense colliding beams with a high gradient [84].*

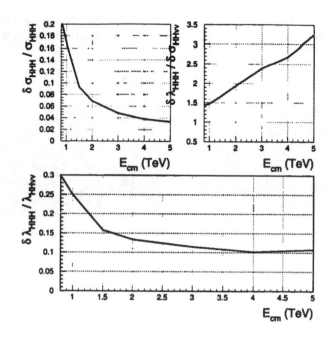

Figure 35. *Analysis of the accuracy with which the triple-Higgs coupling may be measured with a linear e^+e^- collider, as a function of its centre-of-mass energy [101].*

During this epoch of the history of the universe, its energy density would have been dominated by relativistic particles such as photons and neutrinos, in which case the age t of the universe is given approximately by

$$t \propto a^2 \propto \frac{1}{T^2}. \tag{147}$$

The constant of proportionality between time and temperature is such that $t \simeq 1$ second when the temperature $T \simeq 1$ MeV, near the start of cosmological nucleosynthesis. Since typical particle energies in a thermal plasma are $\mathcal{O}(T)$, and the Boltzmann distribution guarantees large densities of particles weighing $\mathcal{O}(T)$, the history of the earlier universe when $T > \mathcal{O}(1)$ MeV was dominated by elementary particles weighing an MeV or more [104].

The landmarks in the history of the universe during its first second presumably included the epoch when protons and neutrons were created out of quarks, when $T \sim 200$ MeV and $t \sim 10^{-5}$ s. Prior to that, there was an epoch when the symmetry between weak and electromagnetic interactions was broken, when $T \sim 100$ GeV and $t \sim 10^{-10}$ s. Laboratory experiments with accelerators have already explored physics at energies $E \lesssim 100$ GeV, and the energy range $E \lesssim 1000$ GeV, corresponding to the history of the universe when $t \gtrsim 10^{-12}$ s, will be explored at the LHC. Our ideas about physics at earlier epochs are necessarily more speculative, but one possibility is that there was

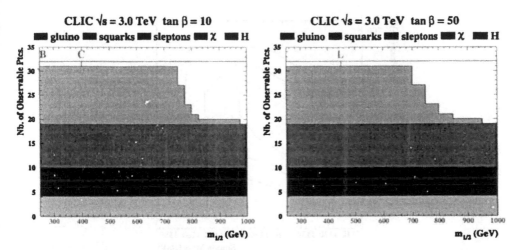

Figure 36. *Estimates of the numbers of MSSM particles that may be detectable at the CLIC 3-TeV linear e^+e^- collider as functions of $m_{1/2}$ along the WMAP lines for $\tan\beta = 10$ and 50 for $\mu > 0$. The locations of updated benchmark points [76] along these WMAP lines are indicated.*

an inflationary epoch when the age of the universe was somewhere between 10^{-40} and 10^{-30} s.

4.2   Density budget of the universe

What does the universe contain? Let us enumerate its composition in terms of the density budget of the universe, measured relative to the critical density: $\Omega_i \equiv \rho_i/\rho_{crit}$.

Inflation [105] suggests that the *total density* of the universe is very close to the critical value: $\Omega_{tot} \simeq 1 \pm O(10^{-4})$, and this estimate is supported by CMB data [74]. I remind you that inflation explains why the universe is so large: the scale size $a \gg \ell_P \sim 10^{-33}$ cm, why the universe is so old: its age $t \gg t_P \sim 10^{-43}$ s, why its geometry is so nearly flat with a Euclidean geometry, and why the universe is so homogeneous on large scales.

It achieves these feats by postulating an epoch of (near-) exponential expansion during the very early universe, making the universe very large and giving it a long time to recollapse (if it ever will). Even the most distant parts of the observable universe would have been very close to each other prior to this inflationary epoch, and so could have synchronized their behaviours. This inflationary expansion would have blown the universe up like an inflated ballon, which seems almost flat to an ant living on its surface. During the inflationary expansion, quantum fluctuations in the inflaton field would have generated small density perturbations (cf. the CMB observations [74]) capable of growing into the structures seen in the universe today [106].

Big-Bang nucleosynthesis suggests that the *baryon density* $\Omega_b \simeq 0.04$ [103], an estimate that has been supported by analyses of the relative sizes of small fluctuations in the CMB

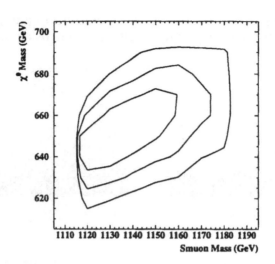

Figure 37. *Like lower-energy e^+e^- colliders, CLIC enables very accurate measurements of sparticle masses to be made, in this case the $\tilde{\mu}$ and the lightest neutralino χ [102].*

at different scales [74].

The baryons are insufficient to explain the *total matter density* $\Omega_m \simeq 0.3$, as estimated independently by analyses of clusters of galaxies and, more recently, by combining the observations of high-redshift supernovae [107] with those of the CMB [74]. The supernovae constrain the density budget of the universe in a way that is almost orthogonal to the CMB constraint, and is very consistent with the prior indications from galaxy clusters [108].

Observations of the structures that have formed at different scales in the universe suggest that most of the missing dark matter is in the form of non-relativistic *cold dark matter*. As already mentioned, the prime candidates for the seeds of these structures are quantum fluctuations in the inflaton field, which would have caused different parts of the universe to expand differently and generated a *Gaussian random field* of density perturbations [106]. If the inflaton energy was roughly constant during inflation, these perturbations would be *almost scale-invariant*, as postulated by astrophysicists. The CMB data are consistent with both these properties. Accepting this scenario, the magnitude of the primordial perturbations would be related to the field energy density μ^4 during inflation:

$$\left(\frac{\delta T}{T}\right) \propto \left(\frac{\delta \rho}{\rho}\right) \propto \mu^2 G_N. \tag{148}$$

Inserting the magnitude of $\delta\rho/\rho \sim 10^{-5}$ obderved by the COBE and subsequent experiments [74], one estimates

$$\mu \simeq 10^{16} \text{ GeV}, \tag{149}$$

comparable with the GUT scale [45].

These primordial perturbations would have produced embryonic potential wells into which the non-relativistic cold dark matter particles would have fallen, while relativistic hot dark matter particles would have escaped. In this way, cold matter particles would have amplified the amplitudes of the primordial density perturbations, while the baryons were still coupled to the relativistic radiation. Then, when the baryonic matter and radiation 're-' combined to form atoms, they would have fallen into the deeper potential wells prepared by the cold dark matter. This theory of structure formation fits remarkably well the data on all scales from over 10^3 Mpc down to ~ 1 Mpc [74, 109].

The theory of structure formation suggests that very little of the dark matter is in the form of *hot dark matter* particles that were relativistic when structures started to form: $\Omega_{hot}h^2 < 0.0076$ [74]. Applying this constraint to neutrinos, for which

$$\Omega_\nu h^2 \simeq \frac{\Sigma_i m_{\nu_i}}{93\text{eV}},\qquad(150)$$

this constraint tells us that $\Sigma_i m_{\nu_i} < 0.7$ eV, a limit that is highly competitive with direct limits [110], as discussed in the next Lecture.

If $\Omega_{tot} \simeq 1$ and the matter density $\Omega_m \sim 0.3$, how do we balance the density budget of the universe? There must be *vacuum energy* Λ with $\Omega_\Lambda \sim 0.7$. All the available cosmological data are consistent with Λ having been constant at redshifts $z \lesssim 1$, as per Einstein's original suggestion of a cosmological constant. However, we cannot yet exclude some slowly varying source of vacuum energy, 'quintessence' with an equation of state parametrized by $w \equiv p/\rho \lesssim -0.8$ [111]. Measurable vacuum energy would provide a second general-relativity observable to explain, in addition to the Planck mass scale m_P. This would provide a tremendous opportunity for any theory of everything including quantum gravity, such as string. The ultimate challenge for theoretical physics may be to calculate Λ.

4.3 Supersymmetric dark matter

The astrophysical cold dark matter could be provided by a neutral, weakly-interacting particle weighing less than about 1 TeV, such as the lightest supersymmetric particle (LSP) χ [63]. This is expected to be stable in the MSSM, and hence should be present in the universe today as a cosmological relic from the Big Bang [63]. Its stability arises because there is a multiplicatively-conserved quantum number called R parity, that takes the values +1 for all conventional particles and -1 for all sparticles [39]. The conservation of R parity can be related to that of baryon number B and lepton number L, since

$$R = (-1)^{3B+L+2S}\qquad(151)$$

where S is the spin. There are three important consequences of R conservation:

1. sparticles are always produced in pairs, e.g., $\bar{p}p \to \bar{q}\bar{g}X$, $e^+e^- \to \tilde{\mu} + \tilde{\mu}^-$,

2. heavier sparticles decay to lighter ones, e.g., $\tilde{q} \to q\tilde{g}, \tilde{\mu} \to \mu\tilde{\gamma}$, and

3. the lightest sparticle (LSP) is stable, because it has no legal decay mode.

This last feature constrains strongly the possible nature of the lightest supersymmetric sparticle [63]. If it had either electric charge or strong interactions, it would surely have dissipated its energy and condensed into galactic disks along with conventional matter. There it would surely have bound electromagnetically or via the strong interactions to conventional nuclei, forming anomalous heavy isotopes that should have been detected.

A priori, the LSP might have been a sneutrino partner of one of the 3 light neutrinos, but this possibility has been excluded by a combination of the LEP neutrino counting and direct searches for cold dark matter. Thus, the LSP is often thought to be the lightest neutralino χ of spin 1/2, which naturally has a relic density of interest to astrophysicists and cosmologists: $\Omega_\chi h^2 = \mathcal{O}(0.1)$ [63].

We see in Fig. 19 that there are significant regions of the CMSSM parameter space where the relic density falls within the preferred range (143). What goes into the calculation of the relic density? It is controlled by the annihilation cross section [63]: $\rho_\chi = m_\chi n_\chi$, where the relic number density n_χ obeys a Boltzmann equation

$$\frac{dn_\chi}{dt} = -3\frac{\dot{a}}{a}n_\chi - \langle\sigma_{ann}v\rangle(n_\chi^2 - n_{eq}^2), \tag{152}$$

where n_{eq} is the thermal equilibrium density, and in many cases of interest one may expand the annihilation rate in even powers of the relative velocity:

$$\langle\sigma_{ann}v\rangle \simeq a + b\cdot x : x \equiv \frac{\langle v^2\rangle}{6}. \tag{153}$$

Usually, the non-relativistic expansion coefficients $a, b = O(1/m_\chi^2)$ and the relic density evolves approximately according to the simplified equation

$$\dot{n}_\chi \simeq \langle\sigma_{ann}v\rangle n_{eq}^2 \tag{154}$$

until a freeze-out temperature $T_f \sim m_\chi/20$, and thereafter

$$\dot{n}_\chi \simeq \langle\sigma_{ann}v\rangle n_\chi^2 \tag{155}$$

to yield a present-day mass density

$$\rho_\chi \simeq 0.8 T_\chi^3 \frac{m_\chi}{a\cdot x_f + \frac{1}{2}b\cdot x_f^2}, \tag{156}$$

where T_χ is the present-day effective relic temperature which is somewhat smaller than the CMB, $x_f \equiv T_f/m_\chi$, and the pre-factor in (156) is an approximate numerical correction. We see from (156) that the relic density typically increases with the relic mass, and this combined with the upper bound in (143) then leads to the common expectation that $m_\chi \lesssim O(1)$ TeV.

However, there are various ways in which the generic upper bound on m_χ can be increased along filaments in the $(m_{1/2}, m_0)$ plane. For example, if the next-to-lightest sparticle (NLSP) is not much heavier than χ: $\Delta m/m_\chi \lesssim 0.1$, the relic density may be suppressed by coannihilation: $\sigma(\chi+\text{NLSP}\rightarrow \ldots)$ [77]. In this way, the allowed CMSSM region may acquire a 'tail' extending to larger sparticle masses. An example of this possibility is the case where the NLSP is the lighter stau: $\tilde{\tau}_1$ and $m_{\tilde{\tau}_1} \sim m_\chi$, as seen in Figs. 19(a) and (b) [78].

Another mechanism for extending the allowed CMSSM region to large m_χ is rapid annihilation via a direct-channel pole when $m_\chi \sim \frac{1}{2}m_{Higgs}$ [79, 112]. This may yield

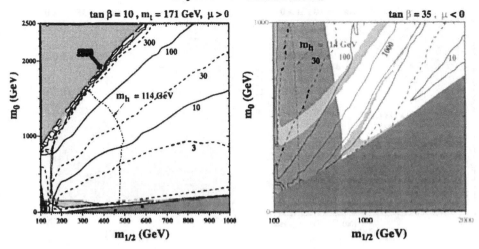

Figure 38. *Contours of the total sensitivity Δ^{Ω} (157) of the relic density in the $(m_{1/2}, m_0)$ planes for (a) $\tan\beta = 10, \mu > 0$ and (b) $\tan\beta = 35, \mu < 0$, for $A_0 = 0$. The light (turquoise) shaded areas are the cosmologically preferred regions with $0.1 \leq \Omega_\chi h^2 \leq 0.3$. In the dark (brick red) shaded regions, the LSP is the charged $\tilde{\tau}_1$, so these regions are excluded. In panel (a), the medium shaded (mauve) region is excluded by the electroweak vacuum conditions.*

a 'funnel' extending to large $m_{1/2}$ and m_0 at large $\tan\beta$, as seen in panels (c) and (d) of Fig. 19 [112]. Yet another allowed region at large $m_{1/2}$ and m_0 is the 'focus-point' region [80], which is adjacent to the boundary of the region where electroweak symmetry breaking is possible. The lightest supersymmetric particle is relatively light in this region.

4.4 Fine tuning

The filaments extending the preferred CMSSM parameter space are clearly exceptional, in some sense, so it is important to understand the sensitivity of the relic density to input parameters, unknown higher-order effects, etc. One proposal is the relic-density fine-tuning measure [81]

$$\Delta^{\Omega} \equiv \sqrt{\sum_i \left(\frac{\partial \ln(\Omega_\chi h^2)}{\partial \ln a_i}\right)^2}, \tag{157}$$

where the sum runs over the input parameters, which might include (relatively) poorly-known standard model quantities such as m_t and m_b, as well as the CMSSM parameters $m_0, m_{1/2}$, etc. As seen in Fig. 38, the sensitivity Δ^{Ω} (157) is relatively small in the 'bulk' region at low $m_{1/2}$, m_0, and $\tan\beta$. However, it is somewhat higher in the $\chi - \tilde{\tau}_1$ coannihilation 'tail', and at large $\tan\beta$ in general. The sensitivity measure Δ^{Ω} (157) is particularly high in the rapid-annihilation 'funnel' and in the 'focus-point' region. This explains why published relic-density calculations may differ in these regions, whereas they agree well when Δ^{Ω} is small: differences may arise because of small differences in the treatments of the inputs.

It is important to note that the relic-density fine-tuning measure (157) is distinct from

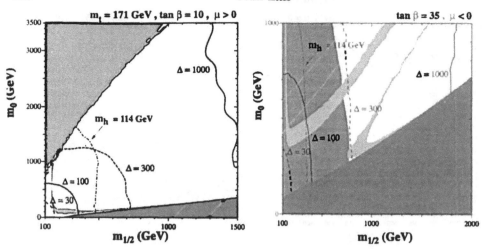

Figure 39. *Contours of the electroweak fine-tuning measure* Δ *(158) in the* $(m_{1/2}, m_0)$ *planes for (a)* $\tan\beta = 10, \mu > 0$ *and (b)* $\tan\beta = 35, \mu < 0$ *for* $A_0 = 0$. *The light (turquoise) shaded areas are the cosmologically preferred regions with* $0.1 \leq \Omega_\chi h^2 \leq 0.3$. *In the dark (brick red) shaded regions, the LSP is the charged* $\tilde{\tau}_1$, *so this region is excluded. In panel (a), the medium shaded (mauve) region is excluded by the electroweak vacuum conditions.*

the traditional measure of the fine-tuning of the electroweak scale [82]:

$$\Delta = \sqrt{\sum_i \Delta_i^2}, \quad \Delta_i \equiv \frac{\partial \ln m_W}{\partial \ln a_i}. \tag{158}$$

Sample contours of the electroweak fine-tuning measure are shown (158) are shown in Figs. 39. This electroweak fine tuning is logically different from the cosmological fine tuning, and values of Δ are not necessarily related to values of Δ^Ω, as is apparent when comparing the contours in Figs. 38 and 39. Electroweak fine-tuning is sometimes used as a criterion for restricting the CMSSM parameters. However, the interpretation of Δ (158) is unclear. How large a value of Δ is tolerable? Different physicists may well have different pain thresholds. Moreover, correlations between input parameters may reduce its value in specific models. Moreover, the regions allowed by the different constraints can be very different from those in the CMSSM when we relax some of the CMSSM assumptions, e.g. the universality between the input Higgs masses and those of the squarks and sleptons.

4.5   Searches for dark matter particles

In the above discussion, we have paid particular attention to the region of parameter space where the lightest supersymmetric particle could constitute the cold dark matter in the universe [63]. How easy would this be to detect?

• One strategy is to look for relic annihilations in the galactic halo, which might produce detectable antiprotons, positrons or γ-rays in the cosmic rays [113]. Unfortunately, the rates for their production are not very promising in the benchmark scenarios we studied [114], when compared with the known and expected backgrounds. Fig. 40 shows that the expected signal for annihilation positrons is far below the expected cosmic-ray background.

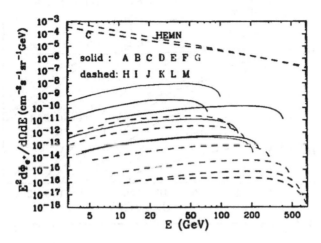

Figure 40. *Spectra of positrons from the annihilations of dark matter particles in the halo of our galaxy, in different benchmark supersymmetric models [114], compared with the background expected from conventional cosmic rays.*

• Alternatively, one might look for annihilations in the core of our galaxy, which might produce detectable γ-rays. As seen in the top panel of Fig. 41, this may be possible in certain benchmark scenarios [114], though the rate is rather uncertain because of the unknown enhancement of relic particles in our galactic core. The experiment that is best-placed to detect these γ-rays may be the GLAST satellite.

• A third strategy is to look for annihilations inside the sun or earth, where the local density of relic particles is enhanced in a calculable way by scattering off matter, which causes them to lose energy and become gravitationally bound [115]. The signature would then be energetic neutrinos that might produce detectable muons. Several underwater and ice experiments are underway or planned to look for this signature, and (particularly for annihilations inside the sun) this strategy looks promising for several benchmark scenarios, as seen in the right panel of Fig. 41 [114]. It will be interesting to have such neutrino telescopes in different hemispheres, which will be able to scan different regions of the sky for astrophysical high-energy neutrino sources.

• The most satisfactory way to look for supersymmetric relic particles is directly via their elastic scattering on nuclei in a low-background laboratory experiment [116]. There are two types of scattering matrix elements, spin-independent – which are normally dominant for heavier nuclei, and spin-dependent – which could be interesting for lighter elements such as fluorine. The best experimental sensitivities so far are for spin-independent scattering, and one experiment has claimed a positive signal [117], in the form of an annual modulation of events in the detector. However, some observers wonder whether this effect could be mimicked by other mechanisms, and this result has failed to be confirmed by a number of other experiments [118]. In the benchmark scenarios the rates are considerably below the present experimental sensitivities [114], but there are

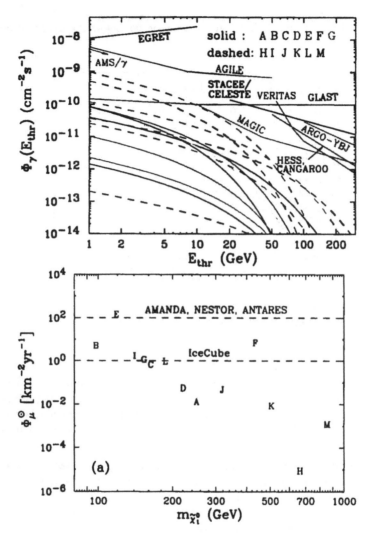

Figure 41. *Top panel: Spectra of photons from the annihilations of dark matter particles in the core of our galaxy, in different benchmark supersymmetric models [114]. Bottom panel: Signals for muons produced by energetic neutrinos originating from annihilations of dark matter particles in the core of the sun, in the same benchmark supersymmetric models [114].*

prospects for improving the sensitivity into the interesting range, as seen in Fig. 42.

What are the prospects for such non-accelerator searches for dark matter, compared with accelerator searches for supersymmetry? There are regions of the supersymmetric parameter space which are accessible to such dark matter experiments, and they have some advantages in the so-called 'focus-point' region close to the boundary of electroweak symmetry breaking at large m_0, where the LSP has a larger higgsino component. If the direct scattering experiments and/or the searches for energetic solar neutrinos can achieve

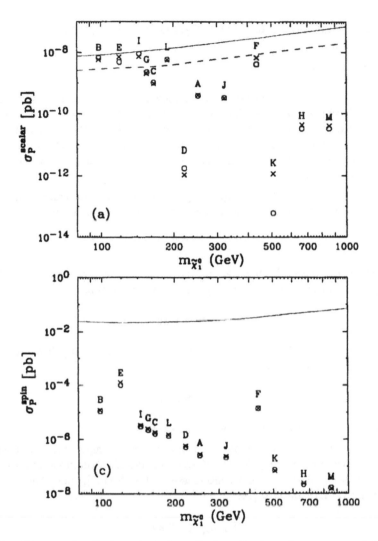

Figure 42. *Top panel: elastic spin-independent scattering of supersymmetric relics on protons calculated in benchmark scenarios [114], compared with the projected sensitivities for CDMS II [119] and CRESST [120] (solid) and GENIUS [121] (dashed). The predictions of the SSARD code (blue crosses) and Neutdriver[122] (red circles) for neutralino-nucleon scattering are compared [114]. The labels A, B, ...,L correspond to the benchmark points in [75, 76]. Bottom panel: prospects for detecting elastic spin-dependent scattering in the benchmark scenarios, which are less bright [114].*

good sensitivity before the LHC gets going, they may stand a chance.

5   Neutrinos and lepton-flavour violation

5.1   Neutrino masses

There is no good reason why either the total lepton number L or the individual lepton flavours $L_{e,\mu,\tau}$ should be conserved. Theorists have learnt that the only conserved quantum numbers are those associated with exact local symmetries, just as the conservation of electromagnetic charge is associated with local U(1) invariance. On the other hand, there is no exact local symmetry associated with any of the lepton numbers, so we may expect non-zero neutrino masses.

However, so far we have only upper experimental limits on neutrino masses [110]. From measurements of the end-point in Tritium β decay, we know that:

$$m_{\nu_e} \lesssim 2.5 \text{ eV}, \tag{159}$$

which might be improved down to about 0.5 eV with the proposed KATRIN experiment [123]. From measurements of $\pi \rightarrow \mu\nu$ decay, we know that:

$$m_{\nu_\mu} < 190 \text{ KeV}, \tag{160}$$

and there are prospects to improve this limit by a factor ~ 20. Finally, from measurements of $\tau \rightarrow n\pi\nu$ decay, we know that:

$$m_{\nu_\tau} < 18.2 \text{ MeV}, \tag{161}$$

and there are prospects to improve this limit to ~ 5 MeV.

Astrophysical upper limits on neutrino masses are stronger than these laboratory limits. As we have already seen, the 2dF data on large-scale structures were used to infer an upper limit on the sum of the neutrino masses of 1.8 eV [109], which has recently been improved using WMAP data to [74]

$$\Sigma_{\nu_i} m_{\nu_i} < 0.7 \text{ eV}, \tag{162}$$

as seen in Fig. 43. This impressive upper limit is substantially better than even the most stringent direct laboratory upper limit on an individual neutrino mass.

Another interesting laboratory limit on neutrino masses comes from searches for neutrinoless double-β decay, which constrain the sum of the neutrinos' Majorana masses weighted by their couplings to electrons [124]:

$$\langle m_\nu \rangle_e \equiv |\Sigma_{\nu_i} m_{\nu_i} U_{ei}^2| \lesssim 0.35 \text{ eV}, \tag{163}$$

which might be improved to ~ 0.01 eV in a future round of experiments.

Neutrinos have been seen to oscillate between their different flavours [125, 126], showing that the separate lepton flavours $L_{e,\mu,\tau}$ are indeed not conserved, though the conservation of total lepton number L is still an open question. The observation of such oscillations strongly suggests that the neutrinos have different masses.

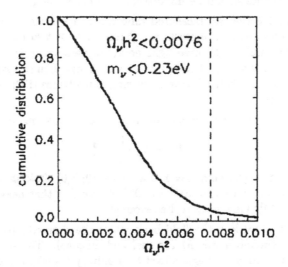

Figure 43. *Likelihood function for $\Omega_\nu h^2$ (related to the sum of neutrino masses) provided by WMAP [74]: the quoted upper limit on m_ν applies if the 3 light neutrino species are degenerate.*

5.2 Models of neutrino masses and mixing

The conservation of lepton number is an accidental symmetry of the renormalizable terms in the standard model Lagrangian. However, one could easily add to the standard model non-renormalizable terms that would generate neutrino masses, even without introducing any new fields. For example, a non-renormalizable term of the form [127]

$$\frac{1}{M}\nu H \cdot \nu H, \tag{164}$$

where M is some large mass beyond the scale of the standard model, would generate a neutrino mass term:

$$m_\nu \nu \cdot \nu : \ m_\nu = \frac{\langle 0|H|0\rangle^2}{M}. \tag{165}$$

However, a new interaction like (164) seems unlikely to be fundamental, and one should like to understand the origin of the large mass scale M.

The minimal renormalizable model of neutrino masses requires the introduction of weak-singlet 'right-handed' neutrinos N. These will in general couple to the conventional weak-doublet left-handed neutrinos via Yukawa couplings Y_ν that yield Dirac masses $m_D = Y_\nu\langle 0|H|0\rangle \sim m_W$. In addition, these singlet neutrinos N can couple to themselves via Majorana masses M that may be $\gg m_W$, since they do not require electroweak symmetry breaking. Combining the two types of mass term, one obtains the seesaw mass matrix [128]:

$$(\nu_L, N) \begin{pmatrix} 0 & M_D \\ M_D^T & M \end{pmatrix} \begin{pmatrix} \nu_L \\ N \end{pmatrix}, \tag{166}$$

where each of the entries should be understood as a matrix in generation space.

In order to provide the two measured differences in neutrino masses-squared, there must be at least two non-zero masses, and hence at least two heavy singlet neutrinos N_i [129, 130]. Presumably, all three light neutrino masses are non-zero, in which case there must be at least three N_i. This is indeed what happens in simple GUT models such as SO(10), but some models [131] have more singlet neutrinos [132]. Here, for simplicity we consider just three N_i.

The effective mass matrix for light neutrinos in the seesaw model may be written as:

$$M_\nu = Y_\nu^T \frac{1}{M} Y_\nu v^2, \tag{167}$$

where we have used the relation $m_D = Y_\nu v$ with $v \equiv \langle 0|H|0 \rangle$. Taking $m_D \sim m_q$ or m_ℓ and requiring light neutrino masses $\sim 10^{-1}$ to 10^{-3} eV, we find that heavy singlet neutrinos weighing $\sim 10^{10}$ to 10^{15} GeV seem to be favoured.

It is convenient to work in the field basis where the charged-lepton masses $m_{\ell^\pm}$ and the heavy singlet-neutrino mases M are real and diagonal. The seesaw neutrino mass matrix M_ν (167) may then be diagonalized by a unitary transformation U:

$$U^T M_\nu U = M_\nu^d. \tag{168}$$

This diagonalization is reminiscent of that required for the quark mass matrices in the standard model. In that case, it is well known that one can redefine the phases of the quark fields [133] so that the mixing matrix U_{CKM} has just one CP-violating phase [134]. However, in the neutrino case, there are fewer independent field phases, and one is left with 3 physical CP-violating parameters:

$$U = \tilde{P_2} V P_0 : P_0 \equiv \text{Diag}\left(e^{i\phi_1}, e^{i\phi_2}, 1\right). \tag{169}$$

Here $\tilde{P_2} = \text{Diag}\left(e^{i\alpha_1}, e^{i\alpha_2}, e^{i\alpha_3}\right)$ contains three phases that can be removed by phase rotations and are unobservable in light-neutrino physics, though they do play a rôle at high energies, V is the light-neutrino mixing matrix first considered by Maki, Nakagawa and Sakata (MNS) [135], and P_0 contains 2 CP-violating phases $\phi_{1,2}$ that are observable at low energies. The MNS matrix describes neutrino oscillations

$$V = \begin{pmatrix} c_{12} & s_{12} & 0 \\ -s_{12} & c_{12} & 0 \\ 0 & 0 & 1 \end{pmatrix} \begin{pmatrix} 1 & 0 & 0 \\ 0 & c_{23} & s_{23} \\ 0 & -s_{23} & c_{23} \end{pmatrix} \begin{pmatrix} c_{13} & 0 & s_{13} \\ 0 & 1 & 0 \\ -s_{13}e^{-i\delta} & 0 & c_{13}e^{-i\delta} \end{pmatrix}, \tag{170}$$

where $c_{ij} \equiv \cos\theta_{ij}$, $s_{ij} \equiv \sin\theta_{ij}$. The three real mixing angles $\theta_{12,23,13}$ in (170) are analogous to the Euler angles that are familiar from the classic rotations of rigid mechanical bodies. The phase δ is a specific quantum effect that is also observable in neutrino oscillations, and violates CP, as we discuss below. The other CP-violating phases $\phi_{1,2}$ are in principle observable in neutrinoless double-β decay (163).

5.3 Neutrino oscillations

The first of the mixing angles in (170) to be discovered was θ_{23}, in atmospheric neutrino experiments. Whereas the numbers of downward-going atmospheric ν_μ were found to

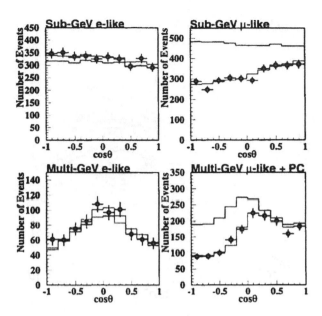

Figure 44. *The zenith angle distributions of atmospheric neutrinos exhibit a deficit of downward-moving ν_μ, which is due to neutrino oscillations [125].*

agree with standard model predictions, a deficit of upward-going ν_μ was observed, as seen in Fig. 44. The data from the Super-Kamiokande experiment, in particular [125], favour near-maximal mixing of atmospheric neutrinos:

$$\theta_{23} \sim 45°, \quad \Delta m_{23}^2 \sim 2.4 \times 10^{-3} \text{ eV}^2. \tag{171}$$

Recently, the K2K experiment using a beam of neutrinos produced by an accelerator has found results consistent with (171) [136]. It seems that the atmospheric ν_μ prefer not to oscillate into ν_e or into light sterile neutrinos, and probably oscillate primarily into ν_τ, though this has yet to be established.

More recently, the oscillation interpretation of the long-standing solar-neutrino deficit has been established, in particular by the SNO experiment. Solar neutrino experiments are sensitive to the mixing angle θ_{12} in (170). The recent data from SNO [126] and Super-Kamiokande [137] prefer quite strongly the large-mixing-angle (LMA) solution to the solar neutrino problem with

$$\theta_{12} \sim 30°, \quad \Delta m_{12}^2 \sim 6 \times 10^{-5} \text{ eV}^2, \tag{172}$$

though they were unable to exclude completely the LOW solution with lower δm^2. However, the KamLAND experiment on reactors produced by nuclear power reactors has recently found a deficit of ν_e that is highly compatible with the LMA solution to the solar neutrino problem [138], as seen in Fig. 45, and excludes any other solution.

Using the range of θ_{12} allowed by the solar and KamLAND data, one can establish a correlation between the relic neutrino density $\Omega_\nu h^2$ and the neutrinoless double-β decay observable $\langle m_\nu \rangle_e$, as seen in Fig. 46 [140]. Pre-WMAP, the experimental limit on $\langle m_\nu \rangle_e$

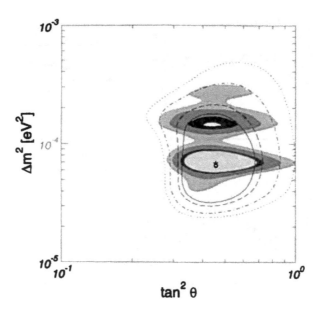

Figure 45. *The KamLAND experiment (shadings) finds [138] a deficit of reactor neutrinos that is consistent with the LMA neutrino oscillation parameters previously estimated (ovals) on the basis of solar neutrino experiments [139].*

could be used to set the bound

$$10^{-3} \lesssim \Omega_\nu h^2 \lesssim 10^{-1}. \tag{173}$$

Alternatively, now that WMAP has set a tighter upper bound $\Omega_\nu h^2 < 0.0076$ (162) [74], one can use this correlation to set an upper bound:

$$\langle m_\nu \rangle_e \lesssim 0.1 \text{ eV}, \tag{174}$$

which is difficult to reconcile with the neutrinoless double-β decay signal reported in [124].

The third mixing angle θ_{13} in (170) is basically unknown, with experiments such as Chooz [141], K2K [142] and Super-Kamiokande only establishing upper limits. *A fortiori*, we have no experimental information on the CP-violating phase δ.

The phase δ could in principle be measured by comparing the oscillation probabilities for neutrinos and antineutrinos and computing the CP-violating asymmetry [143]:

$$
\begin{aligned}
P(\nu_e \to \nu_\mu) - P(\bar{\nu}_e \to \bar{\nu}_\mu) &= 16 s_{12} c_{12} s_{13} c_{13}^2 s_{23} c_{23} \sin\delta \\
&\times \sin\left(\frac{\Delta m_{12}^2}{4E}L\right) \sin\left(\frac{\Delta m_{13}^2}{4E}L\right) \sin\left(\frac{\Delta m_{23}^2}{4E}L\right),
\end{aligned} \tag{175}
$$

as seen in Fig. 47 [144]. This is possible only if Δm_{12}^2 and s_{12} are large enough - as now suggested by the success of the LMA solution to the solar neutrino problem, and if s_{13} is large enough – which remains an open question.

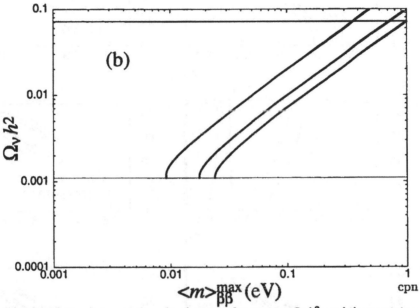

Figure 46. *The correlation between the relic density of neutrinos $\Omega_\nu h^2$ and the neutrino-less double-β decay observable: the different lines indicate the ranges allowed by neutrino oscillation experiments [140].*

A number of long-baseline neutrino experiments using beams from accelerators are now being prepared in the United States, Europe and Japan, with the objectives of measuring more accurately the atmospheric neutrino oscillation parameters, Δm_{23}^2, θ_{23} and θ_{13}, and demonstrating the production of ν_τ in a ν_μ beam.

Beyond these, ideas are being proposed for intense 'super-beams' of low-energy neutrinos, produced by high-intensity, low-energy accelerators such as the SPL [145] proposed at CERN. A subsequent step could be a storage ring for unstable ions, whose decays would produce a 'β beam' of pure ν_e or $\bar{\nu}_e$ neutrinos. These experiments might be able to measure δ via CP and/or T violation in neutrino oscillations [146].

A final step could be a full-fledged neutrino factory based on a muon storage ring, one conceptual layout for which is shown in Fig. 48. This would produce pure ν_μ and $\bar{\nu}_e$ (or ν_e and $\bar{\nu}_\mu$ beams and provide a greatly enhanced capability to search for or measure δ via CP violation in neutrino oscillations [147].

5.4 How to measure the seesaw parameters?

We have seen above that the effective low-energy mass matrix for the light neutrinos in the simplest seesaw model contains 9 parameters: 3 mass eigenvalues, 3 real mixing an-

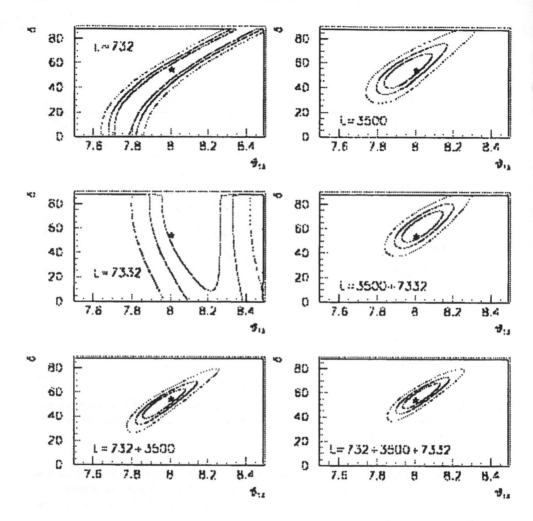

Figure 47. *Possible measurements of θ_{13} and δ that could be made with a neutrino factory, using a neutrino energy threshold of about 10 GeV. Using a single baseline correlations are very strong, but can be largely reduced by combining information from different baselines and detector techniques [144], enabling the CP-violating phase δ to be extracted.*

gles and 3 CP-violating phases. However, these are not all the parameters in the minimal seesaw model. As shown in Fig. 49, this model has a total of 18 parameters [148, 149]. The additional 9 parameters comprise the 3 masses of the heavy singlet 'right-handed' neutrinos M_i, 3 more real mixing angles and 3 more CP-violating phases. Most of the rest of this Lecture is devoted to understanding better the origins and possible manifestations of the remaining parameters, many of which may have controlled the generation of matter in the universe via leptogenesis [150] and may be observable via renormalization in supersymmetric models [151, 149, 152, 153].

To see how the extra 9 parameters appear [149], we reconsider the full lepton sector,

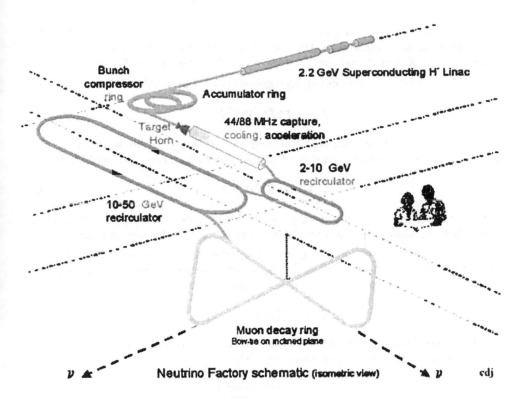

Figure 48. *Conceptual layout for a neutrino factory, based on an intense superconducting proton linac that produces many pions, whose decay muons are captured, cooled in phase space and stored in a 'bow-tie' ring. Their subsequent decays send neutrinos with known energy spectra and flavours to a combination of short- and long-baseline experiments [147].*

assuming that we have diagonalized the charged-lepton mass matrix:

$$(Y_\ell)_{ij} = Y^d_{\ell_i}\delta_{ij}, \tag{176}$$

as well as that of the heavy singlet neutrinos:

$$M_{ij} = M^d_i \delta_{ij}. \tag{177}$$

We can then parametrize the neutrino Dirac coupling matrix Y_ν in terms of its real and diagonal eigenvalues and unitary rotation matrices:

$$Y_\nu = Z^* Y^d_{\nu_k} X^\dagger, \tag{178}$$

where X has 3 mixing angles and one CP-violating phase, just like the CKM matrix, and we can write Z in the form

$$Z = P_1 \bar{Z} P_2, \tag{179}$$

where $\bar{Z}$ also resembles the CKM matrix, with 3 mixing angles and one CP-violating phase, and the diagonal matrices $P_{1,2}$ each have two CP-violating phases:

$$P_{1,2} = \mathrm{Diag}\left(e^{i\theta_{1,3}}, e^{i\theta_{2,4}}, 1\right). \tag{180}$$

In this parametrization, we see explicitly that the neutrino sector has 18 parameters: the 3 heavy-neutrino mass eigenvalues M_i^d, the 3 real eigenvalues of $Y_{\nu_i}^D$, the $6 = 3 + 3$ real mixing angles in X and $\bar{Z}$, and the $6 = 1 + 5$ CP-violating phases in X and $\bar{Z}$ [149].

As we discuss later in more detail, leptogenesis [150] is proportional to the product

$$Y_\nu Y_\nu^\dagger = P_1^* \bar{Z}^* \left(Y_\nu^d\right)^2 \bar{Z}^T P_1, \tag{181}$$

which depends on 13 of the real parameters and 3 CP-violating phases, whilst the leading renormalization of soft supersymmetry-breaking masses depends on the combination

$$Y_\nu^\dagger Y_\nu = X \left(Y_\nu^d\right)^2 X^\dagger, \tag{182}$$

which depends on just 1 CP-violating phase, with two more phases appearing in higher orders, when one allows the heavy singlet neutrinos to be non-degenerate [152].

In order to see how the low-energy sector is embedded in this full parametrization, we first recall that the 3 phases in $\tilde{P}_2$ (169) become observable when one also considers high-energy quantities. Next, we introduce a complex orthogonal matrix

$$R \equiv \sqrt{M^d}^{-1} Y_\nu U \sqrt{M^d}^{-1} [v \sin \beta], \tag{183}$$

which has 3 real mixing angles and 3 phases: $R^T R = 1$. These 6 additional parameters may be used to characterize Y_ν, by inverting (183):

$$Y_\nu = \frac{\sqrt{M^d} R \sqrt{M^d} U^\dagger}{[v \sin \beta]}, \tag{184}$$

giving us the same grand total of $18 = 9 + 3 + 6$ parameters [149]. The leptogenesis observable (181) may now be written in the form

$$Y_\nu Y_\nu^\dagger = \frac{\sqrt{M^d} R M_\nu^d R^\dagger \sqrt{M^d}}{[v^2 \sin^2 \beta]}, \tag{185}$$

which depends on the 3 phases in R, but *not* the 3 low-energy phases $\delta, \phi_{1,2}$, nor the 3 real MNS mixing angles [149]! Conversely, the leading renormalization observable (182) may be written in the form

$$Y_\nu^\dagger Y_\nu = U \frac{\sqrt{M_\nu^d} R^\dagger M^d R \sqrt{M_\nu^d}}{[v^2 \sin^2 \beta]} U^\dagger, \tag{186}$$

which depends explicitly on the MNS matrix, including the CP-violating phases δ and $\phi_{1,2}$, but only one of the three phases in $\tilde{P}_2$ [149].

5.5   Renormalization of soft supersymmetry-breaking parameters

As indicated in Fig. 49, in a supersymmetric seesaw model the soft supersymmetry-breaking parameters m_0^2 and A are renormalized by the neutrino sector, as we now discuss

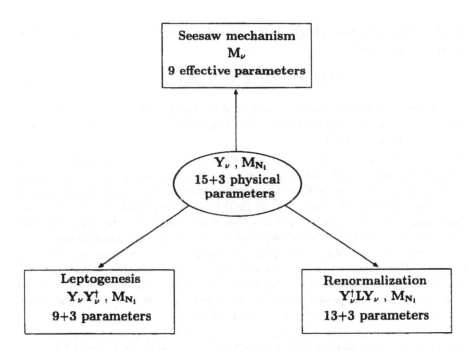

Figure 49. *Roadmap for the physical observables derived from Y_ν and N_i [156].*

in more detail, assuming that the input values at the GUT scale are flavour-independent. If they are not, there will be additional sources of flavour-changing processes, beyond those discussed in this and subsequent sections [58, 154].

In the leading-logarithmic approximation, one finds the following radiative corrections to the soft supersymmetry-breaking terms for sleptons:

$$
\begin{aligned}
\left(\delta m_{\tilde{L}}^2\right)_{ij} &= -\frac{1}{8\pi^2}\left(3m_0^2 + A_0^2\right)\left(Y_\nu^\dagger Y_\nu\right)_{ij} \mathrm{Ln}\left(\frac{M_{\mathrm{GUT}}}{M}\right), \\
\left(\delta A_\ell\right)_{ij} &= -\frac{1}{8\pi^2}A_0 Y_{\ell_i}\left(Y_\nu^\dagger Y_\nu\right)_{ij} \mathrm{Ln}\left(\frac{M_{\mathrm{GUT}}}{M}\right),
\end{aligned}
\tag{187}
$$

where we have intially assumed that the heavy singlet neutrinos are approximately degenerate with $M \ll M_{\mathrm{GUT}}$. In this case, there is a single analogue of the Jarlskog invariant of the standard model [155]:

$$
J_{\tilde{L}} \equiv \mathrm{Im}\left[\left(m_{\tilde{L}}^2\right)_{12}\left(m_{\tilde{L}}^2\right)_{23}\left(m_{\tilde{L}}^2\right)_{31}\right],
\tag{188}
$$

which depends on the single phase that is observable in this approximation. There are other Jarlskog invariants defined analogously in terms of various combinations with the A_ℓ, but these are all proportional [149].

There are additional contributions if the heavy singlet neutrinos are not degenerate:

$$
\left(\bar{\delta} m_{\tilde{L}}^2\right)_{ij} = -\frac{1}{8\pi^2}\left(3m_0^2 + A_0^2\right)\left(Y_\nu^\dagger L Y_\nu\right)_{ij} : L \equiv \mathrm{Ln}\left(\frac{\bar{M}}{M_i}\right)\delta_{ij},
\tag{189}
$$

where $\bar{M} \equiv \sqrt[3]{M_1 M_2 M_3}$, with $\left(\tilde{\delta} A_\ell\right)_{ij}$ being defined analogously. These new contributions contain the matrix factor

$$Y^\dagger L Y = X Y^d P_2 \bar{Z}^T L \bar{Z}^* P_2^* y^d X^\dagger, \tag{190}$$

which introduces dependences on the phases in $\bar{Z} P_2$, though not P_1. In this way, the renormalization of the soft supersymmetry-breaking parameters becomes sensitive to a total of 3 CP-violating phases [152].

As illustrated in Fig. 49, renormalization effects may generate observable rates for flavour-changing lepton decays such as $\mu \to e\gamma, \tau \to \mu\gamma$ and $\tau \to e\gamma$ in supersymmetric models [151, 149, 152, 153], as well as CP-violating observables such as electric dipole moments for the electron and muon. Fig. 50 (left) is a scatter plot of $B(\mu \to e\gamma)$ in one particular texture for lepton mixing, as a function of the singlet neutrino mass M_{N_3}. We see that $\mu \to e\gamma$ may well have a branching ratio close to the present experimental upper limit, particularly for larger M_{N_3}. Analogous predictions for $\tau \to \mu\gamma$ decays are shown in Fig. 50 (right). The branching ratios decrease with increasing sparticle masses, but the range due to variations in the neutrino parameters is considerably larger than that due to the sparticle masses. The present experimental upper limits on $\tau \to \mu\gamma$, in particular, already exclude significant numbers of parameter choices.

The decay $\mu \to e\gamma$ and related processes such as $\mu \to 3e$ and $\mu \to e$ conversion on a heavy nucleus are all of potential interest for the front end of a neutrino factory [157]. Such an accelerator will produce many additional muons, beyond those captured and cooled for storage in the decay ring, which could be used to explore the decays of slow or stopped muons with high statistics. There are several options for studying rare τ decays, such as the B factories already operating or the LHC, which will produce very large numbers of τ leptons via W, Z and B decays.

We now discuss some aspects of CP-violating renormalization effects on the electric dipole moments of the electron and muon. It has often been thought that these are unobservably small in the minimal supersymmetric seesaw model, and that $|d_e/d_\mu| = m_e/m_\mu$. However, d_e and d_μ may be strongly enhanced if the heavy singlet neutrinos are not degenerate [152], and depend on the new phases that contribute to leptogenesis [2], which were discussed above. It should be emphasized that non-degenerate heavy-singlet neutrinos are actually expected in most models of neutrino masses. Typical examples are texture models of the form

$$Y_\nu \sim Y_0 \begin{pmatrix} 0 & c\epsilon_\nu^3 & d\epsilon_\nu^3 \\ c\epsilon_\nu^3 & a\epsilon_\nu^2 & b\epsilon_\nu^2 \\ d\epsilon_\nu^3 & b\epsilon_\nu^2 & e^{i\psi} \end{pmatrix},$$

where Y_0 is an overall scale, ϵ_ν characterizes the hierarchy, a, b, c and d are $\mathcal{O}(1)$ complex numbers, and ψ is an arbitrary phase. For example, there is an SO(10) GUT model of this form with $d = 0$ and a flavour SU(3) model with $a = b$ and $c = d$. The hierarchy of heavy-neutrino masses in such a model is

$$M_1 : M_2 : M_3 = \epsilon_N^6 : \epsilon_N^4 : 1, \tag{191}$$

---

[2]This effect makes lepton electric dipole moments possible even in a two-generation model.

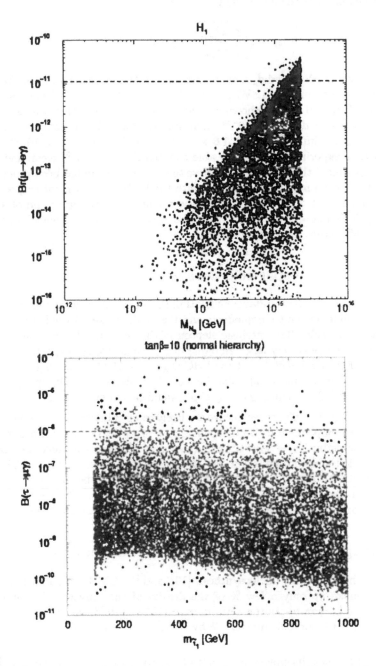

Figure 50. *Scatter plots of the branching ratios for* $\mu \rightarrow e\gamma$ *and* $\tau \rightarrow \mu\gamma$ *in the supersymmetric seesaw model for various values of its unknown parameters [153].*

and indicative ranges of the hierarchy parameters are

$$\epsilon_\nu \sim \sqrt{\frac{\Delta m_{solar}^2}{\Delta m_{atmo}^2}} \, , \; \epsilon_N \sim 0.1 \text{ to } 0.2. \tag{192}$$

Fig. 51 shows how much d_e and d_μ may be increased as soon as the degeneracy between the heavy neutrinos is broken: $\epsilon \neq 1$. We also see that $|d_\mu/d_e| \gg m_\mu/m_e$ when $\epsilon_N \sim 0.1$ to 0.2. Scatter plots of d_e and d_μ are shown in Fig. 52, where we see that values as large as $d_\mu \sim 10^{-27}$ e.cm and $d_e \sim 3 \times 10^{-30}$ e.cm are possible. For comparison, the present experimental upper limits are $d_e < 1.6 \times 10^{-27}$ e.cm [158] and $d_\mu < 10^{-18}$ e.cm. An ongoing series of experiments might be able to reach $d_e < 3 \times 10^{-30}$ e.cm, and a type of solid-state experiment that might be sensitive to $d_e \sim 10^{-33}$ e.cm has been proposed [159]. Also, $d_\mu \sim 10^{-24}$ e.cm might be accessible with the PRISM experiment proposed for the JHF [160], and $d_\mu \sim 5 \times 10^{-26}$ e.cm might be attainable at the front end of a neutrino factory [161]. It therefore seems that d_e might be measurable with foreseeable experiments, whilst d_μ would present more of a challenge.

5.6   (Not so) rare sparticle decays

The suppression of rare lepton-flavour-violating (LFV) μ and τ decays in the supersymmetric seesaw model is due to loop effects and the small masses of the leptons relative to the sparticle mass scale. The intrinsic slepton mixing may not be very small, in which case there might be relatively large amounts of LFV observable in sparticle decays. An example that might be detectable at the LHC is $\chi_2 \to \chi_1 \ell^\pm \ell'^\mp$, where $\chi_1(\chi_2)$ denotes the (next-to-)lightest neutralino [162]. The largest LFV effects might be in $\chi_2 \to \chi_1 \tau^\pm \mu^\mp$ and $\chi_2 \to \chi_1 \tau^\pm e^\mp$ [163], though $\chi_2 \to \chi_1 e^\pm \mu^\mp$ would be easier to detect.

As shown in Fig. 53 [163], these decays are likely to be enhanced in a region of CMSSM parameter space complementary to that where $\tau \to e/\mu\gamma$ decys are most copious. This is because the interesting $\chi_2 \to \chi_1 \tau^\pm \mu^\mp$ and $\chi_2 \to \chi_1 \tau^\pm e^\mp$ decays are mediated by slepton exchange, which is maximized when the slepton mass is close to m_{χ_1}. This happens in the coannihilation region where the LSP relic density may be in the range preferred by astrophysics and cosmology, even if m_{χ_1} is relatively large. Thus searches for LFV $\chi_2 \to \chi_1 \tau^\pm \mu^\mp$ and $\chi_2 \to \chi_1 \tau^\pm e^\mp$ decays are quite complementary to those for $\tau \to e/\mu\gamma$.

5.7   Baryogenesis and the leptogenesis connection

We have seen that Big-Bang nucleosynthesis [103] and the CMB [74] independently imply that baryons make up only a few % of the density of the universe. Numerically, this corresponds to a baryon-to-photon ratio $n_b/n_\gamma \sim 10^{-9} - 10^{-10}$, raising several questions. Why is there so little baryonic matter? Why is there any at all? Why is there apparently no antimatter?

Astronauts did not disappear in a burst of radiation when they landed on the moon, and neither have space probes landing on Mars or an asteroid. The small abundance of antiprotons in the cosmic rays is consistent with their production by primary matter cosmic rays [164], and no antinuclei have been seen [165]. If there were any large concentration of antimatter in our local cluster of galaxies, we would have detected radiation

Electric Dipole Moments

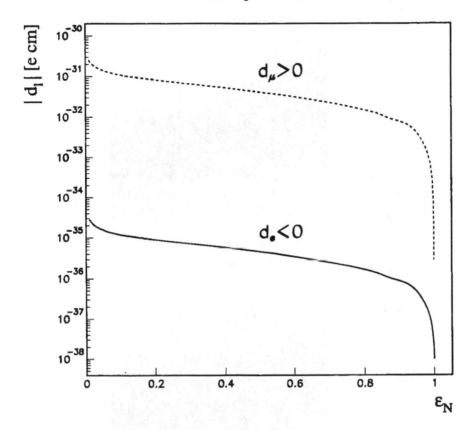

Figure 51. *The electric dipole moments of the electron and muon, d_e and d_μ, may be enhanced if the heavy singlet neutrinos are non-degenerate. The horizontal axis parameterizes the breaking of their degeneracy, and the vertical strip indicates a range favoured in certain models [152].*

from matter-antimatter annihilations at its boundary. The CMB would have been distorted by similar radiation from any matter-antimatter boundary within the observable universe [166]. So it seems that there must be a real cosmological asymmetry betwen matter and antimatter.

This could be explained if, going back to when the universe was less than 10^{-6} s old, it contained about one extra quark for every 10^9 quark-antiquark pairs in the primordial soup. As the universe expanded, most of the quarks would have annihilated with those antiquarks to produce radiation, and the few quarks left over would have survived to

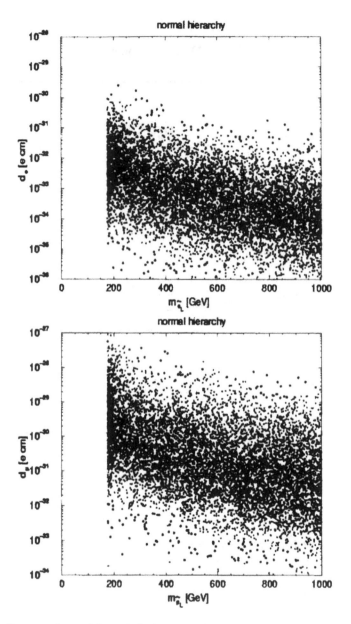

Figure 52. *Scatter plots of d_e and d_μ in variants of the supersymmetric seesaw model, for different values of the unknown parameters [153].*

combine into the baryons seen today. Where did this small quark-antiquark asymmetry originate? Did the Big Bang start off with it, or did the laws of Nature generate it during the subsequent expansion?

The conditions for such cosmological baryogenesis were established by Sakharov in 1967 [167]. There has to be a difference between the interactions of matter and antimatter

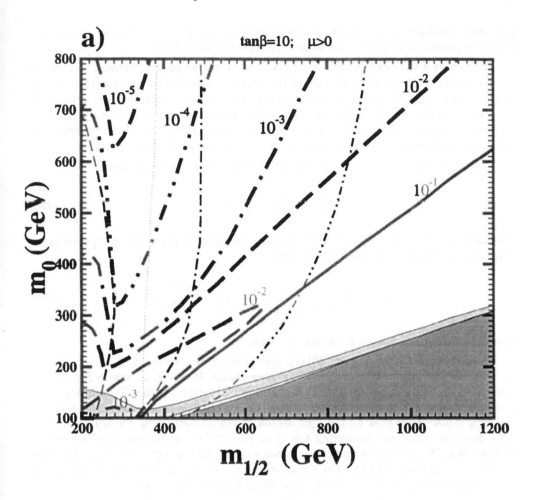

Figure 53. *Contours of the possible ratio of the branching ratios for $\chi_2 \to \chi_1 \tau^{\pm} \mu^{\mp}$ and $\chi_2 \to \chi_1 \mu^{\pm} \mu^{\mp}$ (black lines) and of the branching ratio for $\tau \to \mu\gamma$ (near-vertical grey/blue lines). [163].*

particles, in the form of charge-conjugation (C) violation, which was discovered in the weak interactions in 1957, and CP violation, which was discovered in kaon decays in 1964. There must also have been a departure from thermal equilibrium, which would have been possible during a phase transition, perhaps the electroweak phase transition when $t \sim 10^{-10}$ s or a GUT phase transition when $t \sim 10^{-36}$ s, or at the end of inflation. Finally, there must have been a violation of baryon number, which would have happened through nonperturbative weak interactions at high temperatures [168] and is thought to be a generic feature of GUTs.

Various specific mechanisms for Big-Bang baryogenesis have been proposed, ranging from the out-of-equilibrium decays of GUT bosons [169] or heavy neutrinos [150] to processes around the epoch of the electroweak phase transition. The CP violation in the standard model seems inadequate to generate the required baryon asymmetry, but this might be possible if it is extended to include supersymmetry [170].

The decays of the heavy singlet neutrinos N in the seesaw model provide a mechanism for generating the baryon asymmetry of the universe, namely leptogenesis [150]. In the presence of C and CP violation, the branching ratios for $N \to$ Higgs $+ \ell$ may differ from that for $N \to$ Higgs $+ \bar{\ell}$, producing a net lepton asymmetry in the very early universe. This is then transformed (partly) into a quark asymmetry by non-perturbative electroweak sphaleron interactions during the period before the electroweak phase transition.

The total decay rate of a heavy neutrino N_i may be written in the form

$$\Gamma_i = \frac{1}{8\pi} \left(Y_\nu Y_\nu^\dagger \right)_{ii} M_i. \tag{193}$$

One-loop CP-violating diagrams involving the exchange of heavy neutrino N_j would generate an asymmetry in N_i decay of the form:

$$\epsilon_{ij} = \frac{1}{8\pi} \frac{1}{\left(Y_\nu Y_\nu^\dagger \right)_{ii}} \mathrm{Im} \left(\left(Y_\nu Y_\nu^\dagger \right)_{ij} \right)^2 f \left(\frac{M_j}{M_i} \right), \tag{194}$$

where $f(M_j/M_i)$ is a known kinematic function.

Thus we see that leptogenesis [150] is proportional to the product $Y_\nu Y_\nu^\dagger$, which depends on 13 of the real parameters and 3 CP-violating phases. However, as seen in Fig. 54, the amount of the leptogenesis asymmetry is explicitly independent of the CP-violating phase δ that is measurable in neutrino oscillations [156]. The basic reason for this is that one makes a unitary sum over all the light lepton species in evaluating the asymmetry ϵ_{ij}. This does not mean that measuring δ is of no interest for leptogenesis: if it is found to be non-zero, CP violation in the lepton sector – one of the key ingredients in leptogenesis – will have been established. On the other hand, the phases responsible directly for leptogenesis may contribute to the electric dipole moments of leptons.

In general, one may formulate the following strategy for calculating leptogenesis in terms of laboratory observables [149, 156]:

- Measure the neutrino oscillation phase δ and the Majorana phases $\phi_{1,2}$,

- Measure observables related to the renormalization of soft supersymmetry-breaking parameters, that are functions of $\delta, \phi_{1,2}$ and the leptogenesis phases,

- Extract the effects of the known values of δ and $\phi_{1,2}$, and isolate the leptogenesis parameters.

In the absence of complete information on the first two steps above, we are currently at the stage of preliminary explorations of the multi-dimensional parameter space. As seen in Fig. 54, the amount of the leptogenesis asymmetry is explicitly independent of δ [156]. However, in order to make more definite predictions, one must make some extra hypothesis, of which we now discuss one example.

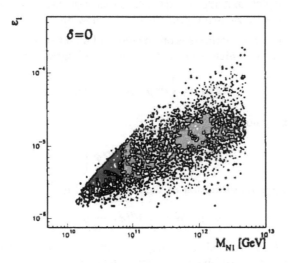

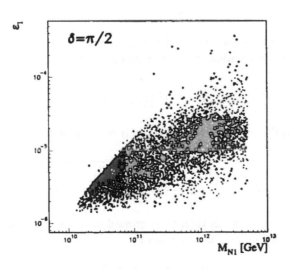

Figure 54. *Comparison of the CP-violating asymmetries in the decays of heavy singlet neutrinos giving rise to the cosmological baryon asymmetry via leptogenesis (top panel) without and (bottom panel) with maximal CP violation in neutrino oscillations [156]. They are indistinguishable.*

5.8   Sneutrino inflation

Could the inflaton postulated in Lecture 4 [105] be the spin-zero supersymmetric partner of one of the heavy singlet neutrinos [171]? This hypothesis makes precise predictions for both the CMB and flavour-violating processes, as we now discuss.

The sneutrino inflation hypothesis is one example of an inflationary model in a generic scalar field theory [172, 4], which may be described by a Lagrangian

$$L(\phi) = \frac{1}{2}\partial^\mu\phi\partial_\mu\phi - V(\phi), \tag{195}$$

yielding a Hubble expansion rate

$$H^2 = \frac{8\pi}{3\pi^2}\left[\frac{1}{2}\dot\phi^2 + V(\phi)\right]. \tag{196}$$

The corresponding equation of motion of the inflaton field is

$$\ddot\phi + 3H\dot\phi + V'(\phi) = 0. \tag{197}$$

The first term in (197) is assumed to be negligible, in which case the equation of motion is dominated by the second (Hubble drag) term, and one has

$$\dot\phi \simeq -\frac{V'}{3H}. \tag{198}$$

In this slow-roll approximation, when the kinetic term in (196) is negligible, and the Hubble expansion rate is dominated by the potential term:

$$H \simeq \sqrt{\frac{1}{3M_P^2}V(\phi)}, \tag{199}$$

where $M_P \equiv 1/\sqrt{8\pi G_N} \simeq 2.4 \times 10^{18}$ GeV. It is convenient to introduce the following slow-roll parameters:

$$\epsilon \equiv \frac{1}{2}M_P^2\left(\frac{V'}{V}\right)^2, \ \eta \equiv M_P^2\left(\frac{V''}{V}\right), \ \xi \equiv M_P^4\left(\frac{VV'''}{V^2}\right). \tag{200}$$

Various observable quantities can then be expressed in terms of ϵ, η and ξ, including the spectral index for scalar density perturbations:

$$n_s = 1 - 6\epsilon + 2\eta, \tag{201}$$

the ratio of scalar and tensor perturbations at the quadrupole scale:

$$r \equiv \frac{A_T}{A_S} = 16\epsilon, \tag{202}$$

the spectral index of the tensor perturbations:

$$n_T = -2\epsilon, \tag{203}$$

and the running parameter for the scalar spectral index:

$$\frac{dn_s}{d\ln k} = \frac{2}{3}\left[(n_s - 1)^2 - 4\eta^2\right] + 2\xi. \tag{204}$$

The amount e^N by which the universe expanded during inflation is also controlled by the slow-roll parameter ϵ:

$$e^N : \quad N = \int H dt = \frac{2\sqrt{\pi}}{m_P} \int_{\phi_{initial}}^{\phi_{final}} \frac{d\phi}{\sqrt{\epsilon(\phi)}}. \tag{205}$$

In order to explain the size of a large feature in the observed universe, one needs $N \sim 40 - 60$.

In the case of a sneutrino inflaton, one would have a $V = \frac{1}{2}m^2\phi^2$, and the slow-roll inflationary parameters would be

$$\epsilon = \frac{2M_P^2}{\phi_I^2}, \eta = \frac{2M_P^2}{\phi_I^2}, \xi = 0, \tag{206}$$

where ϕ_I denotes the *a priori* unknown inflaton field value during inflation at a typical CMB scale k. The overall scale of the inflationary potential is normalized by the WMAP data on density fluctuations:

$$\Delta_R^2 = \frac{V}{24\pi^2 M_P^4 \epsilon} = 2.95 \times 10^{-9} A \quad : \quad A = 0.77 \pm 0.07, \tag{207}$$

yielding

$$V^{\frac{1}{4}} = M_P^4 \sqrt{\epsilon} \times 24\pi^2 \times 2.27 \times 10^{-9} = 0.027 M_P \times \epsilon^{\frac{1}{4}}, \tag{208}$$

corresponding to

$$m^{\frac{1}{2}}\phi_I = 0.038 \times M_P^{\frac{3}{2}} \tag{209}$$

in any simple ϕ^2 inflationary model. Taking

$$N = \frac{1}{4}\frac{\phi_I^2}{M_P^2} \simeq 50, \tag{210}$$

we find

$$\phi_I^2 \simeq 200 \times M_P^2. \tag{211}$$

Inserting this requirement into the WMAP normalization condition (208), we find the following required mass for any quadratic inflaton:

$$m \simeq 1.8 \times 10^{13} \text{ GeV}, \tag{212}$$

which is comfortably within the range of heavy singlet (s)neutrino masses usually considered, namely $m_N \sim 10^{10}$ to 10^{15} GeV [173].

This simple ϕ^2 sneutrino predicts the following values for the primary CMB observables [173]: the scalar spectral index

$$n_s = 1 - \frac{8M_P^2}{\phi_I^2} \simeq 0.96, \tag{213}$$

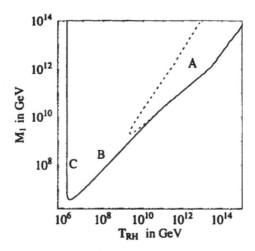

Figure 55. *The solid curve bounds the region allowed for leptogenesis in the (T_{RH}, M_{N_1}) plane in the sneutrino inflation model [173]. In the area bounded by the red dashed curve leptogenesis is entirely thermal.*

the tensor-to scalar ratio

$$r = \frac{32M_P^2}{\phi_I^2} \simeq 0.16, \tag{214}$$

and the running parameter for the scalar spectral index:

$$\frac{dn_s}{d\ln k} = \frac{32M_P^4}{\phi_I^4} \simeq 8 \times 10^{-4}. \tag{215}$$

The sneutrino inflation value of $n_s \simeq 0.96$ appears to be compatible with the data at the 1-σ level, and the ϕ^2 model value $r \simeq 0.16$ for the relative tensor strength is also compatible with the WMAP data. In fact, we note that the favoured individual values for n_s, r and $dn_s/d\ln k$ reported in an independent analysis [174] *all coincide with the ϕ^2 model values, within the latter's errors!*

The next hurdle for the sneutrino inflaton model is to ensure that the reheating temperature after inflation is sufficiently low that one does not overproduce gravitinos after inflation has ended. As seen in Fig. 55, he sneutrino inflation model is indeed quite compatible with a low reheating temperature. This is possible if the couplings of the sneutrino inflaton are small, a possibility also considered for other, more phenomenological reasons [175]. In this case, the combination of constrained couplings and a fixed sneutrino mass enables one to make predictions for charged-lepton flavour-violating processes that are more specific than the generic results presented earlier. For example, as seen in Fig. 56, $\mu \to e\gamma$ decay should appear within a couple of orders of magnitude of the present experimental upper limit [173], though $\tau \to \mu\gamma$ decay may be more difficult to observe.

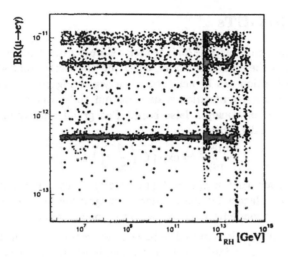

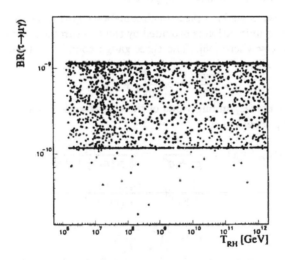

Figure 56. *Calculations of $BR(\mu \rightarrow e\gamma)$ (top) and $BR(\tau \rightarrow \mu\gamma)$ (bottom) in the sneutrino inflation model. Black points correspond to $\sin\theta_{13} = 0.0$, $M_2 = 10^{14}$ GeV, and 5×10^{14} GeV $< M_3 < 5 \times 10^{15}$ GeV. Red points correspond to $\sin\theta_{13} = 0.0$, $M_2 = 5 \times 10^{14}$ GeV and $M_3 = 5 \times 10^{15}$ GeV, while green points correspond to $\sin\theta_{13} = 0.1$, $M_2 = 10^{14}$ GeV and $M_3 = 5 \times 10^{14}$ GeV [173]. We assume for illustration that $(m_{1/2}, m_0) = (800, 170)$ GeV and $\tan\beta = 10$.*

6   Grand unified theories

6.1   The road to GUTs

The philosophy of grand unification [176] is to seek a simple group that includes the untidy separate interactions of the standard model, QCD and the electroweak sector. The hope is that this grand unification can be achieved while neglecting gravity, at least as a first approximation. If the grand unification scale turns out to be significantly less than the Planck mass, this is not obviously a false hope. The grand unification scale is indeed expected to be exponentially large:

$$\frac{m_{GUT}}{m_W} = \exp\left(\mathcal{O}\left(\frac{1}{\alpha_{em}}\right)\right) , \qquad (216)$$

and typical estimates are that $m_{GUT} = \mathcal{O}(10^{16}$ GeV). Such a calculation involves an extrapolation of known physics by many orders of magnitude further than, e.g., the extrapolation that Newton made from the apple to the solar system.

If the grand unification scale is indeed so large, most tests of it are likely to be indirect, such as relations between standard model vector couplings and between particle masses. Any new interactions, such as those that might cause protons to decay or give masses to neutrinos, are likely to be very strongly suppressed.

As already mentioned, a striking piece of circumstantial evidence in favour of the idea of supersymmetric grand unification is provided by the measurements of low-energy gauge couplings at LEP and elsewhere [45]. The three gauge couplings of the standard model are renormalized as follows:

$$\frac{dg_a^2}{dt} = b_a \frac{g_a^4}{16\pi^2} + \dots, \qquad (217)$$

at one-loop order. The corresponding value of the electroweak mixing angle $\sin^2\theta_W(m_Z)$ is given at the one-loop level by:

$$\sin^2\theta_W(m_Z) = \frac{g'^2}{g_2^2 + g'^2} = \frac{3}{5}\frac{g_1^2(m_Z)}{g_2^2(m_Z) + \frac{3}{5}g_1^2(m_Z)} \qquad (218)$$

$$= \frac{1}{1 + 8x}[3x + \frac{\alpha_{em}(m_Z)}{\alpha_3(m_Z)}], \qquad (219)$$

where

$$x \equiv \frac{1}{5}(\frac{b_2 - b_3}{b_1 - b_2}). \qquad (220)$$

One can distinguish the predictions of different GUTs by their different values of the renormalization coefficients b_i, which are in turn determined by the spectra of light particles around the electroweak scale. In the cases of the standard model and the MSSM, these are:

$$\frac{4}{3}N_G - 11 \leftarrow b_3 \rightarrow 2N_G - 9 = -3 , \qquad (221)$$

$$\frac{1}{6}N_H + \frac{4}{3}N_G - \frac{22}{3} \leftarrow b_2 \rightarrow \frac{1}{2}N_H + 2N_G - 6 = +1 , \qquad (222)$$

$$\frac{1}{10}N_H + \frac{4}{3}N_G \leftarrow b_1 \rightarrow \frac{3}{10}N_H + 2N_G = \frac{33}{5} , \qquad (223)$$

$$\frac{23}{218} = 0.1055 \leftarrow x \rightarrow \frac{1}{7} . \qquad (224)$$

If we insert the best available values of the gauge couplings:

$$\alpha_{em} = \frac{1}{128}; \; \alpha_3(m_Z) = 0.119 \pm 0.003, \; \sin^2\theta_W(m_Z) = 0.2315, \tag{225}$$

we find the following value:

$$x = \frac{1}{6.92 \pm 0.07}. \tag{226}$$

We see that experiment strongly favours the inclusion of supersymmetric particles in the renormalization-group equations, as required if the effective low-energy theory is the MSSM (224), as in a simple supersymmetric GUT such as the minimal SU(5) model introduced below.

To examine the indirect GUT predictions for the standard model vector interactions in more detail, one needs the following two-loop renormalization equations:

$$Q \frac{\partial \alpha_i(Q)}{\partial Q} = -\frac{1}{2\pi} \left(b_i + \frac{b_{ij}}{4\pi} \alpha_j(Q) \right) [\alpha_i(Q)]^2 , \tag{227}$$

where the b_i are the one-loop contributions discussed above, and the b_{ij} are the two-loop coefficients:

$$b_{ij} = \begin{pmatrix} 0 & 0 & 0 \\ 0 & -\frac{136}{3} & 0 \\ 0 & 0 & -102 \end{pmatrix} + N_g \begin{pmatrix} \frac{19}{15} & \frac{3}{5} & \frac{44}{15} \\ \frac{1}{5} & \frac{49}{3} & 4 \\ \frac{4}{30} & \frac{3}{2} & \frac{76}{3} \end{pmatrix} + N_H \begin{pmatrix} \frac{9}{50} & \frac{9}{10} & 0 \\ \frac{3}{10} & \frac{13}{6} & 0 \\ 0 & 0 & 0 \end{pmatrix}. \tag{228}$$

These coefficients are all independent of any specific GUT model, depending only on the light particles contributing to the renormalization. Including supersymmetric particles as in the MSSM, one finds [47]

$$b_{ij} = \begin{pmatrix} 0 & 0 & 0 \\ 0 & -24 & 0 \\ 0 & 0 & -54 \end{pmatrix} + N_g \begin{pmatrix} \frac{38}{15} & \frac{6}{5} & \frac{88}{15} \\ \frac{2}{5} & 14 & 8 \\ \frac{11}{5} & 3 & \frac{68}{3} \end{pmatrix} + N_H \begin{pmatrix} \frac{9}{50} & \frac{9}{10} & 0 \\ \frac{3}{10} & \frac{7}{2} & 0 \\ 0 & 0 & 0 \end{pmatrix}, \tag{229}$$

again independent of any specific supersymmetric GUT.

Calculations with these equations confirm that non-supersymmetric models are not consistent with the measurements of the standard model interactions at LEP and elsewhere. However, although extrapolating the experimental determinations of the interaction strengths using the non-supersymmetric renormalization-group equations (228) does not lead to a common value at any renormalization scale, we saw in Fig. 15 that extrapolation using the supersymmetric equations (229) does lead to possible unification at $m_{GUT} \sim 10^{16}$ GeV [45].

Another qualitative success is the prediction of the b quark mass [177, 178]. In many GUTs, such as the minimal SU(5) model discussed shortly, the b quark and the τ lepton have equal Yukawa couplings when renormalized at the GUT sale. The renormalization group then tells us that

$$\frac{m_b}{m_\tau} \simeq \left[\ln \left(\frac{m_b^2}{m_X^2} \right) \right]^{\frac{12}{33-2N_g}} \tag{230}$$

at the one-loop level in a non-supersymmetric GUT. Using $m_\tau = 1.78$ GeV, we predict that $m_b \simeq 5$ GeV, in agreement with experiment[3]. Happily, this prediction remains successful if the effects of supersymmetric particles are included in the renormalization-group calculations, and if one makes a more complete two-loop calculation [179].

6.2 Simple GUT models

The above discussion did not depend on the details of any specific GUT model, but now is the time to get more serious. Any GUT group should have rank 4 or more, and admit complex representations, so that parity and charge conjugation can be violated, as observed in the weak interactions (2).

It is easy to check that the only suitable simple group of rank 4 is SU(5) [176]. Its most useful representations are the complex vector $\underline{5}$ representation denoted by F_α, its conjugate $\underline{\bar{5}}$ denoted by $\bar{F}^\alpha$, the complex two-index antisymmetric tensor $\underline{10}$ representation $T_{[\alpha\beta]}$, and the adjoint $\underline{24}$ representation A_β^α. The latter is used to accommodate the vector bosons of SU(5):

$$
\begin{pmatrix}
 & & & \vdots & \bar{X} & \bar{Y} \\
 & g_{1,\dots,8} & & \vdots & \bar{X} & \bar{Y} \\
 & & & \vdots & \bar{X} & \bar{Y} \\
\hdotsfor{6} \\
X & X & X & \vdots & & \\
 & & & \vdots & W_{1,2,3} & \\
Y & Y & Y & \vdots & &
\end{pmatrix} , \qquad (231)
$$

where the $g_{1,\dots,8}$ are the gluons of QCD, the $W_{1,2,3}$ are weak bosons, and the (X, Y) are new vector bosons, whose interactions we discuss in the next section.

The quarks and leptons of each generation are accommodated in $\underline{\bar{5}}$ and $\underline{10}$ representations of SU(5):

$$
\bar{F} = \begin{pmatrix} d_R^c \\ d_Y^c \\ d_B^c \\ \cdots \\ -e^- \\ \nu_e \end{pmatrix}_L , \quad
T = \begin{pmatrix}
0 & u_B^c & -u_Y^c & \vdots & -u_R & -d_R \\
-u_B^c & 0 & u_R^c & \vdots & -u_Y & -d_Y \\
u_Y^c & -u_R^c & 0 & \vdots & -u_B & -d_B \\
\hdotsfor{6} \\
u_R & u_Y & u_B & \vdots & 0 & -e^c \\
d_R & d_Y & d_B & \vdots & e^c & 0
\end{pmatrix}_L . \qquad (232)
$$

The particle assignments are unique up to the effects of mixing between generations, which we do not discuss in detail here [180].

Beyond SU(5), one can consider simple groups of rank 5. The only one with suitable complex representations is SO(10). In this case, the fermions of each generation are

---

[3]This prediction was made [177] shortly before the b quark was discovered. When we received the proofs of this article, I gleefully wrote by hand in the margin our then prediction, which was already in the text, as 2 *to* 5. This was misread by the typesetter to become 2605: a spectacular disaster!

accommodated in a $\underline{16}$ spinorial representation, which has the following decomposition in terms of SU(5) representations:

$$16 = \underline{10} + \underline{5} + \underline{1}. \tag{233}$$

The first two of these irreducible representations are just those considered earlier in the discussion of the SU(5), with identical assignments for the quarks and leptons, whilst the singlet $\underline{1}$ representation seems tailor-made for a singlet neutrino of the type discussed in the previous Lecture. To break SO(10) down to the standard model gauge group, one could use Higgs bosons in a spinorial $\underline{16}$ representation, and some fermion masses could be generated by a vectorial $\underline{10}$ of SO(10). However, a realistic pattern of symmetry breaking and masses would require additional, larger Higgs representations.

Beyond SO(10), the most suitable simple group of rank 6 is $\mathbf{E_6}$. In this case, the matter particles would be be assigned to a

$$27 = \underline{16} + \underline{10} + \underline{1} \tag{234}$$

representation, with the indicated decomposition into representations of SO(10). This group acquired renewed attention some years ago when it emerged naturally in many compactifications of the heterotic string, about which we hear more in the next Lecture.

We now look briefly at the construction of supersymmetric GUTs, of which the minimal version is based on the group SU(5) [181]. As in the transition from the standard model to the MSSM, one simply extends the conventional GUT multiplets to supermultiplets, so that matter particles are assigned to $\underline{5}$ representations $\bar{F}$ and $\underline{10}$ representations T, one doubles the electroweak Higgs fields to include both $H, \bar{H}$ in $\underline{5}, \underline{\bar{5}}$ representations, and one postulates a $\underline{24}$ representation Φ to break the SU(5) GUT symmetry down to SU(3) × SU(2) × U(1). The superpotential for the Higgs sector takes the general form

$$W_5 = (\mu + \frac{3}{2}\lambda M)H\bar{H} + \lambda H\Phi\bar{H} + f(\Phi). \tag{235}$$

Here, $f(\Phi)$ is chosen so that the vacuum expectation value of Φ has the form

$$\langle 0|\Phi|0\rangle = M \times \mathrm{diag}(1,1,1,-\frac{3}{2},-\frac{3}{2}). \tag{236}$$

The coefficient of the $H\bar{H}$ term has been chosen so that it almost cancels with the term $\propto H\langle 0|\Phi|0\rangle\bar{H}$ coming from the second term in (235), *for the last two components.* In this way, the triplet components of $H, \bar{H}$ acquire large masses $\propto M$, whilst the last two may acquire a vacuum expectation value: $\langle 0|H|0\rangle = \mathrm{column}(0,0,0,0,v)$, $\langle 0|\bar{H}|0\rangle = \mathrm{column}(0,0,0,0,\bar{v})$, once supersymmetry breaking and radiative corrections are taken into account.

In order that $v, \bar{v} \sim 100$ GeV, it is necessary that the residual $H\bar{H}$ mixing term $\mu \lesssim 1$ TeV. Since, as we recall shortly, $M \sim 10^{16}$ GeV, this means that the parameters of W_5 (235) must be tuned finely to one part in 10^{13}. This fine-tuning may appear very unreasonable, but it is technically natural, in the sense that there are no big radiative corrections. Thanks to the supersymmetric no-renormalization theorem for superpotential parameters, we know that $\delta\lambda, \delta\mu = 0$, apart from wave-function renormalization factors. Thus, if we adjust the input parameters of (235) so that μ is small, it will stay small. However, this begs the more profound question: how did μ get to be so small in the first place?

6.3 Baryon decay

Baryon instability is to be expected on general grounds, since there is no exact symmetry to guarantee that baryon number B is conserved, just as we discussed previously for lepton number. Indeed, baryon decay is a generic prediction of GUTs, which we illustrate with the simplest SU(5) model. We see in (231) that there are two species of vector bosons in SU(5) that couple the colour indices (1,2,3) to the electroweak indices (4,5), called X and Y. As we can see from the matter representations (232), these may enable two quarks or a quark and lepton to annihilate. Combining these possibilities leads to interactions with $\Delta B = \Delta L = 1$. The forms of effective four-fermion interactions mediated by the exchanges of massive Z and Y bosons, respectively, are [178]:

$$\left(\epsilon_{ijk} u_{R_k} \gamma_\mu u_{L_j}\right) \frac{g_X^2}{8m_X^2} \left(2 e_R \gamma^\mu d_{L_i} + e_L \gamma^\mu d_{R_i}\right) ,$$

$$\left(\epsilon_{ijk} u_{R_k} \gamma_\mu d_{L_j}\right) \frac{g_Y^2}{8m_X^2} \left(\nu_L \gamma^\mu d_{R_i}\right), \tag{237}$$

up to generation mixing factors.

Since the couplings $g_X = g_Y$ in an SU(5) GUT, and $m_X \simeq m_Y$, we expect that

$$G_X \equiv \frac{g_X^2}{8m_X^2} \simeq G_Y \equiv \frac{g_Y^2}{8m_Y^2}. \tag{238}$$

It is clear from (237) that the baryon decay amplitude $A \propto G_X$, and hence the baryon $B \to \ell +$ meson decay rate

$$\Gamma_B = c G_X^2 m_p^5, \tag{239}$$

where the factor of m_p^5 comes from dimensional analysis, and c is a coefficient that depends on the GUT model and the non-perturbative properties of the baryon and meson.

The decay rate (239) corresponds to a proton lifetime

$$\tau_p = \frac{1}{c} \frac{m_X^4}{m_p^5}. \tag{240}$$

It is clear from (240) that the proton lifetime is very sensitive to m_X, which must therefore be calculated very precisely. In minimal SU(5), the best estimate was

$$m_X \simeq (1 \text{ to } 2) \times 10^{15} \times \Lambda_{QCD} , \tag{241}$$

where Λ_{QCD} is the characteristic QCD scale. Making an analysis of the generation mixing factors [180], one finds that the preferred proton (and bound neutron) decay modes in minimal SU(5) are

$$p \to e^+ \pi^0 , \quad e^+ \omega , \quad \bar{\nu}\pi^+ , \quad \mu^+ K^0 , \quad \ldots$$
$$n \to e^+ \pi^- , \quad e^+ \rho^- , \quad \bar{\nu}\pi^0 , \quad \ldots \tag{242}$$

and the best numerical estimate of the lifetime is

$$\tau(p \to e^+ \pi^0) \simeq 2 \times 10^{31 \pm 1} \times \left(\frac{\Lambda_{QCD}}{400 \text{ MeV}}\right)^4 \ y \tag{243}$$

This is in *prima facie* conflict with the latest experimental lower limit

$$\tau(p \to e^+\pi^0) > 1.6 \times 10^{33} \, y \tag{244}$$

from super-Kamiokande [182].

We saw earlier that supersymmetric GUTs, including SU(5), fare better with coupling unification. They also predict a larger GUT scale [47]:

$$m_X \simeq 10^{16} \text{ GeV}, \tag{245}$$

so that $\tau(p \to e^+\pi^0)$ is considerably longer than the experimental lower limit. However, this is not the dominant proton decay mode in supersymmetric SU(5) [183]. In this model, there are important $\Delta B = \Delta L = 1$ interactions mediated by the exchange of colour-triplet Higgsinos $\tilde{H}_3$, dressed by gaugino exchange [184]:

$$G_X \to \mathcal{O} \left(\frac{\lambda^2 g^2}{16\pi^2} \right) \frac{1}{m_{\tilde{H}_3} \tilde{m}} \tag{246}$$

where λ is a Yukawa coupling. Taking into account colour factors and the increase in λ for more massive particles, it was found [183] that decays into neutrinos and strange particles should dominate:

$$p \to \bar{\nu} K^+ , \quad n \to \bar{\nu} K^0 , \quad \ldots \tag{247}$$

Because there is only one factor of a heavy mass $m_{\tilde{H}_3}$ in the denominator of (246), these decay modes are expected to dominate over $p \to e^+\pi^0$, etc., in minimal supersymmetric SU(5). Calculating carefully the other factors in (246) [183], it seems that the modes (247) may now be close to exclusion at rates compatible with this model [185]. The current experimental limit is $\tau(p \to \bar{\nu} K^+) > 6.7 \times 10^{32} y$. However, there are other GUT models [131] that remain compatible with the baryon decay limits.

6.4 Problems of gravity

It is clear that the GUTs outlined above do not answer all the questions one might ask, which is one of the motivations for tackling string theory, our only candidate for a fundamental theory of Everything including gravity.

The greatest piece of unfinished business for twentieth-century physics is to reconcile general relativity with quantum mechanics. There are aspects of this problem, one being that of the cosmological constant, as discussed above. Another is that of perturbative quantum-gravity effects. Tree-level graviton exchange in $2 \to 2$ scattering, such as $e^+e^- \to e^+e^-$ at LEP, has an amplitude $A_G \sim E^2/m_P^2$, and hence a cross section

$$\sigma_G \sim E^2/m_P^4 . \tag{248}$$

This is very small (negligible!) at LEP energies, reaching the unitarity limit only when $E \sim m_P$. However, when one calculates loop amplitudes involving gravitons, the rapid growth with energy (248) leads to uncontrollable, non-renormalizable divergences. These are of power type, and diverge faster and faster in higher orders of perturbation theory.

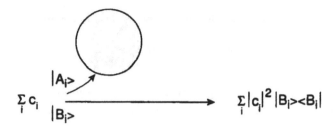

Figure 57. *If a pair of particles $|A\rangle$ $|B\rangle$ is produced near the horizon of a black hole, and one of them ($|A\rangle$, say) falls in, the remaining particle $|B\rangle$ will appear to be in a mixed state.*

There are also non-perturbative problems in the quantization of gravity, that first arose in connection with black holes. From the pioneering work of Bekenstein and Hawking [186] on black-hole thermodynamics, we know that black holes have non-zero entropy S and temperature T, related to the Schwarzschild horizon radius. This means that the quantum description of a black hole should involve mixed states. The intuition underlying this feature is that information can be lost through the event horizon. Consider, for example, a pure quantum-mechanical pair state $|A, B\rangle \equiv \sum_i c_i |A_i\rangle |B_i\rangle$ prepared near the horizon, and what happens if one of the particles, say A, falls through the horizon while B escapes, as seen in Fig. 34. In this case,

$$\sum_i c_i |A_i B_i\rangle \rightarrow \sum_i |c_i|^2 |B_i\rangle\langle B_i| \qquad (249)$$

and B emerges in a mixed state, as in Hawking's original treatment of the black-hole radiation that bears his name [186].

The problem is that conventional quantum mechanics does not permit the evolution of a pure initial state into a mixed final state. This is an issue both for the quantum particles discussed above and for the black hole itself. We could imagine having prepared the black hole by fusing massive or energetic particles in a pure initial state, e.g., by splitting a laser beam and then firing the sub-beams at each other as in a laser device for inertial nuclear fusion.

These problems point to a fundamental conflict between the proudest achievements of early twentieth-century physics, quantum mechanics and general relativity. One or the other should be modified, and perhaps both. Since quantum mechanics is sacred to field theorists, most particle physicists prefer to modify general relativity by elevating it to string theory [6].[4]

6.5   Introduction to string theory

At the level of perturbation theory, the divergence difficulties of quantum gravity can be related to the absence in a point-particle theory of a cutoff at short distances: for

---

[4]It may be that this will eventually also require a modification of the *effective* quantum-mechanical space-time theory, even if the internal formulation of string theory is fully quantum-mechanical, but that is another story [187].

example,

$$\int^{\Lambda \to \infty} d^4 k \left(\frac{1}{k^2}\right) \leftrightarrow \int_{1/\Lambda \to 0} d^4 x \left(\frac{1}{x^6}\right) \sim \Lambda^2 \to \infty . \tag{250}$$

Such divergences can be alleviated or removed if one replaces point particles by extended objects. The simplest possibility is to extend in just one dimension, leading us to a theory of strings. In such a theory, instead of point particles describing one-dimensional world lines, we have strings describing two-dimensional world sheets. Most popular have been closed loops of string, whose simplest world sheet is a tube. The 'wiring diagrams' generated by the Feynman rules of conventional point-like particle theories become the 'plumbing circuits' generated by the junctions and connections of these tubes of closed string. One could imagine generalizing this idea to higher-dimensional extended objects such as membranes describing world volumes, etc., and we return later to this option.

On a historical note, string models first arose from old-fashioned strong-interaction theory, before the advent of QCD. The lowest-lying hadronic states were joined by a very large number of excited states with increasing masses m and spins J:

$$J = \alpha' m^2 , \tag{251}$$

where α' was called the 'Regge slope'. One interpretation of this spectrum was of $\bar{q}q$ bound states in a linearly-rising potential, like an elastic string holding the constituents together, with tension $\mu = 1/\alpha'$. It was pointed out that such an infinitely (?) large set of resonances in the direct s-channel of a scattering process could be dual (equivalent) to the exchange of a similar infinite set in the crossed channel. Mathematically, this idea was expressed by the Veneziano [188] amplitude for $2 \to 2$ scattering, and its generalizations to $2 \to n$ particle production processes. Then it was pointed out that these amplitudes could be derived formally from an underlying quantum theory of string [189]. However, this first incarnation of string theory was not able to accommodate the point-like partons seen inside hadrons at this time – the converse of the quantum-gravity motivation for string theory mentioned at the beginning of this section. Then along came QCD, which incorporated these point-like scaling properties and provided a qualitative understanding of confinement, which has now become quantitative with the advent of modern lattice calculations. Thus string theory languished as a candidate model of the strong interactions.

It was realized early on that unitarity required the existence of closed strings, even in an *a priori* open-string theory. Moreover, it was observed that the spectrum of a closed string included a massless spin-2 particle, which was an embarrassment for a theory of the strong interactions. However, this led to the idea [6] of reinterpreting string theory as a theory of Everything, with this massless spin-2 state interpreted as the graviton and the string tension elevated to $\mu = O(m_P^2)$.

As already mentioned, one of the primary reasons for studying extended objects in connection with quantum gravity is the softening of divergences associated with short-distance behaviour. Since the string propagates on a world sheet, the basic formalism is two-dimensional. Accordingly, string vibrations may be described in terms of left- and right-moving waves:

$$\phi(r, t) \to \phi_L(r - t), \phi_R(r + t) . \tag{252}$$

If the string has no boundary, as for a closed string, the left- and right-movers are independent. When quantized, they may be described by a two-dimensional field theory. Compared to a four-dimensional theory, it is relatively easy to make a two-dimensional field theory finite. In this case, it has conformal symmetry, which has an infinite-dimensional symmetry group in two dimensions. However, as you already know from gauge theories, one must be careful to ensure that this classical symmetry is not broken at the quantum level by anomalies. If the quantum string theory is to be consistent in a flat background space-time, the conformal anomaly fixes the number of left- and right-movers each to be equivalent to 26 free bosons if the theory has no supersymmetry, or 10 boson/fermion supermultiplets if the theory has $N = 1$ supersymmetry on the world sheet. There are other important quantum consistency conditions, and it was the demonstration by Green and Schwarz [6] that certain string theories are completely anomaly-free that opened the floodgates of theoretical interest in string theory as a theory of Everything [6].

Among consistent string theories, one may enumerate the following. The *bosonic string* exists in 26 dimensions, but this is not even its worst problem! It contains no fermionic matter degrees of freedom, and the flat-space vacuum is intrinsically unstable. *Superstrings* exist in 10 dimensions, have fermionic matter and also a stable flat-space vacuum. On the other hand, the ten-dimensional theory is left-right symmetric, and the incorporation of parity violation in four dimensions is not trivial. The *heterotic string* [190] was originally formulated in 10 dimensions, with parity violation already incorporated, since the left- and right movers were treated differently. This theory also has a stable vacuum, but suffers from the disadvantage of having too many dimensions. *Four-dimensional heterotic strings* may be obtained either by compactifying the six surplus dimensions: $10 = 4 + 6$ compact dimensions with size $R \sim 1/m_P$ [191], or by direct construction in four dimensions, replacing the missing dimensions by other internal degrees of freedom such as fermions [192] or group manifolds or ...? In this way it was possible to incorporate a GUT-like gauge group [193] or even something resembling the standard model [194].

What are the general features of such string models? First, they predict there are no more than 10 dimensions, which agrees with the observed number of 4! Secondly, they suggest that the rank of the four-dimensional gauge group should not be very large, in agreement with the rank 4 of the standard model! Thirdly, the simplest four-dimensional string models do not accommodate large matter representations [195], such as an $\underline{8}$ of SU(3) or a $\underline{3}$ of SU(2), again in agreement with the known representation structure of the standard model! Fourthly, simple string models predict fairly successfully the mass of the top quark. This is because the maximum generic value of a Yukawa coupling λ_t is of the same order as the gauge coupling g. Applied to the top quark, this suggests that

$$m_t = \lambda_t \langle 0|H|0 \rangle = O(g) \times 250 \text{ GeV} . \tag{253}$$

Moreover, the renormalization-group equation for λ_t exhibits an approximate infra-red fixed point, as seen in Fig. 35. This means that a large range of Yukawa coupling inputs at the Planck scale yield very similar physical values of $m_t \lesssim 190$ GeV. Fifthly, string theory makes a fairly successful prediction for the gauge unification scale in terms of m_P. If the intrinsic string coupling g_s is weak, one predicts [196]

$$M_{\text{GUT}} = O(g) \times \frac{m_P}{\sqrt{8\pi}} \simeq \text{few} \times 10^{17} \text{ GeV} , \tag{254}$$

where g is the gauge coupling, which is $\mathcal{O}(20)$ higher than the value calculated from the bottom up in Lecture 4 on the basis of LEP measurement of the gauge couplings. On

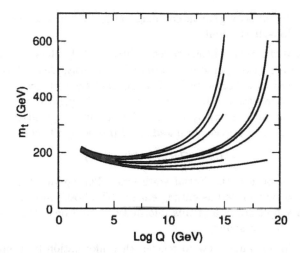

Figure 58. *The approximate infra-red fixed point of the renormalization-group equation for m_t means that a wide range of input Yukawa couplings at the GUT or string scale lead to similar physical values of $m_t \lesssim 190$ GeV.*

the one hand, it is impressive that the bottom-up extrapolation over 14 decades agrees to within 10 % (on a logarithmic scale) with the top-down calculation (253). Nevertheless, it would be nice to obtain closer agreement, and this provides the major motivation for considering strongly-coupled string theory, which corresponds to a large internal dimension $\ell > m_{\text{GUT}}^{-1}$, as we discuss next.

6.6 Beyond string

Current developments involve going beyond string to consider higher-dimensional extended objects, such as generalized membranes with various numbers of internal dimensions. These may be obtained as solitons (non-perturbative classical solutions) of string theory, with masses

$$m \propto \frac{1}{g_s} \tag{255}$$

analogously to monopoles in gauge theory. It is evident from (255) that such membrane-solitons become light in the limit of strong string coupling $g_s \to \infty$.

It was observed some time ago that there should be a strong-coupling/weak-coupling duality [197] between elementary excitations and monopoles in supersymmetric gauge theories. These ideas have recently been confirmed in a spectacular solution of $N = 2$ supersymmetric gauge theory in four dimensions [198]. Similarly, it has recently been shown that there are analogous dualities in string theory, whereby solitons in some strongly-coupled string theory are equivalent to light string states in some other weakly-coupled string theory. Indeed, it appears that all string theories are related by such dualities. A particularity of this discovery is that the string coupling strength g_s is related to an extra dimension, in such a way that its size $R \to \infty$ as $g_s \to \infty$. This then leads to the idea of an underlying 11-dimensional framework called M theory [199] that reduces to the different string theories in different strong/weak-coupling linits, and reduces to eleven-dimensional supergravity in the low-energy limit.

A particular class of string solitons called D-branes [200] offers a promising approach to the black hole information paradox mentioned previously. According to this picture, black holes are viewed as solitonic balls of string, and their entropy simply counts the number of internal string states [187, 201]. These are in principle countable, so string theory may provide an accounting system for the information contained in black holes. Within this framework, the previously paradoxical process (249) becomes

$$|A, B\rangle + |BH\rangle \rightarrow |B'\rangle + |BH'\rangle , \qquad (256)$$

and the final state is pure if the initial state was. The apparent entropy of the final state in (249) is now interpreted as entanglement. The 'lost' information is in principle encoded in the black-hole state, and this information could be extracted if we measured all properties of this ball of string.

In practice, we do not know how to recover this information from macroscopic black holes, so they appear to us as mixed states. What about microscopic black holes, namely fluctuations in the space-time background with $\Delta E = O(m_P)$, that last for a period $\Delta t = O(1/m_P)$ and have a size $\Delta x = O(1/m_P)$? Do these steal information from us [202], or do they give it back to us when they decay? Most people think there is no microscopic leakage of information in this way, but not all of us [187] are convinced. The neutral kaon system is among the most sensitive experimental areas [203, 204, 205] for testing this speculative possibility.

A final experimental comment concerns the magnitude of the extra dimension in M theory: LEP data suggest that it may be relatively large, with size $L_{11} \gg 1/m_{\text{GUT}} \simeq 1/10^{16}$ GeV $\gg 1/m_P$ [206]. Remember that the naïve string unification scale (254) is about 20 times larger than m_{GUT}. This may be traced to the fact that the gravitational interaction strength, although growing rapidly as a power of energy (248), is still much smaller than the gauge coupling strength at $E = m_{\text{GUT}}$. However, if an extra space-time dimension appears at an energy $E < m_{\text{GUT}}$, the gravitational interaction strength grows fast, as indicated in Fig. 59. unification with gravity around 10^{16}GeV then becomes possible, *if* the gauge couplings do not also acquire a similar higher-dimensional kick. Thus we are led to the startling capacitor-plate framework for fundamental physics shown in Fig. 60.

Each plate is *a priori* ten-dimensional, and the bulk space between then is *a priori* eleven-dimensional. Six dimensions are compactified on a scale $L_6 \sim 1/m_{\text{GUT}}$, leaving a theory which is effectively five-dimensional in the bulk and four-dimensional on the walls. Conventional gauge interactions and observable matter particles are hypothesized to live on one capacitor plate, and there are other hidden gauge interactions and matter particles living on the other plate. The fifth dimension has a characteristic size which is estimated to be $O(10^{12}$ to 10^{13} GeV$)^{-1}$, and physics at large distances (smaller energies) looks effectively four-dimensional. Supersymmetry breaking is may originate on the hidden capacitor plate in this scenario, and to be transmitted to the observable wall by gravitational strength interactions in the bulk.

The phenomenological richness of this speculative M-theory approach is only beginning to be explored, and it remains to be seen whether it offers a realistic phenomenological description. However, it does embody all the available theoretical wisdom as well as offer-

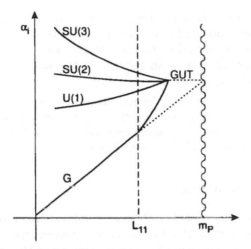

Figure 59. *Sketch of the possible evolution of the gauge couplings and the gravitational coupling G: if there is a large fifth dimension with size $\gg m_{GUT}^{-1}$, G may be unified with the gauge couplings at the GUT scale.*

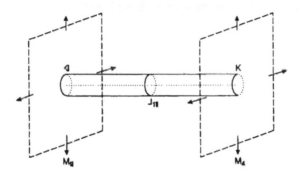

Figure 60. *The capacitor-plate scenario favoured in eleven-dimensional M theory. The eleventh dimension has a size $L_{11} \gg M_{GUT}^{-1}$, whereas dimensions 5, ..., 10 are compactified on a small manifold K with characteristic size $\sim M_{GUT}^{-1}$. the remaining four dimensions form (approximately) a flat Minkowski space M_4.*

ing the prospect of unifying all the observable gauge interactions with gravity at a single effective scale $\sim m_{GUT}$. As such, it constitutes our best contemporary guess about the theory of everything within and beyond the standard model.

One of the possibilities thrown up by M theory is that the extra dimensions required for the consistency of string theory might actually be much larger than the $\ell \sim 1/m_P$ that was popular originally. If at least one of the extra dimensions could be $\gg 1/m_{GUT}$, how much larger could it be? One early suggestion was that one or more extra dimensions might be as large as a TeV^{-1}, in which case they might be responsible for supersymmetry breaking [60]. In this case, experiments at the LHC might reveal the Kaluza-Klein excitations of standard model particles, depending which if any of them propagate into the extra dimensions. In some such scenarios, the lightest Kaluza-Klein excitation could even be a candidate for cold dark matter. Alternatively, it has been suggested that there might be one or more much bigger dimensions, potentially even at the millimetre scale. This suggestion was originally motivated by the hierarchy problem, though I have never been convinced that extra-dimensional models do more than reformulate the hierarchy problem, albeit in a creative way that might lead to new understanding.

As you might gather from the space devoted in these lectures to large extra dimensions, I am not a big fan. I have not yet seen a convincing motivation for them. On the other hand, why not? It is a profound scientific question just how large they might be, and thinking through their theory and phenomenology is surely fun.

6.7   Concluding remarks

In these lectures, I have concentrated on the the possible extensions of the standard model that seem to me best motivated: the Higgs boson or whatever replaces it, supersymmetry to tame the hierarchy problem, neutrino masses, mixing and the ensuing lepton-flavour violation, grand unification and string theory. Neutrino physics has already taken us beyond the standard model, and I am sure that the LHC will take us beyond it in a different direction. Cosmology and astrophysics cry out for physics beyond the standard model, and also offer new ways to explore it. Bring it on!

Bibliography

[1] J. R. Ellis, Lectures at 1998 CERN Summer School, St. Andrews, *Beyond the Standard Model for Hill Walkers*, arXiv:hep-ph/9812235.

[2] J. R. Ellis, Lectures at 2001 CERN Summer School, Beatenberg, *Supersymmetry for Alp Hikers*, arXiv:hep-ph/0203114.

[3] J. R. Ellis, Lectures at the 2002 PSI Summer School, Zuoz, *Limits of the Standard Model*, arXiv:hep-ph/0211168.

[4] J. R. Ellis, Lectures at the 2003 ANU Summer School, Canberra, *Particle Physics and Cosmology*, arXiv:astro-ph/0305038.

[5] LEP Electroweak Working Group, http://lepewwg.web.cern.ch/LEPEWWG/Welcome.html.

[6] J. Scherk and J. H. Schwarz, Nucl. Phys. **B81**, 118 (1974); M. B. Green and J. H. Schwarz, Phys. Lett. **149B**, 117 (1984) and **151B**, 21 (1985); M. B. Green, J. H. Schwarz and E. Witten, *Superstring Theory*, (Cambridge Univ. Press, 1987).

[7] S.L. Glashow, Nucl. Phys. **22** (1961) 579; S. Weinberg, Phys. Rev. Lett. **19** (1967) 1264; A. Salam, Proc. 8th Nobel Symposium, Stockholm 1968, ed. N. Svartholm (Almqvist and Wiksells, Stockholm, 1968), p. 367.

[8] P. W. Higgs, Phys. Lett. **12**, 132 (1964), Phys. Rev. Lett. **13**, 508 (1964) and Phys. Rev. **145**, 1156 (1966); F. Englert and R. Brout, Phys. Rev. Lett. **13**, 321 (1964); see also C. Quigg, *Gauge Theories of the Strong, Weak and Electromagnetic Interactions* (Benjamin-Cummings, Reading, 1983).

[9] D. Brandt, H. Burkhardt, M. Lamont, S. Myers and J. Wenninger, Rept. Prog. Phys. **63**, 939 (2000).

[10] M. Veltman, Nucl. Phys. B **123**, 89 (1977); M.S. Chanowitz, M. Furman and I. Hinchliffe, Phys. Lett. B **78**, 285 (1978).

[11] M. Veltman, Acta Phys. Pol. **8**, 475 (1977).

[12] D. A. Ross and M. J. Veltman, Nucl. Phys. B **95**, 13 (1975).

[13] C. T. Hill, Phys. Lett. B **266** (1991) 419; for a recent review, see: C. T. Hill and E. H. Simmons, arXiv:hep-ph/0203079.

[14] For a historical reference, see: E. Farhi and L. Susskind, Phys. Rept. **74**, 277 (1981).

[15] S. Dimopoulos and L. Susskind, Nucl. Phys. B **155** (1979) 237; E. Eichten and K. Lane, Phys. Lett. B **90** (1980) 125.

[16] J. R. Ellis, M. K. Gaillard, D. V. Nanopoulos and P. Sikivie, Nucl. Phys. B **182** (1981) 529.

[17] S. Dimopoulos and J. R. Ellis, Nucl. Phys. B **182** (1982) 505.

[18] G. Altarelli and R. Barbieri, Phys. Lett. B **253** (1991) 161; M. E. Peskin and T. Takeuchi, Phys. Rev. Lett. **65** (1990) 964.

[19] J. R. Ellis, G. L. Fogli and E. Lisi, Phys. Lett. B **343**, 282 (1995).

[20] G. Altarelli, F. Caravaglios, G. F. Giudice, P. Gambino and G. Ridolfi, JHEP **0106** (2001) 018 [arXiv:hep-ph/0106029].

[21] For a recent reference, see: K. Lane, *Two lectures on technicolor*, arXiv:hep-ph/0202255.

[22] B. Holdom, Phys. Rev. D **24** (1981) 1441.

[23] J. R. Ellis, M. K. Gaillard and D. V. Nanopoulos, Nucl. Phys. B **106**, 292 (1976).

[24] LEP Higgs Working Group for Higgs boson searches, OPAL Collaboration, ALEPH Collaboration, DELPHI Collaboration and L3 Collaboration, *Search for the Standard Model Higgs Boson at LEP*, CERN-EP/2003-011.

[25] J. Erler, Phys. Rev. D **63**, 071301 (2001) [arXiv:hep-ph/0010153].

[26] For a review, see: T. Hambye and K. Riesselmann, arXiv:hep-ph/9708416.

[27] J. R. Ellis and D. Ross, Phys. Lett. B **506**, 331 (2001) [arXiv:hep-ph/0012067].

[28] G. Isidori, G. Ridolfi and A. Strumia, Nucl. Phys. B **609** (2001) 387 [arXiv:hep-ph/0104016].

[29] M. Carena *et al.* [Higgs Working Group Collaboration], arXiv:hep-ph/0010338.

[30] ATLAS Collaboration, *ATLAS detector and physics performance Technical Design Report*, CERN/LHCC 99-14/15 (1999); CMS Collaboration, Technical Proposal, CERN/LHCC 94-38 (1994).

[31] S. R. Coleman and J. Mandula, Phys. Rev. **159** (1967) 1251.

[32] Y. A. Golfand and E. P. Likhtman, JETP Lett. **13** (1971) 323 [Pisma Zh. Eksp. Teor. Fiz. **13** (1971) 452].

[33] A. Neveu and J. H. Schwarz, Nucl. Phys. B **31** (1971) 86.

[34] P. Ramond, Phys. Rev. D **3** (1971) 2415.

[35] D. V. Volkov and V. P. Akulov, Phys. Lett. B **46** (1973) 109.

[36] J. Wess and B. Zumino, Phys. Lett. B **49** (1974) 52; Nucl. Phys. B **70** (1974) 39.

[37] J. Wess and B. Zumino, Nucl. Phys. B **78** (1974) 1.

[38] S. Ferrara, J. Wess and B. Zumino, Phys. Lett. B **51**, 239 (1974); S. Ferrara, J. Iliopoulos and B. Zumino, Nucl. Phys. B **77**, 413 (1974).

[39] P. Fayet, as reviewed in *Supersymmetry, Particle Physics And Gravitation*, CERN-TH-2864, published in *Proc. of Europhysics Study Conf. on Unification of Fundamental Interactions*, Erice, Italy, Mar 17-24, 1980, eds. S. Ferrara, J. Ellis, P. van Nieuwenhuizen (Plenum Press, 1980).

[40] D. Z. Freedman, P. van Nieuwenhuizen and S. Ferrara, Phys. Rev. D **13** (1976) 3214; S. Deser and B. Zumino, Phys. Lett. B **62** (1976) 335.

[41] S. W. Hawking, *Is The End In Sight For Theoretical Physics?* Phys. Bull. **32** (1981) 15.

[42] L. Maiani, *Proceedings of the 1979 Gif-sur-Yvette Summer School On Particle Physics*, 1; G. 't Hooft, in *Recent Developments in Gauge Theories, Proceedings of the Nato Advanced Study Institute, Cargese, 1979*, eds. G. 't Hooft *et al.*, (Plenum Press, NY, 1980); E. Witten, Phys. Lett. B **105**, 267 (1981).

[43] Y. Okada, M. Yamaguchi and T. Yanagida, Prog. Theor. Phys. **85** (1991) 1; J. R. Ellis, G. Ridolfi and F. Zwirner, Phys. Lett. B **257** (1991) 83; H. E. Haber and R. Hempfling, Phys. Rev. Lett. **66** (1991) 1815.

[44] H. P. Nilles, Phys. Rept. **110**, 1 (1984); H. E. Haber and G. L. Kane, Phys. Rept. **117**, 75 (1985).

[45] J. Ellis, S. Kelley and D. V. Nanopoulos, Phys. Lett. B **260**, 131 (1991); U. Amaldi, W. de Boer and H. Furstenau, Phys. Lett. B **260**, 447 (1991); P. Langacker and M. x. Luo, Phys. Rev. D **44**, 817 (1991); C. Giunti, C. W. Kim and U. W. Lee, Mod. Phys. Lett. A **6**, 1745 (1991).

[46] H. Georgi, H. Quinn and S. Weinberg, Phys. Rev. Lett. **33**, 451 (1974).

[47] L. E. Ibanez and G. G. Ross, Phys. Lett. B **105**, 439 (1981); S. Dimopoulos, S. Raby and F. Wilczek, Phys. Rev. D **24**, 1681 (1981).

[48] R. Haag, J. T. Lopuszanski and M. Sohnius, Nucl. Phys. B **88** (1975) 257.

[49] F. Iachello, Phys. Rev. Lett. **44** (1980) 772.

[50] For an early review, see: P. Fayet and S. Ferrara, Phys. Rept. **32** (1977) 249.

[51] B. A. Campbell, S. Davidson, J. R. Ellis and K. A. Olive, Phys. Lett. B **256** (1991) 457; W. Fischler, G. F. Giudice, R. G. Leigh and S. Paban, Phys. Lett. B **258** (1991) 45.

[52] P. Fayet and J. Iliopoulos, Phys. Lett. B **51** (1974) 461.

[53] M. Dine, N. Seiberg and E. Witten, Nucl. Phys. B **289** (1987) 589.

[54] L. O'Raifeartaigh, Nucl. Phys. B **96** (1975) 331; P. Fayet, Phys. Lett. B **58** (1975) 67.

[55] J. Polonyi, Hungary Central Inst. Res. preprint KFKI-77-93 (1977).

[56] E. Cremmer, B. Julia, J. Scherk, S. Ferrara, L. Girardello and P. van Nieuwenhuizen, Nucl. Phys. B **147** (1979) 105.

[57] K. Inoue, A. Kakuto, H. Komatsu and S. Takeshita, Prog. Theor. Phys. **68** (1982) 927 [Erratum-ibid. **70** (1982) 330]; L.E. Ibáñez and G.G. Ross, Phys. Lett. B **110** (1982) 215; L.E. Ibáñez, Phys. Lett. B **118** (1982) 73; J. Ellis, D.V. Nanopoulos and K. Tamvakis, Phys. Lett. B **121** (1983) 123; J. Ellis, J. Hagelin, D.V. Nanopoulos and K. Tamvakis, Phys. Lett. B **125** (1983) 275; L. Alvarez-Gaumé, J. Polchinski, and M. Wise, Nucl. Phys. B **221** (1983) 495.

[58] J. R. Ellis and D. V. Nanopoulos, Phys. Lett. B **110** (1982) 44; R. Barbieri and R. Gatto, Phys. Lett. B **110** (1982) 211.

[59] S. L. Glashow, J. Iliopoulos and L. Maiani, Phys. Rev. D **2** (1970) 1285.

[60] For a phenomenological review, see: I. Antoniadis and K. Benakli, Int. J. Mod. Phys. A **15** (2000) 4237 [arXiv:hep-ph/0007226].

[61] J. R. Ellis and S. Rudaz, Phys. Lett. B **128** (1983) 248.

[62] J. R. Ellis, T. Falk, G. Ganis, K. A. Olive and M. Schmitt, Phys. Rev. D **58** (1998) 095002 [arXiv:hep-ph/9801445].

[63] J. Ellis, J.S. Hagelin, D.V. Nanopoulos, K.A. Olive and M. Srednicki, Nucl. Phys. B **238** (1984) 453; see also H. Goldberg, Phys. Rev. Lett. **50** (1983) 1419.

[64] G. Altarelli, T. Sjöstrand and F. Zwirner, eds., *Proceedings of the Workshop on Physics at LEP 2*, CERN Report 96-01 (1996).

[65] Joint LEP 2 Supersymmetry Working Group, *Combined LEP Chargino Results, up to 208 GeV*, http://lepsusy.web.cern.ch/lepsusy/www/inos_moriond01/charginos_pub.html.

[66] Joint LEP 2 Supersymmetry Working Group, *Combined LEP Selectron/Smuon/Stau Results, 183-208 GeV*, http://lepsusy.web.cern.ch/lepsusy/www/sleptons_summer02/slep_2002.html.

[67] J. Ellis, K. A. Olive, Y. Santoso and V. C. Spanos, arXiv:hep-ph/0303043.

[68] M. S. Alam *et al.*, [CLEO Collaboration], Phys. Rev. Lett. **74**, 2885 (1995), as updated in S. Ahmed et al., CLEO CONF 99-10; BELLE Collaboration, BELLE-CONF-0003, contribution to the 30th International conference on High-Energy Physics, Osaka, 2000. See also K. Abe *et al.*, [Belle Collaboration], arXiv:hep-ex/0107065; L. Lista [BaBar Collaboration], arXiv:hep-ex/0110010; C. Degrassi, P. Gambino and G. F. Giudice, JHEP **0012**, 009 (2000) [arXiv:hep-ph/0009337]; M. Carena, D. Garcia, U. Nierste and C. E. Wagner, Phys. Lett. B **499**, 141 (2001) [arXiv:hep-ph/0010003].

[69] J. R. Ellis, G. Ganis, D. V. Nanopoulos and K. A. Olive, Phys. Lett. B **502**, 171 (2001) [arXiv:hep-ph/0009355].

[70] G. W. Bennett *et al.* [Muon g-2 Collaboration], arXiv:hep-ex/0401008 and references therein.

[71] M. Davier, S. Eidelman, A. Hocker and Z. Zhang, arXiv:hep-ph/0208177; see also K. Hagiwara, A. D. Martin, D. Nomura and T. Teubner, arXiv:hep-ph/0209187; F. Jegerlehner, unpublished, as reported in M. Krawczyk, arXiv:hep-ph/0208076.

[72] M. Knecht and A. Nyffeler, Phys. Rev. D **65**, 073034 (2002) [arXiv:hep-ph/0111058]; M. Knecht, A. Nyffeler, M. Perrottet and E. De Rafael, Phys. Rev. Lett. **88**, 071802 (2002) [arXiv:hep-ph/0111059]; M. Hayakawa and T. Kinoshita, arXiv:hep-ph/0112102; I. Blokland, A. Czarnecki and K. Melnikov, Phys. Rev. Lett. **88**, 071803 (2002) [arXiv:hep-ph/0112117]; J. Bijnens, E. Pallante and J. Prades, Nucl. Phys. B **626**, 410 (2002) [arXiv:hep-ph/0112255].

[73] L. L. Everett, G. L. Kane, S. Rigolin and L. Wang, Phys. Rev. Lett. **86**, 3484 (2001) [arXiv:hep-ph/0102145]; J. L. Feng and K. T. Matchev, Phys. Rev. Lett. **86**, 3480 (2001) [arXiv:hep-ph/0102146]; E. A. Baltz and P. Gondolo, Phys. Rev. Lett. **86**, 5004 (2001)

[arXiv:hep-ph/0102147]; U. Chattopadhyay and P. Nath, Phys. Rev. Lett. **86**, 5854 (2001) [arXiv:hep-ph/0102157]; S. Komine, T. Moroi and M. Yamaguchi, Phys. Lett. B **506**, 93 (2001) [arXiv:hep-ph/0102204]; J. Ellis, D. V. Nanopoulos and K. A. Olive, Phys. Lett. B **508**, 65 (2001) [arXiv:hep-ph/0102331]; R. Arnowitt, B. Dutta, B. Hu and Y. Santoso, Phys. Lett. B **505**, 177 (2001) [arXiv:hep-ph/0102344] S. P. Martin and J. D. Wells, Phys. Rev. D **64**, 035003 (2001) [arXiv:hep-ph/0103067]; H. Baer, C. Balazs, J. Ferrandis and X. Tata, Phys. Rev. D **64**, 035004 (2001) [arXiv:hep-ph/0103280].

[74] C. L. Bennett *et al.*, arXiv:astro-ph/0302207; D. N. Spergel *et al.*, arXiv:astro-ph/0302209; H. V. Peiris *et al.*, arXiv:astro-ph/0302225 and references therein.

[75] M. Battaglia *et al.*, Eur. Phys. J. C **22**, 535 (2001) [arXiv:hep-ph/0106204].

[76] M. Battaglia, A. De Roeck, J. R. Ellis, F. Gianotti, K. A. Olive and L. Pape, arXiv:hep-ph/0306219.

[77] S. Mizuta and M. Yamaguchi, Phys. Lett. B **298**, 120 (1993) [arXiv:hep-ph/9208251]; J. Edsjo and P. Gondolo, Phys. Rev. D **56**, 1879 (1997) [arXiv:hep-ph/9704361].

[78] J. Ellis, T. Falk and K. A. Olive, Phys. Lett. B **444**, 367 (1998) [arXiv:hep-ph/9810360]; J. Ellis, T. Falk, K. A. Olive and M. Srednicki, Astropart. Phys. **13**, 181 (2000) [arXiv:hep-ph/9905481]; M. E. Gómez, G. Lazarides and C. Pallis, Phys. Rev. D **61**, 123512 (2000) [arXiv:hep-ph/9907261] and Phys. Lett. B **487**, 313 (2000) [arXiv:hep-ph/0004028]; R. Arnowitt, B. Dutta and Y. Santoso, Nucl. Phys. B **606**, 59 (2001) [arXiv:hep-ph/0102181].

[79] M. Drees and M. M. Nojiri, Phys. Rev. D **47**, 376 (1993) [arXiv:hep-ph/9207234]; H. Baer and M. Brhlik, Phys. Rev. D **53**, 597 (1996) [arXiv:hep-ph/9508321] and Phys. Rev. D **57**, 567 (1998) [arXiv:hep-ph/9706509]; H. Baer, M. Brhlik, M. A. Diaz, J. Ferrandis, P. Mercadante, P. Quintana and X. Tata, Phys. Rev. D **63**, 015007 (2001) [arXiv:hep-ph/0005027]; A. B. Lahanas, D. V. Nanopoulos and V. C. Spanos, Mod. Phys. Lett. A **16**, 1229 (2001) [arXiv:hep-ph/0009065].

[80] J. L. Feng, K. T. Matchev and T. Moroi, Phys. Rev. Lett. **84**, 2322 (2000) [arXiv:hep-ph/9908309]; J. L. Feng, K. T. Matchev and T. Moroi, Phys. Rev. D **61**, 075005 (2000) [arXiv:hep-ph/9909334]; J. L. Feng, K. T. Matchev and F. Wilczek, Phys. Lett. B **482**, 388 (2000) [arXiv:hep-ph/0004043].

[81] J. R. Ellis and K. A. Olive, Phys. Lett. B **514**, 114 (2001) [arXiv:hep-ph/0105004].

[82] J. Ellis, K. Enqvist, D. V. Nanopoulos and F. Zwirner, Mod. Phys. Lett. A **1**, 57 (1986); R. Barbieri and G. F. Giudice, Nucl. Phys. B **306**, 63 (1988).

[83] S. Matsumoto *et al.* [JLC Group], *JLC-1*, KEK Report 92-16 (1992); J. Bagger *et al.* [American Linear Collider Working Group], *The Case for a 500-GeV e⁺e⁻ Linear Collider*, SLAC-PUB-8495, BNL-67545, FERMILAB-PUB-00-152, LBNL-46299, UCRL-ID-139524, LBL-46299, Jul 2000, arXiv:hep-ex/0007022; T. Abe *et al.* [American Linear Collider Working Group Collaboration], *Linear Collider Physics Resource Book for Snowmass 2001*, SLAC-570, arXiv:hep-ex/0106055, hep-ex/0106056, hep-ex/0106057 and hep-ex/0106058; TESLA Technical Design Report, DESY-01-011, Part III, *Physics at an e⁺e⁻ Linear Collider* (March 2001).

[84] R. W. Assmann *et al.* [CLIC Study Team], *A 3-TeV e^+e^- Linear Collider Based on CLIC Technology*, ed. G. Guignard, CERN 2000-08; for more information about this project, see: http://ps-div.web.cern.ch/ps-div/CLIC/Welcome.html.

[85] Neutrino Factory and Muon Collider Collaboration, http://www.cap.bnl.gov/mumu/mu_home_page.html.

[86] European Muon Working Groups, http://muonstoragerings.cern.ch/Welcome.html; B. Autin, A. Blondel and J. R. Ellis, *Prospective study of muon storage rings at CERN*, CERN-99-02; C. Blochinger *et al.*, *Physics opportunities at $\mu^+\mu^-$ Higgs factories*, arXiv:hep-ph/0202199.

[87] G. L. Kane, J. Lykken, S. Mrenna, B. D. Nelson, L. T. Wang and T. T. Wang, Phys. Rev. D **67**, 045008 (2003) [arXiv:hep-ph/0209061].

[88] D. R. Tovey, Phys. Lett. B **498**, 1 (2001) [arXiv:hep-ph/0006276].

[89] F. E. Paige, hep-ph/0211017.

[90] S. Abdullin *et al.* [CMS Collaboration], arXiv:hep-ph/9806366; S. Abdullin and F. Charles, Nucl. Phys. B **547**, 60 (1999) [arXiv:hep-ph/9811402].

[91] D. Denegri, W. Majerotto and L. Rurua, Phys. Rev. D **60**, 035008 (1999).

[92] I. Hinchliffe, F. E. Paige, M. D. Shapiro, J. Soderqvist and W. Yao, Phys. Rev. D **55**, 5520 (1997).

[93] F. Gianotti *et al.*, *Physics potential and experimental challenges of the LHC luminosity upgrade*, arXiv:hep-ph/0204087.

[94] J. R. Ellis, S. Heinemeyer, K. A. Olive and G. Weiglein, Phys. Lett. B **515**, 348 (2001) [arXiv:hep-ph/0105061].

[95] J. R. Ellis, S. Heinemeyer, K. A. Olive and G. Weiglein, JHEP **0301**, 006 (2003) [arXiv:hep-ph/0211206].

[96] A. De Roeck, J. R. Ellis and F. Gianotti, *Physics motivations for future CERN accelerators*, arXiv:hep-ex/0112004.

[97] J. Erler, S. Heinemeyer, W. Hollik, G. Weiglein and P. M. Zerwas, Phys. Lett. B **486**, 125 (2000) [arXiv:hep-ph/0005024].

[98] G. A. Blair, W. Porod and P. M. Zerwas, Phys. Rev. D63, 017703 (2001). [arXiv:hep-ph/0007107].

[99] CTF3 home page, http://ctf3.home.cern.ch/ctf3/M_Root/Minutes.htm.

[100] CLIC Physics Study Group, http://clicphysics.web.cern.ch/CLICphysics/ and Yellow Report in preparation.

[101] M. Battaglia, E. Boos and W. M. Yao, *Proc. of the APS/DPF/DPB Summer Study on the Future of Particle Physics (Snowmass 2001)* ed. N. Graf, eConf C010630, E3016 (2001) [arXiv:hep-ph/0111276].

[102] M. Battaglia and M. Gruwe, arXiv:hep-ph/0212140.

[103] For a recent review, see: K. A. Olive, arXiv:astro-ph/0202486.

[104] E. W. Kolb and M. S. Turner, *The Early Universe* (Addison-Wesley, Redwood City, USA, 1990).

[105] D. H. Lyth and A. Riotto, Phys. Rept. **314**, 1 (1999) [arXiv:hep-ph/9807278].

[106] J. M. Bardeen, P. J. Steinhardt and M. S. Turner, Phys. Rev. D **28**, 679 (1983).

[107] A. G. Riess *et al.* [Supernova Search Team Collaboration], Astron. J. **116**, 1009 (1998) [arXiv:astro-ph/9805201]; S. Perlmutter *et al.* [Supernova Cosmology Project Collaboration], Astrophys. J. **517**, 565 (1999) [arXiv:astro-ph/9812133]; Perlmutter, S. & Schmidt, B. P. 2003 arXiv:astro-ph/0303428; J. L. Tonry *et al.*, arXiv:astro-ph/0305008.

[108] N. A. Bahcall, J. P. Ostriker, S. Perlmutter and P. J. Steinhardt, Science **284**, 1481 (1999) [arXiv:astro-ph/9906463].

[109] O. Elgaroy *et al.*, Phys. Rev. Lett. **89**, 061301 (2002) [arXiv:astro-ph/0204152].

[110] K. Hagiwara *et al.* [Particle Data Group Collaboration], Phys. Rev. D **66**, 010001 (2002).

[111] L. M. Wang, R. R. Caldwell, J. P. Ostriker and P. J. Steinhardt, Astrophys. J. **530**, 17 (2000) [arXiv:astro-ph/9901388].

[112] J. R. Ellis, T. Falk, G. Ganis, K. A. Olive and M. Srednicki, Phys. Lett. B **510**, 236 (2001) [arXiv:hep-ph/0102098].

[113] J. Silk and M. Srednicki, Phys. Rev. Lett. **53**, 624 (1984).

[114] J. Ellis, J. L. Feng, A. Ferstl, K. T. Matchev and K. A. Olive, arXiv:astro-ph/0110225.

[115] J. Silk, K. A. Olive and M. Srednicki, Phys. Rev. Lett. **55**, 257 (1985).

[116] M. W. Goodman and E. Witten, Phys. Rev. D **31**, 3059 (1985).

[117] R. Bernabei *et al.* [DAMA Collaboration], Phys. Lett. B **436**, 379 (1998).

[118] D. Abrams *et al.* [CDMS Collaboration], arXiv:astro-ph/0203500; A. Benoit *et al.* [EDEL-WEISS Collaboration], Phys. Lett. B **513**, 15 (2001) [arXiv:astro-ph/0106094].

[119] R. W. Schnee *et al.* [CDMS Collaboration], Phys. Rept. **307**, 283 (1998).

[120] M. Bravin *et al.* [CRESST Collaboration], Astropart. Phys. **12**, 107 (1999) [arXiv:hep-ex/9904005].

[121] H. V. Klapdor-Kleingrothaus, arXiv:hep-ph/0104028.

[122] G. Jungman, M. Kamionkowski and K. Griest, Phys. Rept. **267**, 195 (1996) [arXiv:hep-ph/9506380];
http://t8web.lanl.gov/people/jungman/neut-package.html.

[123] A. Osipowicz *et al.* [KATRIN Collaboration], arXiv:hep-ex/0109033.

[124] H. V. Klapdor-Kleingrothaus *et al.*, Eur. Phys. J. A **12**, 147 (2001) [arXiv:hep-ph/0103062]; see, however, H. V. Klapdor-Kleingrothaus *et al.*, Mod. Phys. Lett. A **16**, 2409 (2002) [arXiv:hep-ph/0201231].

[125] Y. Fukuda *et al.* [Super-Kamiokande Collaboration], Phys. Rev. Lett. **81**, 1562 (1998) [arXiv:hep-ex/9807003].

[126] Q. R. Ahmad *et al.* [SNO Collaboration], Phys. Rev. Lett. **89**, 011301 (2002) [arXiv:nucl-ex/0204008]; Phys. Rev. Lett. **89**, 011302 (2002) [arXiv:nucl-ex/0204009].

[127] R. Barbieri, J. R. Ellis and M. K. Gaillard, Phys. Lett. B **90**, 249 (1980).

[128] M. Gell-Mann, P. Ramond and R. Slansky, Proceedings of the Supergravity Stony Brook Workshop, New York, 1979, eds. P. Van Nieuwenhuizen and D. Freedman (North-Holland, Amsterdam); T. Yanagida, Proceedings of the Workshop on Unified Theories and Baryon Number in the Universe, Tsukuba, Japan 1979 (edited by A. Sawada and A. Sugamoto, KEK Report No. 79-18, Tsukuba); R. Mohapatra and G. Senjanovic, Phys. Rev. Lett. **44**, 912 (1980).

[129] P. H. Frampton, S. L. Glashow and T. Yanagida, arXiv:hep-ph/0208157.

[130] T. Endoh, S. Kaneko, S. K. Kang, T. Morozumi and M. Tanimoto, arXiv:hep-ph/0209020.

[131] J. R. Ellis, J. S. Hagelin, S. Kelley and D. V. Nanopoulos, Nucl. Phys. B **311**, 1 (1988).

[132] J. R. Ellis, M. E. Gómez, G. K. Leontaris, S. Lola and D. V. Nanopoulos, Eur. Phys. J. C **14**, 319 (2000).

[133] J. R. Ellis, M. K. Gaillard and D. V. Nanopoulos, Nucl. Phys. B **109**, 213 (1976).

[134] M. Kobayashi and T. Maskawa, Prog. Theor. Phys. **49**, 652 (1973).

[135] Z. Maki, M. Nakagawa and S. Sakata, Prog. Theor. Phys. **28**, 870 (1962).

[136] Y. Oyama, arXiv:hep-ex/0210030.

[137] S. Fukuda *et al.* [Super-Kamiokande Collaboration], Phys. Lett. B **539**, 179 (2002) [arXiv:hep-ex/0205075].

[138] K. Eguchi *et al.* [KamLAND Collaboration], Phys. Rev. Lett. **90**, 021802 (2003) [arXiv:hep-ex/0212021].

[139] S. Pakvasa and J. W. Valle, arXiv:hep-ph/0301061.

[140] H. Minakata and H. Sugiyama, arXiv:hep-ph/0212240.

[141] M. Apollonio *et al.* [CHOOZ Collaboration], Phys. Lett. B **466**, 415 (1999) [arXiv:hep-ex/9907037].

[142] M. H. Ahn *et al.*, [K2K Collaboration], arXiv:hep-ex/0402017.

[143] A. De Rújula, M.B. Gavela and P. Hernández, Nucl. Phys. B **547**, 21 (1999) [arXive:hep-ph/9811390].

[144] A. Cervera *et al.*, Nucl. Phys. B **579**, 17 (2000) [Erratum-ibid. B **593**, 731 (2001)].

[145] B. Autin *et al.*, *Conceptual design of the SPL, a high-power superconducting H^- linac at CERN*, CERN-2000-012.

[146] P. Zucchelli, Phys. Lett. B **532**, 166 (2002).

[147] M. Apollonio *et al.*, *Oscillation physics with a neutrino factory*, arXiv:hep-ph/0210192; and references therein.

[148] J. A. Casas and A. Ibarra, Nucl. Phys. B **618**, 171 (2001) [arXiv:hep-ph/0103065].

[149] J. R. Ellis, J. Hisano, S. Lola and M. Raidal, Nucl. Phys. B **621**, 208 (2002) [arXiv:hep-ph/0109125].

[150] M. Fukugita and T. Yanagida, Phys. Lett. B **174**, 45 (1986).

[151] S. Davidson and A. Ibarra, JHEP **0109**, 013 (2001).

[152] J. R. Ellis, J. Hisano, M. Raidal and Y. Shimizu, Phys. Lett. B **528**, 86 (2002) [arXiv:hep-ph/0111324].

[153] J. R. Ellis, J. Hisano, M. Raidal and Y. Shimizu, Phys. Rev. D **66**, 115013 (2002) [arXiv:hep-ph/0206110].

[154] A. Masiero and O. Vives, New J. Phys. **4** (2002) 4.

[155] C. Jarlskog, Phys. Rev. Lett. **55** (1985) 1039; Z. Phys. C **29** (1985) 491.

[156] J. R. Ellis and M. Raidal, Nucl. Phys. B **643**, 229 (2002) [arXiv:hep-ph/0206174].

[157] J. Aystö *et al.*, *Physics with low-energy muons at a neutrino factory complex*, arXiv:hep-ph/0109217.

[158] B. C. Regan, E. D. Commins, C. J. Schmidt and D. DeMille, Phys. Rev. Lett. **88** (2002) 071805.

[159] S. K. Lamoreaux, arXiv:nucl-ex/0109014.

[160] M. Furusaka *et al.*, JAERI/KEK Joint Project Proposal *The Joint Project for High-Intensity Proton Accelerators*, KEK-REPORT-99-4, JAERI-TECH-99-056.

[161] J. Äystö *et al.*, *Physics with Low-Energy Muons at a Neutrino Factory Complex*, CERN-TH/2001-231, hep-ph/0109217; and references therein.

[162] I. Hinchliffe and F. E. Paige, Phys. Rev. D **63** (2001) 115006 [arXiv:hep-ph/0010086].

[163] D. F. Carvalho, J. R. Ellis, M. E. Gómez, S. Lola and J. C. Romao, arXiv:hep-ph/0206148.

[164] T. Maeno *et al.* [BESS Collaboration], Astropart. Phys. **16**, 121 (2001) [arXiv:astro-ph/0010381].

[165] M. Aguilar *et al.* [AMS Collaboration], Phys. Rept. **366**, 331 (2002) [Erratum-ibid. **380**, 97 (2003)].

[166] A. G. Cohen, A. De Rujula and S. L. Glashow, Astrophys. J. **495**, 539 (1998) [arXiv:astro-ph/9707087].

[167] A. D. Sakharov, Pisma Zh. Eksp. Teor. Fiz. **5**, 32 (1967)

[168] G. 't Hooft, Phys. Rev. D **14**, 3432 (1976) [Erratum-ibid. D **18**, 2199 (1978)].

[169] M. Yoshimura, Phys. Rev. Lett. **41**, 281 (1978) [Erratum-ibid. **42**, 746 (1979)].

[170] See, for example: M. Carena, M. Quiros, M. Seco and C. E. Wagner, Nucl. Phys. B **650**, 24 (2003) [arXiv:hep-ph/0208043].

[171] H. Murayama, H. Suzuki, T. Yanagida and J. Yokoyama, Phys. Rev. Lett. **70**, 1912 (1993); H. Murayama, H. Suzuki, T. Yanagida and J. Yokoyama, Phys. Rev. D **50**, 2356 (1994) [arXiv:hep-ph/9311326].

[172] W. H. Kinney, Phys. Rev. D **58**, 123506 (1998) arXiv:astro-ph/9806259; arXiv:astro-ph/0301448.

[173] J. R. Ellis, M. Raidal and T. Yanagida, arXiv:hep-ph/0303242.

[174] V. Barger, H. S. Lee and D. Marfatia, arXiv:hep-ph/0302150.

[175] P. H. Chankowski, J. Ellis, S. Pokorski, M. Raidal and K. Turzyński, in preparation.

[176] H. Georgi and S.L. Glashow, Phys. Rev. Lett. **32** (1974) 438.

[177] M. Chanowitz, J. Ellis and M.K. Gaillard, Nucl. Phys. B**128** (1977) 506.

[178] A.J. Buras, J. Ellis, M.K. Gaillard and D.V. Nanopoulos, Nucl. Phys. B**135** (1978) 66.

[179] D.V. Nanopoulos and D.A. Ross, Phys. Lett. **118B** (1982) 99.

[180] J. Ellis, M.K. Gaillard and D.V. Nanopoulos, Phys. Lett. **91B** (1980) 67.

[181] S. Dimopoulos and H. Georgi, Nucl. Phys. B**193** (1981) 150.

[182] Super-Kamiokande collaboration, M. Shiozawa et al., Phys. Rev. Lett. **81** (1998) 3319.

[183] J. Ellis, D.V. Nanopoulos and S. Rudaz, Nucl. Phys. B**202** (1982) 43; S. Dimopoulos, S. Raby and F. Wilczek, Phys. Lett. **112B** (1982) 133.

[184] S. Weinberg, Phys. Rev. D**26** (1982) 287, N. Sakai and T. Yanagida, Nucl. Phys. B**197** (1982) 533.

[185] B. Bajc, A. Melfo, G. Senjanovic and F. Vissani, arXiv:hep-ph/0402122 and references therein.

[186] J. Bekenstein, Phys. Rev. D**12** (1975) 3077; S. Hawking, Comm. Math. Phys. **43** (1975) 199.

[187] J. Ellis, N.E. Mavromatos and D.V. Nanopoulos, Mod.Phys.Lett. A**10** (1995) 425 and references therein

[188] G. Veneziano, Nuovo Cimento **57A** (1968) 190 and Phys.Rep. C**9** (1974) 199.

[189] Y. Nambu, *Proc. Int. Conf. on Symmetries and Quark Models*, Wayne State University (1969); P. Goddard, J. Goldstone, C.Rebbi and C. Thorn, Nucl. Phys. B**181** (1981) 502.

[190] D.J. Gross, J.A. Harvey, E. Martinec and R. Rohm, Phys. Rev. Lett. **54** (1985) 502 and Nucl. Phys. B**256** (1985) 253, B**267** (1985) 75.

[191] P. Candelas, G. Horowitz, A. Strominger and E. Witten, Nucl. Phys. B**258** (1985) 46.

[192] H. Kawai, D. Lewellen and S.H.H. Tye, Nucl. Phys. **B287** (1987) 1;
I. Antoniadis, C. Bachas and C. Kounnas, Nucl. Phys. **B289** (1987) 87;
I. Antoniadis and C. Bachas, Nucl. Phys. **B298** (1988) 586.

[193] I. Antoniadis, J. Ellis, J.S. Hagelin and D.V. Nanopoulos, Phys. Lett. **B194** (1987) 231
and **B231** (1989) 65.

[194] A. Faraggi, D.V. Nanopoulos and K. Yuan, Nucl. Phys. **B335** (1990) 347.

[195] H. Dreiner, J. Lopez, D.V. Nanopoulos and D.B. Reiss, Phys. Lett. **B216** (1989) 283.

[196] I. Antoniadis, J. Ellis, R. Lacaze and D.V. Nanopoulos, Phys. Lett. **B268** (1991) 188.

[197] C. Montonen and D. Olive, Phys. Lett. **72B** (1977) 117.

[198] N. Seiberg and E. Witten, Nucl. Phys. **B431** (1994) 484.

[199] For a review, see: Miao Li, hep-th/9811019.

[200] J. Polchinski, Phys. Rev. Lett. **75** (1995) 4724.

[201] A. Strominger and C. Vafa, Phys. Lett. **B379** (1996) 99.

[202] S. Hawking, Commun. Math. Phys. **87** (1983) 395 and Phys. Rev. **D37** (1988) 904.

[203] J. Ellis, J.S. Hagelin, D.V. Nanopoulos and M. Srednicki, Nucl. Phys. **B241** (1984) 381.

[204] J. Ellis, J. Lopez, N.E. Mavromatos and D.V. Nanopoulos, Phys. Rev. **D53** (1996) 3846.

[205] CPLEAR collaboration, R, Adlet et al., and J. Ellis, J. Lopez, N.E. Mavromatos and D.V.
Nanopoulos, Phys. Lett. **B364** (1995) 239.

[206] P. Horava and E. Witten, Nucl. Phys. **B460** (1996) 506 and Nucl. Phys. **B475** (1996) 94;
P. Horava, Phys. Rev. **D54** (1996) 7561.

The LHC accelerator and its challenges

Rüdiger Schmidt

CERN, Switzerland

1 Introduction

The motivation to construct the Large Hadron Collider (LHC) at CERN comes from fundamental questions in particle physics. The first problem of particle physics today is the problem of mass: is there an elementary Higgs boson? The primary task of the LHC is to make an initial exploration in the 1 TeV range. The major LHC experiments, ATLAS and CMS, should be able to accomplish this for any Higgs mass in this range. To get into the 1 TeV scale, a proton collider with two beams of 7 TeV/c momenta is being constructed in the 27 km long LEP tunnel. A magnetic field of 8.33 Tesla is necessary to achieve the required deflection of 7 TeV/c protons that can only be generated with superconducting magnets. The machine is also designed for collisions of heavy ions (for example lead) at very high centre of mass energies. A dedicated heavy ion detector, ALICE, will be built to exploit the unique physics potential of nucleus-nucleus interactions at LHC energies. The fourth detector, LHCb, will be built for precision measurements of CP-Violation and rare decays of B Mesons. The physics motivation for the LHC and the construction of the experiments are discussed in detail in the other lectures.

The LHC accelerator has been under preparation since the beginning of the eighties, with a research and development program for superconducting dipole magnets and the first design of the machine parameters and lattice. The CERN Council approved the LHC in 1994. At that time it was proposed to build the machine in two energy stages due to limited funding. Strong support for the LHC from outside the CERN member states was found (Canada, India, Japan, USA and Russia will contribute with manpower and money), and the CERN Council decided in 1996 to approve the LHC to be built in only one stage with 7 TeV beam energy. The LHC accelerator is being constructed in collaboration with laboratories from both member and non-member states and regular beam operation shall start in 2007.

Particle physics requires a luminosity for the LHC in the order of 10^{34} cm^{-2} s^{-1}, a significant challenge for the accelerator. Assuming a total cross section of about 100 mbarn for proton-proton collisions, the event rate for this luminosity is in the order of 10^9

events/second. The challenge for the LHC experiments to cope with this rate is discussed in other lectures.

For the construction of an accelerator with the complexity of the LHC many disciplines in physics and engineering are required: electromagnetism and relativity, thermodynamics, mechanics, quantum mechanics, physics of nonlinear systems, solid state and surface physics, particle physics and radiation physics, and vacuum physics. Engineering disciplines mainly involved in the LHC are civil and mechanical engineering, cryogenics, electrical engineering, automation and computing.

CERN and the LHC are introduced in the first lecture. The basis of accelerator physics required to derive the LHC main parameters is briefly discussed. The layout of the machine, the main parameters and the limitations are presented.

The first part of the second lecture is devoted to LHC technology. The second part addresses the operation of the accelerator and the associated risks due to an unprecedented amount of energy stored in the magnet system and in the beams.

The lectures are intended for non-specialists. It is not possible to discuss all systems for the LHC in these lectures. For more information, the LHC accelerator is well documented in many publications. A design report was published in 1995 (LHC Study Group 1995), and an updated design report is in preparation. There are also many textbooks on accelerator physics, for example the proceedings of the CERN Accelerator School (Turner S. 1994). A reference book on accelerator physics and technology is the "Handbook of accelerator physics and engineering" (Chao A., Tigner M. 2002).

2   CERN and the LHC

CERN is the leading European institute for particle physics. It is located close to Geneva across the French Swiss border. There are 20 European member states, 5 observer states, and many other states participating in the research at CERN.

CERN has a unique expertise in the construction of hadron accelerators. The CERN Proton Synchrotron (CPS) was built in the fifties and is still operating. The first proton-proton collider, the ISR (Intersecting Storage Ring), was built at CERN. Two continuous proton beams were colliding at a momentum of about 30 GeV/c. Later, bunched proton and antiproton beams were accelerated from 26 GeV/c to 315 GeV/c and brought into collision at the SPS (Super Proton Synchrotron) (Schmidt R., Harrison, M. 1990), where the Z_0 and W bosons were discovered. Most of this infrastructure will be re-used for the LHC project.

The LHC will be installed in a tunnel with a circumference of 26.8 km that was previously used for the Large Electron Positron collider (LEP). LEP was operating between 1989 and 2000. The maximum momentum of 104 GeV/c (centre-of-mass energy about 208 GeV) was reached after an upgrade of the RF system by installing a large number of superconducting cavities. In 2001, the LEP equipment was removed preparing the tunnel for the LHC. Installation of LHC equipment in the former LEP tunnel started in 2003. Major CERN milestones relevant to the LHC project are given in Table 1.

Beams for the LHC machine will be prepared and accelerated by existing particle sources and pre-accelerators (Benedikt M. et al. 2001). The injector complex includes

1982	First studies for the LHC project
1983	Z_0 discovered at SPS proton antiproton collider
1985	Nobel Prize for S. van der Meer and C. Rubbia for the Z_0 discovery
1989	Start of LEP operation (Z-factory)
1994	Approval of the LHC by the CERN Council
1996	Final decision to start the LHC construction
1996	LEP operation at 100 GeV/c (W-factory)
2000	End of LEP operation
2002	LEP equipment removed (second life for sc cavities ?)
2003	Start of the LHC installation
2005	Start of hardware commissioning
2006	Injection of beam into 1/8 of LHC
2007	Commissioning with beam

Table 1. *History of CERN related to the LHC Project.*

many accelerators at CERN (see Figure 1): Linacs, Booster, LEIR (Low Energy Ion Ring) as an ion accumulator, CPS and the SPS. The beams will be injected from the SPS into the LHC at 450 GeV/c and accelerated to 7 TeV/c in about 30 min, and then collide for many hours.

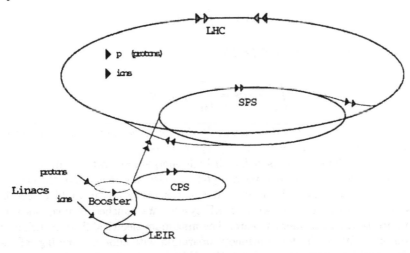

Figure 1. *Schematic view of the LHC injector complex. The beam is injected from the SPS into the LHC at a momentum of 450 GeV/c. Booster and CPS are used to prepare the bunches and to accelerate to the injection momentum of the SPS (26 GeV/c). LEIR will be used for accumulating and cooling of ions.*

The LHC beam parameters, beam sizes and beam intensities are basically determined by the performance of the injector complex. The pre-accelerators are operational and the modifications required to achieve the LHC beam parameters are essentially finished. The main part of civil engineering for the LHC is the construction of two large underground caverns for ATLAS and CMS, two transfer tunnels, each of 2.5 km length from SPS to LHC, and two tunnels for the beam dumping system.

3   Accelerator physics primer

The basic questions addressed in this chapter are: Why are protons (and not electrons) accelerated in the LHC? Why in the LEP tunnel? Why superconducting magnets?

3.1   Acceleration and deflection of charged particles

The force on a particle with charge q is given by the Lorentz equation with the electrical field $\mathbf{E}$, the magnetic field $\mathbf{B}$ and the particle velocity $\mathbf{v}$:

$$\mathbf{F} = q \cdot (\mathbf{E} + \mathbf{v} \times \mathbf{B}) . \tag{1}$$

An increase of energy is only possible by electrical fields, and not by magnetic fields that are used to deflect charged particles. The energy gain for a particle is proportional to the force along the path:

$$\Delta E = \int \mathbf{F} \cdot \mathbf{ds} . \tag{2}$$

The energy gain in an electrical field is:

$$\Delta E = \int_{s_1}^{s_2} \mathbf{F} \cdot \mathbf{ds} = \int_{s_1}^{s_2} q \cdot \mathbf{E} \cdot \mathbf{ds} = q \cdot U . \tag{3}$$

For an acceleration to 7 TeV/c a voltage U of 7 TV is required. Since it is not possible to accelerate particles to an energy above, say, some MeV in a constant potential, radio-frequency (RF) acceleration is used for all high energy accelerators. In a RF cavity a time-varying electrical field accelerates charged particles that must enter the cavity at the correct phase. A second particle coming later at the wrong phase would be decelerated. The consequence of acceleration with an RF system are bunched beams, since it is not possible to accelerate a continuous beam. The maximum electrical field in such cavities is in the order of 20-30 MV/m for continuous operation with superconducting RF cavities. Using pulsed cavities, gradients of about 40 MV/m have been achieved.

In circular accelerators (synchrotrons) the particles pass many times through the cavity and are accelerated during each passage. Dipole magnets keep the particles during the acceleration on an (approximately) circular path. The typical frequency of an RF cavity is between some 10 MHz and some 100 MHz. The LHC RF system operates at 400 MHz (Boussard D., Linnecar T. 1999). For comparison, accelerating to 7 TeV with a linear accelerator would require a length of about 350 km (!) (assuming 20 MV/m) for each beam.

Deflection of charged particles in a magnetic field is also determined by the Lorentz force:

$$\mathbf{F} = m \cdot a = q \cdot (\mathbf{v} \times \mathbf{B}) \, . \tag{4}$$

Assuming that a particle moves on a circle with radius ρ in a homogenous magnetic field perpendicular to the velocity, the Lorentz force is equal to the centrifugal force:

$$q \cdot v \cdot B = m \cdot \frac{v^2}{\rho} \, . \tag{5}$$

The radius of the accelerator is determined by the particle energy and the strength of the magnetic field:

$$\rho = \frac{E}{c \cdot q \cdot B} \, . \tag{6}$$

The bending radius of the magnets is determined by the LHC tunnel. At injection for 450 GeV/c the magnetic field is about 0.54 T. During acceleration, the field in the dipole magnets is increased to 8.33 T for the maximum momentum of 7 TeV/c. Such high field can only be achieved with superconducting magnets.

Compared to a magnetic field of the LHC dipoles of 8.33 T, a typical electrical field that can be applied for the deflection of charged particles in an accelerator is limited to about 10^7 V/m. For these numbers, the force of the magnetic field is about 300 times stronger compared to the force of the electrical field. In high energy accelerators only magnetic fields are used for particle deflection (apart from some few exceptions, for example the separation of two counter-rotating beams with opposite charge). The gravitational force is many orders of magnitude smaller than the electromagnetic force and can be neglected.

3.2 Synchrotron radiation

Electrons and protons moving in a circular accelerator emit synchrotron radiation due to the (longitudinal) acceleration by electrical fields, and due to the (transverse) acceleration when the particles are deflected. The emitted power for longitudinal acceleration is very small and can be neglected. The emitted power for a particle that is deflected is given by:

$$P = \frac{e_0^2 \cdot c}{6 \cdot \pi \cdot \epsilon_0 \cdot (m_0 c^2)^4} \cdot \frac{E^4}{\rho^2} \, , \tag{7}$$

where E is the energy, e_0 the elementary charge for an electron / proton, m_0 the rest mass, $\epsilon_0 = 8.85 \cdot 10^{-12}$ Vs/Am (dielectric constant) and ρ the deflection radius.

Due to the large difference in rest mass, the energy loss for electrons is many orders of magnitude larger than for protons. As an example, an electron at LEP at 100 GeV loses $3.8 \cdot 10^9$ eV per turn. A proton in the LHC at 7 TeV loses $8.1 \cdot 10^3$ eV per turn. The total power loss due to synchrotron radiation of the LEP beam with 10^{12} particles was in the order of $1.3 \cdot 10^7$ W. For the LHC beam with $3 \cdot 10^{14}$ protons the power is 2700 W. Although this is very small compared to LEP, the power is dissipated into a cryogenic system. The cryogenic system, described later, keeps the superconducting magnets at a temperature of 1.9 K using superfluid helium. The energy dissipation by synchrotron radiation must be taken into account for the cooling capacity of the refrigerators.

Electrons at 7 TeV in the LHC would lose their entire energy within the first deflecting dipole magnet due to synchrotron radiation!

4 LHC layout and particle oscillations

The LHC has an eight-fold symmetry with eight arc sections, and eight straight sections with experiments and systems for machine operation (see Figure 2). Two counter-rotating proton beams will circulate in separate beam pipes installed in the twin-aperture magnets and cross at four points. The total path length is the same for both beams. The beams will collide at a small angle in the centre of the experimental detectors (ATLAS, ALICE, CMS and LHCb), which are installed in four of the straight sections. In the insertions for ALICE and LHCb the injection elements are installed. The other insertions are dedicated to machine operation, two for beam cleaning, one for the beam dumping system, and one for RF and beam instrumentation.

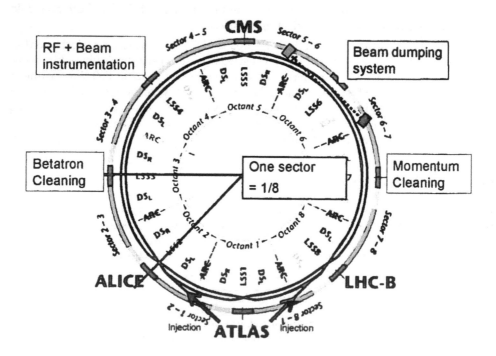

Figure 2. *Schematic layout of the LHC machine.*

The particle transport through the arcs requires both dipole magnets for deflection and quadrupole magnets for focusing. Without quadrupole magnets, two particles with slightly different angles would move apart within a very short time. Quadrupole magnets focus the particles in a way similar to lenses used in light optics. From Maxwell's equation it can be shown that a quadrupole magnet is focusing in one plane, and defocusing in the other plane. It can be demonstrated that for focusing the beams in both planes, a succession of focusing and defocusing quadrupole magnets with drift spaces in between is required, the so-called FODO structure.

Each of the LHC arcs consist of 23 regular cells, each cell with six dipole magnets to deflect the particles and two quadrupole magnets with opposite polarities to focus the beams in both planes (see Figure 3). The accelerator lattice is determined by the positions of dipole and quadrupole magnets along the accelerator. The lattice, together with the magnet strengths, determines the beam optics.

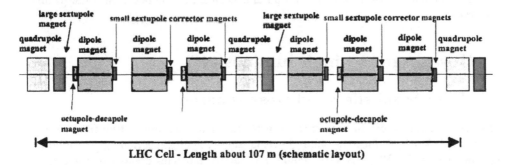

LHC Cell - Length about 107 m (schematic layout)

Figure 3. *Schematic layout of one LHC cell with dipole, quadrupole and corrector magnets seen from the inside of the LHC (not to scale).*

Using an appropriate coordinate transformation, it can be shown that with quadrupole and dipole magnets only, the particle trajectories can be stable (Turner S. 1994). In this coordinate system the particle motion is described as an harmonic oscillation due to the linear driving force of a quadrupole magnet (betatron oscillation). An important parameter of circular accelerators is the betatron tune (the number of oscillations of one particle in one turn, for each plane). For each plane, the transverse beam size along the accelerator is given by $\sigma = \sqrt{\epsilon \cdot \beta}$. The emittance ϵ_x (and ϵ_y for the vertical plane) is the area of the phase space. It is constant along the circumference and mainly determined by the chain of injectors. $\beta_x(s)$ and $\beta_y(s)$ are the beta-functions. For each plane, the beam size is proportional to the square of the β-function. The β-functions depend on the beam optics and vary along the accelerator. For the calculation of these functions the MAD computer codes are being used (Grote H., Schmidt F. 2003).

Due to magnetic field errors of higher order the movement becomes nonlinear. In the presence of nonlinearities, particle trajectories can become unstable. This can happen after a few turns, but also after many turns (even after, say, 10^6 turns). The stability of a particle trajectory depends on several parameters and decreases as the oscillation amplitude increases. The dynamic aperture is the limit of the stability region (Schmidt F. 1998) and depends on the strength of the field errors. Without any field error, the dynamic aperture would be outside the vacuum chamber. One of the challenges for the magnet production is to minimize the nonlinearities such that the dynamic aperture becomes acceptable (Brüning O., Fartoukh S. 2001). The field error of a dipole magnet is defined as the relative difference between the field at a give radius and the field in the centre. For the LHC, field errors are in the order of some 10^{-4} at a radius of 17 mm. This is still not sufficient for beam stability. Sextupole, octupole and decapole corrector magnets at the end of the dipoles correct for residual errors. For the, calculation of the dynamic aperture, particle tracking with computer codes such as MAD or SIXTRACK (Schmidt F. 1994) is performed.

To correct errors in the focusing for particles that have a slightly different momentum, typically in the order of $\Delta p/p = 10^{-3}$, sextupole magnets are installed (chromaticity correction). Since sextupole magnets are nonlinear elements, the dynamic aperture depends not only on the field errors, but also on the arrangement of the sextupole magnets for correction of chromatic effects as well as on parameters such as the betatron tunes (Schmidt F. 1998).

5 How to get to high luminosity?

5.1 Beam-beam effect and crossing angle

When two bunches cross in the centre of a physics detector, only a tiny fraction of the particles collide to produce the wanted events, since the cross-section σ_{hep} to produce a particle such as the Higgs is very small. The rate of particles created in the collisions is given by the product of cross-section and luminosity: $N = \sigma_{hep} \cdot L$. To exploit the LHC energy range, a luminosity in the order of 10^{34} cm^{-2}s^{-1} is required, which exceeds the luminosity of today's proton colliders by more than one order of magnitude. The luminosity L increases with the number of protons in each bunch, and decreases with the beam size:

$$L = \frac{N^2 \cdot f \cdot n_b}{4 \cdot \pi \cdot \sigma_x^* \cdot \sigma_y^*} \ . \tag{8}$$

where N is the number of particles in each bunch, f the revolution frequency (11246 Hz, determined by the circumference of the LHC tunnel) and n_b the number of bunches. $\sigma_{x,y}^*$ are the beam sizes in the horizontal and vertical plane at the collision point. The beam size is given by $\sigma_{x,y} = \sqrt{\beta_{x,y} \cdot \epsilon_{x,y}}$, with ϵ the transverse emittance and β the beta function at the interaction point. The transverse particle distribution is approximately Gaussian. At the collision point the beams are round. This equation is valid for head-on collisions when the bunch length is small compared to the beta function at the collision point.

The emittance of a proton beam decreases during acceleration: $\epsilon = \epsilon^*/\gamma$ (with $\gamma = E/m_0 c^2$). The beam size decreases from its value at injection σ_0 with the energy E, with E_0 the energy at injection:

$$\sigma = \sigma_0 \cdot \sqrt{\frac{E_0}{E}} \ . \tag{9}$$

The normalized emittance ϵ^* is given by the injectors and maintained during acceleration, if the machine is operating correctly. The value of the beta function at the collision point is 0.5 m, compared to a typical value of β in the arc of about 100 m. At 7 TeV, with the normalized emittance of 3.75 μm the beam size at the collision point is about 16 μm, and in the arcs about 1 mm.

The Gaussian distribution of an intense proton bunch creates a highly non-linear electromagnetic field, which modifies the trajectory of particles in the counter-rotating beam (beam-beam effect) (Evans L. 2003) (Herr W. 2003). The force on the particle is proportional to the number of protons in the other bunch, and limits the bunch intensity to about 1-2 $\cdot 10^{11}$ protons.

Inserting these numbers with the nominal bunch intensity of 1.1 $\cdot 10^{11}$ gives for one bunch in each beam a luminosity of $3.5 \cdot 10^{30}$ cm^{-2}s^{-1}. In order to achieve the required luminosity of 10^{34} cm^{-2}s^{-1}, each beam will have 2808 bunches.

5.2   Luminosity insertions

The nominal bunch spacing is 25 ns. This corresponds to a distance between two bunches of about 8 m. The beams collide in the interaction points to provide luminosity for the LHC experiments. All other interactions have to be avoided, since the beam-beam effect is detrimental for beam stability. In the insertions with experiments, the beams collide at an angle, in order to separate the bunches as efficiently as possible (see Figure 4). The crossing angle slightly reduces the luminosity. In the arcs and in the insertions without experiments, the beams are guided in two separate vacuum chambers.

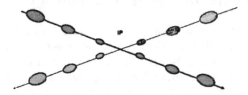

Figure 4. *Schematic view of two beams crossing at an angle. The beams have the smallest transverse size at the crossing point.*

A schematic layout of the insertions for the high luminosity experiments ATLAS and CMS is shown in Figure 5 (Ostojic R., Taylor T., Weisz S. 1997). At the collision point the two beams with dimensions of about 16 μm are crossing at an angle of about 300 μrad in order to avoid parasitic beam crossings. An inner triplet of superconducting quadrupole magnets with an aperture of about 70 mm is installed at a distance of about 20 m from the collision point. The superconducting quadrupoles must accommodate separated beams at injection, provide high field gradients (up to 250 T/m) and low multipole errors for colliding beams, and sustain considerable heat loads from secondary particles generated in the two high luminosity experiments (Lamm M. et al. 2002) (Muratore J.F. et al. 2002). For the high luminosity insertions normal conducting magnets are used to further separate the beams. For the insertions with ALICE and LHCb superconducting separation dipoles and sections with quadrupole magnets share the available space with the injection equipment. When the separation between the beams has reached 194 mm (Figure 5), a

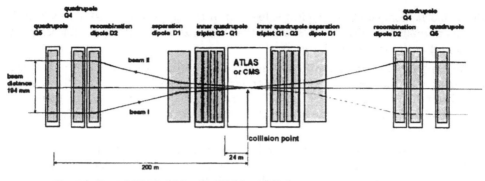

Example for an LHC insertion with ATLAS or CMS

Figure 5. *Schematic layout of an insertion for a high luminosity experiment*

superconducting dipole magnet (D2) guides the beams into the separate vacuum chambers of the outer superconducting quadrupoles (Q4 and Q5).

6   LHC parameters and challenges

6.1   LHC parameters and magnet technology

The main LHC parameters are summarized in Table 2. Two counter-rotating proton beams will be brought into collisions. Contrary to electron-positron or proton-antiproton colliders, the beams need opposite deflecting magnetic field in the arcs. The consequence is to build either two separate superconducting dipole magnets, or twin aperture magnets. Due to space constraints in the tunnel and for reasons of economy, the LHC uses superconducting twin aperture magnets.

During acceleration from 450 GeV/c to 7 TeV/c the emittance decreases proportional to the energy. The beam size is largest at injection. This determines the required size of the vacuum chamber. The cold bore of the superconducting magnets has a diameter of 56 mm. At injection, the beam fills a significant fraction of the vacuum chamber.

6.2   Challenges when operating with high beam current

To achieve the required luminosity at 7 TeV, high intensity beams with a beam current of about 0.5 A per beam are accelerated. The energy stored in each beam is about 350 MJ (Figure 6), two orders of magnitude more than for any other accelerator (Assmann R.W. et al. 2003). Since the transverse dimensions of the LHC beams are very small, the transverse energy density is even a factor of 1000 above the transverse energy density in machines such as CERN-SPS, TEVATRON at Fermilab and HERA at DESY. Three of the eight insertions are reserved for machine protection systems.

After a physics run or in case of failure, safe deposition of the beams is challenging. When operating at 7 TeV with nominal beam intensity, a very small fraction of the beam

Top energy	TeV	7
Injection energy	TeV	0.45
Dipole field at top energy	Tesla	8.33
Number of main dipole magnets		1232
Number of main quadrupole magnets		430
Number of corrector magnets		about 8000
Luminosity	$cm^{-2}s^{-1}$	10^{34}
Coil aperture	mm	56
Centre-to-centre distance between apertures	mm	194
Particles per bunch		$1.1 \cdot 10^{11}$
Number of bunches		2808
Bunch spacing	ns	25

Table 2. *Some machine parameters*

(in the order of 10^{-7}) lost in a superconducting dipole magnet would cause a quench (resistive transition in superconducting magnet). Significant particle losses are only acceptable in the cleaning insertions with normal conducting magnets. The efficiency of beam cleaning must be better than 99.98%. Only a very small fraction of the particles should be lost in the arcs and other insertions.

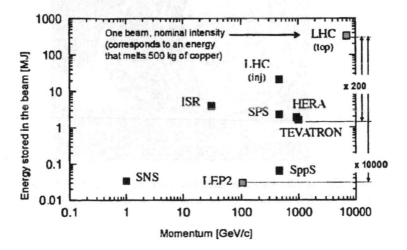

Figure 6. *Energy stored in the beams for different accelerators (courtesy R.Assmann).*

Vacuum stability has to be ensured in the beam vacuum of the arcs with a cold bore at a temperature of 1.9 K and in the insertions with sections at 1.9 K, 4.5 K and 300 K. A beam screen operating between 5 K and 20 K will be inserted in the cold bore around most of the circumference, to intercept the heat load from synchrotron radiation and other effects. The screen has a copper layer to reduce the impedance, and is also required to avoid any ion-induced instability.

A critical factor is the electron cloud instability that has been observed in several accelerators (Zimmermann F. 2002) (Ruggiero F., Rumolo G., Thomashausen J., Zimmermann F. 2002) (Cornelis K., Arduini G., Rumolo G., Zimmermann F. 2002). Depending on bunch intensity and spacing, a build-up of secondary electrons from photons emitted by the beam during the passage of the bunch train could produce heating of the chamber wall and could even cause beam instabilities. Various cures have been devised, such as minimizing the beam screen reflectivity and lowering the production yield for secondary electrons (Hilleret N. et al. 2002). After an initial time of operation the accumulated dose of electrons should have sufficiently "scrubbed" the surfaces thus reducing the secondary emission yield (SEY) to below the threshold. Increasing the bunch spacing from 25 ns to 75 ns is expected to suppress the electron cloud.

Figure 7 shows the beam in a standard arc vacuum chamber for 7 TeV. The cold bore is shown together with the beam screen and its capillaries to control the temperature between 5 K and 20 K.

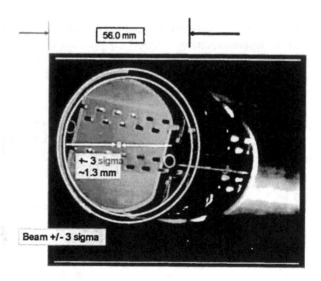

Figure 7. *View of the vacuum chamber with beam screen. The dot on the left represents the beam size at 7 TeV. The diameter of the cold bore is 56 mm.*

6.3   Challenges for LHC technology

Figure 8 shows a view of a typical tunnel cross-section in the arcs, with some of the major systems:

- The high field superconducting magnets operating at 1.9 K.

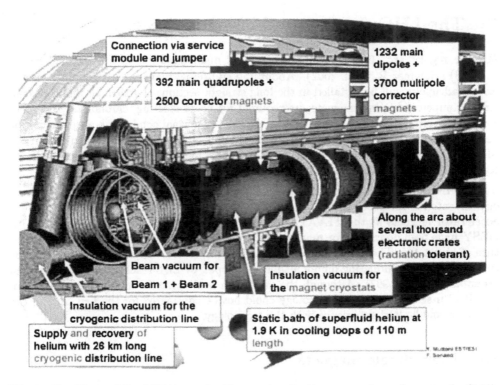

Figure 8. *View of the LHC tunnel with superconducting magnets and cryogenic distribution line.*

- The large cryogenics system distributed around the entire accelerator (Lebrun Ph. 2003). Supply and recovery of helium is done with a 26 km long cryogenic distribution line.

- Four vacuum systems: one for each beam, one insulation vacuum system for the magnets, and one insulation vacuum for the cryogenic distribution line.

- Along the arc, several thousand crates with radiation tolerant electronics are required (Rausch R. 2002), for quench protection, orbit corrector power converters and instrumentation. Instruments are for beam diagnostics, and for the vacuum and cryogenics system.

- At the end of the long arc cryostat and at the end of each other cryostat, current feedthroughs are installed in feed-boxes. Feeding the current from ambient temperature into the magnets operating at 1.9 K includes as novel technology the industrial use of High Temperature Superconducting material (Ballarino A. 2002).

7   The LHC magnet and cryogenic system

For the regular arc, 1232 main dipole and 392 main quadrupoles are required (Rossi L. 2003) (Burgmer R. et al. 2002). About 150 additional main superconducting dipoles and quadrupoles will be installed in the long straight sections. Further, several thousand smaller superconducting magnets are required for beam steering, chromaticity compensation and correction of multipole errors. About 140 normal conducting magnets will be installed in the LHC ring, mainly in the cleaning insertions, and more than 600 in the SPS-LHC transfer lines. Field errors for all magnets must be kept below specified limits to ensure sufficient dynamic aperture (Brüning O., Fartoukh S. 2001).

The LHC magnet coils are made of niobium-titanium (NbTi) cables. The technology, invented in the 1960s at the Rutherford-Appleton Laboratory, UK, was first used for accelerator magnets at the TEVATRON at Fermilab. The TEVATRON and HERA at DESY have been operating for many years with superconducting magnets at a temperature of about 4 K. The heavy ion collider RHIC at BNL uses the same technology. Some time ago, a superconducting dipole magnet developed by LBL reached 13 T using NbSn3 (Scanlan R. et al. 1997). Although the maximum field with this material is higher than for NbTi magnets, this new technology could not be considered for the LHC due to the high cost and limited experience with the technology.

7.1   Main dipole magnets

The main dipole magnets must reach a field of at least 8.33 T, and possibly 9 T after a few training quenches.

The field quality must be excellent and the magnet must be bent by about 5 mrad to follow the tunnel curvature. The geometry of the dipole magnet with the attached corrector magnets must closely follow the specifications. The magnets must be produced in time, delivered to CERN, installed in the cryostats, tested at cold temperatures, and finally installed into the LHC tunnel.

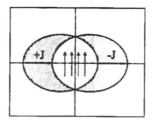

Figure 9. *Ideal current distribution for creating dipole field.*

A current density with a $cos(\theta)$ distribution produces a perfectly homogenous dipole field (Figure 9) (Mess K.H., Schmüser P., Wolff S. 1996). In practice, the ideal current distribution is approximated by NbTi conductors, thus the field has higher order multipolar field components (Figure 10). The coils of the LHC dipoles are designed with six blocks. For 8.33 T, each cable carries a current of about 12 kA. In first order, the dipole field has sextupolar, decapolar, etc. components. Mechanical imperfections during construction can lead to other multipole errors.

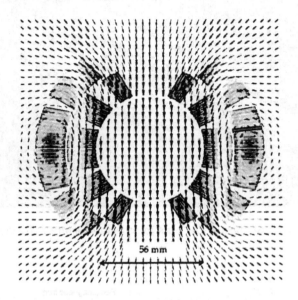

Figure 10. *Cross section of one coil in the LHC dipole magnet with the field map. The color represents the field strength, from red (adjacent to inner ring) for 8.3 T to blue/black (vortices at left and right of image) for low field. The cables have a width of 15 mm and are 2 mm thick (Russenschuck S. 1999).*

Figure 11 shows how the twin aperture dipole magnet is wound. Conductors run parallel to the beam along the vacuum chamber. The arrows indicate the direction of the current. In order to deflect the counter-rotating proton beams accordingly, the field direction in the apertures is opposite.

The cross-section of the dipole magnet installed in a cryostat is shown in Figure 12. The two beam tubes are separated by 194 mm. The cold mass inside the pressure vessel is filled with liquid helium at a pressure slightly more than one bar and cooled using a heat exchanger pipe at 1.9 K. A vacuum vessel, radiative insulation and a thermal shield at 55-75 K reduce heat in-leaks to the minimum.

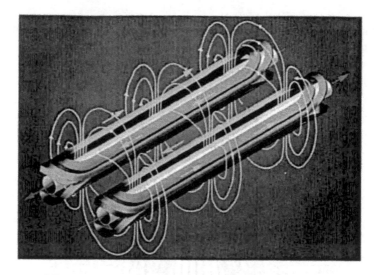

Figure 11. *Schematic view of the coils for a twin aperture LHC dipole magnet.*

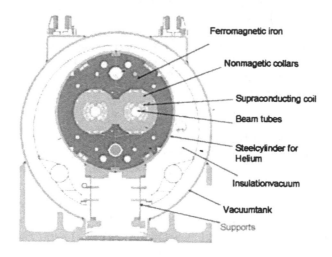

Figure 12. *Cross section of twin aperture LHC dipole magnet and cryostat.*

7.2   Superconducting cables

The high critical field of superconductor type II allows the construction of high field magnets. In 1962 superconducting NbTi wires became commercially available. The state of the superconductor can only be maintained below the "critical" values of three parameters: magnetic field, current density and temperature. Two of these parameters are determined by the material that is only superconducting below the critical temperature T_c

(at zero magnetic field), and below the critical field B_c (at very low "zero" temperature). The production process of the superconducting wires determines the third parameter, the critical current density J_c.

For the cable to be superconducting, the working point must be below the critical surface (see Figure 13). If one of the parameters exceeds the critical surface, the magnet quenches and becomes resistive. Superconducting wires are usually qualified with the critical current density for a field of 6 T at a temperature of 4.5 K. Typical values for NbTi at 4.5 K are $J_c = 2000$ A/mm^2 and 6 T.

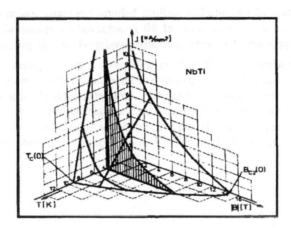

Figure 13. *Critical surface for a superconductor. For 7 TeV, the LHC dipoles operate at 1.9 K and 8.3 T.*

For NbTi magnets operating at 4.2 K with a magnetic field at 9 T-10 T, the current density is limited to unacceptably low values. Therefore the dipole magnets and most other magnets in the LHC will operate at 1.9 K, which allows much higher current density. Compared to operating at a temperature of 4.2 K, the magnetic field can be about 2.7 T higher.

The coils are wound with Rutherford type cables (Figure 14) (Adam J.D. et al. 2002). Such cable is made up of 28 (cable for the inner coil) and 36 (cable for the outer coil) individual wires (Rossi L. 2003). The typical current of such a wire at a field of 8 T and a temperature of 1.9 K is 800 A.

Figure 14. *Rutherford cable for the LHC dipole magnets.*

7.3   Cryogenic system

The LHC magnets operate in a static bath of pressurised superfluid helium at 1.9 K, cooled
by continuous heat exchange with flowing saturated superfluid helium. One cryogenic
loop extends along a lattice cell of 107 m, supplying the magnets in one cell. In each
sector, many loops are connected to the 3.3 km cryogenic distribution line (QRL) that is
connected to one of the eight cryogenic plants.

In addition to the four existing cryogenic plants from LEP (each with a cooling power
of 18 kW at 4.5 K), four new plants with about the same cooling power are being installed.

For the production of saturated superfluid helium, the compression of high flow-rates
of helium vapor over a pressure ratio of 80 is achieved by means of multi-stage cold
hydrodynamic compressors that were specially developed for this purpose.

Below a temperature of 2.17 K helium undergoes a second phase transition and becomes superfluid (helium II). The viscosity of superfluid helium enables it to penetrate
the magnet windings and make use of its very large specific heat, about 2000 times that
of the cable per unit volume. The huge thermal conductivity at moderate heat flux, 1000
times of OFHC copper, allows to transport the heat efficiently over distances up to a few
tens of meters (Lebrun Ph. 2003). Helium II has the highest thermal conductivity of any
material.

8 Operation and machine protection

Operation of the LHC will be strongly constrained due to the risks from the large stored energy in the magnet system and the beams. In the first part of this chapter the operational cycle is introduced. Safe operation of the cryogenics, vacuum, magnet and powering system is already required for the commissioning of the LHC hardware systems starting in 2005. Regular beam operation will start in 2007. Operation of the delicate hardware in presence of the high intensity beam is discussed in the second part of this chapter.

8.1 LHC magnetic cycle

The beams are accelerated in the SPS from 26 GeV/c to 450 GeV/c, and then transferred to the LHC. This energy corresponds to a magnetic field of the LHC dipoles of about 0.54 T (Figure 15) (Bottura L. et al. 1998). During the injection phase, 12 batches per beam (one batch has either 216 or 288 bunches) from the SPS are injected into the LHC (Figure 16). Injection of the two beams will take about 10 min. Then the field of the LHC dipole magnets is ramped within 25 min to 8.33 T corresponding to a momentum of 7 TeV/c. Normally, the beams will collide for several hours (physics fill). At the end of the fill, the beams will be dumped and the magnets will ramp down to prepare for the next injection. Before a new injection, the field is slightly lowered, and then ramped to injection level.

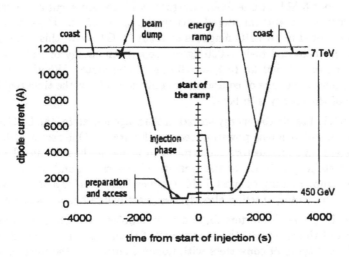

Figure 15. *Current versus time for the nominal magnetic cycle of the LHC (courtesy L. Bottura).*

8.2 Powering operation and quench protection

The energy stored in a dipole magnet is approximately proportional to the volume V inside the magnet aperture times the square of the magnetic field:

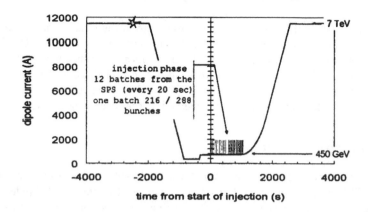

Figure 16. *LHC cycle and the injection of 12 batches from the SPS for each beam.*

$$E_{magnet} = 2 \cdot \frac{B^2 \cdot V}{\mu_0}. \tag{10}$$

This yields about 5 MJ for one dipole magnet at 12 kA. An accurate calculation with the magnet inductance L and the stored energy $E_{magnet} = 0.5 \cdot L \cdot I^2$ yields 7.6 MJ. The magnetic energy stored in all LHC magnets is about 10 GJ. With this amount of energy it is possible to heat up and melt 12000 kg of copper. An energy of 10 GJ is equivalent to the energy content of 500 kg fuel, or 400 kg of chocolate. Whether an accidental energy release leads to damage of equipment mainly depends on the time required for the transformation of the energy into heat.

In order to safely handle the energy stored in the magnet system, the LHC magnets are powered in several independent powering sectors (Figure 2). One sector includes several subsectors, the arc, in some sectors the triplets to focus the beams into the interaction point, and the matching sections for the transition between arc and triplets. This allows for progressive commissioning of most LHC systems (magnets, cryogenics, vacuum and powering), sector by sector, starting already two years before beam operation.

There are eight power converters for the main dipoles, each powering 154 dipoles in one arc. Control of the current with an accuracy of about one ppm is required (King Q. et al. 2001). Several types of converters with nominal current in the range between 60 A to 13 kA have been developed and are under construction by industry.

During the magnetic ramp the energy is taken from the powering grid. The time for normal discharge (ramping down the magnets) is about the same as for ramping up and the energy is delivered back to the grid.

As it has been discussed above, the operational margin of a superconducting magnet has to be respected. With a given magnetic field and current density in the cable, the temperature must not exceed the critical temperature, otherwise the magnet quenches. There are several mechanisms that could lead to a quench. Quenches are initiated by an energy deposition in the order of mJ (this corresponds to the energy of 1000 protons at 7 TeV deposited into a small volume). Other mechanisms for inducing a quench are movements of the superconductor by several μm (friction and heat dissipation), or a failure in the cooling system.

Beam losses must be kept below the quench level. At injection energy, the magnet would quench after an instantaneous loss of about $5 \cdot 10^9$ protons, at top energy of only about $5 \cdot 10^6$ protons (see Figure 17) (Jeanneret J.B. 2003).

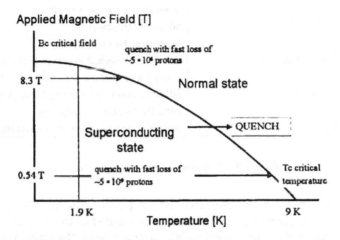

Figure 17. *Operational margin of a dipole magnet at injection and at 7 TeV.*

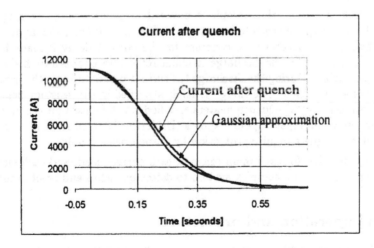

Figure 18. *Current decay in a dipole magnet after a quench.*

Without protection, after a quench, the temperature in the resistive zone would increase within less than one second to 1000 K, and the magnet would be destroyed. Since quenches cannot be avoided, sophisticated protection systems are required (Dahlerup-Petersen K. et al. 1999).

Beam lifetime	Beam power into accelerator components for one beam	Comment
100 h	1 kW	healthy operation
10 h	10 kW	operation acceptable
		collimators must absorb large part of beam losses
12 min	100 kW	operation only possibly for short time,
		collimators must be very efficient
1 sec	330 MW	failure of equipment
		beam must be dumped
15 turns	several 100 GW	failure of D1 normal conducting dipole magnet
		detect beam losses, beam dump as fast as possible
1 turn	several TW	failure at injection or during beam dump,
		potential damage of equipment,
		passive protection relies on collimators

Table 3. *Lifetime of the LHC beams.*

Firstly, the quench has to be detected and the power converter has to be switched off. The energy of the quenching magnet is then distributed in the magnet coils by firing quench heaters, mounted on the magnet coils that force-quench the superconductor. After such a quench at 7 TeV, the current in a dipole magnet decays with a time constant of about 200 ms (see Figure 18). The temperature of the magnet coil increases by some ten Kelvin.

If the magnet in an electrical circuit is powered in series with other magnets, the energy stored in all magnets needs to be extracted. For the LHC arc, 154 dipoles magnets are in one circuit. The natural time constant for the current decay is many hours. To reduce the time constant for the discharge to about 100 s, a resistor is switched into series with the magnet string. During the discharge the current passes through the bypass diode of the magnet that quenched. The energy is safely absorbed in the resistors, heating eight tons of steel to about 300 °C. Quench heaters and diodes will be used for the protection of the LHC main dipole and quadrupole magnets (Dahlerup-Petersen K. et al. 2001). Many other stand-alone magnets require protection by quench heaters only.

Individual protection for each main magnet, each current lead, and each circuit with corrector magnets requires a complex system to detect quenches and react accordingly.

8.3   Beam operation and protection

It is assumed that beams with nominal parameters collide at 7 TeV. A single beam lifetime larger than 100 hours corresponds to particle losses equivalent to a power of about 1 kW for each beam. This is far below the cooling capacity of the cryogenic system, even if all power would go into the 1.9 K system.

During colliding beam operation with a luminosity of $10^{34} cm^{-2} s^{-1}$, the lifetime of the beam is dominated by collisions. 10^9 protons / second are lost per beam and per experiment in the high luminosity insertions. This corresponds to a power of about several kW locally deposited in the high luminosity insertions. Part of the power goes into the close-by superconducting quadrupoles that have been designed to accept high heat load. Heavy shielding for these insertions is required.

For decreased lifetime, particles must be captured by collimators in the cleaning insertions to limit particle losses and the associated heat load into the superconducting magnets. Below a lifetime of about 12 min, the heat load on the collimators would become unacceptable and the beams must be dumped.

In case of equipment or operational failures, the lifetime could further decrease. A failure of a normal conducting magnet D1 in two of the insertions with physics experiments (ATLAS and CMS) is most critical. The deflection by the D1 magnet installed at a location with a large β-function of more than 4000 m leads to a fast change of the closed orbit. Since the collimators are very close to the beam, protons in the tails of the distribution would touch the collimator jaws already after several turns. A failure of the normal conducting D1 magnets leads to losses within very short time, which corresponds to 15 turns (one turn = 89 μs). The beams must be dumped as fast as possible.

9 LHC Protection systems

9.1 Beam losses into material

Beam impact in material produces particle cascades. The temperature increases with the energy deposition that depends mainly on the material and on the number and energy of the particles. The energy deposition of a particle as a function of material and the geometry is calculated with programs such as FLUKA (Fasso A. et al. 2003). Using the specific heat, the temperature increase in the material is calculated.

Already small beam losses into superconducting magnets could lead to a quench. Quench recovery to re-establish the conditions for beam operation will take several hours. For large beam impact the material could be damaged. It could melt or lose its performance (decrease of mechanical strength) leading to long and costly repairs. For example, the exchange of one superconducting magnet would take about 5 weeks.

A simple approximation for the temperature increase in material for a 7 TeV beam impact is given in the following example: For copper, the maximum longitudinal energy deposition for one 7 TeV proton at about 25 cm inside the material is $E_{dep} = 1.5 \cdot 10^{-5}$ J/kg (calculation with FLUKA). The energy required to heat and melt copper is $E = 6.3 \cdot 10^5$ J/kg. Assuming a pencil beam, the number of particles required to damage copper is in the order of $N = 4.2 \cdot 10^{10}$. For graphite, the number is about one order of magnitude larger. The 7 TeV beam with $3 \cdot 10^{14}$ protons could melt about 500 kg of copper.

9.2 Beam dumping system

The only element in the LHC that can absorb the full energy of the 7 TeV beam without being damaged is the beam dump block. At the end of a physics fill or in case of failure the beams are extracted into this block: Fast kicker magnets with 15 modules and a rise-time of about 3 μs deflect the beam by an angle of 260 μrad (Figure 19). The extracted beam is further deflected by septum magnets towards the beam dump block.

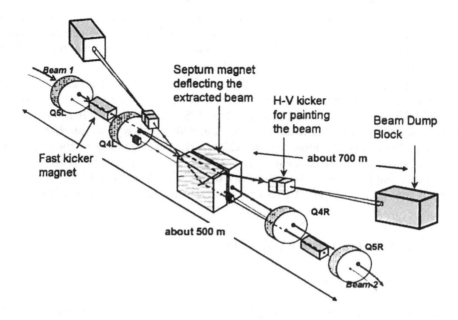

Figure 19. *Layout of the beam dumping system (courtesy M. Gyr).*

The average beam size at 7 TeV is about σ=0.2 mm. The beam dump block has a graphite core and a concrete shielding (Figure 20). Even this block could not stand such a small beam. The beam size increases in the 700 m long extraction line. In addition, the beam is blown up by two additional sets of kicker magnets deflecting the beam in horizontal and vertical direction. The rise-time of these dilution magnets is about one turn. The projection of the beam on the beam dump block is shown in (Figure 21) (Bruno L., Péraire S. 1999) (Zazula J., Péraire S. 1997). For nominal beam parameters, the maximum temperature in the beam dump block is expected to be in the order of 700 °C.

During the rise of the extraction kicker, several bunches would be deflected by an angle between zero and 260 μrad. These bunches would oscillate with large amplitudes and could hit machine elements. In order to avoid quenches or even damage, the circulating beam has a particle free abort gap with a length of 3 μs corresponding to the kicker rise-time. The extraction kicker is triggered to increase the field during this gap when there are no particles.

A failure that cannot be excluded is spontaneous firing of one beam dump kicker module. The other 14 kicker modules would be immediately triggered after such failure, but about 94 bunches deflected up to an amplitude of 4 σ (σ is the beam size) would not be extracted correctly.

- The bunches having received the smallest kick would travel through the machine, come back after one turn and then be deflected by the dump kicker in the second turn. These bunches could have a large offset at the second deflection, and absorbers must ensure that no equipment is damaged.

- About 10 bunches would receive a kick such that they reach the cleaning insertion and hit the collimators (Assmann R.W. et al. 2003).

- Bunches that are deflected by a larger angle would hit a beam absorber in the beam dump insertion (TCDQ, installed in front of the Q4R).

- Some 40 bunches would hit an absorber in front of the septum magnet (TCDS).

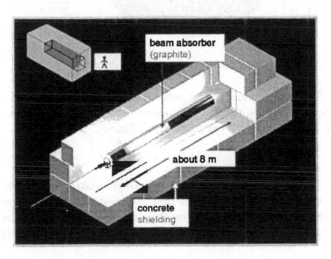

Figure 20. *Layout of the beam dump block (courtesy L.Bruno).*

9.3 Beam cleaning system

Collimators in the beam cleaning system are absorbers that should capture all particles with large amplitudes that could be lost around the accelerator, in particular into the superconducting magnets. The LHC will be the first accelerator requiring collimators to define the mechanical aperture throughout the entire cycle. For efficient beam cleaning, the collimators are adjusted to a position of 5-9 σ away from the beam centre (Figure 22). For operating at 7 TeV with nominal beam parameters, more than 99.98% of the protons in the beam halo should be captured in the cleaning insertions to minimise protons impacting on the cold magnets (Assmann R.W. et al. 2003). For the first year(s) of operation, the parameters are relaxed due to operation with beams below nominal intensity.

In case of equipment failures, collimators will be the first device to intercept the beam and must absorb part of the energy until the beams are extracted.

In one of the two insertions for beam cleaning, collimators in locations with non-zero dispersion catch protons with momentum deviations that are too large. In the

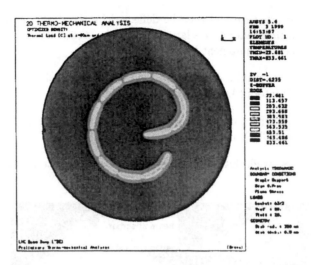

Figure 21. *Temperature map of the beam dump block after full beam impact. The maximum temperature in the center of the spiral (indicated in red) is about 800 °C. The diameter of the block is about 70 cm (courtesy L. Bruno).*

other insertion, collimators capture protons with large betatron amplitudes. Downstream from a primary collimator (the collimator that is closest to the beam), several secondary collimators catch the protons scattered by the primary collimator. If the beam lifetime would exceptionally drop to 0.2 h, the power deposition in the cleaning section could reach 0.5 MW.

Collimator jaws are blocks of solid materials. Primary collimators intercept the pri-

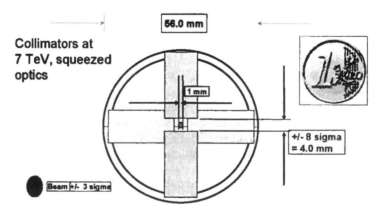

Figure 22. *Typical setting of the collimators for nominal luminosity at 7 TeV. The opening between the jaws has the same size as Spain on a one Euro coin (Assmann R. W. 2003).*

mary halo with an impact parameter in the order of some μm (Figure 23). They scatter the protons into the so-called secondary halo that is intercepted by secondary collimators downstream, with an impact parameter of about 0.2 mm. Most of the protons are absorbed, but there is some leakage to a tertiary halo.

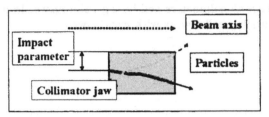

Figure 23. *Impact of beam halo on collimator (courtesy R. Assmann) (Assmann R. W. 2003).*

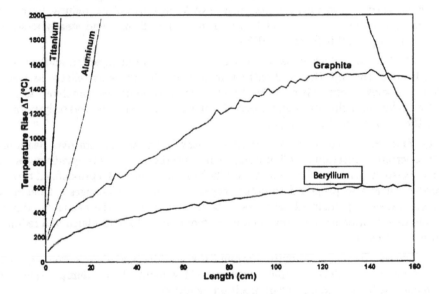

Figure 24. *Temperature increase after about 20 bunches at 7 TeV hit the collimators versus length of the jaw (courtesy A. Ferrari) and (Assmann R. W. et al. 2003).*

Initially it has been suggested to use copper or aluminium for the collimator jaws. Extensive studies of possible failure scenarios showed that up to about 10 bunches at 7 TeV can impact on a collimator jaw in case of beam dump pre-firing, or for a failure at injection. The increase of the temperature (see Figure 24) for a material such as aluminium or copper is not acceptable. Only very light material with low number of protons in the nucleus (low-Z material) survives. This is the motivation to use graphite as the work-horse for the collimator jaws during the LHC start-up.

A drawback of graphite is its high impedance. For beam stability, conducting materials are preferred. More exotic materials with lower impedance were also considered: copper loaded graphite, graphite that is partially copper plated, etc. Studies are ongoing to relax possible constraints due to the impedance being too large. Beryllium is not favored since it has insufficient mechanical properties and it is toxic.

The collimators for cleaning and protection will be installed in several phases. For the start-up of the LHC, all collimators will have graphite jaws. In a second phase, it is planned to install additional collimators with metal jaws.

9.4   Strategy for machine protection

Beam losses due to a failure can occur in a single turn, or during many turns and are therefore classified as one turn failures (ultra-fast losses), multiturn failures (either very fast losses in less that 5 ms, or fast losses in more than 5 ms), and steady losses (one second or more) (Schmidt R. et al. 2003b).

Ultra-fast losses: Such losses could be due to a failure at injection and extraction. The hardware of the injection and extraction systems should be as reliable as possible. However, it is known from other accelerators that failures cannot be completely excluded. Protection relies on collimators and beam absorbers that need to be correctly positioned close to the beam to capture the particles.

Very fast losses: For failures that lead to very fast beam movements. Beam losses close to aperture limitations (collimators, beam absorbers, low-beta quadrupoles, etc.) will be detected by fast beam loss monitors. The beam is dumped if a pre-defined threshold is exceeded. It is also proposed to detect rapid beam position changes. If beam orbit movements exceed a predefined value, say, about 0.2 mm / ms, the beams are dumped. This could detect failures earlier than beam loss monitors and is considered as a redundant system for protection.

Fast Losses: Beam loss monitors around the machine and signals from equipment in case of hardware failure for many systems will generate beam dump requests and complement fast beam loss and beam position monitors.

Steady losses: the beam losses and heat load at collimators will be monitored. A beam dump is also being considered if the beam lifetime becomes unacceptably low.

For detecting the beam losses, ionisation chambers installed close to the collimators and other aperture limitations will monitor the flux of secondary particles continuously (Gschwendtner E. et al. 2002). Although failures of magnets (quenches) or power converters could lead to fast beam losses within ten turns (Kain V. et al. 2002), enough time is available for detection and dumping the beams.

The beam dump requests from the beam loss monitors and from many other systems (RF, vacuum, powering, personnel access, ...) will be collected by the beam interlock system (Schmidt R. et al. 2003a). The interlock system including electronics distributed around the LHC will transmit such requests within less than 60 μs to the beam dumping system.

10 Outlook

Construction of the Large Hadron Collider is well under way, and planning of the commissioning has started. Civil construction for the LHC is about to be completed. 30-40 different types of superconducting magnets and some types of normal conducting magnets are being produced. Series production of the magnets is ongoing in European industry, and in collaboration with institutes in Canada, India, Japan, Russia and the USA. The LHC is the first large CERN accelerator project with a strong participation of external institutes to the machine construction. Essential components both for the accelerator and for the transfer lines are provided by collaborating institutes.

The LHC String (an arc cell with two quadrupole and six dipole magnets) demonstrated that a system of LHC superconducting magnets at a temperature of 1.9 K can be reliably operated (Saban R. et al. 2002). The String operated for several years, the magnets were quenched more than 100 times and many thermal cycles were performed.

The commissioning of the first extraction system to send the beam from the SPS into the transfer line started recently. Most of the magnets for the transfer lines have been built at BINP / Novosibirsk and are at CERN ready for installation. Commissioning of the first transfer line is foreseen for 2004.

Installation of the LHC components into the former LEP tunnel has started. Services like electricity, ventilation etc. are being installed. Installation of the cryogenic distribution line started in summer 2003. Magnet installation will start in 2004. Injection into the first sector of the LHC is foreseen for 2006. It is planned to finish the machine installation and to start operation with colliding beams in 2007.

Before the first protons will be injected into the LHC, many of the technical systems will have been commissioned: cryogenics, vacuum, powering, quench protection and interlocks. A period of about two years is foreseen for this so-called hardware commissioning. It is anticipated that at the beginning of beam commissioning the majority of the teething troubles of the complex hardware have been overcome.

The status of the entire production is regularly updated[1]. Later, this will also include the status of magnet installation.

After a meeting of the external Machine Advisory Committee (Machine Advisory Committee 2002) the chairman concluded: "The LHC is a global project with the worldwide high-energy physics community devoted to its progress and results. As a project, it is much more complex and diversified than the SPS or LEP or any other large accelerator project constructed to date".

Acknowledgments

The material presented in these lectures reflects a tremendous amount of excellent work by many colleagues, inside and outside CERN. In addition, the help of those who helped writing this report is gratefully acknowledged. I also appreciate my colleagues for providing me with a number of figures.

---

[1]http://lhc-new-homepage.web.cern.ch/lhc-new-homepage/DashBoard/ index.asp

Finally I would also like to thank the organizers of the school for their hospitality and creating a stimulating environment. In particular, thanks to P. Soler for his help and encouragement during the write-up.

References

Adam J.D. et al. (2002), 'Status of the LHC superconducting cable mass production', *IEEE Trans. Appl. Supercond.* **12**(1), 1056–1062.

Assmann R.W. (2003), Private communication.

Assmann R.W. et al. (2003), Designing and building a collimation system for the high-intensity LHC beam, *in* 'PAC 2003', Portland, USA. May 2003.

Ballarino A. (2002), HTS current leads for the LHC magnet powering system, Technical Report CERN-LHC-Project-Report-608, CERN, Geneva.

Benedikt M. et al. (2001), Performance of the LHC injectors, *in* '18th Int. Conf. High Energy Accelerators, HEACC 2001', Tsukuba, Japan.

Bottura L. et al. (1998), LHC main dipoles proposed baseline current ramping, Technical Report CERN-LHC-Project-Report-172, CERN. 23 March 1998.

Boussard D., Linnecar T. (1999), The LHC superconducting RF system, Technical Report CERN-LHC-Project-Report-316, CERN, Geneva.

Brüning O., Fartoukh S. (2001), Field quality specification for the LHC main dipole magnets, Technical Report CERN-LHC-Project-Report-501, CERN.

Bruno L., Péraire S. (1999), Design studies of the LHC beam dump, Technical Report LHC Project Note 196, CERN.

Burgmer R. et al. (2002), 'Launching the series fabrication of the LHC main quadrupoles', *IEEE Trans. Appl. Supercond.* **12**(1), 287–290.

Chao A., Tigner M. (2002), *Handbook of accelerator physics and engineering*, World Scientific Publishing Co, Singapore.

Cornelis K., Arduini G., Rumolo G., Zimmermann F. (2002), Electron cloud instability in the SPS, *in* 'EPAC 2002, 3-7 June, 2002', La Villette, Paris, France.

Dahlerup-Petersen K. et al. (1999), The protection system for the superconducting elements of the large hadron collider at cern, *in* '18th Biennial Particle Accelerator Conference PAC 1999', New York City, NY, USA. 29 March-2 April, 1999.

Dahlerup-Petersen K. et al. (2001), Energy extraction in the CERN LHC: a project overview, *in* '13th Int. Pulsed Power Conf. PPPS2001', Nevada, USA.

Evans L. (2003), Beam physics at LHC, Technical Report CERN-LHC-Project-Report-635, CERN, Geneva.

Fasso A. et al. (2003), The physics models of FLUKA: status and recent development, *in* 'CHEP 2003', La Jolla, California.

Grote H., Schmidt F. (2003), MAD-X, an upgrade from MAD8, *in* 'PAC 2003, 12-16 May, 2003', Portland, OR, USA.

Gschwendtner E. et al. (2002), Beam loss detection system of the LHC ring, *in* 'EPAC 2002, 3-7 June, 2002', La Villette, Paris, France.

Herr W. (2003), Features and implications of different LHC crossing schemes, Technical Report CERN-LHC-Project-Report-628, CERN, Geneva.

Hilleret N. et al. (2002), The variation of the secondary desorption yield of copper during electron bombardment, *in* 'EPAC 2002, 3-7 June, 2002', La Villette, Paris, France.

Jeanneret J.B. (2003), Collimation at LHC, *in* 'HALO 2003', Montauk, Long Island, New York. 19-23 May 2003.

Kain V. et al. (2002), Equipment failure, *in* 'EPAC 2002, 3-7 June, 2002', La Villette, Paris, France.

King Q. et al. (2001), The all-digital approach to LHC power converter current control, *in* 'ICALEPS 2001', San Jose, CA, USA.

Lamm M. et al. (2002), Production and test of the first LQXB inner triplet quadrupole at Fermilab, *in* 'EPAC 2002, 3-7 June, 2002', La Villette, Paris, France.

Lebrun Ph. (2003), Large cryogenic helium refrigeration system for the LHC, Technical Report CERN-LHC-Project-Report-629, CERN, Geneva.

LHC Study Group (1995), The Large Hadron Collider: Conceptual Design, Technical Report CERN/AC/95-05 (LHC), CERN.

Machine Advisory Committee (2002), Recommendations of the 11th meeting of the machine advisory committee for the LHC. March 2002.

Mess K.H., Schmüser P., Wolff S. (1996), *Superconducting accelerator magnets*, World Scientific, Singapore.

Muratore J.F. et al. (2002), Test results for initial production LHC insertion region dipole magnets, *in* 'EPAC 2002, 3-7 June, 2002', La Villette, Paris, France.

Ostojic R., Taylor T., Weisz S. (1997), System layout of the low-β insertions for the LHC experiments, Technical Report CERN-LHC-Project-Report 129, CERN.

Rausch R. (2002), Electronic components and systems for the control of the LHC machine, *in* 'EPAC 2002, 3-7 June, 2002', La Villette, Paris, France.

Rossi L. (2003), The LHC superconducting magnets, Technical Report CERN-LHC-Project-Report-660, CERN, Geneva.

Ruggiero F., Rumolo G., Thomashausen J., Zimmermann F., ed. (2002), *ECLOUD'02 - Mini-Workshop on Electron-Cloud Simulations for p and e$^+$ Beams*, CERN Yellow Report CERN-2002-001, CERN. http://slap.cern.ch/collective/ecloud02/.

Russenschuck S. (1999), ROXIE: A computer code for the integrated design of accelerator magnets, Technical Report CERN-LHC-Project-Report-276, CERN, Geneva.

Saban R. et al. (2002), First results and status of the LHC Test String 2, *in* 'EPAC 2002, 3-7 June, 2002', La Villette, Paris, France.

Scanlan R. et al. (1997), 'Preliminary test results of a 13 Tesla niobium tin dipole', *Applied Superconductivity* pp. 1503-1506. Netherlands, 30 June - 3 July 1997.

Schmidt F. (1994), SIXTRACK version 1.2: user's reference manual, Technical Report CERN-SL-94-56-AP, CERN.

Schmidt F. (1998), Nonlinear single particle issues for the LHC at CERN, *in* 'HEACC 1998, 7-12 September, 1998', Dubna, Russia.

Schmidt R. et al. (2003a), Beam interlocks for LHC and SPS, *in* 'ICALEPCS 2003', Gyeongju, Korea. 13-17 October 2003.

Schmidt R. et al. (2003b), Beam loss scenarios and strategies for machine protection at the LHC, *in* 'HALO 2003', Montauk, Long Island, New York. 19-23 May 2003.

Schmidt R., Harrison, M. (1990), The performance of proton antiproton colliders, *in* 'EPAC 1990, 12-16 June 1990', Nice, France.

Turner S., ed. (1994), *Proceedings 5th General CERN Accelerator School (CAS), CERN 94-01, January 1994*, Vol. 2. http://www.cern.ch/Schools/CAS/.

Zazula J., Péraire S. (1997), Design studies of the LHC beam dump, Technical Report CERN-LHC-Project-Report 112, CERN.

Zimmermann F. (2002), Two-stream effects in present and future accelerators, *in* 'EPAC 2002, 3-7 June, 2002', La Villette, Paris, France.

General Purpose Detectors at the LHC

Tejinder S. Virdee

CERN, Switzerland, and Imperial College, London, UK

1 Introduction

Although the Standard Model of particle physics has so far been tested to exquisite precision it is considered to be an effective theory up to some scale $\Lambda \sim$ TeV. The prime motivation of the LHC is to discover the nature of electro-weak symmetry breaking. The Higgs mechanism is the currently favoured mechanism. However, there are alternatives that invoke more symmetry, such as supersymmetry, or invoke new forces or constituents such as strongly broken electroweak symmetry breaking, technicolour etc. An as yet unknown mechanism is also possible. Furthermore, there are high hopes for discoveries that could pave the way towards a unified theory. These discoveries could take the form of supersymmetry or extra dimensions, the latter requiring modification of gravity at the TeV scale. There is something magic about the TeV energy scale. The energy of the LHC has been chosen in order to study this energy scale. The wide range of physics that is potentially possible requires a very careful design of the detectors.

At the LHC, the SM Higgs provides a good benchmark to test the performance of a General Purpose Detector (GPD). The decay modes generally accepted to be promising for detection of the Standard Model (SM) Higgs boson are listed in Table 1. The current lower limit on the mass of the Higgs boson from LEP is 114.5 GeV. Fully hadronic final states dominate the branching ratios (BR) but unfortunately these cannot be used to discover the Higgs boson at the LHC due to the large QCD backgrounds. Hence the search is conducted using final states containing isolated leptons and photons despite the smaller branching ratios. In the mass interval 114 – 130 GeV the two-photon decay is the only channel likely to give a significant signal. The Higgs boson should be detectable via its decay into two Z bosons if its mass is larger than about 130 GeV (one of the Z's may be virtual). For $2m_Z < m_H < 600$ GeV the ZZ decay mode resulting in 4 leptons is the mode of choice. In the region $700 < m_H < 1000$ GeV the cross-section decreases so that higher BR modes involving jets or E_T^{miss} have to be employed.

Region 1: Intermediate mass region (LEP limit 114.5 GeV $< m_H < 2\, m_Z$)
$m_H < 120$ GeV: pp $\to$ WH $\to$ $\ell\nu$ bb or tt H $\to$ $\ell\nu$X bb
$m_H < 150$ GeV: H $\to$ $\gamma\gamma$, $Z\gamma$
$130 < m_H < 2\, m_Z$: qq $\to$ qqH with H $\to$ $\tau\tau$, H $\to$ WW etc
$130 < m_H < 2\, m_Z$: H $\to$ WW* $\to$ $\ell\nu\ell\nu$
$130 < m_H < 2\, m_Z$: H $\to$ ZZ* $\to$ $\ell\ell\ell\ell$
Region 2: High mass region ($2\, m_Z < m_H < 700$)
H $\to$ ZZ $\to$ $\ell\ell\ell\ell$
Region 3: Very high mass region ($700 < m_H < 1$ TeV)
H $\to$ ZZ $\to$ $\ell\ell\nu\nu$, H $\to$ ZZ* $\to$ $\ell\ell$jet-jet
H $\to$ WW $\to$ $\ell\nu$jet-jet

Table 1. *The Promising Higgs Boson Decay Modes.*

The dominant Higgs-boson production mechanism, for masses up to $\sim$ 700 GeV, is gluon–gluon fusion via a t-quark loop. The W–W or Z–Z fusion mechanism becomes important for the production of higher-mass Higgs bosons. Here, the quarks that emit the Intermediate Vector Bosons (IVBs) have transverse momenta of the order of the W and Z masses. The detection of the resulting high-energy jets in the forward regions ($2.0 \lesssim |\eta| \lesssim 5.0$ where η is pseudorapidity) can be used to tag the reaction, improving the signal-to-noise ratio and extending the mass range over which the Higgs can be discovered.

2 Requirements at the LHC

2.1 Search for the Higgs boson and detector requirements

We now consider in detail channels that are demanding of detector performance and extract the detector performance requirements.

The $\gamma\gamma$ Decay of the Higgs Boson

The natural width of the Higgs boson in the *intermediate-mass* region ($m_Z < m_H < 2m_Z$) is small ($\sim$ 10 MeV) and the observed width of a H$\to$ $\gamma\gamma$ signal will be entirely dominated by the instrumental two-photon mass resolution. For the best mass resolution, a good photon energy and angular resolution is needed.

At LHC the longitudinal position of the interaction vertices will be distributed with a root mean squared (r.m.s.) spread of about 53 mm. At high luminosity, the position of the correct interaction vertex has to be determined reliably and with sufficiently high efficiency (each bunch crossing contains $\sim$ 17 interactions). In order to reconstruct the angle of emission of the photons with the required precision, there are two possibilities: the photon shower position can be measured at two depths in the electromagnetic calorimeter or the vertex can be localized using the difference in 'hardness' of tracks in the minimum bias events as compared to those arising from Higgs boson production.

A good mass resolution would not be essential were it not for a substantial background.

There are two major sources of background, an irreducible one from events containing two real photons and a reducible one in which one or both of the photons is fake.

The reducible backgrounds are from the bremsstrahlung process, gq → γq followed by q → γq, and the jet processes gq → γq (q → jet) and pp → jet-jet where the jet or jets are mis-identified as single photons. The di-jet cross-section is about seven orders of magnitude larger than the di-photon cross-section, so the probability of jets faking isolated photons must be reduced to a very low level if the di-jet background is not to swamp the H → γγ signal. The fake photons come predominantly from single π^0s, carrying a large fraction of the jet p_T, which deposit all of their energy in a small area of the electromagnetic calorimeter. Much of the rejection power against jets comes from the fact that the probability for the fragmentation that leads to a jet dominated by a single π^0 is small. A further rejection factor can be obtained by making an isolation cut on the photon candidates, but the remaining background is still uncomfortably close to the irreducible two-photon background.

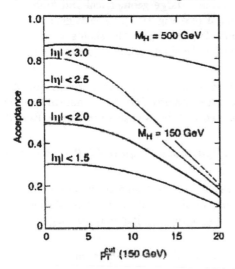

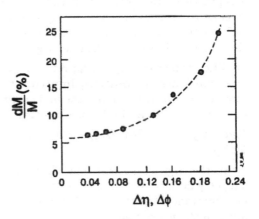

Figure 2. *The fractional jet-jet mass resolution as a function of calorimeter tower size.*

Figure 1. *The geometric acceptance for leptons in the reaction $H \rightarrow ZZ^* \rightarrow 4\ell^{\pm}$*

In view of the large uncertainties in calculating the jet background, it is essential to be able to reduce the jet background by a further factor of about ten. This can only be achieved by the recognition of π^0s as two-photon clusters. Extensive simulations and some measurements suggest that an average rejection factor of about three is possible on each photon candidate using electromagnetic calorimeters or preshower detectors with fine lateral granularity.

The process H → γγ can be selected in the trigger with reasonable efficiency by requiring two isolated high- p_T photons with a threshold $p_T \gtrsim 20$ GeV.

The detector requirements can be summarized as: good electromagnetic energy resolution, measurement of photon direction or correct localization of the primary interaction vertex, π^0 rejection and efficient photon isolation.

The $H \to ZZ^* \to 4\ell^\pm$ Decay of the Higgs Boson

If the mass of the Higgs boson is less than $2m_Z$, it can decay by emitting one real and one virtual Z. In this final state, the leptons typically have low p_T, of the order of 5 – 50 GeV (all four leptons should be isolated). The natural width of such a Higgs boson is small and the observed width of the signal will be dominated by the detector resolution.

The major source of background comes from $t\bar{t}$ production with charged leptons from the t-quark decay (isolated) and second-generation b-quark decays (non-isolated). This background can be reduced by requiring one lepton pair, consistent with arising from a Z so it is desirable to have a di-lepton mass resolution that matches Γ_Z. Another source of background is from events with a $Zb\bar{b}$ final state. Isolation cuts are efficient against both the above backgrounds for rejecting leptons from b-quark decays. Further rejection power can be obtained if precision vertexing is available, rejecting leptons that originate from displaced secondary vertices or have significant impact parameters.

The expected number of signal events is small so a large geometrical and kinematic acceptance for the detection of leptons is essential (see Figure 1). The trigger for this channel will require one or more high- p_T leptons. Typical trigger thresholds are $p_T > 30$ GeV for the single-electron trigger, $p_T > 20$ GeV for the single-muon trigger and $p_T > 15$ GeV for each lepton in di-lepton triggers.

The detector requirements can be summarized as: good dimuon (or di-electron) mass resolution, implying good momentum (energy) resolution for low momenta muons (electrons), a large geometric acceptance and efficient lepton isolation at high luminosities.

Search for the SM Higgs Boson in the High mass Region ($700 < m_H < 1$ TeV)

In this region, the reactions likely to provide the clearest signals involve neutrinos and/or jets from W or Z decays. Hence, good missing E_T and di-jet mass resolution will be important. The jets from W and Z will be boosted and the jets may be close to each other in the $\eta - \phi$ space.

In addition to the energy resolution, the mass resolution depends on the angular resolution, $d\theta$, in defining the jet axes and is given by:

$$\frac{dM}{M} = \frac{p_T}{M}d\theta . \tag{1}$$

Highly boosted di-jets (e.g. from boosted Zs from $H \to ZZ$) will have a significant contribution from the angular error. Hence, fine lateral granularity in the hadron calorimeter is required to resolve the two boosted jets from W/Z decays and attain a good mass resolution. A lateral granularity of $\Delta\eta\Delta\phi \sim 0.1 \times 0.1$ would be sufficient as can be seen in Figure 2.

In experiments at e^+e^- machines, the jet energy resolution can be improved by using the centre of mass energy to constrain the energies of jets, provided the jet directions are measured relatively precisely.

The detector requirements can be summarized as: good missing-E_T and di-jet mass resolution This requires hadron calorimeters with a large geometric coverage and with fine lateral segmentation.

Higgs Production using Vector Boson Fusion

There is a sizeable production cross-section (~10%) for Higgs boson via the vector boson fusion process qq→qqH. The vector bosons (IVB) are radiated from the initial state quarks, which get a typical p_T kick of $m_{IVB}/2$. As these quarks carry high energy, they fragment as narrow jets in the forward region (i.e. at high rapidities). A large rapidity coverage is therefore required to detect these forward jets to 'tag' this IVB fusion processes. Most of these jets are detected by the very forward calorimeters. Since these jets are highly boosted, their transverse size is similar to that of a high-energy hadron shower.

Tagging jets are also needed for the study of strong $W^{\pm}W^{\pm}$, W^+W^-, WZ and ZZ scattering. Typical cuts at LHC would be : high p_T isolated charged leptons ($p_T \geq 40$ or 100 GeV), no central jets (with $E_T \geq 30$ or 60 GeV in $|\eta| \leq 3$) and a single tagging jet (with $E \geq 0.8$ to 1.5 TeV in $3 \leq |\eta| \leq 5$).

The detector requirements can be summarized as: hadron calorimeter geometric coverage up to $|\eta| < 5$ with sufficiently fine lateral granularity to reconstruct narrow jets above large background from MB events in the forward region $(3< |\eta| <5)$.

New Massive Vector Bosons

The detector requirements for high momenta can be ascertained by considering decays of high mass objects such as $Z' \rightarrow e^+e^-$ or $\mu^+\mu^-$. For masses above 1 TeV the main background comes from the Drell-Yan process. The discovery of objects like Z' will be rate limited and hence a good mass resolution for both the electron and muon channels is desirable. Good momentum resolution is required only in the central region ($|\eta| \leq 2$) since the rapidity plateau shrinks with increasing mass. Ways of distinguishing between different models involve the measurement of the natural width and the forward-backward asymmetry both of which require sufficiently good high p_T momentum resolution ($\Delta p_T/p_T \approx 0.3$ for $p_T \approx 1$ TeV) to determine the sign of the leptons.

The detector requirements can be summarized as: good high p_T momentum resolution that is sufficient to unambiguously determine the charge of leptons in the region $|\eta| <2$.

Supersymmetric Higgs Bosons

In addition to the channels used in the search for a SM Higgs boson the other channels that have been studied are listed in Table 2.

$h, H, A \rightarrow \tau^+\tau^- \rightarrow (e/\mu)^+ + h^- + E_T^{miss}$ or $\rightarrow e^+ + \mu^- + E_T^{miss}$ or $\rightarrow h^+ + h^- + E_T^{miss}$
$H^+ \rightarrow \tau^+\nu$ from $t\bar{t}$
$H^+ \rightarrow \tau^+\nu$ and $H^+ \rightarrow t\bar{b}$ for $M_H > M_{top}$
$A \rightarrow Zh$ with $h \rightarrow b\bar{b}$; $A \rightarrow \gamma\gamma$
$H, A \rightarrow \chi_2^0\chi_2^0, \tilde{\chi}_i^0\chi_j^0, \tilde{\chi}_i^+\tilde{\chi}_j^-$
$H^+ \rightarrow \tilde{\chi}_2^+\chi_2^0$

Table 2. *List of channels studied for SUSY Higgs boson in addition to those for SM Higgs boson.*

It can be seen that most decay chains contain the lightest SUSY particle, leading to significant E_T^{miss} in the final state and an abundance of b-jet or τ-jet production.

Hence, in addition to the requirements mentioned above, the additional requirements of efficient b/τ-jet tagging and triggering on τ's are important.

3   The Large Hadron Collider project

As outlined above, a search for new particles has to be made at a centre-of-mass energy of $\sim$ 1 TeV. An exploratory machine is therefore required to cover the diverse possible signatures and mechanisms outlined above. A hadron (proton-proton) collider is such a machine as long as the proton energy is high enough and the luminosity is sufficiently large. The most interesting and easily detectable final states involve leptons and photons with a low $\sigma \cdot$BR. The hadron colliders can provide these conditions but at the expense of 'clean' experimental conditions.

The energy and luminosity requirements can be ascertained by considering the reaction in which a 1 TeV Higgs boson is produced via the WW fusion production mechanism. Each radiated W should have an energy of about 0.5 TeV implying that the radiating quark should carry an energy of $\sim$ 1 TeV and hence the proton about 6 $\times$ 1 TeV. This roughly defines the energy of the beam: the LHC accelerates protons to 7 TeV. The luminosity can be estimated using the same reaction H $\rightarrow$ ZZ $\rightarrow$ 4 charged leptons. In order to register 10 such events/year the luminosity has to be $L = 10/(\sigma \cdot \text{BR} \cdot \Delta t) = 10/(10^{-37} \times 10^{-3} \times 10^7) \sim 10^{34}$ cm^{-2} s^{-1}.

3.1   The parameters of LHC

As the LHC accelerator has to be installed in an already built tunnel, and the energy is defined to be 7 TeV; the dipole bending magnetic field can be calculated from the formula $p(\text{TeV}) = 0.3B(\text{T}) \times R(\text{km})$, giving $B = 8.4$ T. The LHC machine comprises about 1230 dipole magnets, with radio-frequency (R.F.) cavities giving an acceleration kick of $\sim$ 0.5 MeV per turn. The luminosity is given by:

$$L = \frac{\gamma f \, k_B \, N_p^2}{4\pi \, \varepsilon_n \, \beta^*} F \,, \tag{2}$$

where f is the revolution frequency, k_B is the number of bunches, N_p is the number of protons/bunch, ε_n is the normalized transverse emittance, β^* the betatron function and F the reduction factor due to the crossing angle. The values of these parameter and others are shown in Table 3.

The first pp collisions at the LHC are expected to occur in mid-2007, although the commissioning of the machine will start earlier. The current timeline for the first year of operation of the machine and the experiments, could be as follows (t_0=April 2007):

Energy at collision	E	7	TeV
Dipole field at 7 TeV	B	8.33	T
Luminosity	L	10^{34}	$cm^2 s^1$
Bunch separation		24.95	ns
No. of bunches	k_b	2835	
No. particles per bunch	N_p	1.1	10^{11}
Normalized transverse emittance (r.m.s.)	ε_n	3.75	μm
Collisions			
β−value at IP	β^*	0.5	m
r.m.s. beam radius at IP	σ^*	16	μm
Luminosity lifetime	τ_L	10	hr
Number of evts/crossing	n_c	17	

Table 3. *The parameters of the LHC.*

- 1st Beam to 1st Collisions (Pilot Run):

  - Single Beam: t_0 to t_0 + 2 mo. (months). Set-up the machine for safe operation.
  - Colliding Beams:

    * t_0 + 2 mo. to t_0 + 3 mo.: One bunch on one bunch to 43 × 43 bunches.
    * t_0 + 3 mo. to t_0 + 4 mo.: Increase number of bunches to attain L ∼few × 10^{32} cm^{-2} s^{-1}.
    * Experiments: Synchronization, set-up for physics running, 'pilot' physics.

  - Shutdown (3 mo.).

- Physics Run:

  - t_0 > 7 mo. until 5-10 fb^{-1} have been accumulated for ATLAS and CMS each.

4 The experimental challenge at the LHC

As outlined above, in the search for high-mass objects with rare signatures, high $\sqrt{s}$ and high luminosity are required. The main LHC machine parameters (proton-proton mode) are a centre of mass energy of $\sqrt{s}$ = 14 TeV, design luminosity of L=10^{34} cm^{-2} s^{-1}and a bunch crossing interval of 25 ns. The large proton-proton inelastic cross-section ($\approx$ 70 mb) leads to some 10^9 interactions/sec. These parameters lead to formidable experimental challenges (Ellis N and Virdee T S, 1994).

At $\sqrt{s}$ = 14 TeV the total pp cross-section is σ_{tot} ∼ 105 mb, comprising $\sigma_{elastic}$ ∼ 28 mb, $\sigma_{single\ diffractive}$ ∼ 12 mb, $\sigma_{double\ diffractive}$ ∼ few mb and $\sigma_{inelastic}$ ∼ 65 mb. Hence, the detectors see an inelastic event rate of:

$$\text{Event Rate} = L \cdot \sigma = 10^{34} \times 65 \times 10^{-27} s^{-1} = 6.5 \times 10^8 s^{-1} \ .$$

Since only 2835 bunches out of the possible 3564 bunches are full, the number of events per crossing ~ 20 at design luminosity.

The event selection (trigger) must reduce the billion interactions/s to $\approx 10^2$ events/s for storage. The short bunch-crossing period has implications for the design of the readout and trigger systems. It is clearly not feasible to make a trigger decision in the time between bunch crossings, yet new events occur every crossing and a trigger decision has to be made for every crossing. This requires relatively *pipelined* trigger processing and readout, where many bunch crossings are processed concurrently by a chain of processing elements. The Level-1 trigger decision takes $\approx 3 \mu s$ so the data must be stored in pipelines for $\approx 3 \mu s$.

At design luminosity a mean of ≈ 20 minimum bias events will be superposed on the event of interest. Around 1000 charged tracks emerge from the interaction region every 25 ns. Thus, the products of an interaction under study may be confused with those from other interactions in the same bunch crossing. This problem, known as *pileup*, clearly becomes more severe if detectors with a response time longer than 25 ns are used. The effect of pileup can be reduced by using highly granular detectors with good time resolution, giving low *occupancy* (fraction of detector elements that contain information) at the expense of having large numbers of detector channels.

The high particle fluxes emanating from the interaction region lead to high radiation levels, requiring radiation hard detectors and front-end electronics. The expected radiation levels in CMS for the various detectors are given in Tables 4 and 5 for an integrated luminosity of 5×10^5 pb^{-1}, corresponding to the first 10 years of running.

It is clear from the above that the LHC detectors cannot be just larger versions of the previous generation of HEP detectors. A major R&D effort in particle detectors was required during the last decade to select and develop detectors and electronics that could survive and operate reliably in the harsh environment of the LHC. For each technology chosen for a given detector, several others were investigated. Development of new particle detectors takes a long time and goes through many phases starting from the idea or concept, intensive R&D, prototyping, systems integration, installation and commissioning and finally data taking. This can be illustrated using one of the many detector technologies in the LHC experiments.

Radius (cm)	Fluence of fast hadrons (10^{14} cm^{-2})	Dose (kGy)	Charged Particle Flux (cm^{-2}s^{-1})
4	32	840	10^8
11	4.6	190	
22	1.6	70	6.10^6
75	0.3	7	
115	0.2	1.8	3.10^5

Table 4. *Hadron fluence and radiation dose in different radial layers of the CMS Tracker (barrel part) for an integrated luminosity of 500 fb^{-1} ($\sim$ 10 years).*

Pseudorapidity η	ECAL Dose (kGy)	HCAL Dose (kGy)	ECAL Dose Rate (Gy/h)
0 - 1.5	3	0.2	0.25
2.0	20	4	1.4
2.9	200	40	14
3.5	-	100	-
5	-	1000	-

Table 5. *Radiation dose in CMS Calorimeters for an integrated luminosity of 500 fb^{-1} ($\sim$10 years).*

We have chosen the electromagnetic calorimeter of CMS, using lead tungstate scintillating crystals, as an illustrative example.

- Idea: in 1992 some yellowish crystal samples of a few cc were shown - the final volume required is 10 m^3!!

- R&D: 1993-1998: much work was carried out to increase the size of crystals (both length and cross-section), requiring improvements of growing techniques for heavier and heavier crystals, the transparency of crystals, the radiation hardness requiring optimizing the stoechiometry (fraction of lead oxide and tungsten oxide), purity of raw materials (balance cost versus level of purity), compensation of defects by specific doping (production crystals are now doped with Yttrium and Niobium).

- Prototyping: 1994-2002: performance of larger and larger matrices of crystals was studied in test beams. The information obtained was fed back into the R&D e.g. radiation damage at low dose rate ($\sim$ 1 Gy/hr) was observed in 1997.

- Mass manufacture: 1998-2006: a sizeable and high-yield crystal growing capability had to be put in place – around 130 ovens had to be refurbished with computer-control. The diameter of the boules has been steadily increased so that now two crystals per boule can be obtained.

- Systems integration: 2003-2006: the crystals are assembled into the specially developed light and thin mechanical structures, the electronics, cooling pipes and signal and voltage cables have to be integrated into the each one of 36 phi 'super-modules' comprising 4 'baskets' each, or endcap 'Dees'.

- Data Taking: 2007.

It can be seen that almost one and half-decades will have passed from the concept to physics data taking!

5 Physics tools and algorithms

Experimentally, the programme of physics outlined above involves the study of hard particle interactions, determining the identity of the resulting particles and measuring their momenta or energies with as high a precision as possible. Some thirty years ago a single detection device, the bubble chamber, was sufficient to reconstruct the full event information. At the current high centre of mass energies, no single detector can accomplish this even though the number of particles whose identity and momenta can be usefully determined is limited [electrons, muons, photons, jets, b-jets, tau-jets and missing transverse energy $E_T^{miss}(\nu)$]. This leads to a familiar cylindrical onion-like structure of present day high energy physics experiments. Each layer is designed to perform a specific task and together they allow identification and measurement of the momenta and/or energies of all particles produced in LHC proton-proton collisions.

Starting from the interaction vertex, the momenta of charged particles is determined in the inner tracker, which is usually immersed in a solenoidal magnetic field. Identification of b-jets can be accomplished by placing high spatial resolution detectors such as silicon pixel or microstrip detectors close to the interaction point. Following the tracking detectors are calorimeters, which measure the energies and identify electrons, photons, single hadrons or jets of hadrons. Only muons and neutrinos penetrate through the calorimeters. The muons are identified and measured in the outermost sub-detector, the muon system, which is usually immersed in a magnetic field. The presence of neutrinos is deduced from the apparent imbalance of transverse momentum or energy.

5.1   Particle detection

Moderately relativistic charged particles, other than electrons, lose energy in matter through their Coulomb interaction with the atomic electrons. The energy transferred to the electrons causes them either to be ejected from the parent atom (*ionisation*) or to be excited to a higher level (*excitation*). Particle detection is based on one or both of these electromagnetic processes. For example in a silicon microstrip detector a charged particle traversing the diode creates electron-hole pairs. The drift of the electrons and holes induces a signal on the electrodes (microstrips). Similarly in lead tungstate crystals, the passage of a charged particle excites the atoms in the material, which de-excite by emitting photons (scintillation) that are detected in a photodetector placed at the end of the crystal.

5.2   Characteristics of minimum bias events

In order to extract interesting physics signals at the LHC, the characteristics of the background or pileup events, termed as 'minimum bias', have to be understood. The kinematic properties of minimum bias events are illustrated in Figure 3. It can be seen that there are ~ 7 charged tracks per unit of rapidity in the central region. These tracks have a mean p_T of ~ 300 MeV/c and rarely do these events contain tracks with $p_T > 2$ GeV/c when compared with events containing hard interactions (e.g. Higgs boson production). These facts can be used to establish 'isolation' criteria. Isolation is one of the most powerful event selection tools at Hadron Colliders. Particles such as electrons, muons or photons emanating from a fundamental parton-parton interaction will tend to be isolated i.e. they are produced with no other accompanying particles. The isolation criterion can be established by requiring that no other particle with $p_T \sim 1$-3 GeV/c lies within a cone of $\Delta R \sim$ 0.2-0.4 surrounding the particle of interest. The rejection power against π^0s in jets and the accidental loss at high luminosity are shown in Figure 4.

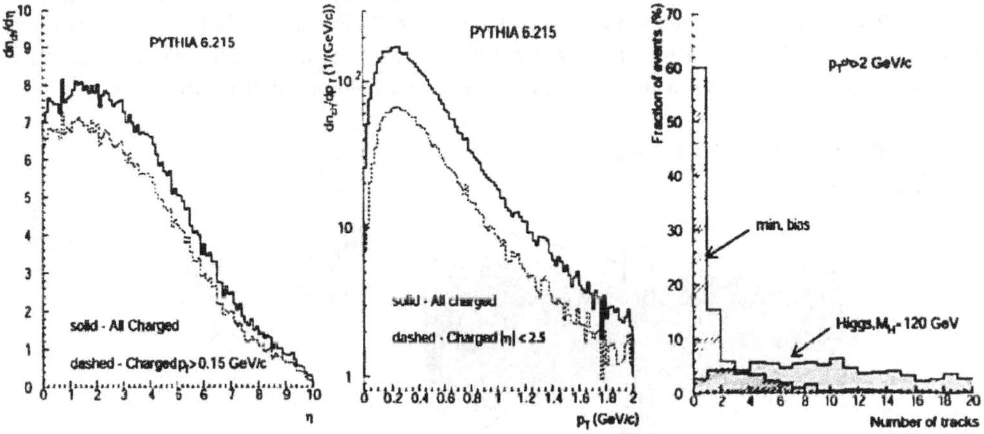

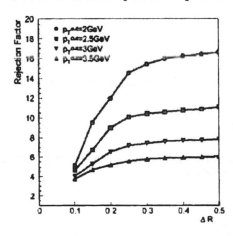

Figure 3. *Kinematic characteristics of minimum bias events.*

Furthermore, in H→ $\gamma\gamma$ events, charged tracks with $p_T > 2\text{GeV}/c$ can be used to localize the primary vertex since the probability of such a track originating from a minimum bias event is very small, thus minimizing the probability of making an error in the localization of the Higgs primary vertex.

It is also instructive to look at the energy flow due to minimum bias events. At design luminosity (10^{34} cm^{-2} s^{-1}), the transverse energy flow in an area of $\Delta\eta\Delta\phi \sim 0.1 \times 0.1$ is illustrated in Figure 5. This area corresponds to one that is typically used to measure the energy of an isolated electron or photon in the electromagnetic calorimeter. It can be seen that in central rapidities no particle enters such a region in 70% of the cases.

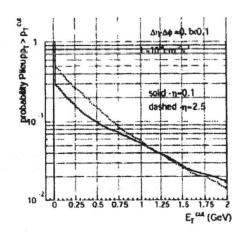

Figure 4. *The rejection power against π^0s in jets and the accidental loss at high luminosity.*

Figure 5. *The transverse energy flow in an area of $\Delta\eta\Delta\phi \sim 0.1 \times 0.1$.*

The large channel count is the key to high track reconstruction efficiency (ϵ_{trk}). Figure 6(a) shows a H→ ZZ → 4ℓ event superposed on 30 minimum bias events in CMS. The strong curling of the low p_T tracks is evident leaving the outer regions rather free of tracks. Figure 6(b) shows the same event once a cut of $p_T \geq 2$ GeV is applied. Figures 6(c) and 6(d) show a z = 12.5 cm slice of the same event in the Si tracker. The event looks much simpler and indicates how a high ε_{trk} can be attained even at high luminosities.

Overlap of 30 min bias events
H→ZZ (Z →µµ)

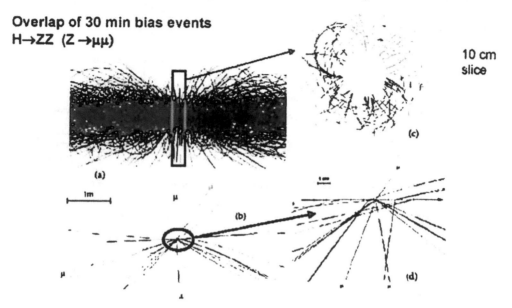

Figure 6. *A H→ ZZ → 4ℓ event superposed on 30 minimum bias events in CMS (see text).*

5.3   Measurement of momentum

Consider the motion of a charged particle in a uniform solenoid magnetic field (Figure 7). The radius of curvature, r, is given by:

$$r = \frac{p_T}{0.3B} , \qquad (3)$$

where r is measured in m, B is the magnetic field strength measured in T and p_T is the momentum perpendicular to B and measured in GeV/c.

The angle θ is given by

$$\sin\frac{\theta}{2} = \frac{L}{2r} . \qquad (4)$$

If r ≫ L then:

$$\frac{\theta}{2} \approx \frac{L}{2r} \Rightarrow \theta \approx \frac{0.3BL}{p_T} . \qquad (5)$$

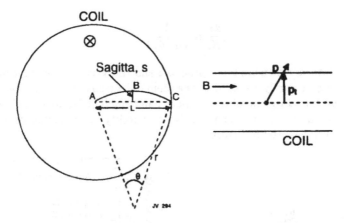

Figure 7. *The trajectory of a charged particles in a magnetic field.*

Therefore the sagitta, s, is given by:

$$s = r - r\cos\left(\frac{\theta}{2}\right) = \frac{r\theta^2}{8} \approx \frac{0.3BL^2}{8p_T} .$$ (6)

As an example, s ≈ 3.75 cm for p_T=1 GeV/c, L=1 m and B=1 T. Suppose the sagitta is measured using points A, B and C. Then the relative momentum resolution can be estimated as:

$$\frac{dp_T}{p_T} = \frac{\sigma_s}{s} = \frac{\sqrt{3}\sigma_x}{2s} = \frac{\sqrt{3}}{2}\sigma_x\frac{8p_T}{0.3BL^2} ,$$ (7)

where σ_s is the error on the sagitta and σ_x is the single point error.

Hence, the momentum resolution will degrade linearly with increasing p_T but will improve for higher field and larger radial size of the tracking cavity. The latter improvement is quadratic in L! The next question that can be asked is the arrangement of N measuring points. Uniform spacing is best for minimizing the effect of multiple scattering and the resolution is given by:

$$\frac{dp_T}{p_T} = \frac{\sigma_s p_T}{0.3BL^2}\sqrt{\frac{720}{N+4}} .$$ (8)

For example, dp_T/p_T ≈ 0.5% for p_T= 1 GeV/c, L = 1 m, B = 1 T, σ_x= 200 μm and N=10. However, in a real tracker the errors due to multiple scattering need to be included.

Multiple Scattering

The electric field close to an atomic nucleus may give a large acceleration to a charged particle. This will result in a change of direction for a heavy charged particle ($m \gg m_e$). For small particle-nucleus impact parameters a single large angle scatter is possible. This is described by Rutherford scattering and the angle θ is given by

$$\frac{d\sigma}{d\Omega} \propto \frac{1}{\sin^4 \theta/2} \cdot \tag{9}$$

Larger impact parameters are more probable and the scattering angle will be smaller as the nuclear charge is partly screened by the atomic electrons. Hence in a relatively thick material there will be a large number of random and small deflections. This is described by multiple Coulomb scattering. The relative probability of scattering as a function of scattering angle is illustrated in Figure 8.

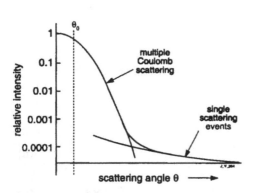

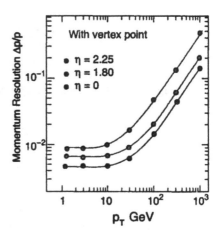

Figure 8. *The relative probability of scattering as a function of scattering angle.*

Figure 9. *Illustrative momentum resolution in an early version of CMS.*

The r.m.s. of the scattering angle is given by

$$\theta_0 \approx \frac{13.6 \text{ MeV}}{\beta pc} Z_{inc} \sqrt{\frac{L}{X_0}} , \tag{10}$$

where L is the thickness of the material in the tracker and X_0 is the radiation length of the material (both measured in m). The apparent sagitta due to multiple scattering is given by:

$$s_{ms} = \frac{L\theta_0}{4\sqrt{3}} . \tag{11}$$

If the extrapolation error from one measuring plane to the next is larger than the point resolution, i.e. $\theta_0 \Delta r > \sigma_x$, then the momentum resolution will be degraded. The relative momentum resolution due to multiple scattering is then given by

$$\frac{s_{ms}}{s} = \frac{dp}{p}\bigg|_{ms} \approx 0.05 \frac{1}{B\sqrt{LX_0}} , \tag{12}$$

since

$$s = \frac{0.3BL^2}{8p} .$$

Hence the relative momentum resolution is independent of p and is proportional to 1/B. For example, $dp/p \approx 0.5\%$ for argon gas with L=1 m and B=1 T. The estimated momentum resolution in an early version of CMS is illustrated in Figure 9. The momentum resolution is independent of momentum in the range where the multiple scattering error dominates (up to $\approx$ 20 GeV/c). The resolution, $\Delta p/p$, above 20 GeV/c is proportional to p.

5.4 Measurement of energy

Neutral and charged particles incident on a block of material deposit their energy through creation and destruction processes. The deposited energy is rendered measurable by ionisation or excitation of the atoms of matter in the active medium. The active medium can be the block itself (*totally active or homogeneous calorimeter*) or a sandwich of dense absorber and light active planes (*sampling calorimeter*). The measurable signal is usually linearly proportional to the incident energy.

An example of the phenomena (bremsstrahlung and pair production) involved in electromagnetic showers is illustrated in Figure 10.

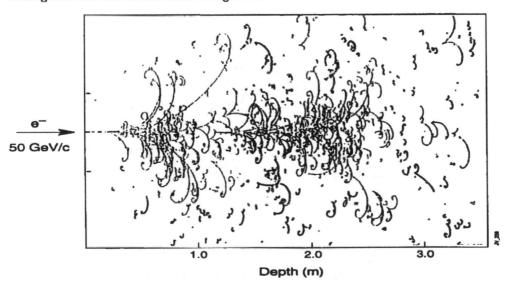

Figure 10. *An example of a 50 GeV electron shower in the Ne/H$_2$ (70%/30%) filled Big European Bubble Chamber (BEBC) at CERN. The radiation length is $\approx$ 34 cm.*

The Electromagnetic cascade

Longitudinal Development of the Electromagnetic Cascade

A high energy electron or photon incident on a thick absorber initiates a cascade of secondary electrons and photons via bremsstrahlung and pair production, as illustrated in Figure 11. With increasing depth, the number of secondary particles increases while their mean energy decreases. The multiplication continues until the energies fall below the critical energy, ε, the energy at which the loss due to ionization is equal to that from radiation process. Ionization and excitation rather than generation of more shower particles dominate further dissipation of energy.

Consider a simplified model of development of an electromagnetic shower initiated by an electron or a photon of an energy E. A universal description, independent of material, can be obtained if the development is described in terms of scaled variables:

$$t = \frac{x}{X_0} \quad \text{and} \quad y = \frac{E}{\varepsilon} \; .$$

Since in 1 X_0 an electron loses about $2/3^{rd}$ of its energy and a high energy photon has a probability of 7/9 of pair conversion, we can naively take 1 X_0 as a generation length. In each generation the number of particles increases by a factor of 2. After t generations the energy and number of particles is:

$$e(t) = \frac{E}{2^t} \quad \text{and} \quad n(t) = 2^t \; .$$

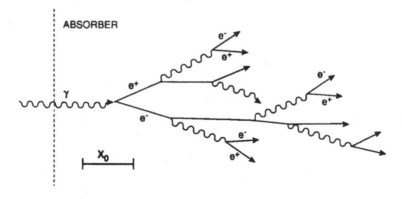

ABSORBER

Figure 11. *Schematic development of an electromagnetic shower.*

At shower maximum, where $e \approx \varepsilon$, the number of particles is

$$n(t_{max}) = \frac{E}{\varepsilon} \quad \text{and} \quad t_{max} = \ln \frac{E}{\varepsilon} = \ln y \; .$$

Hence the longitudinal depth required to contain high energy electromagnetic cascades increases only logarithmically, whereas for magnetic spectrometers this size increases as $\sqrt{p}$ for constant dp/p. The longitudinal development (Figure 12) is rather similar for different materials if the depth is expressed in units of X_0.

Lateral Development of the Electromagnetic Shower

The lateral spread of an e.m. shower is determined by multiple scattering of electrons away from the shower axis. In different materials the lateral extent of electromagnetic (e.m.) showers scales fairly accurately with the Molière radius ($R_M \sim 7A/Z$ g.cm^{-2}). An

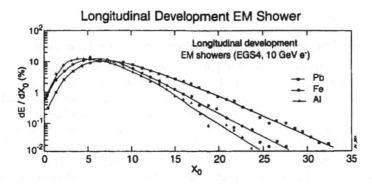

Figure 12. *Simulation of longitudinal development of 10 GeV electron showers in Al, Fe and Pb.*

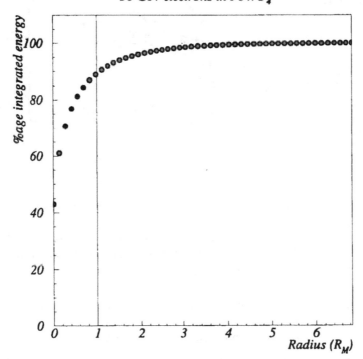

Figure 13. *The percentage of energy contained in a cylinder of lead tungstate of different radii.*

infinite cylinder with a radius of $\approx 1\ R_M$ contains $\approx 90\%$ of the shower energy. For lead tungstate, and a depth of 26 X_0, the amount of energy contained in a cylinder of a given radius is shown in Figure 13. The fact that e.m. showers are very narrow at the start, can be used to distinguish single photons from π^0s (see Section 5.13).

The hadronic cascade

hadronic cascade **The Longitudinal Development of the Hadronic Cascade**

A situation analagous to that for e.m. showers exists for hadronic showers. The interaction responsible for shower development is the strong interaction rather than electromagnetic. The interaction of the incoming hadron with absorber nuclei leads to multiparticle production. The secondary hadrons in turn interact with further nuclei leading to a growth in the number of particles in the cascade. Nuclei may break up leading to spallation products. The cascade contains two distinct components, namely the electromagnetic one (π^0s etc.) and the hadronic one ($\pi^\pm$, n, etc) one. This is illustrated in Figure 14. In contrast to electromagnetic showers there is a considerable variation from one hadronic shower to another, as illustrated in Figure 15.

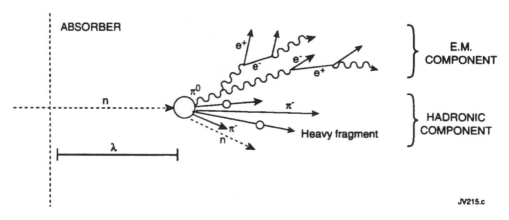

Figure 14. *Schematic development of hadronic showers.*

The multiplication continues until pion production threshold is reached. The average number, n, of secondary hadrons produced in nuclear interactions is given by $n \alpha \ln E$ and grows logarithmically. The secondaries are produced with a limited transverse momentum of the order of 300 MeV.

It is convenient to describe the average hadronic shower development using scaled variables:

$$\nu = x/\lambda \quad \text{and} \quad E_{th} \approx 2m_\pi = 0.28 \text{ GeV}, .$$

where λ is the nuclear interaction length and is the scale appropriate for longitudinal and lateral development of hadronic showers. The generation length can be taken to be λ. Note that $\lambda \approx 35A^{1/3}$ g.cm^{-2}. Furthermore, if it is assumed that $\langle n \rangle$ secondaries/primary are produced for each generation and that the cascade continues until no more pions can be produced, then in generation ν:

$$e(\nu) = \frac{E}{\langle n \rangle^\nu} , \tag{13}$$

$$e(\nu_{max}) = E_{th} \quad \therefore E_{th} = \frac{E}{\langle n \rangle^{\nu_{max}}} , \tag{14}$$

270 GeV Incident Pions in Copper

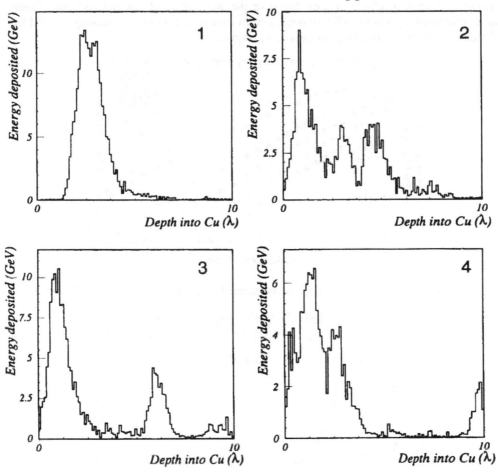

Figure 15. *A simulation of the development of four representative pion showers in a block of copper.*

$$n^{\nu_{max}} = \frac{E}{E_{th}} \Rightarrow \nu_{max} = \frac{\ln(E/E_{th})}{\ln\langle n \rangle} . \tag{15}$$

The number of independent particles in the hadronic cascades compared to electromagnetic ones is smaller by E_{th}/ϵ and hence the intrinsic energy resolution will be worse at least by a factor $\sqrt{E_{th}/\epsilon} \approx 6$. The average longitudinal energy deposition profiles are characterised by a sharp peak near the first interaction point (from π^0s) followed by a exponential fall-off with scale λ. This is illustrated in Figure 16. The maximum occurs at $t_{max} \approx 0.2 \ln E + 0.7$ (E in GeV).

A parameterisation for the depth required for almost full containment (95%) is given by $L_{0.95}(\lambda) \approx t_{max} + 2\lambda_{att}$, where $\lambda_{att} \approx \lambda E^{0.13}$. Figure 16 shows that over 9λ are

required to contain almost all the energy of high energy hadrons. The peaks arise from
energy deposited locally by π^0s produced in the interactions of charged hadrons. These
interactions take place at differing depths from shower to shower. The energy carried by
π^0s also varies considerably from shower to shower.

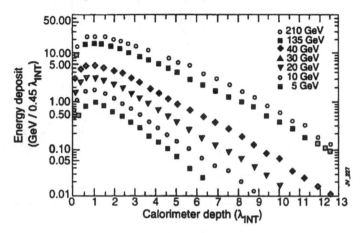

Figure 16. *Longitudinal profile of energy deposition for pion showers of different energies.*

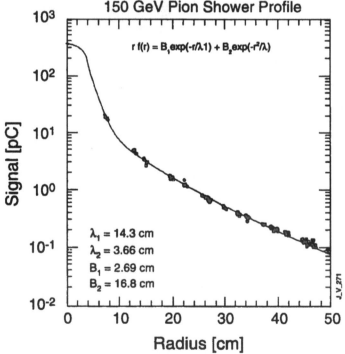

Figure 17. *The lateral profile of energy deposition of pion showers.*

The Lateral Development of the Hadronic Cascade

The secondary hadrons are produced typically with $\langle p_T \rangle \approx 300$ MeV. This is comparable to the energy lost in 1λ in most materials. At shower maximum, where the mean energy of the particles is $E_{th} \approx 280$ MeV, the radial extent will have a characteristic scale of $R_\pi \approx \lambda$. High energy hadronic showers show a pronounced core, caused by the π^0 component with a characteristic transverse scale of R_M, surrounded by an exponentially decreasing halo with scale λ. This is illustrated in Figure 17 for a lead/scintillating fibre calorimeter.

Energy Resolution

The energy resolution of calorimeters is usually parameterised as :

$$\frac{\sigma}{E} = \frac{a}{\sqrt{E}} \oplus \frac{b}{E} \oplus c .$$

where the right hand side is the square root of the quadratic sum of the three terms.

The first term, with coefficient a, is the *stochastic or sampling* term and accounts for the statistical fluctuation in the number of primary and independent signal generating processes, or any further process that limits this number. An example of the latter is the conversion of light into photo-electrons by a photo-device.

The second term, with coefficient b, is the *noise* term and includes:

- the energy equivalent of the electronics noise and

- the fluctuation in energy carried by particles, other than the one(s) of interest, entering the measurement area. This is usually labeled pileup.

The last term, with coefficient c, is the *constant* term and accounts for:

- imperfect quality of construction of the calorimeter,

- non-uniformity of signal generation and/or collection,

- cell-to-cell inter-calibration error,

- the fluctuation in the amount of energy leakage from the front or the rear (though somewhat increasing with energy) of the volume used for the measurement of energy,

- the contribution from the fluctuation in the e.m. component in hadronic showers.

The tolerable size of the three terms depends on the energy range involved in the experiment. The above parametrisation allows the identification of the causes of resolution degradation. The quadratic summation implies that the three types of contributions are independent, which may not always be the case.

5.5 Electron reconstruction

The amount of material in the ATLAS and CMS trackers averages over 50% of X_0. Electrons radiate photons (bremsstrahlung) as they traverse this material. Furthermore, as the electrons 'bend' in the magnetic field the resulting energy deposits appear as a spray, extended in the ϕ-direction in case of the barrel parts of the electromagnetic calorimeter as illustrated in Figure 18. A proper measurement of the electron energy requires the inclusion of the energy of the bremsstrahlung photons. This is clearly important as e.g. an electron with a p_T=35 GeV emits photons that carry on average $\sim$ 45 % of the initial electron energy. Dedicated algorithms are needed to include as much of the initial energy of the electron. An example of the algorithm used in CMS is outlined (see Figure 18). The energy of the electron is determined by adding energies of several clusters, nearby in ϕ-space, taking advantage of the fine η-ϕ granularity of the crystals and the knowledge of the lateral shape in the η-direction of e.m. showers. Figure 18 illustrates the effect on the energy resolution of 'bremsstahlung-recovery'. The achieved energy resolution for electrons with p_T=35 GeV can be quantified by $\sigma_{Gauss}/\mu = 1.06\%$ and $\sigma_{eff}/\mu = 2.24\%$, where $\mu = \langle E_{meas}/E_{true} \rangle$. One σ_{eff} is the width that contains 68.3% of the events in the distribution.

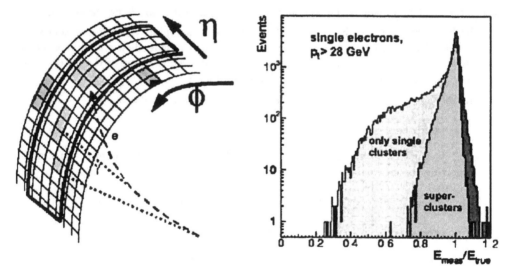

Figure 18. *Left - as electrons 'bend' in the magnetic field, the resulting energy deposits appear as a spray. Right - energy resolution before and after bremsstrahlung recovery'.*

5.6   Gamma reconstruction

Reconstructing the energy of unconverted high energy photons is relatively straightforward. In the case of CMS, the energy deposited in a matrix of 5 × 5 crystals yields a very good energy resolution. This area corresponds to a relatively small size in η-ϕ space ($\sim$ 0.09 × 0.09).

However, a substantial fraction of the photons convert in the material of the trackers. Nevertheless, for high energy photons that convert, standard electron reconstruction algorithms can be employed. The reason is illustrated in Figure 19 (Left), taken from the CMS ECAL Technical Design Report (TDR), that shows a scatter plot of the radial point of conversion and the separation of the resulting electrons. Because the energies of γs from Higgs boson decays is high, and the momenta of the electrons and positrons from conversions is also high, the separation (in ϕ) for conversions occurring at large radial distances is small ($\sim$ 5 cm equivalent to the lateral size of two crystals). In fact, for most of the conversions, except for those occurring essentially in the pixels detector, the separation remains smaller than 7 crystals. Hence the use of a matrix of 5 × 9 crystals or superclusters, as outlined above, results in a relatively good energy resolution. Figure 19 (Right) shows that the energy resolution for the converted photons is slightly worse.

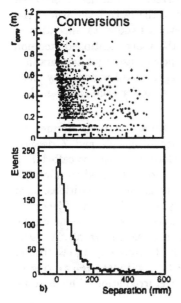
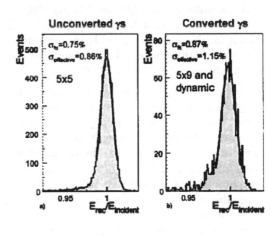

Figure 19. *Left - Radial point of conversion and the separation of the resulting electrons in CMS. Right - The energy resolution for unconverted and converted photons.*

5.7 ECAL calibration

One of the key issues in achieving a good energy resolution for high energy electrons and photons is cell-to-cell inter-calibration of the electromagnetic calorimeter. The inter-calibration error enters almost directly into the 'constant term' as most of the energy of an e.m. shower goes into a single crystal for CMS or single cell for ATLAS.

In CMS, the calibration procedure proceeds through several phases outlined below.

Only a small subset of the 36 'Supermodules' in the barrel region and 'Dees' in the forward region will be inter-calibrated in the test beams. It is hoped that after transport into the experiment pit this calibration will yield an inter-calibration error of less than 2%.

For the other Supermodules and Dees, the starting calibration will be set by using the laboratory measurements such as crystal light yield, APD response, etc. From comparative measurements made in test beams this procedure is known to yield an inter-calibration error of about 5%.

At LHC startup, a method has been developed that uses energy from minimum bias events to measure the energy flow into each crystal. Using rotational (ϕ) symmetry, each crystal at the same η should, on the average, have the same energy flow. Hence the 60,000 or so constants in the barrel part can be reduced to about 170. From a simulation using 18 million minimum-bias events, it has been shown that an inter-calibration error of about 2% can be reached. The limitation of this method is the non-homogeneity of the tracker material, which is lumpy in ϕ. With increased knowledge of the distribution of the tracker material, the precision may be improved to about 1%. However, account has to be taken of drawbacks such as non-Gaussian noise (spikes, etc) and of the large lever arm of the extrapolation from hundreds of MeV to tens of GeV. A Level-1 bandwidth allocation of 1 kHz, handled in the High Level Trigger farm, should allow the above mentioned precision (2%) in a matter of a few hours. Use of higher energy deposits, using jet triggers in a similar fashion and only considering deposits in the non-trigger-biased region, may allow to overcome the drawbacks mentioned above. The energy scale for each ring can then be established using $Z \rightarrow e^+e^-$ decays.

Once the tracker is fully operational, the precise calibration can be performed by comparing the measurement of momentum of electrons (from e.g. W and Z decays) in the tracker and their energy in the electromagnetic calorimeter, the so-called E/p matching. However, electrons that have not radiated much should be used to minimize corrections. This sample can be obtained by imposing strict isolation conditions, such as a high threshold (close to 1) of a quantity like the energy in 5×5 crystals (energy in a supercluster of crystals). It is hoped that an inter-calibration error of 0.3% should be attainable after accumulating data over a period of a month or so (see Figure 20). An issue is the deconvolution of calibration constants from the energy deposited by an electron in several crystals. Furthermore, it is imperative that there be continuous monitoring of the variation of crystal transparency due to even the low-level irradiation. This is to be accomplished by injecting laser light into each crystal.

5.8 Jets at LHC

Hadronic calorimeters, in conjunction with the electromagnetic calorimeters, are primarily used to measure the energies of jets. The most important quantities that characterize their performance are:

- Jet energy resolution and energy linearity.

- Missing transverse energy resolution.

The performance of the calorimeter system depends on the 'physics' of an object such as a jet and on detector related effects.

In the category of physics effects there are:

- The intrinsic fluctuation in the fragmentation process. For example the energy of jets is often estimated by adding the calorimeter energy contained in a cone, with half angle ΔR, where $\Delta R = \sqrt{\Delta\eta^2 + \Delta\phi^2}$ in pseudorapidity (η) and ϕ-space, and whose axis is centred on a seed cell with an energy above a pre-defined threshold. The energy (and indeed the particle content) flowing into a cone of a given ΔR fluctuates from one jet to another with the same energy.

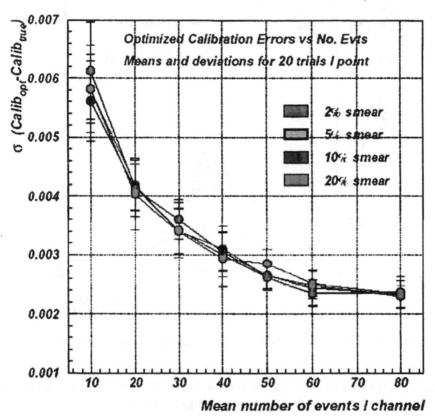

10x10 array, 40GeV electrons, L3 Iterative Algorithm

Figure 20. *The inter-calibration precision attained versus the number of events/crystal.*

- Initial and final state radiation.

- The fluctuation in energy pileup from the underlying event and/or overlapping minimum bias events at hadron colliders, especially at high luminosities.

In the category of detector related effects there are:

- Different calorimeter response to hadrons and electrons and photons.

- Non-linearities, effects of dead material, cracks between calorimeters, longitudinal leakage, electronics noise, lateral segmentation, etc.

- Magnetic field, which deflects low momenta charged tracks out of the jet 'cone'.

The impact of these effects on the performance of the calorimeter system with respect to jets can be minimized by careful design. However, simultaneously achieving the best possible electromagnetic energy resolution (< 1% at 100 GeV) and the best possible jet energy resolution (<4% at 100 GeV) is very difficult because minimizing the impact of the

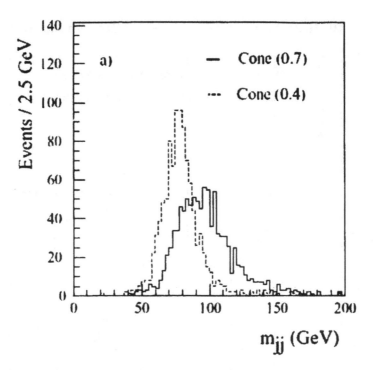

Figure 21. *Reconstructed W mass with 50 pileup events for two cone sizes.*

first effect requires a compensating calorimeter (either built-in or attained by software). Such calorimeters have poor electromagnetic energy resolution.

As mentioned above, many effects limit the precision with which jet energy can be measured and hence in both ATLAS and CMS more importance is placed on achieving a Gaussian jet-energy response and linearity than on the exact value of the stochastic term in the energy resolution function.

Jet Reconstruction

The cone algorithm is often used to reconstruct the energy of jets. An example of the performance of such an algorithm, taken from the ATLAS Physics TDR, uses the W ($\rightarrow$ jet jet) mass, reconstructed using jet energy collected in a cone of ΔR=0.4 or 0.7. A cone size of 0.7 is best to collect the energy of the fragmenting parton at low luminosities. However, when running at high luminosities, and with an average of 50 overlapping minimum bias events ($\sim$ 2 crossings integrated) the picture is somewhat different, where a cone size of ΔR=0.4 gives a better estimate of the energy (see Figure 21). Clearly, conversion from jet energy to parton energy depends on the cone size and on the number of overlapping events, i.e. on the instantaneous luminosity. Other more sophisticated algorithms can also be used.

The reconstruction of two jets from W/Z decays will play an important role in many physics channels. ATLAS has made a study of the W $\rightarrow$ jj mass resolution in different p_T^W ranges, namely < 50 GeV, 100<p_T < 200 GeV from top production or W+jets events and 200< p_T < 700 GeV from Higgs boson decays (m_H=1 TeV). Three different methods are used:

- Method 1: the mass is calculated from the four-momenta of the two massless jets.

- Method 2: the mass is calculated from the 4-momentum of each calorimeter tower ($m_{tower} = 0$) inside the two jets.

- Method 3: same as Method 2 but the energy is collected in a single cone to treat decays with severe overlap.

The results, using the fixed cone algorithm, are shown in Table 6. In the first region the jets are well apart ($\Delta R > 3$) and the reconstructed W mass is dominated by the jet energy resolution resulting in a mass resolution of 9.5 GeV at low luminosity and 13.8 GeV with pileup (after applying a > 1 GeV cut on the energies of towers included in the energy sum). The reconstructed mass is compatible with the generated W mass. In the second region the average distance between jets is $\Delta R \sim 1.5$. The energy and hence the mass resolution improves because the energies of the jets are higher. However, a low tail appears at low masses due to a bias in the reconstruction of the angle between the two jets when the opening angle is small. In the high p_T^W region, for the Ws arising from 1 TeV H decays, due to the boost the most probable separation is $\Delta R=0.4$ and hence the two jets will overlap. The cone size has to be decreased to 0.3 in order to efficiently separate the two jets. It can be reduced and resolution improved by iterating the jet direction resulting in the energy being shared between the two jets.

p_T^W (GeV)	ΔR	σ_{LowL} (GeV)	σ_{HiL} (GeV)
$p_T < 50$	0.4	9.5	13.8
$100 < p_T < 200$	0.4	7.7	12.9
$200 < p_T < 700$	0.3	5.0	6.9

Table 6. *W mass resolution in ATLAS.*

Jet Energy Measurement using Energy Flow

Considerable improvement in the jet energy measurement and resolution when using reconstructed tracks (the so called 'energy flow' algorithm) has been demonstrated in several experiments at LEP, Tevatron and HERA. The possibility to resolve and identify calorimeter clusters using fine transverse and/or longitudinal granularity is the key feature. As an example in CMS, the energy flow algorithm uses as a starting point the fixed cone algorithm but the information provided by the tracker is used in addition. Several steps can be envisaged:

(i) include the energy (in fact the measured momentum, assuming zero mass particle) of the reconstructed charged tracks that are swept out of the cone by the strong magnetic field;

(ii) replace the energy, measured in the calorimeter system, of a charged track-calorimeter matched cluster by the momentum measured in the tracker;

(iii) for clusters with poor charged track(s), the calorimeter matched energy is replaced by the measured momenta. The energy is estimated from a library of responses to charged particle showers generated by MC or created in test beams.

The results are shown in Figure 22. Considerable improvement is observed when the out of cone charged tracks are included (labeled 'calo + tracks (out)'). Further improvement is observed when steps (ii) and (iii) are implemented (labeled 'calo + tracks (all)'). The improvement is striking especially for jets at low E_T and in the ratio E_{Treco}/E_{TMC} which is close to 1.

Such algorithms will be particularly useful when reconstructing di-jet masses, especially when low E_T jets are present. Although the improvement in mass resolution is not striking, the estimated effective mass is much closer to the generated mass. Jets were generated from a test Z-like object with a mass of 100 GeV with E_T jets > 20 GeV. When the information from the tracker is used, the ratio of mean reconstructed to mean generated mass goes up from 0.88 to 1.01 and the mass resolution improves from 13.7% to 11.9%.

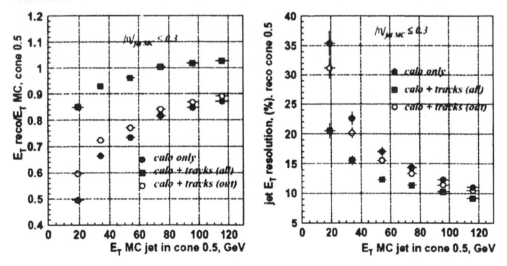

Figure 22. *The fraction of jet energy contained in the cone and the energy resolution in CMS for various conditions (see text).*

5.9   Missing E_T

A limited geometric calorimetric coverage leads to fake E_T^{miss} due to energy escaping outside the active part of the detector. Figure 23 illustrates the effect on minimum bias events, at design luminosity, of limiting the geometric coverage to $|\eta|$ <3, 5 or 10 is illustrated. CMS and ATLAS have coverage up to $|\eta|$ < 5.

The detector performance in terms of E_T^{miss} resolution has been studied by using $H/A \rightarrow \tau\tau$ events. The x- and y-components of the E_T^{miss} vector are obtained from the transverse energies deposited in all the calorimeter towers. An energy cut-off is applied to minimize the impact of electronics noise and pile-up. For such events, at the particle level, the resolution of each of the components degrades from 2.3 GeV to 8.3 GeV if calorimeter coverage is reduced from $|\eta| < 5$ to $|\eta| < 3$. In ATLAS, the resolution of each component of the E_T^{miss} vector is found to be about 7 GeV with full simulation (see Figure 24). The E_T^{miss} resolution can be parameterized as $\sigma(E_T^{miss}) = a \times \sqrt{\Sigma E_T}$, where ΣE_T is the scalar sum of E_T in the event. The parameter a is found to be around 0.45 for ATLAS and 0.65 for CMS, reflecting the poorer jet energy resolution in CMS.

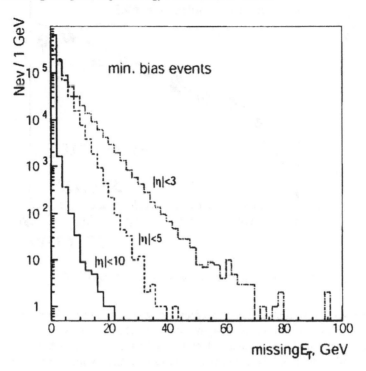

Figure 23. *The effect on E_T^{miss} of limiting the calorimeter geometric coverage.*

5.10   Identification of b-jets

As b quarks have a large mass, the charged tracks in their decays have a higher p_T with respect to the b-jet direction than the typical 0.3 GeV for lighter quarks. Furthermore, bottom hadrons have a relatively long (≈ 1.5 ps) lifetime. Hence, for the identification of b's one looks for:

- electrons or muons within a jet arising from the semi-leptonic decay of a b-quark with relatively high p_T;

- one or more charged tracks within a jet with a significant impact parameter (defined to be the distance of closest approach of the track from the primary vertex);

- a secondary vertex consistent with the flight path of a B-meson (see Figure 25).

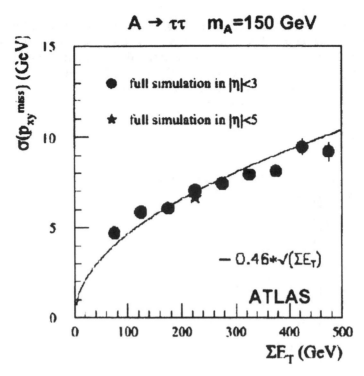

Figure 24. *x- or y- component E_T^{miss} in ATLAS.*

The second and third methods require measuring layers close to the interaction vertex. Both CMS and ATLAS have up to 3 layers of pixel detectors between r = 4 and 12 cm. The harsh and congested environment close to the interaction point has necessitated the development and use of Si pixel detectors instead of the Si microstrip detectors used at LEP. The precision with which the impact parameter, and secondary vertex, can be measured is determined by:

- the closeness of the first measuring layer from the interaction vertex,

- the number of measured points close to the interaction vertex,

- the spatial resolution of the measured points and

- the amount of material in these layers leading to the degradation in the significance of impact parameter due to multiple scattering.

The estimated impact parameter resolution for 10 GeV tracks is around 15μm. In ATLAS, b-identification proceeds as follows: a likelihood of the jet being a b-jet is formed by first estimating the significance, S_i, of the impact parameter (measured impact parameter/impact parameter resolution) for each track in the jet. Then a quantity is formed, labeled relative significance, $r_i = f_b(S_i)/f_u(S_i)$, using the probability distributions shown in Figure 26. Finally, a jet weight is estimated from $W = \Sigma \log r_i$. By keeping jets above

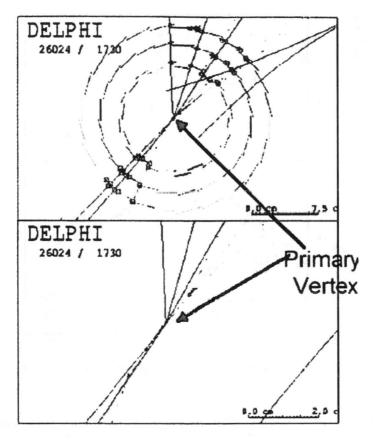

Figure 25. *Decay of a B-hadron in DELPHI.*

some value of W (which can be varied) the efficiency for different samples can be computed. For a sample of b-jets from Higgs bosons of a mass of 100 GeV or 400 GeV, a typical jet rejection against u, d and s quarks or gluons of a factor of 100 can be attained for an efficiency of b-tagging of 50 % (shown in Figure 27). In case of real data, since the jet type will not be a priori known, the rejection will have to be optimized for each specific background under study. The rejection at low p_T for a fixed ε_b worsens (a factor of 30 at 20 GeV rising to $\sim$ 500 at 100 GeV) due to a decrease in the number of charged particles (after cuts e.g. $p_T > 1$ GeV) coupled with the fact that the tracks tend to be softer, leading to a worse impact parameter resolution caused by multiple scattering. The rejection tends to get somewhat worse at $p_T > 100$ GeV due to increased charged track multiplicity (reducing the discrimination between b- and u-jets) and difficulties of pattern recognition in the denser environment, amongst other effects.

5.11 Identification of jets from hadronic τ-decays

At LHC, taus can arise from decays of massive objects such as the A or H bosons. The taus emerge from these decays with a high boost. This, coupled with a low charged

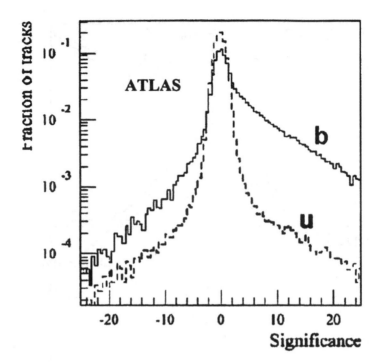

Figure 26. *Significance distribution (signed impact parameter divided by its error for b- and u-jets).*

particle multiplicity in τ-decay (in 77% of the cases only one charged track is produced), leads to a τ-jet that is narrow. The following variables can be used to distinguish τ-jets from normal QCD jets:

- R_{em} – the jet radius computed using only the e.m. cells.

- Isolation criterion, ΔE_T^{12} – the fraction of E_T in e.m. and hadronic calorimeters that is contained in a region defined by $0.1 < \Delta R < 0.2$ around the barycentre of the cluster.

- N_{tr} – number of charged tracks with a p_T above a given threshold pointing to the calorimeter cluster within $\Delta R = 0.3$.

In a study for the search of $A \to \tau\tau$, taken from ATLAS, a jet is identified as a τ-jet if it satisfies the cuts given in Table 7. Also given in the table are the cumulative efficiencies of the cuts on τ-jets from A bosons, QCD jets and jets from the main background processes. After applying these cuts, the QCD jet rejection versus τ-jet efficiency is shown in Figure 28 for different p_T ranges.

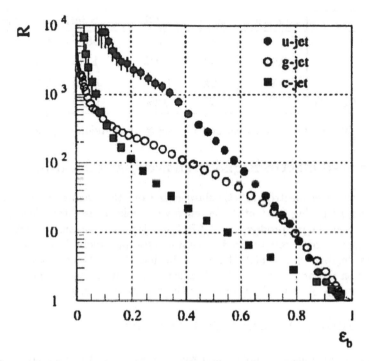

Figure 27. *Background rejection as a function of b-jet efficiency in ATLAS.*

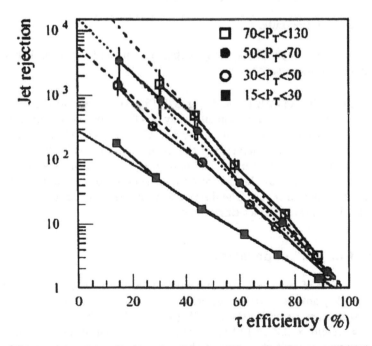

Figure 28. *Jet rejection as a function of τ efficiency in ATLAS.*

	Cut Value	A→ $\tau\tau$	QCD jets	b-jets	t-$\bar{t}$ events
$\langle p_T(\tau) \rangle$ **GeV**		73	44	58	65
$\mathbf{R}_{em}$	< 0.07	45%	1.1%	1.9%	1.3%
ΔE_T^{12}	< 0.1	32%	0.6%	0.9%	0.7%
$\mathbf{N}_{tr}$ (**p$_T$ > 2 GeV**)	1 or 3	25%	0.2%	0.2%	0.2%

Table 7. τ *identification criteria and their cumulative efficiency.*

5.12   Isolated electromagnetic shower-jet separation

The largest source of electromagnetic showers is from the fragments of jets, especially π^0s. A leading π^0 taking most of the jet energy can fake an isolated photon. There are large uncertainties in jet production at LHC and jet-fragmentation. Furthermore, the ratio of production of di-jets to irreducible di-photon background at LHC is $\approx 2 \times 10^6$ and γ-jet/irreducible $\gamma\gamma$ is ≈ 800. Hence, a rejection of ≈ 5000 against jets is needed. A certain rejection factor (≈ 20) can be obtained by simply asking for e.m. showers with a transverse energy greater than some threshold with the energy measured in a small region.

Jets can be distinguished from single electromagnetic showers by:

- demanding an energy smaller than some threshold in the hadronic compartment behind the electromagnetic one ($E_T^{had} < E_T^{cut}$),

- using other isolation cuts,

- demanding a lateral profile of energy deposition in the ECAL consistent with that from an electromagnetic shower.

Using these criteria, ATLAS (Liquid Argon TDR) estimates that the rejection factor against jets can be ≈ 1500 for a photon efficiency of 90%. This is illustrated in Figure 29, where the effect of various cuts is shown: a) $E_T^{had} < 0.5$ GeV in HCAL in $\Delta\eta\Delta\phi = 0.2 \times 0.2$, b) e.m. isolation ($R_{isol}$)– more than 90% of the energy is contained in the central 3×5 e.m. cells compared with that in the central 7×7 e.m. cells, c) lateral shower profile ($R_{lateral}$)– look for an e.m. core such that the central 4 towers contain more than 65% of the shower energy, d) shower width in $\eta(\sigma_\eta)$. The distribution for jets is shown as dashed histogram whereas the full histograms depict single photons.

5.13   Photon – π^0 separation

After the application of the above criteria, only jets resulting in leading π^0s ($z \sim 1$) can fake genuine single photons. Further rejection can only be achieved by the recognition of two e.m. showers close to each other. CMS (see the ECAL TDR) uses the fine lateral granularity (≈ 2.2 cm $\times$ 2.2 cm) of their crystals and a neural network algorithm that compares the energy deposited in each of the 9 crystals in a 3×3 crystal array with that expected from a single photon. Variables are constructed from the 9 energies, x and y

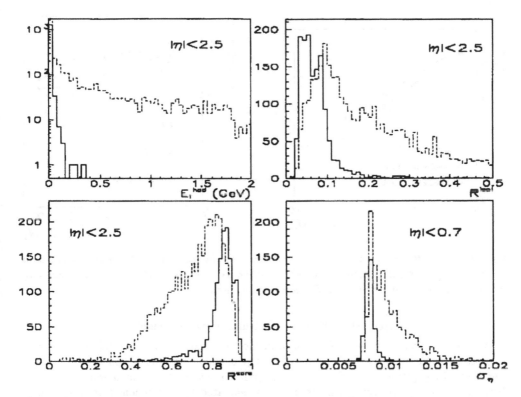

Figure 29. *The distributions used to cut against jets. Solid histogram is for photons and the dashed one for jets. See text for explanation.*

position of impact and a pair measuring the shower width. The fraction of π^0s rejected is shown in Figure 30.

The narrowness of the e.m. shower in the early part can be used to reject events consisting of two close-by e.m. showers. Planes of fine pitch orthogonal strips after a pre-shower, placed at a depth of $\approx 2.5X_0$, can also be used to distinguish π^0s from single photons. Results using 2 mm pitch strips are shown in Figure 30.

5.14 Identification of muons

Muons are identified by their penetration through the material of calorimeters which absorb the electrons, photons and hadrons. About two metres equivalent of iron is needed to absorb most of the energy of the hadrons. This corresponds to over 10λ of material. Insufficient depth of material can allow debris from hadronic showers to emerge and hence cause false identification of a hadron as a muon. The added confusion can also lead to difficulty in matching muon-tracks in jets and increase in trigger rate.

Two configurations are possible for measurement of the momentum of muons: tracking in an air-filled magnetic field or in magnetized iron.

Extra material is usually required when muons are tracked through an air-filled field in order to decrease hadronic punchthough from the calorimeters. However, the magnetic

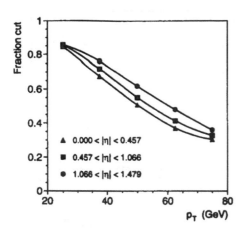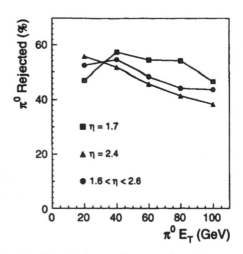

Figure 30. *Left- Fraction of $\pi^0 s$ rejected using the lateral shape of energy deposit as a function of p_T. Right - Variation of π^0 rejection as a function of η using two planes of orthogonally oriented 2 mm pitch Si strips after 2 and 3 X_0.*

field can help by sweeping the soft debris that may 'punch-through'.

Muons of energies above a few hundred GeV generate their own background when traversing magnetized iron. The critical energy of muons in iron is 350 GeV and hard bremsstrahlung (sometimes labeled catastrophic energy loss) can spoil muon tracking. When tracking in iron, several muon stations separated by a sufficient thickness of iron are required so as to kill the e.m. shower before the following station is reached.

6   The proton-proton experiments at the LHC

The p-p General Purpose Detectors (GPDs) at the LHC follow closely the onion-like structure discussed in the introduction. The design requirements can be summarized as follows:

(i) Very good muon identification and momentum measurement to allow efficient muon triggering and the measurement of the sign of ~ 1 TeV muons;

(ii) Very good energy resolution electromagnetic calorimetry;

(iii) Powerful inner tracking, with a factor 10 better momentum resolution than at LEP;

(iv) Hermetic and large geometric coverage calorimetry for good missing E_T resolution;

(v) An affordable cost.

The single most important aspect of the overall detector design is the magnetic field configuration for the measurement of muon momenta. The choice strongly influences the rest of the detector design. The two basic configurations are solenoidal and toroidal. The closed configuration of a toroid does not provide magnetic field for inner tracking. Since a detector without magnetic inner tracking cannot adequately study a number of important physics topics, an additional inner solenoid is required to supplement a toroid. Large bending power is needed to measure precisely high momentum muons or other charged tracks. This forces a choice of superconducting technology for solenoids whereas both superconducting (air or iron core) and warm (iron core) are possible for toroids. Below, we discuss the design, the advantages and the drawbacks of each of the configurations (Fournier D and Serin L, 1996 and Virdee T S, 1999).

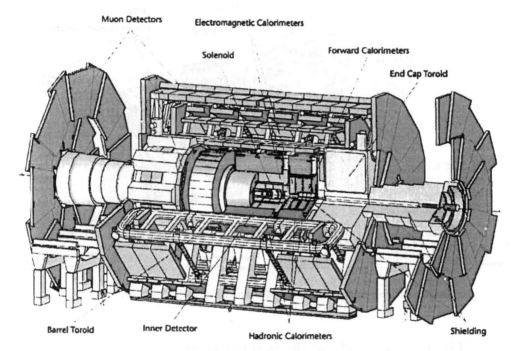

Figure 31. *The 3-D view of the ATLAS detector displaying the various sub-detectors.*

6.1 ATLAS

The overall detector layout is shown in Figure 31. The magnet configuration uses large superconducting air-core toroids consisting of independent coils arranged with an eight-fold symmetry outside the calorimetry. The magnetic field for the inner tracking is provided by an inner thin superconducting solenoid generating a field of 2 T.

The inner detector is contained in a cylinder of length 6.8 m and radius 1.15 m. It consists of a combination of 'discreet' high-resolution Si pixel and microstrip detectors in the inner part and 'continuous' straw-tube tracking detectors with transition radiation capability in the outer part of the tracking volume. Highly granular liquid-argon (LAr) e.m. sampling calorimetry covers the pseudorapidity range $|\eta| <3.2$. In the endcaps, the LAr technology is used for the hadronic calorimeter. The forward LAr calorimeters, extending the coverage to $|\eta|=4.9$ are also housed in the same cryostat. The barrel part of the hadronic calorimetry is provided by an iron-scintillator tile sampling calorimeter using wavelength shifting fibres (WLS). The calorimetery is surrounded by the muon spectrometer. The air-core toroid system encloses a large field volume. The muon chambers, grouped into three stations, are placed in the open and light structure to minimize effects of multiple scattering. The muon spectrometer defines the overall dimensions of the ATLAS detector with a diameter of 22 m and a length of 46 m. The weight of the detector is about 7000 tons.

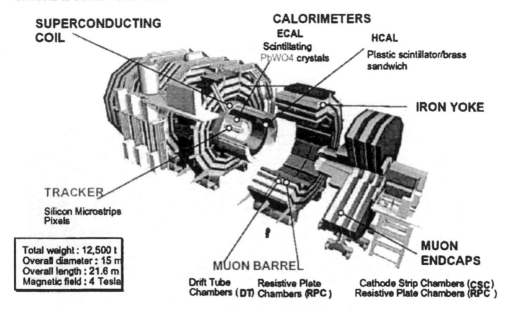

Figure 32. *An exploded view of the CMS detector.*

6.2   The Compact Muon Solenoid (CMS)

The overall layout is shown in Figure 32. At the heart of CMS sits a superconducting solenoid. In order to achieve a good momentum resolution within a compact spectrometer without making stringent demands on muon-chamber resolution and alignment, a high magnetic field is required. CMS has a long (13 m), large bore (ϕ=5.9 m) and high field (4 T) solenoid. The field is large enough to saturate 1.5 m of iron, which is thick enough to accommodate four muon stations to ensure robustness and full geometric coverage. Each muon station consists of many measuring planes. These consist of aluminium drift tubes in the barrel region and Cathode Strip Chambers (CSCs) in the endcap region.

The bore of the magnet is also large enough to accommodate the inner tracker and the calorimetry inside the coil. The tracking volume is given by a cylinder of length 6 m and a diameter of 2.6 m. In order to deal with high track multiplicities, tracking detectors with small cell sizes are used. Solid-state and gas microstrip detectors provide the required granularity and precision. Pixel detectors placed close to the interaction region improve the measurement of the track impact parameter and secondary vertices. The electromagnetic calorimeter (ECAL) uses lead tungstate ($PbWO_4$) crystals. A preshower system is installed in front of the endcap ECAL for π^0 rejection. The ECAL is surrounded by a brass-scintillator sampling hadronic calorimeter. The light is channeled by clear fibres fused to wavelength shifting fibres embedded in scintillator plates. The light is detected by photodetectors (Hybrid Photo-Diodes) that can provide gain and operate in high axial magnetic fields (proximity focussed hybrid photodiodes). Coverage up to rapidities of 4.7 is provided by a Cu-quartz fibre calorimeter. The Cerenkov light emitted in the quartz fibres is detected by photomultipliers. The forward calorimeters ensure full geometric coverage for transverse energy measurement. The overall dimensions of the CMS detector are: a length of 21.6 m, a diameter of 14.6 m and a total weight of 12500 tons.

6.3 Muon systems

ATLAS

The number of field lines crossed by a muon track in toroids is constant. In the endcap region, the magnetic field increases as $1/R$. Hence, toroids have the property that the transverse momentum resolution is constant over a wide range of pseudo-rapidity. The integral $\int B \cdot dl$ ($\propto 1/\sin\theta$) compensates for the Lorentz boost in the forward direction. In an air-core toroid, a good stand alone momentum resolution can be reached as long as the quantity BL^2 is large enough. Two drawbacks of the toroidal configuration are:

- the bending does not take place in the transverse plane and hence benefit cannot be drawn from the precise knowledge of the beam-beam crossing point (20 μm at LHC), and

- a solenoid is needed to provide field for the inner tracker, opening the debate of whether the coil should be in place before or after the electromagnetic calorimetry.

The design criteria for the muon system can be obtained by requiring that an unambiguous determination is made of the sign for muons of 1 TeV. This implies that $\Delta p/p \approx 10\%$. The sagitta, s, for a track of momentum p in a uniform magnetic field is given by $s = 0.3BL^2/8p$. In the case of ATLAS, where $B \approx 0.6$ T, $L \approx 4.5$ m $s \approx 0.5$ mm for $p_\mu = 1$ TeV, this implies that the sagitta has to be measured with a precision of $\approx 50\mu$m. For a muon system as large as in ATLAS, precision of this nature presents special challenges of spatial and alignment precision. From the term BL^2 it is clear that a large magnet is required. However, it is not easy to generate a high field over a large volume. By considering Ampere's theorem we can estimate the current required, Now

$$2\pi RB = \mu_0 nI \Rightarrow nI = 20 \times 10^6 At ,$$

$$i.e. \quad 2.5 \times 10^6 At \quad \text{for 8 coils} ,$$

where n is the number of turns (t) and I is the current (A). ATLAS employ $2 \times 2 \times 30$ turns, leading to a current of I=20 kA. Such currents can only be considered in the context of superconducting magnets.

The basis of measurement in the barrel part of the ATLAS spectrometer is to measure a position on the muon trajectory before and after the magnet, and a third point between the other two.

$$\frac{\Delta p}{p} = 26.7\sigma \sqrt{\left(\frac{1}{2N_1} + \frac{1}{N_2}\right)} \frac{p}{BL^2}(\%) , \tag{16}$$

where p is in GeV, B in T, and σ and L are in m. In ATLAS, $\sigma \approx 70~\mu$m, $N_i \approx 6$, $B \approx 0.6$ T, $L \approx 4.5$ m, implying $\Delta p/p \approx 0.8$ % at 100 GeV, which is very close to the value found in simulations.

The momentum resolution is limited by energy loss fluctuation in the calorimeters at small momenta and by detector resolution at high momenta, whereas the multiple scattering effect is approximately momentum independent, as can be seen in Figure 33. The momentum resolution is typically 2-3% over most of the kinematic range, apart from very high momenta, where it increases to ≈ 10% at $p_T = 1$ TeV. The assembly of the barrel toroid coils at CERN can be seen in Figure 34.

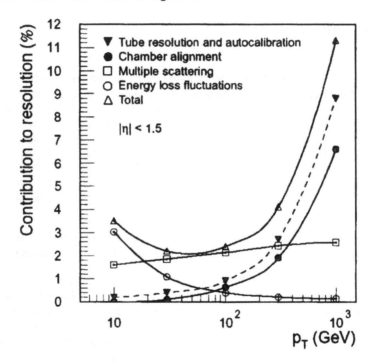

Figure 33. *Various contributions to muon momentum resolution in ATLAS.*

CMS

A large $\int B \cdot dL$ can be obtained for a modest size using high field solenoids. The bending, which takes place in the transverse plane, starts at the primary vertex. For tracks that pass through the end of the solenoid, the momentum resolution worsens as $L_C \tan \theta / r_C$, where L_C and r_C are the length and radius of the solenoid. The effect can be attenuated by choosing a favourable length/radius ratio. For the CMS coil, where a high magnetic field is chosen, the challenge lies in the production of a reinforced superconducting cable that can take an outwards pressure of about 60 atm.

The field generated in a solenoid is given by $B = \mu_0 nI$. For CMS, with B=4 T, n = 2168, implying that $I \approx 20$ kA, again requiring superconducting technology.

Centrally produced muons are measured three times: in the inner tracker, after the coil and in the return flux. In the stand alone mode (no inner tracking) the momentum

Figure 34. *Assembly of the ATLAS toroid coils at CERN.*

resolution in magnetized iron and in the multiple scattering dominated region is given by $\Delta p/p = 44\%/ \left(B\sqrt{L} \right) \sim 20\%$ in CMS. However, the momentum resolution is considerably better if the beam spot and the angle is measured straight after the coil are used (labeled 'muon system only' in Figure 35). Furthermore, if multiple scattering and energy loss are neglected, then the muon trajectory beyond the return yoke extrapolates back to the beam-line due to the compensation of the bending before and after the coil. This fact can be used to improve the momentum resolution at high momenta (labeled 'full system' in Figure 35). The sagitta is given by the perpendicular distance between the outermost inner tracking points and the line joining the beam to the muon beyond the return yoke. The assembly of the magnet yoke in the surface experiment hall can be seen in Figure 36.

6.4   Inner tracking

The most powerful way to 'see' the event topology is by using the inner tracker. The role of the innner tracker is to measure the momentum and impact parameter of charged tracks with minimal disturbance. The figures of merit are the track finding efficiency, the momentum resolution and the secondary vertex resolution. As described earlier the inner tracker plays a crucial role in the identification of electrons, taus and b-jets.

During the 60's the bubble chamber was the detector of choice for tracking. Immersed in a magnetic field the momenta of charged tracks could be measured. Usually the identity of the charged particles could also be deduced from the density of bubbles. Energy-momentum conservation was usually applied to determine the kinematic proper-

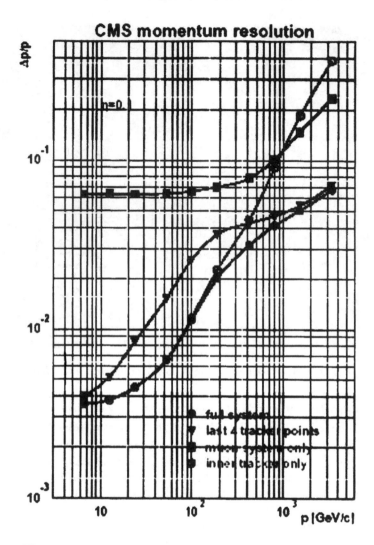

Figure 35. *The muon momentum resolution in CMS using the inner tracker and the muon system.*

ties of all the participating particles. However, it was superseded by electronic detectors as HEP moved to the study of lower cross-section phenomena. The bubble chambers had a low repetition rate and lacked sufficient triggering capability. Recently, large detectors such as ALEPH and DELPHI at LEP have used 'electronic bubble chambers' in the form of Time Projection Chambers (TPCs). These give 3-D spatial information with high granularity and some particle identification capability is in-built using dE/dx measurements. However, these are not used in the LHC GPDs as the electron drift time is long (25-45 μs). They are suitable for LEP as the event rate is low and the bunch crossing interval is large. The tracking detectors at the LHC have to deal with very high particle rates ($\approx 4 \times 10^{10}$ particles/s emerging from the interaction point) and very short bunch

Figure 36. *Assembly of the yoke of the CMS magnet. The inner and outer vacuum tanks of the solenoid can be seen.*

crossing time (25 ns). Furthermore, the target momentum resolution for 100 GeV tracks is almost an order of magnitude smaller at the LHC than at LEP. Hence, Si pixel and Si microstrip detectors, and short drift-time gaseous detectors (straw tubes) are used.

The particle fluence per crossing versus radius at high luminosity is shown in Figure 37. Three regions can be delineated.

- Pixel detectors are placed closest to the interaction vertex where the particle flux is the highest. The typical size of a pixel is ~ 125 μm $\times 125$ μm, giving an occupancy of about 10^{-4} per pixel per LHC crossing.

- In the intermediate regions, the particle flux is low enough to enable the use of Si microstrip detectors with typical cell size of 10 cm $\times$ 75 μm, leading to an occupancy of $\approx 1\%$ per LHC crossing.

- In the outermost regions of the inner tracker, the particle flux has dropped sufficiently to allow use of larger pitch Si microstrip or gaseous straw tube detectors. Typical cell size in CMS is 25 cm $\times 80$ μm, giving an occupancy of a few percent.

ATLAS Inner Tracker

Close to the interaction vertex are 3 barrel layers of hybrid pixels detectors at r = 5, 10 and 13 cm and 3 endcap disks on each side. The size of the pixels is 50 $\times$ 300 μm^2 in r-ϕ and z respectively and the pixel count is some 70 million. The surface area is 2.3 m^2 and provides coverage up to $|\eta| < 2.5$. The pixel detector is described as 'hybrid', as the readout chips are 'bump-bonded' to the silicon pixel elements. The chips contain

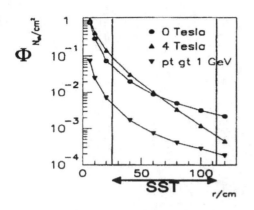

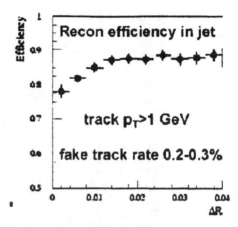

Figure 37. *Particle fluence per crossing versus radius at high luminosity.*

Figure 38. *Charged reconstruction efficiency in jets as a function of distance ΔR from the jet-axis in ATLAS.*

individual circuits for each pixel element, including buffering to store data while awaiting the Level-1 trigger. The silicon microstrip detectors, with a total area of 60 m^2 and mounted on carbon-fibre cylinders and placed at radii between 25 and 50 cm, provide 4 precision space points (4 layers between 30 cm and 50 cm) in the barrel region and 9 points in the endcaps on each side. The silicon microstrip detectors are surrounded by straw-tubes, with transition radiation capability for electron identification, that provide 'continuous tracking' for powerful pattern recognition, especially at low luminosities. Up to 36 points per track can be registered in straws that have a diameter of 4 mm. The transition radiation photons are generated in polyethelene and detected in Xe loaded gas (70%Xe-20%CO$_2$-10%CF$_4$).

The momentum and impact parameter resolutions are parameterized by:

- $\sigma(1/p_T) \sim 0.36 + 13/(\, p_T\sqrt{\sin\theta}\,)$ in (TeV^{-1}) with p_T in TeV/c
- $\sigma(d_0) \sim 11 + 73/(\, p_T\sqrt{\sin\theta}\,)$ in μm with p_T in GeV/c

The track reconstruction efficiency inside jets for tracks with $p_T > 2$ GeV is shown in Figure 38 for a fake rate of < 0.3%.

CMS Inner Tracker

The layout of the CMS tracker is shown in Figure 39. Close to the interaction vertex are 3 layers of hybrid pixels detectors at r = 4, 7 and 11 cm. The size of the pixels is $125 \times 125\ \mu$m^2. The Si microstrip detectors, placed in the barrel region at r between 20 and 115 cm, provide 10 precision points. The forward region has two pixel and nine micro-strip layers in each of the two End-Caps. The silicon micro-strip barrel is separated into an Inner and an Outer Barrel. In order to avoid excessively shallow track crossing angles, the Inner Barrel is shorter that the Outer Barrel, and there are an additional three Inner Disks in the transition region between Barrel and End-Caps, on each side of the Inner Barrel. The total area of the silicon detectors is 220 m^2.

The momentum resolution of 100 GeV pions is shown in Figure 40.

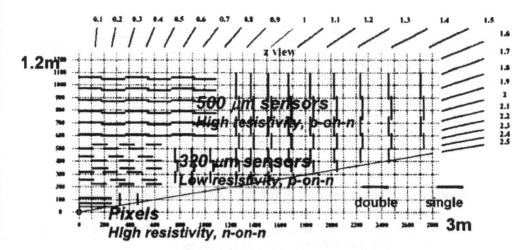

Figure 39. *The layout of the CMS Inner Tracker.*

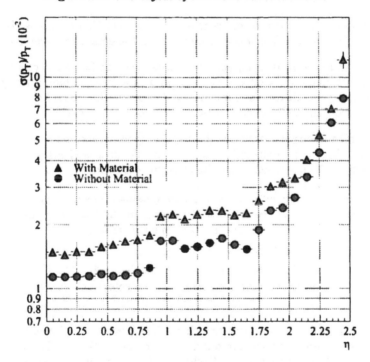

Figure 40. *Momentum resolution in CMS inner tracker.*

6.5 Electronics for LHC experiments

The main components of electronics systems are: front-end, signal processing, data transmission, and power supplies, services, etc.

The features that differentiate the electronics of the LHC experiments from, e.g. LEP

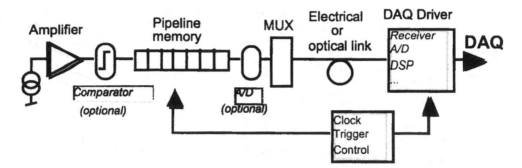

Figure 41. *A generic readout system for a p-p experiment at the LHC.*

experiments are: high speed signal processing, signal pileup, high radiation levels, larger number of channels (large data volume), and new technologies.

For example, the challenges for the inner tracker electronics are:

- Signals are small and fast response must be preserved. Hence, long leads cannot be used and the preamplifiers must be mounted on the detectors themselves.

- The data must be held in pipeline memories awaiting Level-1 decision. It is not feasible to transfer data off the detector at a rate of 40 million events/s for millions of channels. Hence, the pipeline memories must be located on the detectors. Consideration has to be given to how the signals are taken out.

- The several millions of channels will dissipate a considerable amount of heat (power dissipation has to be kept as low as possible; the goal is $\leq$ a few mW/channel). This leads to the question of how the electronics are cooled.

The above leads to difficult engineering and systems challenges. All this has to be accomplished whilst keeping the amount of material in the tracker to the minimum to minimize multiple scattering and conversion or bremsstrahlung.

Electronics of Sub-detectors

The generic LHC readout system is illustrated in Figure 41. The characteristics and requirements for the electronics for the various sub-detectors can be summarized as follows:

- *Tracking*: large number of microstrip channels (e.g.$\approx$ 10's of millions of microstrips), limited energy precision and limited dynamic range ($<$ 8-bits). The power dissipation/channel has to be low ($\approx$ mW/ch) and the electronics have to withstand very high radiation levels (neutron fluence of 10^{15} n/cm^2, integrated doses of 10's of Mrads).

- *Calorimetry*: medium number of channels ($\approx 10^5$), high measurement precision (12-bits), large dynamic range (16-bits), very good linearity and very good stability in time. The power constraints and the radiation levels (neutron fluence of 10^{13} n/cm^2, integrated doses of 100's of krads in the barrel region) are not as stringent as for the tracker.

- *Muon system*: the large surface area that needs to be instrumented means that the electronics are distributed over a large area and the radiation levels are low.

The functions that are common to all systems are amplification, analogue to digital conversion, association to beam crossing, storage prior to trigger, deadtime-free readout, zero suppression and formatted storage prior to access by the data acquisition, calibration control and monitoring.

An example electronics chain, taken from the CMS silicon strip tracker, is shown in Figure 42. Each microstrip is read out by a charge sensitive amplifier with a shaping time of 50 ns. The output voltage is sampled at the beam crossing rate of 40 MHz and the samples are stored in an analogue pipeline for up to 3.2 μs (Level-1 latency). Following a trigger, a weighted sum of 3 consecutive samples is formed in an analog circuit. This confines the signal to a single bunch crossing and gives the pulse height. The buffered pulse height data are multiplexed out on optical fibres as light signals where the light intensity is the output of a laser modulated by the pulse height for each strip. The light signals, transported over optical cables to another cavern containing services and electronics, are transformed into electrical pulses by a Si photodiode and digitized. After some digital processing (zero suppression etc.) the data are formatted and placed into dual port memories for access by the data acquisition.

The calorimeter and muon systems have also to generate the primitives (energy or momentum values) for the first-level trigger.

6.6   The calorimeters

CMS Calorimeters

The most demanding physics channel for an electromagnetic calorimeter at the LHC is the two-photon decay of an intermediate-mass Higgs boson. The background is large and the signal width is determined by the calorimeter performance. The best possible performance in terms of energy resolution is only possible using fully active calorimeters such as inorganic scintillating crystals. CMS has chosen lead tungstate ($PbWO_4$)scintillating crystals for its e.m. calorimeter. These crystals have short radiation (0.89 cm) and Molière (2.2 cm) lengths, fast (80 % of the light is emitted within 25ns) and radiation hard (up to 10 Mrad). However, the relatively low light yield (3000 γ/MeV) requires use of photo-detectors with intrinsic gain that can operate in a magnetic feld, silicon avalanche photo-diodes (APDs have been chosen for the barrel crystals and vacuum phototriodes for the endcap crystals). In addition, the sensitivity of both the crystals and the APD response to temperature changes requires temperature stability (goal is 0.1 $^{\circ}$C). The calorimeter comprises over 75000 crystals with a size of around 22 $\times$ 22 mm^2 ($\Delta\eta\Delta\phi \sim 0.016 \times 0.016$ in the barrel region).

In the 2003 test beam running, using 0.25 μm CMOS front-end electronics, the energy resolution has been measured to be close to the design value of:

$$(\sigma/E)^2 \sim \left(2.7\%/\sqrt{E}\right)^2 + (0.5\%)^2 + (0.150/E)^2, \quad \text{where E is in GeV}.$$

The CMS hadronic calorimeter comprises a brass-scintillator tile sampling calorimeter with a lateral granularity of $\Delta\phi\Delta\eta \sim 0.09 \times 0.09$. The brass and scintillator plates

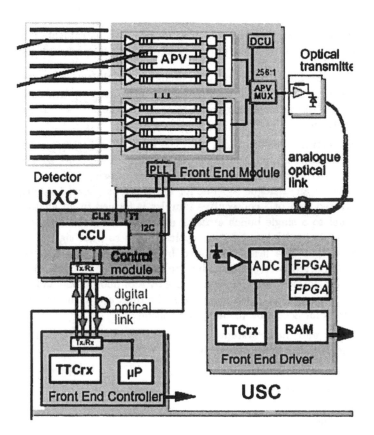

Figure 42. *Schematic of the CMS Tracker readout and control system.*

have a thickness of 5.0 and 0.4 cm respectively. The scintillation light is readout out by wavelength shifting fibres with the light channeled on clear fibres to hybrid photo-diodes placed axially to the 4 T magnetic field. The energy resolution of the hadronic calorimeter, combined with lead tungstate crystals, is measured to be

$$(\sigma/E)^2 \sim \left(130\%/\sqrt{E}\right)^2 + (6.5\%)^2, \quad \text{where E is in GeV}.$$

The forward calorimeter is constructed out of diffusion welded steel plates with grooves in which 0.8 mm quartz fibres are inserted. The Cerenkov light generated by relativistic particles in showers is transported to photomultipliers in shielded housing at the rear of the calorimeter. The pion energy resolution of the forward calorimeter is measured to be around 40% at 50 GeV and is predicted to be $\sim$ 20% at 1 TeV. The barrel and endcap hadronic calorimeters have been assembled in the CMS-SX5 surface assembly Hall (see Figure 43).

ATLAS Calorimeters

A view of the ATLAS calorimeters is shown in Figure 44. The e.m. and endcap hadronic calorimeters use liquid argon as the active medium. Calorimeters using liquid filled ionization chambers as detection elements have several important advantages. The

Figure 43. *The CMS HCAL half-barrel in CMS-SX5 surface assembly Hall.*

absence of internal amplification of charge results in a stable calibration over long periods of time, provided that the purity of the liquid is sufficiently good. The considerable flexibility in the size and the shape of the charge collecting electrodes allows high granularity both longitudinally and laterally. Conventionally, liquid ionization chambers are oriented perpendicularly to the incident particles. However, in such a geometry it is difficult to realize fine lateral segmentation with small size towers (which in addition need to be projective) and to implement longitudinal sampling without introducing insensitive regions, a large number of penetrating interconnections and long cables, (which necessarily introduce electronics noise and lead to significant charge transfer time). To overcome these shortcomings, ATLAS introduced a novel absorber-electrode configuration, known as the 'accordion' (Figure 45), in which the particles traverse the chambers at angles around 45°.

The accordion geometry provides complete ϕ-symmetry without azimuthal cracks. The thickness of lead has been optimized as a function of η in terms of performance in energy resolution. The LAr gap has a constant thickness of 2.1 mm in the barrel because

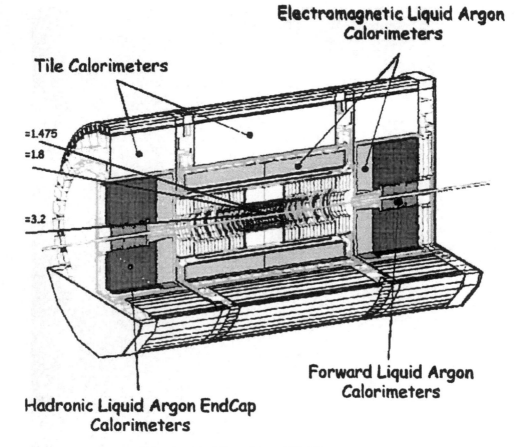

Figure 44. *View of the ATLAS calorimeters.*

the amplitude of the accordion waves increases with radius. Over the region of precision physics ($|\eta| < 2.5$) the e.m. calorimeter is segmented into 3 longitiudinal sections. In the barrel, the first section is equipped with narrow strips, with 4 mm pitch and a size of $\Delta\eta\Delta\phi = 0.003 \times 0.1$, that act as a preshower detector, the second with cells of 0.025×0.025 and the third with cells of 0.05×0.025. The signals are extracted at the inner and outer faces and sent to preamplifiers located outside the cryostats close to the feedthroughs. The preamplifier output is formed by bipolar shapers, sampled every 25 ns and stored in analogue memories during the Level-1 trigger latency.

The electromagnetic calorimeter is a lead/liquid argon detector covering $|\eta| < 3.2$. Over the range $|\eta| < 1.8$ it is preceded by a presampler detector installed immediately behind the cryostat wall, and is used to correct for the energy lost in the material upstream of the calorimeter, enhance γ/π^0 and e/π separation.

The performance of the electromagnetic calorimeter can be deduced from Figure 46. Some corrections (see Figure 47) for impact point have to be made in order to take account of the lateral containment and the ϕ-modulation introduced by the accordion geometry. The barrel e.m. calorimeter is now assembled and inserted into the cryostat

Figure 45. *View of the accordion geometry.*

(see Figure 48).

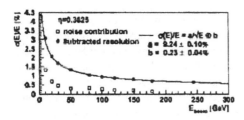

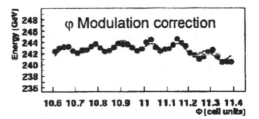

Figure 46. *The measured energy resolution of the ATLAS e.m. calorimeter.*

Figure 47. *The ϕ-modulation of response in the ATLAS e.m. calorimeter.*

The hadronic calorimeter in the barrel region is an iron-scintillator tile sampling calorimeter (see Figure 49). Again, in a novel way, the tiles are placed radially and staggered in depth. The structure is periodic along z. The tiles are 3 mm thick and the total thickness of the iron plates in one period is 14 mm. Two sides of the scintillating tiles are read out by wavelength shifting fibres into two separate photomultipliers. The wedges of the barrel hadronic calorimeter have been built and have been trial assembled into rings at CERN.

A combined test of the e.m. LAr and hadronic tile calorimeter have been performed in the test beams and the energy resolution can be parameterised as:

$$(\sigma/E)^2 = \left(70\%/\sqrt{E}\right)^2 + (3.3\%)^2 + (1.8/E)^2 \quad \text{where E is in GeV}.$$

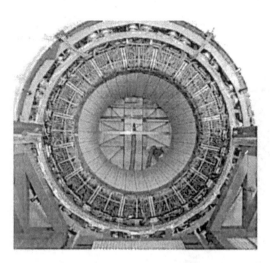

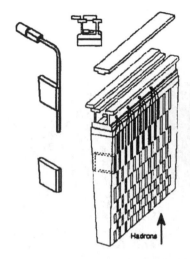

Figure 48. *The ATLAS barrel e.m. LAr calorimeter inserted into the cryostat.*

Figure 49. *The schematic of the ATLAS barrel Tilecal hadron calorimeter.*

6.7 Muon detectors

Two kinds of muon detectors are used at LHC, serving complementary roles. These are gaseous drift chambers that provide accurate position measurement for momentum determination and 'trigger' chambers, such as resistive plate chambers (RPCs) or thin gap chambers (TGCs), that have a short response time ($\leq$25 ns) for precise bunch crossing identification but a less accurate position measurement. The former category of detectors can also provide a first level-trigger on muons. In LHC GPDs, the rate in the barrel region ($\approx$ 10 Hz/cm^2) is two orders of magnitude smaller than in the endcaps. This rate is due mainly to hits induced by photons from neutron capture. The neutrons are evaporation neutrons produced by breakup of nuclei in hadronic showers. In the endcap region, drift chambers are replaced by faster chambers such as cathode strip chambers (CSCs). Since the dominant background is neutron induced, usually affecting two detecting layers, each of the muon stations comprises several ($\approx$6) layers of detectors.

ATLAS Muon System

The layout of the ATLAS muon system in shown in Figure 50. In the barrel region, tracks are measured in chambers arranged in 3 cylindrical layers (stations) around the beam-axis; in the transition region and the endcaps the chambers are installed vertically, also in three stations. The precision measurement of the track is performed in Monitored Drift Tubes (MDTs) except in the forward region where it is performed by CSCs. The MDTs are constructed out of long aluminium tubes (3 cm diameter) at 3 bar pressure.

The ionization electrons drift to the anode wire with a maximum drift time of 500 ns. Two groups of 3 planes of the tubes, separated by ∼ 20 cm, form a station (see Figure 51). The spatial resolution afforded by each tube has been measured to be about 80 μm. The Level-1 trigger is provided by 3 stations of RPCs in the barrel region and TGCs in the endcap region, located as shown in Figure 50.

There are some 1200 MDTs, 1140 RPCs, 1600 TGCs and 30 CSCs to be constructed, totalling a surface area of over 12000 m^2. The construction is well advanced in many laboratories around the world and an endcap MDT station is shown in Figure 51.

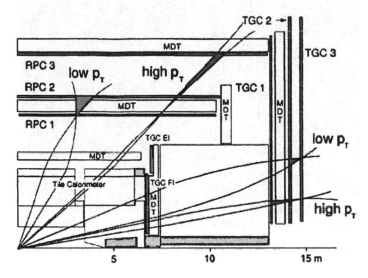

Figure 50. *The layout of the ATLAS Muon Chambers.*

CMS Muon System

The layout of the CMS Muon system is shown in Figure 52. Centrally produced muons are measured three times, in the inner tracker, after the coil and in the return flux. They are then identified and measured in four identical muon stations (MB) inserted in the return yoke. Special care has been taken to avoid pointing cracks and to maximize the geometric acceptance. Each muon station consists of twelve planes of aluminium drift tubes designed to give a muon vector in space, with 100 μm precision in position and better than 1 mrad in direction. The four muon stations include RPC triggering planes that also identify the bunch crossing and enable a cut on the muon transverse momentum at the first trigger level. The endcap muon system also consists of four muon stations (ME). Each station consists of six planes of Cathode Strip Chambers. The chambers are arranged such that all muon tracks traverse four stations at all rapidities, including the transition region between the barrel and the endcaps. The last muon stations are after a total of ≥ 20λ of absorber so that only muons can reach them. The four muon stations lead to a redundant and robust muon system.

The construction of the muon chambers is well advanced in many laboratories around the world and installation has started in SX5 (see Figure 53).

Figure 51. *An ATLAS endcap Monitored Drift Tube Station.*

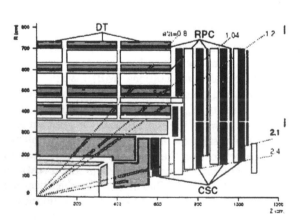

Figure 52. *The layout of the CMS Muon Chambers.*

Figure 53. *Installation of CMS CSCs on an endcap yoke disk.*

7 Trigger and data acquisition System

The bunch crossing rate at the LHC is 40 MHz. The role of the trigger and data acqusition system is to look at all the bunch crossings, select about one hundred of these containing the most interesting events and record their data on permanent storage for offline analysis. This is a daunting task because the selection process:

- must be highly efficient. The vast majority of the events have to be rejected. However, none of the few expected rare events should be missed.

- should not introduce any bias.

- should cause as little dead-time as possible.

- must use data from the same crossing for all sub-detectors. This requires synchronisation of millions of channels.

- needs an information super-highway, as the 20 or so interactions every 25 ns lead to the generation of 40,000 Gbits/s. The data flow has to be reduced as quickly as possible by high selectivity.

- is carried out in real time i.e. one cannot go back and recover lost events. Furthermore, it is essential to monitor the selection process.

A typical trigger and data acquisition system consists of four parts: the detector electronics, the calorimeter and muon first level trigger processors, the readout network and an on-line event filter system (processor farm).

The Level-1 Trigger System uses custom processors (ASICs and increasingly FPGAs) and is required to reduce the bunch-crossing rate of 40 MHz to an event rate of 100 kHz. Upon receipt of a Level-1 trigger, after a fixed time interval between 2-3 μs, the data from the pipelines are first transferred to 'derandomizing' memories that can accept the very high instantaneous input rate (the Level-1 can accept several events within the space of $\approx$ 10 crossings even though the average rate is much lower). These memories are emptied into readout buffers: usually many individual channels are multiplexed over a single readout link. After further signal processing (e.g. digitisation, deconvolution), zero suppression and/or data compression takes place before the reduced amount of data are placed in dual-port memories for access by the DAQ system. Each physics event ($\approx$1 Mbytes large) is contained in about 500 front-end *Readout* buffers. To further analyse the event, it is necessary to transfer the data from the 500 Readout units to a single processor, running the appropriate physics selection algorithm. The input rate of 100 kHz is thus reduced to the 10^2 Hz of sustainable physics. The 'event building' is performed using a data switch; in fact, use will be made of switching technologies from the telecommunications industry. The most important elements, and also the most difficult ones to develop, are the front-end buffers, the switch that will connect these memories to the processor farm and the physics selection algorithms.

The CMS and ATLAS architectures are illustrated in Figure 54. ATLAS uses a more conventional architecture with a dedicated Level-2 processor before the switch network. The use of a processor farm for all selection beyond Level-1, produces maximal benefit and can be taken from the evolution of computing technology. Flexibility is maximized since there are no built-in design or architectural limitations; there is complete freedom in what data to access and in the sophistication of algorithms. Evolution is possible, allowing response to unforeseen backgrounds.

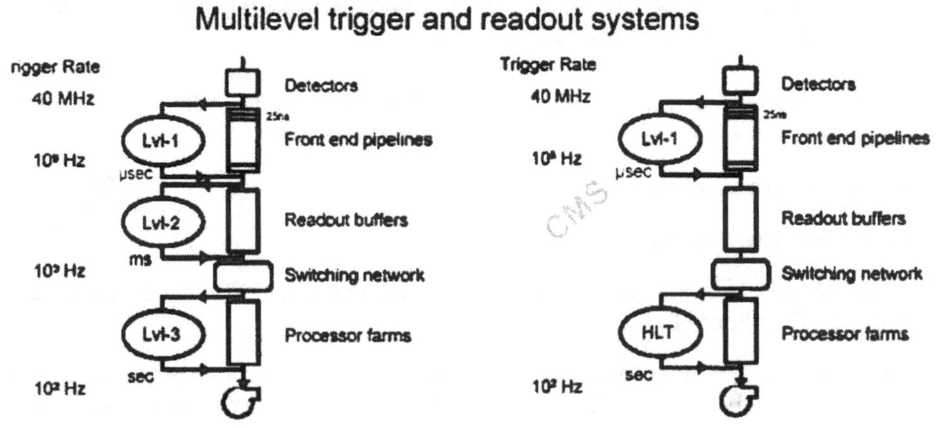

Multilevel trigger and readout systems

Figure 54. *The functional view of the ATLAS and CMS trigger and data acquisition systems.*

7.1 Level-1 trigger

Track stubs in the muon system or energy deposits in the calorimeters are used to create the so-called trigger objects: isolated e.m. clusters, muons, jets, E_t^{miss}, etc. The selection is based on e.m. and/or hadronic clusters and/or muons with transverse energy and momentum above certain pre-loaded thresholds in trigger processors. Reduced granularity and reduced resolution data are used to form trigger objects. For example, in CMS ECAL, information from groups of 25 crystals are combined to form one trigger-tower and 8-bit resolution is used instead of full 12-bit information for energy in the trigger towers.

An example from ATLAS of how an isolated e.m. calorimeter is selected is shown in Figure 55. There are some 4000 ECAL and geometrically matching 4000 HCAL trigger towers each giving an 8-bit value every 25 ns. The Level-1 selection proceeds as follows: the energy in a 'hot' trigger tower is combined with that from the hottest one in either η or ϕ (Figure 55). The transverse energy of the sum should be greater than some E_T^{cut}. For a cluster to be labeled 'isolated', the transverse energy in the 12 towers surrounding the central 2×2 has to be smaller than some threshold E_T^{em}. Furthermore, the transverse energy leaking into the HCAL (sum energy in the 16 HCAL towers behind) for an e.m. shower has to be smaller than some threshold E_T^{had}.

The above algorithm is applied for each of the 4000 window positions, representing a massive computing task. Pipelined and parallel processing is employed. Pipelined processing means that the logic is organised in a chain of operations to be performed one after the other for each crossing. Each processing element in the chain performs its function in 25 ns and passes its result to the next element in the chain. Data corresponding to successive bunch crossings follow each other down the processing 'pipe'. Parallel processing

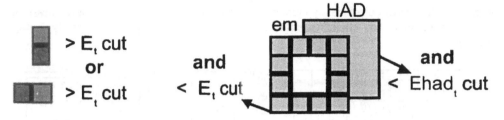

Figure 55. *Selection of isolated e.m. clusters at the Level-1 Trigger in ATLAS.*

means that many processing elements act in parallel, for example, performing the same operations on different data.

The above algorithms can be extended to triggers on taus or jets. For a τ-trigger, the vertical and horizontal transverse energy sum of the two (2×1) or four (2×2) trigger towers in both ECAL and HCAL should be greater than some cut E_T^{cut}. The isolation is applied only in the ECAL. For jet triggers, the transverse energy is summed in 4×4 trigger towers in both the ECAL and the HCAL. A sliding window can be employed centred on blocks of 2×2 towers.

For muon triggering, 'roads' are defined from one station to the next. The width of the road depends on the desired p_T threshold. The calculation allows for magnetic deflection and multiple scattering. An example, from CMS, is shown in Figure 56.

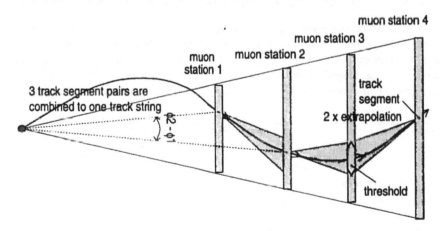

Figure 56. *Schematic of Muon Level-1 trigger in CMS using 'roads'.*

Each barrel station consists of two sets of 4 planes of drift tubes measuring the r-ϕ co-ordinate. The two sets are separated by about 20 cm and, hence, a 'primitive' giving the track direction can be created. This is used to project to the next station and, if a consistent primitive is found there, the process is continued. This allows muon finding and a yes or no answer for the trigger, depending on the momentum of the muon.

The efficiency and trigger rates for single e.m clusters for CMS are illustrated in Figure 57. Cuts may be placed on the isolation, on the hadronic/electromagnetic fraction (HoE), and on fine-grain (FG) lateral shape in the ECAL (which acts as a sort of local isolation). The turn-on efficiency curves are also shown. Clearly, for the lowest possible rate the turn-on in the efficiency curve should be as steep as possible (ideally a step function).

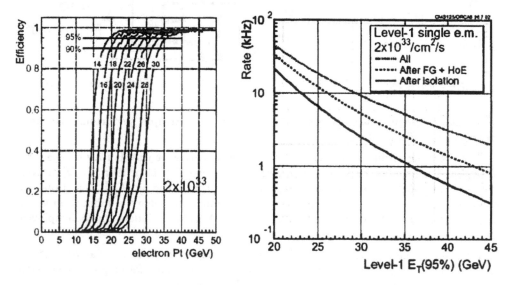

Figure 57. *The efficiency and trigger rates for single e.m clusters in CMS.*

For events within the geometric acceptance of the detector, the trigger 'cocktail' and the rates at startup luminosity for a CMS DAQ staged to 50% of its capability are detailed in Table 8. The cumulative rate of 16 kHz is predicted. A margin for 3 is taken for uncertainties. The staged Level-1 input rate is set at 50 kHz. Both ATLAS and CMS have designed their systems to handle at least 100 kHz of input rate from Level-1.

Trigger	Threshold (GeV or GeV/c)	Rate (kHz)	Cumulative Rate (kHz)
Inclusive isolated electron/photon	29	3.3	3.3
Di-electrons/di-photons	7	1.3	4.3
Inclusive isolated muon	14	2.7	7.0
Di-muons	3	0.9	7.9
Single tau jet	86	2.2	10.1
Two tau jets	59	1.0	10.9
1 jet, 3 jets, 4 jets	177, 86, 70	3.0	12.5
Jet * E_T^{miss}	88 * 46	2.3	14.3
Electron * Jet	21 * 45	0.8	15.1
Minimum Bias (calibration)		0.9	16.0
Total			**16.0**

Table 8. *Trigger rates in CMS running at low luminosity for 95% efficiency of selection.*

7.2   Higher level triggers and data acquisition

The output rate (100 kHz) from the Level-1 trigger has to be reduced to 100 Hz by using more complex algorithms. In CMS, all this will be carried out in a processing farm of mass-market processors using code that is as close as possible to the offline code. This will allow offline code development to be rapidly exploited in the trigger. The farm will consist of about ~1000 computers with the capability of a few times 1000 Gips (Giga instructions per second). The aim is to bring to the farm the full event information.

Both CMS and ATLAS have made detailed studies at low (2×10^{33} cm^{-2} s^{-1}) and design (10^{34} cm^{-2} s^{-1}) luminosities of the performance of the High Level Triggers. These studies involve full detector simulation of many millions of events using GEANT3. Simulated digitization, including both in-time and out-of-time pileup was performed. Digitization and reconstruction were done within the OO environment and C++ code.

Various strategies guide the development of HLT code: regional reconstruction or reconstruction on demand. Rather than reconstruct all possible objects in an event, whenever possible, only those objects and regions of the detector that are needed are reconstructed. Events are to be discarded as soon as possible. This leads to the idea of partial reconstruction and to the notion of virtual trigger levels: at Level-2, the calorimeter and muon information is used, at Level-2.5, tracker pixel information is used and at Level-3 is where the full event information is used (including the complete tracker).

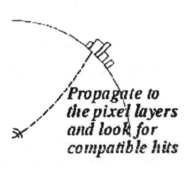
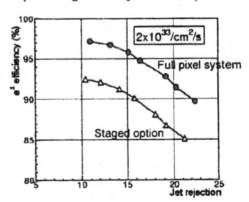

Figure 58. *Left - electron identification using the pixels. Right - electron efficency versus jet rejection.*

The HLT for a given event runs on a single processor, which deals with one event at a time. We consider one example of HLT algorithms, that for electrons and photons in CMS. The steps are as follows.

Reconstruct cluster (or supercluster) in ECAL matched to Level-1 trigger using full granularity information. At Level-1 the electron and photon trigger rate is entirely dominated by the decay of π^0s in jets.

The most important step in the electron selection comes at Level-2.5, where superclusters are 'propagated' back in the field from the ECAL to the pixel detector layers and matching hits are sought. In the pixel region, most electrons have not yet radiated significantly and photons do not have a large probability to convert. Searching for 2 out of

3 possible hits provides a large rejection factor with a small efficiency loss (see Figure 58). The efficiency for electron identification and jet rejection are also shown in Figure 58. The unmatched clusters become photon candidates, the rate of which is reduced by a much higher threshold than used in the electron channel.

The electron and photon rates output by HLT at low luminosity in CMS are listed in Table 9. A loose calorimetric isolation has been applied to the photon streams but none (beyond Level-1) to the electron streams. To control the two-photon rate, the thresholds have been raised to $E_T^1 > 40$ GeV, $E_T^2 > 25$ GeV (equal to the final offline cuts envisaged for H$\rightarrow \gamma\gamma$).

The final rates to storage for all the other channels are given in Table 10, corresponding to low luminosity. For CMS, the efficiency of the whole selection process for selected physics channels is given in Table 11.

	Signal	**Background**	**Total**
Inclusive Electrons	W$\rightarrow e\nu$: 10 Hz	$\pi^{+/-}/\pi^0$ overlap: 5 Hz	33 Hz
		π^0 conversions: 10 Hz	
		$b/c \rightarrow e$: 8 Hz	
Di-electrons	Z$\rightarrow ee$: 1 Hz	$\sim$0	1 Hz
Inclusive photons	2 Hz	2 Hz	4 Hz
Di-photons	$\sim$ 0	5 Hz	5 Hz
TOTAL			**43 Hz**

Table 9. *The electron and photon rates output by HLT at low luminosity in CMS.*

A key issue for the HLT selection is the CPU power required for the execution of the algorithms. In the case of CMS, the time taken by the selection algorithms has been measured on a Pentium-III 1 GHz processor, and the results vary from a very fast $\sim$ 50 ms for jet reconstruction to $\sim$ 700 ms for muon reconstruction. The weighted average is found to be $\sim$ 300 ms per event passing the Level-1. Using Moore's Law (doubling of CPU power every 18 months) a factor 8 increase in computing power (from end 2002) leads to a requirement of $\sim$ 1000 dual-CPU PCs for the HLT farm to be ready for LHC startup in 2007.

There is a considerable advantage in carrying out the whole of the HLT process on PCs. The full event record is available and great flexibility and room is afforded to improve the selection of the various physics channels as well as adjusting to unforeseen circumstances resulting from bad beam conditions, high background levels or new physics channels not previously studied.

Trigger	Threshold (GeV or GeV/c)	Rate (Hz)	Cumulative Rate (Hz)
Inclusive Electrons	29	33	33
Di-electrons	17	~ 1	34
Inclusive photons	80	4	38
Di-photons	40, 25	5	43
Inclusive Muons	19	25	68
Di-muons	7	4	72
Inclusive τ-jets	86	3	75
Di- τ jets	59	1	76
1 jet * Etmiss	180 * 123	5	81
1 jet OR 3 jets OR 4 jets	657, 247, 113	9	89
Electron * jet	19, 45	2	90
Inclusive b-jets	237	5	95
Calibration, etc (10%)		10	105
TOTAL			**105 Hz**

Table 10. *The breakdown of the final estimated rate for offline storage in CMS.*

Channel	Efficiency (%)
$H(115 \text{ GeV/c}^2) \rightarrow \gamma\gamma$	77
$H(150 \text{ GeV/c}^2) \rightarrow ZZ^* \rightarrow \mu\mu\mu\mu$	98
$H(160 \text{ GeV/c}^2) \rightarrow WW^* \rightarrow \mu\nu\mu\nu$	92
$A/H(200 \text{ GeV/c}^2) \rightarrow \tau\tau$	45
SUSY (~ 0.5 TeV/c^2 sparticles)	~ 60
With R parity-violation	~ 20
$W \rightarrow e\nu$	42
$W \rightarrow \mu\nu$	69
Top $\rightarrow \mu X$	72

Table 11. *Efficiency of the whole selection process for some physics channels.*

8 Physics reach in the first year

In order to outline the discovery reach in the first year, the timeline of the anticipated integrated luminosity has to be examined. At a luminosity of $\sim 10^{33}$ cm^{-2} s^{-1}, it is estimated that 25 pb^{-1}/day can be integrated, leading to ~ 0.7 fb^{-1} per month. It is assumed that:

- a 14-hr run is followed by a refilling time of 10-hrs,

- the luminosity lifetime is 20-hours and

- the efficiency of operation is 2/3.

The Higgs physics reach is illustrated in Figure 59. The vertical axis gives the amount of integrated luminosity, for ATLAS and CMS combined, that is needed to make a 5σ discovery of a Standard Model Higgs boson. A sizeable integrated luminosity is needed before an 'attack' can be made on finding the Higgs boson. However, assuming that 2 fb^{-1} per experiment can be integrated over the first 3 months of physics data-taking (at 10^{33} cm^{-2} s^{-1}, $\sim$ 80 fills) then the Higgs boson can be discovered in the mass regions 140-155 GeV and 190-450 GeV using the 'gold-plated' mode (H $\rightarrow$ ZZ$^{(*)}$ $\rightarrow$ 4ℓ). Although there is great discovery potential in the H$\rightarrow$WW mode, a very good understanding of the background is required as there is no resonance peak in this channel. The horizontal lines illustrate the potential for integrated luminosities, for ATLAS and CMS combined, of 4 fb^{-1} and 20 fb^{-1}.

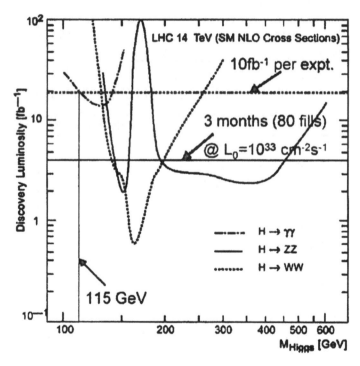

Figure 59. *The Higgs physics reach of ATLAS and CMS combined.*

A topic that potentially can lead to early discoveries is Supersymmetry. Squarks and gluinos are produced with strong cross-sections and, hence, large numbers can be produced even at modest luminosities. Hence, some SUSY particles can be found quickly. This is illustrated in Figure 60 that shows the squark and gluino reach for CMS at integrated luminosities of 1, 10 and 100 fb^{-1}. The chosen point in the SUSY parameter space is close to or beyond the reach of the Tevatron experiments.

It should be remarked that the potential for b-physics can be exploited right from the startup.

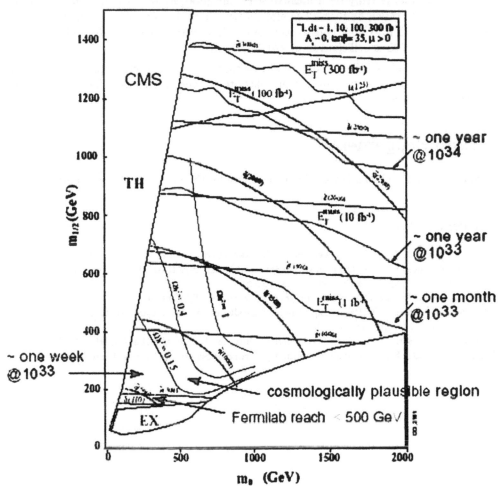

Figure 60. *The squark and gluino reach of CMS.*

9   Conclusion

Physics at the LHC will be extremely rich. The SM Higgs boson (if there) will be found and attention will turn to the measurement of its properties and couplings. Low energy supersymmetry (if there) should also be found relatively easily and attention will turn to determining the SUSY 'zoology' and numerous measurements of the properties of these particles. The large centre of mass energy should also enable search for compositeness, new bosons and extra dimensions.

Much R&D and prototyping has been carried out during the last decade in order to develop detectors that could cope with the harsh conditions anticipated in the pp LHC experiments. These detectors are not just bigger versions of the current detectors but are substantially different, innovative and at the frontier of technology. The construction of these detectors is truly on a massive scale and production of sub-detectors is on an industrial scale. Much hardware has already been built and assembled. The LHC GPD detectors are well into their construction and will be ready for first beam. The machine schedule calls for first collisions in mid-2007.

The ATLAS and CMS detectors should be capable of discovering whatever Nature has in store at the TeV energy scale and are likely to provide answers to some of the biggest question in physics.

Acknowledgements

I would like to thank the Organizing Committee for the invitation to give these lectures.

References

ATLAS Outreach, http://atlas.web.cern.ch/Atlas/Welcome.html

ATLAS Technical Design Reports, http://atlas.web.cern.ch/Atlas/internal/tdr.html

CMS Outreach, http://cmsdoc.cern.ch/outreach/

CMS Technical Design Reports, http://cmsdoc.cern.ch/LHCC.html

Ellis N and Virdee T S, 1994, Experimental Challenges in High Luminosity Collider Physics, Ann Rev Nucl Part Sci **44** 609.

Fournier D and Serin L, 1996, Experimental Techniques, European School of HEP, CERN 96-04.

Virdee T S, 1999, Experimental Techniques, European School of Physics, CERN 99-04.

The LHC Computing Grid

H F Hoffmann

CERN, Switzerland

1 Introduction

Particle physics has extended its reach into the microcosm by an order of magnitude every 10 to 15 years over the past 40 years. Important steps were the discovery of the antiproton[1], the J/Ψ^2 and the W, Z particles[3] and the precise determination of the standard model parameters. The LHC, to become operational in 2007, will increase this range with respect to Tevatron and LEP, again by an order of magnitude to around 10^{-18} m and to objects of up to 2000 GeV mass. It will exceed the discovery potential of the original CERN PS accelerator, measured in centre of mass (cms) energy, by a factor of 2000.

Whereas the early discoveries were still performed with visual observation of individual events, now electronic recording of large numbers of events is customary. Increasing amounts of detailed parameters are measured and online selection and filtering is used, followed by sophisticated offline reconstruction, simulation and analysis programs. The LHC general purpose experiments today, ATLAS and CMS, have in excess of 100 million electronic channels and are able to digest $\sim 10^9$ events/s, each with an average multiplicity of ~ 100 generated secondary particles.

In fact, the influx of data into ATLAS or CMS[4] at preamplifier level and assuming everything was digitized, runs at an equivalent PetaByte/s in-flow of data of which ~ 100 MByte/s are recorded offline, or ~ 100 events/s. This corresponds to an online selection and reduction of the available data by a factor of 10^7. The recording rate for the LHC heavy ion detector reaches 1.25 GBytes/s originating from the extremely high multiplicity of the individual ion-ion collisions.

For many decades, particle physics has depended on and profited from the advances of electronics, as well as from the progress of information and communication technologies (ICT). In some specific cases, such as data acquisition and filtering as well as in scientific collaboration and analysis tools, particle physics even "drives" the wide use of these technologies by applying the novel technologies extensively.

---

[1]http://www.nobel.se/physics/laureates/1959/
[2]http://www.bnl.gov/bnlweb/history/Nobel/Nobel_76.html
[3]http://www.nobel.se/physics/laureates/1984/rubbia-lecture.pdf
[4]http://greybook.cern.ch/

CERN and its collaboration institutes perform a continuous technology watch named "PASTA"[5] in the fields of semiconductor technology, secondary storage, mass storage, networking, data management and storage management technologies and high performance computing solutions. An important feature of these studies is to identify the best cost/performance solutions.

A final important ingredient to the subject of the lecture is the collaborative nature of particle physics. At this time, CERN collaborates with almost 300 institutes and more than 4000 scientists in Europe and in excess of 200 institutes and another almost 2000 scientists elsewhere in the world. The ATLAS or CMS collaborations are about one third of this each and probably the largest long-term scientific collaborations ever existing until today, with well defined and commonly pursued scientific objectives. In the Grid context, we shall name such collaborations "virtual organisations", distant groups of people collaborating towards a common objective.

2   The LHC challenge

The present perception of the physics at the LHC is summarised in Figure 1.

Between the total cross section of proton–proton collisions at LHC energies and the production cross section of a Higgs particle lie nine orders of magnitude of event rate. Multiplying with the branching ratio into detectable decay channels of the Higgs particle, LHC experiments are conceived for selectivity of 1 in 10^{12} for important physics processes.

The difficulty to achieve this selectivity is augmented by the operating mode of the LHC collider. In order to achieve the highest possible and acceptable luminosity the beams in the accelerator are bunched with a bunch frequency of 40 MHz and an average of 25 collisions occurring in any bunch-bunch collision in the centre of the experiment. Therefore any of the "interesting" events will be contaminated by a number of less interesting events which have to be separated out[6,7]. This scientific environment is an unprecedented challenge for offline computing and analysis.

The Technical Proposals of the LHC experiments, the first comprehensive papers describing the complete scope of the detectors and published in 1994[8] give little information on computing and application software issues. At that time the focus of the collaborations was to identify the most promising hardware components, validated by acceptance and performance calculations of the important physics channels. Only a few pages dealt with the online data acquisition and filtering, offline reconstruction and analysis software and computing hardware requirements. Because of the exponential change (Moore's law) of the performance of computing hardware, no serious estimates of the cost of the necessary computing facilities were given.

Following the approval of the LHC project, the computing needs of the accelerator and the four experiments (ALICE, ATLAS, CMS, LHCb) started to be considered again in

---

[5]http://lcg.web.cern.ch/LCG/PEB/PASTAIII/pasta2002Report.htm

[6]Technical Design Reports (TDR) on Higher Level Triggers, Data Acquisition and Physics for each LHC experiment to be found in respective websites (http://greybook.cern.ch/)

[7]LHC Design Report, 1995: CERN/AC/95-05 (LHC); 20 October 1995; The LHC Study Group and http://cern.ch/lhc

[8]LHC Experiments Technical Proposals, 1994: CERN/LHC/94-43; LHCC/P2, 15 December 1994.

1996. In a common effort between ATLAS and CMS, the MONARC project (MONARC, 2000) provided key information on the design and operation of the worldwide distributed Computing Models for the LHC experiments.

These computing models envisaged thousands of physicists involved in data analysis

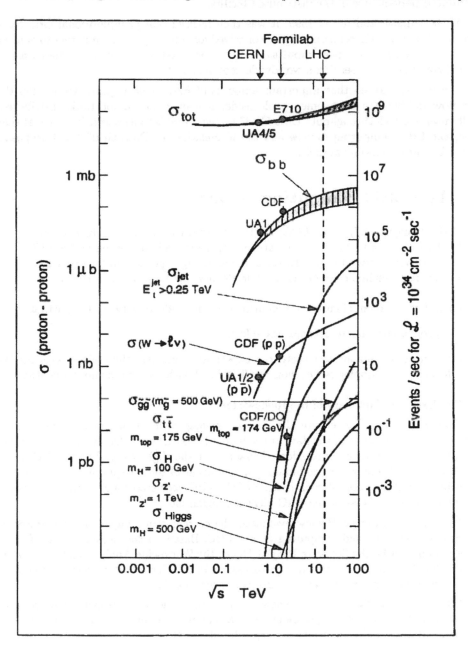

Figure 1. *Proton-proton cross sections as a function of centre of mass (cms) energy.*

at institutes around the world making use of a complex, tiered set of compute and data servers connected by powerful wide-, regional- and local area networks. Each of the experiments envisaged storing and partially distributing data volumes of Petabytes per year. The MONARC project provided for a simulation tool to assess and optimise the cumulated performance of the distributed facility.

In the summer and fall of 1999, it was then realised that Computational Grid technology (Foster and Kesselmann, 1998), conceived for coupling supercomputers to achieve maximum performance supercomputing, could be extended and adapted to data-intensive, high throughput tasks serving a worldwide community.

It was also realised that important amounts of R&D were required to develop the workflow and resource management tools needed to manage such a worldwide, distributed, multi-user and high throughput "Data Grid" system. Two activities were launched at that time, the LHC Computing Review and the formulation of "Data Grid" R&D proposals with EU and US funding agencies.

3   The LHC computing review

The LHC Computing Review (LCG, 2001) evaluated the status of LHC computing and established plans and prospects for data management and computing at the LHC. The Review involved representative software and computing experts from the experiments and the Information Technologies Division. The Review was supposed to establish:

- solid requirements figures and identified common efforts between the partners;

- a proposal on how to organise the efforts;

- the need to collaborate closely, the experiments amongst themselves, with the Information Technologies Division at CERN and with national computer centres.

The Executive Summary of the Review stated:

"The requirements to ensure the storage, management, simulation, reconstruction, distribution and analysis of the data of the four LHC experiments (ALICE, ATLAS, CMS and LHCb) constitute an unprecedented challenge to the High Energy Physics (HEP) and Information Technology (IT) communities. Fulfilment of these requirements will, on the same basis as the successful construction of the LHC accelerator and particle detectors, be crucial for the success of the LHC physics programme.

The LHC Computing Review evaluated the current situation, plans and prospects of data management and computing at the LHC. Based on the detailed work of three independent panels, the Software Project Panel, the Worldwide Analysis and Computing Model Panel and the Management and Resources Panel, this report reaches the following main conclusions and makes associated recommendations".

After critical assessment, the review accepted the scale of the resource requirements as submitted by the four experiments and as summarised in Table 1. A selection of recommendations and conclusions are reproduced below.

Parameter	Unit	ALICE		ATLAS	CMS	LHCb	TOTAL
		p-p	Pb-Pb				
# assumed Tier1 not at CERN		4		6	5	5	
# assumed Tier2 not at CERN*					25		
Event recording rate	Hz	100	50	100	100	200	
Running time per year	M sec-onds	10	1	10	10	10	
Events/year	Giga	1	0.05	1	1	2	
Storage for real data	PB	1.2	1.5	2.0	1.7	0.45	6.9
RAW SIM Event size	MB	0.5	600	2	2	0.2	
Events SIM/year	Giga	0.1	0.0001	0.12	0.5	1.2	
Number of reconst. passes	Nb	2		2-3	2	2-3	
Storage for simul. data	PB	0.1	0.1	1.5	1.2	0.36	3.2
Storage for calibration	PB	0.0	0.0	0.4	0.01	0.01	0.4
Total tape storage / year		4.7		10.4	10.5	2.8	28.5
Total disk storage		1.6		1.9	5.9	1.1	10.4
Time to reconstruct 1 event	k SI-95 sec	0.4	100	0.64	3	0.25	
Time to simulate 1 event	k SI-95 sec	3	2250	3	5	1.5	
CPU reconstruction, calib.		65	525	251	1040	50	1931
CPU simulation	k SI-95	19	269	30	587	660	1564
CPU analysis		880		1479	1280	215	3854
Total CPU at CERN T0+T1		824		506	820	225	2375
Total CPU each Tier1 (Avg.)	k SI-95	}234		}209	204	}140	}787
Total CPU each Tier2 (Avg.)*					43		
Total CPU		1758		1760	2907	925	7349
WAN, Bandwidths							
Tier0 - Tier1 link, 1 expt.	Mbps	1500		1500	1500	310	4810
Tier1 - Tier2 link		622		622	622		

* for all except CMS, the Tier1 and Tier2 needs are merged together.

Table 1. *Consolidated Computing Resources planned by the four LHC Experiments in 2007 (or the first full year with design luminosity).*

A multi-Tier hierarchical model similar to that developed by the MONARC project should be the key element of the LHC computing model. In this model, for each experiment, raw data storage and reconstruction will be carried out at a Tier0 centre. Analysis, data storage, some reconstruction, Monte-Carlo data generation and data distribution will mainly be the task of several (national or supra-national) "Regional" Tier1 centres, followed by a number of (national or infra-national) Tier2 centres to which the end users would be connected appropriately. The CERN-based Tier0+Tier1 hardware for all LHC experiments should be installed as a single partitionable facility.

Grid Technology will be used to attempt to contribute solutions to this model that provide a combination of efficient resource utilisation and rapid turnaround time.

Estimates of the required bandwidth of the wide area network between Tier0 and the Tier1 centres arrive at 1.5 to 3 Gbps for a single experiment. The traffic between other pairs of nodes in the distributed systems will be comparable.

Concerning software, joint efforts and common projects between the experiments and CERN/IT were recommended to minimise costs and risks. In more detail the following points were raised:

- Data Challenges of increasing size and complexity should be performed as planned by all the experiments until LHC start-up.

- CERN should sponsor a coherent programme to ease the transition of the bulk of the physics community from Fortran to Object Oriented (OO, C++) programming.

- Further identified areas of concern were the limited maturity of current planning and resource estimates, the development and support of simulation packages and the support and future evolution of analysis tools.

Concerning resources, the complete distributed facility was estimated in terms of materials cost and of personnel and material cost at CERN. The PASTA extrapolation was used to assess the cost. The CERN material effort was about one third of the total.

The Core Software teams of all four experiments were regarded as currently seriously understaffed, as well as the CERN staff for developing the software, control and management tools needed for the distributed facility.

The approach to Maintenance and Operation of the LHC computing system included the strategy of rolling replacement within a constant budget. Maintenance and Operation would require within each three-year operating period an amount roughly equal to the initial investment assuming that the computing needs increase along with the cost/performances advances.

The construction of a common prototype of the distributed computing system should be launched urgently as a joint project of the four experiments and CERN/IT, along with the major Regional Centres. It should grow progressively in complexity and scale to reach ~50% of the overall computing and data handling structure of one LHC experiment in time to influence the acquisitions of the full-scale systems.

4   "Data Grid" projects

The vision of Grids is the long standing wish to connect computing resources together in a transparent way when they are required for a given task, or expressed in the words of Larry Smarr and Charles Catlett, 1992: "Eventually, users will be unaware they are using any computer but the one on their desk, because it will have the capabilities to reach out across the internet and obtain whatever computational resources are necessary."

The word "Grid" is derived from the electrical power grids which provide for a transparent service, namely to make available at standardised sockets electrical power at given tension and a limited amount of current alternating at 50 Hz. The user does not need to understand the complexity of the power stations and their power distribution system and their delivery to the socket with given parameters. In Ian Foster's and Carl Kesselman's definition, the "Grid infrastructure will provide us with the ability to dynamically link together resources as an ensemble to support the execution of large-scale, resource-intensive, and distributed applications."

In the layered view originally introduced for networking, the following layers could be associated with an integrated science environment (snapshot from DataGrid):

- "Problem Solving Environment" layer: domain specific application interfaces for scientists;

- Grid Middleware layer:

  - Grid Application: job, data and metadata management;

  - Collective Services: Grid scheduler, replica manager, information and monitoring services;

  - Underlying Grid Services: SQL database, computing element, storage element services, replica catalogue, authorisation, authentication, accounting, service index;

  - Fabric services: resource, configuration, node and installation and storage managements, as well as monitoring and fault tolerance;

- Networking and Computing Fabric layers.

The challenge of the task is to identify an architecture of the tasks which allows factorizing them into individual services whilst preserving simplicity, overall efficiency and transparent use. This is the reason for CERN and its collaborating institutes to foresee ample testing opportunities involving the whole community.

Several Data Grid projects were launched subsequently, covering different aspects of the problem. In Europe, and supported by the EU, DataGrid (large datasets, high throughput, many users), DataTag (quality of service and high bandwidth in Grids), crossgrid (more applications) and EGEE (Grid infrastructure to enable e-science in Europe); in European nations, to name some of the major ones, GridPP (UK), GridKa (DE), INFNGrid (I), Nordugrid (Scandinavian Countries), Virtual Labs (NL) and finally in the US: GriPhyn, PPDG and iVDGL, named together: "Trillium".

The projects where CERN was or is deeply involved will be described here. The European DataGrid[9] project was elaborated, proposed and approved in the years 1999 and 2000. It was funded by the European Union and the participating institutes with the aim of setting up a computational and data-intensive Grid of resources for the analysis of data from scientific exploration.

The main goal of the DataGrid initiative was to develop and test the technological infrastructure that will:

---

[9]http://cern.ch/edg

- Enable the implementation of scientific "collaboratories" or "virtual organizations" where researchers and scientists will perform their activities regardless of geographical location and where they will collaborate and share resources and facilities for agreed common objectives;

- Allow interaction with colleagues from sites all over the world as well as the sharing of data and instruments on a scale not yet previously attempted;

- Devise and develop scalable software solutions and testbeds in order to handle many PetaBytes of distributed data, tens of thousand of computing resources (processors, disks, etc.), and thousands of simultaneous users from multiple research institutions.

The DataGrid initiative was led by CERN together with five other main partners and fifteen associated partners. The project brought together the following European leading research agencies: the European Space Agency (ESA), France's Centre National de la Recherche Scientifique (CNRS), Italy's Istituto Nazionale di Fisica Nucleare (INFN), the Dutch National Institute for Nuclear Physics and High Energy Physics (NIKHEF) and UK's Particle Physics and Astronomy Research Council (PPARC). The fifteen associated partners come from the Czech Republic, Finland, France, Germany, Hungary, Italy, the Netherlands, Spain, Sweden and the United Kingdom.

The project covered three years, from 2001 to 2003, with over 200 scientists and researchers involved. The EU funding amounted to 9.8 million euros, mostly for people. The contributions from the partners exceeded that amount.

The first and main challenge facing the project was the sharing of huge amounts of distributed data over the currently available network infrastructure. The DataGrid project relied upon emerging Computational Grid technologies[10] that were expected to make feasible the creation of a giant computational environment out of a distributed collection of files, databases, computers, scientific instruments and devices.

The DataGrid project was divided into twelve Work Packages distributed over four Working Groups: Testbed and Infrastructure, Applications, Computational & DataGrid Middleware, Management and Dissemination. Figure 2 illustrates the structure of the project and the interactions between the work packages.

More details on the functions and relations of the various middleware packages can be found in the DataGrid tutorials[11].

DataGrid has obtained most of its objectives by the end of 2003. The important "deliverables" were:

- Proof of viability of the Grid concept;

- Provision of training (>200 persons trained) for the applications addressed, namely for particle physics, bio-medical and earth observation communities, based on extensive testbeds;

- Delivery of a tested and stable middleware package for the LHC Computing Grid, the delivered package containing basic elements of the Globus toolkit, contributed

---

[10]http://www.globus.org
[11]http://eu-datagrid.web.cern.ch/eu-datagrid/Tutorial/tutorial.htm

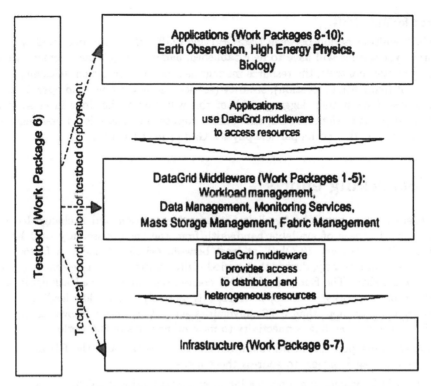

Organisation of the technical work packages in the DataGrid project

Figure 2. *Interaction between Work Packages.*

by Trillium issuing the "Virtual Data Toolkit" (VDT) together with Globus and a substantial number of components delivered by DataGrid.

DataGrid delivered these items before and during 2003.

In 2003, and within the now active 6th Framework Programme of the EU, the possibility arose of a new Grid project named EGEE[12] for "Enabling Grids for E-Science in Europe". A number of the DataGrid participants and many further institutes from all over Europe joined in elaborating the proposal. In the meantime the proposal is basically accepted for an initial period of two years.

The desired goal of the project is to create a general European Grid infrastructure of production quality on top of present and future EU Research and education networks. The project will build on the major investment by the EU and EU member states in Grid Technology and will exploit the international connections with the US and Asia Pacific. It is, in particular, accepted to use the particle physics community for the early adoption and for most of the initial tests and debugging to be performed during the first two years of the project. To this end, the project foresees a close collaboration with the LHC Computing Grid Project described below. The most probable start-up time for EGEE is

---

[12]http://cern.ch/egee

envisaged for April 2004.

In the transition period between DataGrid and EGEE the basic functionality for the LCG prototype service will have to be established, namely the job management services (submit, monitor, control), the resource management services (location, matching), data management (replication, location), security (services as they evolve) and operations and fault diagnosis (accounting, logging). All of this will have to be done in close technical partnership with Globus, VDT and other projects, for example the e-science core programme in the UK and its particle physics Grid project GridPP.

5   Networking efforts

In the past 20 years networks for digital information transfer and exchange have had a phenomenal increase of available bandwidth from the then customary $\sim$10 Kb/s to today's equally customary 10 Gb/s, at least between privileged nodes, a factor of 10^6 increase. More importantly, during that period digital networks have become a generally accepted commodity. The European GÉANT connections between the National Research and Education Networks (NRENs) in Europe is probably the world's leading research network, connecting more than 3100 universities and R&D centres in over 32 countries across Europe and providing connectivity to most other parts of the world.

However, Grids pose particular requirements to networks, and the EU and the US funded the DataTag[13] project to address those needs.

The DataTAG project is to create a large-scale intercontinental Grid testbed that focuses upon advanced networking issues and interoperability between these intercontinental Grid domains, hence extending the capabilities of each and enhancing the worldwide programme of Grid development.

The project addresses the issues which arise in the sector of high performance Grid and inter-Grid networking, including sustained and reliable high performance data replication, end-to-end advanced network services, and novel monitoring techniques.

The Transport Control Protocol (TCP), which is today the most common solution for reliable data transfer over the Internet, is not well adapted to the needs of Grid operations and needs to be tuned and adapted to large sustained data-flows. The DataTag project is ongoing.

6   The LHC Computing Grid Project (LCG)

In September 2001, CERN Council approved the first phase of the LHC Computing Grid[14] project based on the needs established in the Computing Review discussed above and aimed at full exploitation of LHC data by all participating institutes and physicists. The Council followed the proposal of the CERN Management of a two-phased LHC Computing Grid project resulting from discussions with the experiments and collaborating institutes subsequent to the review, and its proposals for action. It was stated in more detail that:

---

[13]http://cern.ch/datatag
[14]http://cern.ch/lcg

- The size and complexity of this task required a worldwide development effort and the deployment of a worldwide distributed computing system, based on tiered regional centres, to which all institutes and funding agencies participating in the LHC experimental programme were asked to contribute;

- The envisaged common exploitation by the LHC experiments of the distributed computing system required coherence, common tests and developments, and close co-operation between all nodes of the computing system amongst each other and in particular with the CERN installations.

The LHC Computing Grid Project was conceived as a two-phased approach to the problem, covering the years 2001 to 2008 or, in more detail:

- Phase 1: Development and prototyping at CERN and in Member States and Non Member States from 2001 to 2005, requiring expert manpower and some investment to establish a distributed production prototype at CERN and elsewhere that would be operated as a platform for the data challenges of the experiments. The experience acquired towards the end of this phase would allow the elaboration of a Technical Design Report, which will serve as a basis for agreeing the relations between the distributed Grid nodes and their co-ordinated deployment and exploitation.

- Phase 2: Installation and operation of the full world-wide initial production Grid system in the years 2006 to 2008, requiring continued manpower efforts and substantial material resources.

Further, a formal project structure was proposed that would ensure the achievement of the required functionality and performance of the overall system with an efficient use of the allocated resources. Participation in the project structure by the LHC experiments and the emerging regional centres would ensure the formulation of a work plan addressing the fundamental needs of the LHC experimental programme. The formulation of the work plan as a set of work packages, schedules and milestones would facilitate contributions by collaborating institutes and by pre-existing projects, in particular the EU DataGrid and other Grid projects. Appropriate liaisons with these pre-existing projects as well as with industry would be put in place to promote efficient use of resources, avoid duplication of work and preserve possibilities for technology transfer.

It was proposed that CERN should lead the project and would perform the necessary CERN-based development and prototyping activities, given the special roles as host lab. CERN hosts the Tier-0 and Tier 1 centre, where storage of the raw data emerging from the four detectors will be kept, and their first pass reconstruction will be performed, as well as providing analysis capacity for people at CERN and as data source for the rest of the Grid infrastructure.

The project was clearly seen to provide an excellent training ground for external staff from participating countries and institutes who would acquire substantial expertise, which would subsequently become available to their institutes or home countries. CERN would finally concentrate all its available computing efforts into that project and execute it in close collaboration with the LHC experiments and the collaborating computer centres.

The LCG project aims at creating a global virtual computing centre for particle physics, focusing at the service aspects for LHC physics and its world-wide community of physicists.

LCG has four working areas, namely the:

- Application area for the LHC experiments, including the development of the software environment, the joint projects between the experiments, the data management and the distributed analysis;

- Grid deployment area, establishing and managing the Grid Service, middleware certification, security, operations, registration, authorization and accounting, to be performed together with the EGEE project described above and possibly other complementary Grid projects elsewhere in the world;

- Grid technology or middleware area, providing a base set of Grid middleware – acquisition, development, integration, testing and support, acquiring software from Grid projects or industry;

- Fabric area with large cluster management, data recording, cluster technology, networking and computing service at CERN.

The latter three areas would proceed gradually, offering increasingly difficult services, starting with simulation, then batch analysis and later interactive analysis. Middleware acquired from DataGrid and VDT would be used first, then re-engineered middleware from projects such as EGEE or others based on increased experience with testbeds exercising intensive large scale batch work (the "data-challenges" of the LHC experiments) and later interactive analysis.

At this time and concerning the Grid deployment area, LCG1, the first LCG production Grid, 24 hrs/day and 7 days/week, is operating and undergoing extensive tests. Releases with upgraded functionality will succeed each other until early next year where LCG2 will offer a stable environment for the planned data challenges of the LHC experiments. The main goals for the time between now, autumn 2003, and early spring 2004 will be to establish a production quality service for the data challenges, robust, fault tolerant, of sufficient capacity and well integrated into the distributed LHC physics computing services around the world.

For the Grid Technology area, the important event in autumn will be the publishing of the ARDA (Architectural Roadmap Towards Distributed Analysis) study. The results of this study will give rise to new prototype Grid testbeds, probably based on variants of the Globus 3 toolkit offering "Grid services".

For the CERN fabric area, a 1.1 GB/s data recording rate has been sustained over a period of 8 hours and 0.9 GB/s over 3 days, matching the required recording rate for Phase II of the project, namely data taking for all LHC experiments combined during LHC operation. For wide area networks a new internet land speed record was set by transmitting at a rate of $\sim$ 1 GB/s from Chicago to Geneva for a sustained transmission of a Terabyte of data. The other focus of the Fabric area activities is to prepare the CERN computer centre for the increase in number of computers and their power requirements which rise roughly linearly with their performance.

Finally, in the application area, the scope of the LHC experiments common software is now almost completely defined. The areas addressed are:

- The Software Process and Infrastructure (SPI) package addressing the librarian, QA, testing and developer tools, documentation and training;

- The Persistency Framework (POOL), the hybrid data store, object oriented derived from ROOT combined with a relational database management system;

- The Core Tools and Services (SEAL), the foundation and utilities libraries, basic framework services, system services, object dictionary and whiteboard, Grid enabled services and mathematical libraries;

- The Physics Interfaces (PI), the interfaces and tools by which the physicists use the software directly, interactive and distributed analysis, visualization and Grid portals;

- The Simulation with the packages GEANT 4 and FLUKA, the simulation framework, the generator services and the physics validation. The present effort in this field is directed towards integrating these application tools "upwards" into the application software of the individual experiments and "downwards" into the Grid services and via the Grid services and the networks into the distributed computing fabrics around the world.

7 Particle physics and e-science, e-infrastructure, cyberinfrastructure

Science that is competitive at a global scale will require coordinated resource sharing, collaborative processing and analysis of huge amounts of data produced and stored by many scientific laboratories belonging to several countries and institutions. Such sciences have complete mechanisms for collecting the relevant raw data, turning raw data into validated, calibrated pieces of information. Such pieces of information are ordered, completed and validated and described coherently. Finally, in the comparison with underlying theoretical understanding, such data confirm or not the current understanding and complement and extend the knowledge produced by that science. With the exponential growth of the amounts of data in many sciences this process is performed relying on ICTs. Sciences that analyse and understand all their data in this manner are e-sciences.

In the words of the then Director of the UK Research Councils, Sir John Taylor, "e-science is about global collaboration in key areas of science and the next generation of infrastructure that will enable it".

Particle physics in the LHC context is probably the most advanced e-science at this time. In many countries, Grids and e-science are seen as the next step towards a comprehensive "cyberinfrastructure", as the environment is named by the US-NSF. They state that the cost and complexity of the 21st century science requires the creation of an advanced and coherent global information infrastructure (infostructure). The process, in their view, will be driven by global applications, changing the way science is done. Particle physics with the LHC challenge is such an early driving science and the US is interested to join forces globally to advance the "infostructure" using this particular global scientific effort.

In a similar way, and possibly even ahead of the US, the EU has made e-infrastructure one of their key actions in the 6th Framework Programme, supporting early production Grids such as EGEE as e-infrastructure and the "complex problem solving" initiative to advance the scope of the comparatively simple particle physics to more complex sciences and their enabling application technologies and underlying advanced architecture, design and development of the next generation Grids.

The UK e-science initiative has been instrumental in advancing e-science applications and collaborations, and serves as a model for Europe and its individual nations. CERN and the LHC community are grateful to their support and the speaker is grateful to the organizers for the opportunity to present results of that support.

Acknowledgement

Numerous colleagues have contributed to the content of these notes, in particular persons from the European DataGrid (http://cern.ch/edg), the LHC Computing Grid, LCG (http://cern.ch/lcg), and the EGEE (http://cern.ch/egee) projects. At the respective web sites ample and comprehensive information can be found. I would like to mention, in particular, Fabrizio Gagliardi (CERN), Les Robertson (CERN), Manuel Delfino (CERN), Alois Putzer (Kirchhoff Institut, Heidelberg) and A Reinefeld (Zuse Institut, Berlin). All of those have significantly contributed to my understanding of the subject and, more importantly, to the mentioned projects.

References

I. Foster and C. Kesselmann (1998), "The GRID: Blueprint for a New Computing Infrastructure", Morgan Kaufmann Publishers, San Francisco, 1998.
   Compare also: UNICORE (http://www.fz-juelich.de/unicore/) and e-Science Programme in the UK (http://www.escience-grid.org.uk/).

LCG (2001): S Bethke, (chair) MPI Munich, M Calvetti, INFN Florence, H F Hoffmann, CERN, D Jacobs, CERN, M Kasemann, FNAL, D Linglin, IN2P3, "Report of the Steering Group of the LHC Computing Review", CERN/LHCC/2001-004, CERN/RRB-D 2001-3; 22 February 2001.

MONARC (2000): M Aderholz (MPI) *et al,* "Models of Networked Analysis at Regional Centres for LHC Experiments" (MONARC), Phase 2 report; KEK preprint 2000-8, CERN/LCB 2000-001, April 2000. http://www.cern.ch/MONARC/

Higgs and new physics searches

Michael Andrew Parker

University of Cambridge, UK

1   Introduction

The Standard Model (SM) of particle physics represents an amazing triumph of theory and experiment. The agreement of the predictions with the high precision data from LEP and elsewhere is excellent for all the observables so far tested. The theory itself is economical, with a small set of matter fields, and includes the startling predictions of the unification of the weak and electromagnetic interactions, and the existence of the Higgs field giving masses to matter. The only unknown parameter in the model is the mass of the Higgs itself, and the precision data indicates that it should be discovered at a mass not too much above the existing search bound.

However, particle physicists do not rest on their laurels for long. The theory is widely regarded as flawed, and an enormous range of hypotheses have been put forward to correct its perceived weaknesses. These lectures will focus on two of the most popular extensions to the SM - supersymmetry and extra space dimensions.

The long wait since the last major physics discoveries, the W and Z bosons, has allowed beyond the standard model physics to evolve without experimental constraints for perhaps too long. As Richard Lewontin, writing on biology in the New York Review of Books put it in 1995:

"the problem is not a want of a theory, but a want of evidence. If scientific advance really came from theorizing, natural scientists would have long ago wrapped up their affairs and gone on to more interesting matters."

In considering the models on offer, we must consider not only the elegance of the theory, but also the experiments that can choose between them. The existing evidence for each hypothesis will be considered, and most importantly the experimental tests which are now planned. The Large Hadron Collider is likely to provide the first solid evidence of physics beyond the SM, and its search potential will be considered in some detail.

2   Reaching beyond the standard model

While the SM is rather economical theoretically, it still contains at least 19 parameters which are input from experimental measurements. These include the lepton and quark masses, the strengths of the forces, the electroweak mixing angle, the parameters of the CKM matrix, including the CP-violating phase, and the parameters of the Higgs sector. The SM also fails to explain some striking facts. For example, the charge on the electron is the same magnitude as that on the proton to very high precision (otherwise electrostatic forces would dominate gravity in the Universe). Yet there is no link in the theory between the quark and lepton sectors. The existence of 3 families of quarks and leptons is also unpredicted. These facts, among others, point strongly to the existence of a framework underlying the SM in which these parameters are naturally related.

In order to search for new physics, two broad classes of experiment are possible. Direct searches attempt to produce new particles in their real states - this allows their masses and properties to be determined experimentally, but of course the reach of such a search is limited by the available centre-of-mass energy. This approach has been the driving force behind LEP, the Tevatron, the LHC and the planned Linear Collider. Alternatively, one can search for manifestations of new physics through virtual diagrams. Such searches are not limited by accelarator energy. For example, the proton can decay to $e^+\pi^0$ via the subprocess $d + u \rightarrow \bar{\bar{s}} \rightarrow e^+\bar{u}$ in supersymmetric models with R-parity violation. Measurements of proton decay at a mass scale of 1 GeV can therefore constrain physics models including squarks with masses of several TeV. However, the interpretation of a deviation from the SM in such a process is highly model dependent. Hence the principal use of such results has been to constrain the existence of new physics using null results.

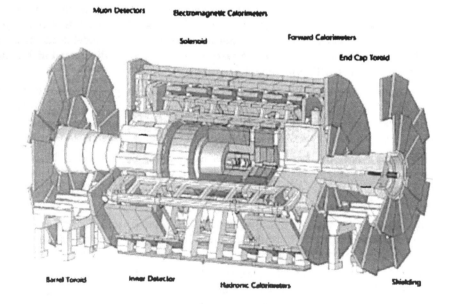

Figure 1. *The ATLAS Detector.*

The LEP experiments have performed direct searches for new particles in the 100 GeV range, setting many limits such as (Hagiwara 2002):

- SM Higgs mass > 114.3 GeV,

- heavy charged lepton masses > 100.8 GeV,

- heavy neutral lepton masses > 90.3 GeV,

- new quark masses > 46.0 GeV.

These, and many other detailed results, illustrate the power of e^+e^- machines in new physics searches. However, in order to proceed with this technique, at least an order of magnitude increase in energy is required. Since pointlike cross-sections drop with the square of the energy, two orders of magnitude increase in luminosity is required. A linear collider running with LEP's luminosity at 1 TeV would aquire only one event per day. Reaching such high luminosities, with a machine in which the beam is not stored but created afresh for each pulse, is a major challenge.

By contrast, the Large Hadron Collider runs with stored proton beams, making its design luminosity of 10^{34}cm^{-2}s^{-1} (relatively) easy to obtain. With an expected total cross section of 100 mb, this gives the astonishing event rate of 10^9 events per second! Fortunately, most of these are low energy scatters which are removed by simple triggers. The principal difficulty is that interesting high energy collisions are overlayed by the pileup of around 18 other events. This complicates the reconstruction of the interesting events. However, the tracks from the pileup can be separated from those from the signal events by their low transverse momentum.

Two general purpose detectors are under construction for the LHC, ATLAS (Figure 1) and CMS (Figure 2). The CMS detector follows the conventional layout for a collider detector, with a single solenoid magnet providing both the field for the momentum measurement in the inner tracker, and (by the return field) the bending for the momentum measurement in the outer muon system. This simplifies some aspects of the design, but it also involves some compromises. The calorimeter system is constrained by the geometry of the solenoid field, and more importantly, the bending power of the field is reduced for particles emitted near the beam axis, since the particles travel almost parallel to the field lines diverging from the end of the solenoid. CMS overcomes this by the use of a 4T central field, making the magnet a very challenging engineering problem.

ATLAS has chosen a different approach. In order to provide a uniform bending field in the muon system, a toroidal field is created outside the calorimeter, while a 2T solenoid provides the field in the tracking region. This layout avoids constraints on the geometry of the calorimeter (other than cost), and decouples the design of the tracker from that of the muon system.

Simulations indicate that both designs will provide a very similar discovery potential. In fact, the long approval process for these detectors virtually guarantees such an outcome, as review committees seek to optimise the detector performance by understanding and eliminating any gross differences. In these lectures, I shall discuss the performance of ATLAS in more detail, but it may be taken as read that CMS will be able to compete on all measurements.

3   Higgs searches

From the point of view of an experimentalist, the SM is an excellent model. All the available experimental data is explained to high precision. The theory has been checked at distance scales of $1/M_W = 2.5 \times 10^{-18}$ m. Only one state, the Higgs, and only one unknown parameter, the Higgs mass m_H, are unaccounted for.

The Higgs field itself is a completely new beast in the quantum jungle, being a scalar field, present throughout the vacuum. The mechanism by which it gives rise to particle masses can be envisaged by the analogy of a dipole in a ferromagnet. At high temperatures, the dipoles of the ferromagnet point in random directions, giving a symmetric field with a zero expectation value. The energy of a test dipole does not depend on its orientation. However, at low temperature, the ferromagnet undergoes a phase change, and all its dipoles spontaneously align along a common direction. The field is no longer symmetric, it has a non-zero expectation value, and the energy of test dipoles will be at a minimum if they are parallel with the field. The same process is thought to occur with the Higgs field in the vacuum, except that the direction of the field is along a direction of the SU(2) quantum number, rather than a direction in space. This means that particles with different SU(2) states are affected differently by the presence of the Higgs field, leading ultimately to the W and Z boson masses.

There is no contradiction between the best fit Higgs mass of 81 GeV and the search

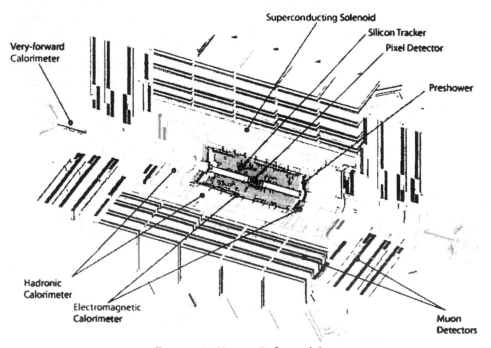

Figure 2. *The CMS Detector.*

limit of 114 GeV, since the Higgs has only a logarithmic effect on the likelihood, as shown in Figure 3 (LEP Electroweak Working Group, 2002).

The fit gives an upper limit, at 95% confidence, that $m_H < 193$ GeV. The Higgs mass is therefore very well constrained, and one would expect a discovery to be simply a matter of time.

3.1 Higgs searches at LEP

The LEP experiments detected a hint of a signal just before the machine was closed, with a preferred Higgs mass of 115 GeV. Figure 4 (LEP Working Group for Higgs Boson Searches, 2003) shows a plot of the reconstructed masses of the Higgs candidates in channel $e^+e^- \rightarrow HZ \rightarrow 4$ jets, for one particular set of cuts. The SM background and the predicted

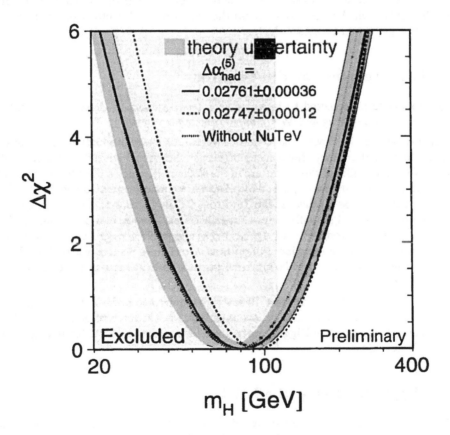

Figure 3. *The χ^2 for a fit to all precision electroweak data as a function of the Higgs mass. The fit is performed with only SM fields.*

signal for $m_H = 115$ GeV are displayed. The interpretation of such data is always difficult, and the LEP community did an exemplary job in analysing the significance of their data. Confidence limits were calculated for a variety of variables, and for different sets of cuts, and there was much discussion of the statistical interpretation. It should be noted that the derivation of confidence limits is a matter of some debate in expert circles, and that confidence levels from different experiments cannot be combined in general. In dealing with such problems, the advice of James (James F. 1986) should always be followed. It is mandatory to publish the number of events in the signal (if necessary divided into bins), the corresponding background estimates, and any conversion factors such as integrated luminosity necessary to obtain physical units. This contains all the new information from the experiment, and crucially can be combined with other experiments in a meaningful way. It is optional to publish an upper limit, and you may derive that limit by any method you choose, as long as the method is explained.

The LEP data in Figure 4 corresponds to 17 data events, with 15.8 background and 8.4 signal expected. Clearly this does not constitute a discovery. But the full data set contained many more variables, which could be combined with the mass information, and other cuts could be used to suppress the background. Nonetheless, no statistically significant signal could be established.

3.2   Higgs searches at the Tevatron

Higgs production is expected to be copious at the Tevatron, mediated by diagrams involving virtual top quarks and W's. Some examples are shown in Figure 5.

The problem at the Tevaton is to find the signal for Higgs production. To understand this, consider the dominant Higgs decay modes, as shown in Figure 6 (Spira et al 1995). At the lowest mass, 80% of the events are in the $b\bar{b}$ channel, which suffers from a large QCD background. At higher masses, the WW, ZZ and $t\bar{t}$ channels start to contribute. In the mass range that can be reached at the Tevatron, the only alternative to the $b\bar{b}$ channel is the WW decay mode. This suffers from large QCD background, and poor mass resolution when a W decays hadronically. In the case of semileptonic decays, the reconstruction is hampered by the missing neutrino. All of these factors have been considered by the CDF and D0 teams (Carena M. et al 2000), resulting in a best estimate of the discovery reach shown in Figure 7.

This analysis shows that at least 10 fb^{-1} are required to make a discovery above the existing LEP limit. Run II, which is now underway, was expected to deliver about 2 fb^{-1}. This makes is likely that the Higgs discovery will be made at the LHC (assuming of course that the Higgs exists).

3.3   Higgs searches at the LHC

Although ATLAS and CMS are designed to search for any new physics, the detector designs have been driven by the demands of the Higgs search. Over the full mass range considered interesting by theory (from 100-1000 GeV), the decay modes of the Higgs change as new channels open. Also, the event rate drops at high mass as the limit of the incident parton energy is approached. This forces the experiments to consider

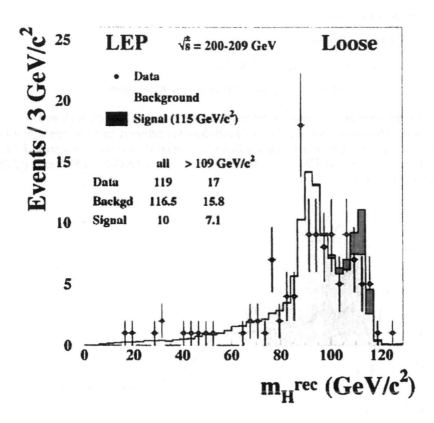

Figure 4. *The mass distribution of Higgs candidates seen at LEP (points) together with the SM background and the signal expected for a Higgs mass of 115 GeV.*

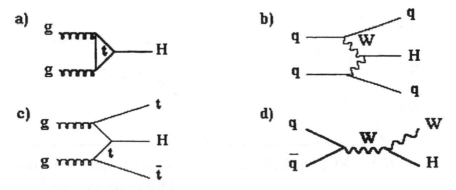

Figure 5. *Some diagrams contributing to Higgs production at the Tevatron. a) gluon fusion, b) W fusion, c) top associated production, d) WH associated production.*

search channels with higher branching ratios, even though they may be more difficult to reconstruct. Figure 8 (Spira *et al* 1995) shows the branching ratios of interest at the LHC. The channels offering significant discovery potential are:

- $H \rightarrow \gamma\gamma$ for $m_H < 150$ GeV.
- $H \rightarrow b\bar{b}$ for $m_H < 160$ GeV.
- $H \rightarrow ZZ \rightarrow llll$ for $120 < m_H < 700$ GeV.
- $H \rightarrow ZZ \rightarrow ll\nu\nu$ and $H \rightarrow WW \rightarrow l\nu jj$ for $m_H > 700$ GeV.

Each of these channels presents a different experimental challenge, and will be discussed in turn. The components of the ATLAS detector relevant to each signal will be discussed as well, showing how the design is linked to the needs of the Higgs search. Fuller accounts of the ATLAS and CMS detectors can be found in (ATLAS 1999) and (CMS 1997-2002) respectively.

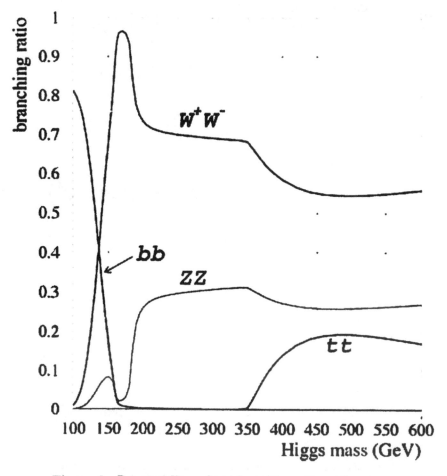

Figure 6. *Principal Higgs decay branching ratios at low mass.*

3.3.1 $H \to \gamma\gamma$

This channel is based on the small branching ratio (a few times 10^{-3}) of the Higgs to photons, produced by loop diagrams. It is of interest because it is the only channel at very low mass which does not suffer from a very large QCD background from jets. Nonetheless it requires the detection of a relatively small resonance in the photon pair mass on a significant background from π^0 pairs, whose decay photons overlap in the calorimeter. The $H \to \gamma\gamma$ channel makes the most stringent demands on the electromagnetic calorimeters of ATLAS and CMS, and both have adopted high technology solutions: liquid argon for ATLAS, and scintillating lead-tungsten crystals for CMS.

The requirements set by ATLAS for the calorimetry include

- rapidity coverage to a pseudorapidity of ± 5, to give good reconstrution of missing energy signatures.

- an electromagnetic energy resolution of $\Delta E/E \leq 10\%/\sqrt{E}$, with a systematic uncertainty $< 1\%$ (E in GeV). This is needed in particular for the $H \to \gamma\gamma$ channel.

- an uncertainty on the energy scale of better than 0.1%, in order to make precision mass measurements.

- high granularity, in order to correctly associate energy deposits with tracks. Liquid argon has the relatively low Moliere radius of 7.2 cm, setting the limit on what can be acheived.

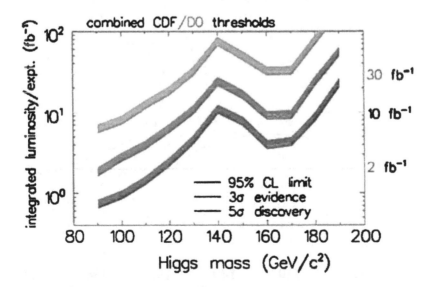

Figure 7. *The discovery potential for the Higgs at the Tevatron, as a function of Higgs mass. The contours show the reach for signals with 3 and 5 σ significance as well as the region which can be excluded with 95% confidence.*

- total thickness > 24 radiation lengths, to prevent energy leakage.

- must survive high radiation levels at the LHC. Liquid argon is inert, and the liquid, which acts as both detector and absorber, can be replaced.

Liquid argon is a noble gas with a high electron density and it allows ionized electrons to travel without trapping. It therefore can act both as an absorber for electromagnetic showers, and as the detector medium. The liquid surrounds a set of electrodes with an accordion geometry, which apply the necessary electric field to collect the ionization, and pass the collected charge to the back of the calorimeter structure for measurement. This structure avoids dead regions of mechanical support and gives a very uniform response. The electromagnetic calorimeter is backed by a hadronic calorimeter constructed of scintillating plastic tiles interleaved with steel, with a total weight of over 3000 tons, and around 10,000 readout channels. This massive device measures jet energies, distinguishes between electromagnetic and hadronic showers, and is crucial for the missing energy resolution.

The signal expected in ATLAS in this channel is shown in Figure 9 (ATLAS 1999) for $m_H = 120$ GeV. The large background from photon pairs in the SM is subtracted,

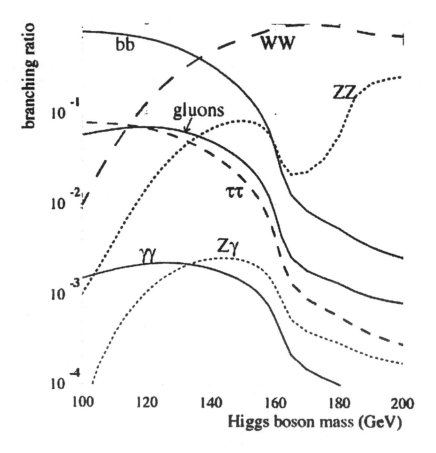

Figure 8. *Higgs decay branching ratios as a function of mass.*

leaving a clearly resolved signal peak, thanks to the excellent calorimeter resolution. An example of one event is shown in Figure 10, for $m_H = 100$ GeV, at design luminosity. The two calorimeter clusters from the photon pair can be seen easily. One cluster is split into two components, since its parent photon converted to an e^+e^- pair in the material of the tracker. The conversion of photons is a source of low energy tails in the mass resolution, in the case that one of the electrons has low momentum and impacts too far from the main cluster. The high energy tracks pointing to this cluster can be seen by eye among the many low transverse momentum tracks produced by the underlying event and the pileup.

3.3.2 $H \to b\bar{b}$

The channel $H \to b\bar{b}$ can be detected only if the tracking detector (called the Inner Detector or ID in ATLAS) can tag b-quark decays with sufficient efficiency and background rejection. The layout of the ATLAS ID is shown in Figure 11. The detector has three major subsystems: the pixel detector, the semiconductor tracker (SCT) and the transition radiation tracker (TRT). The whole ID is enclosed in a solenoid with a 2T central field.

The pixel detector comprises a barrel section and endcap disks. The innermost layer is devoted to b physics. The pixel size of 50×300 μm gives a resolution of 14 μm in the r-ϕ direction and 87 μm in the z direction. The detector can operate successfully just outside the beampipe thanks to its low occupancy of 10^{-4} per pixel and 10^{-2} per readout column.

Beyond the pixel detector the occupancy drops as the radius increases, and so the detector element size can increase. The region from 30 to 60 cm in radius is occupied

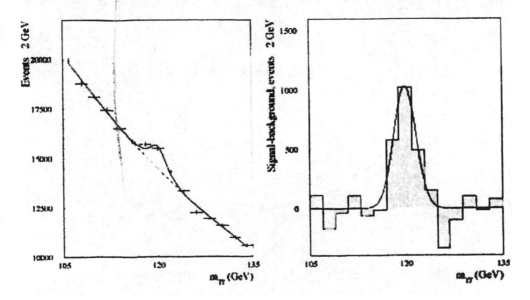

Figure 9. *The Higgs signal expected in ATLAS in the γγ channel, for $m_H = 120$ GeV, before and after background subtraction.*

by the SCT, consisting of silicon detector modules. Each module is constructed from 4 silicon wafers, each 6 cm long. Each side of the module has two detectors bonded together, providing 768 strips, each 12 cm long, with an intrinsic resolution of 20 μm. The two sides are offset by an angle of 40 mrad, allowing the z coordinate of tracks to be reconstructed using stereo. With over 4000 modules in the final system, around 12 kW of heat is dissipated in the detector. This threatens the dimensional stability of the support structure, if the cooling system does not maintain a uniform temperature. In addition, both silicon and pixel detector wafers suffer from radiation damage, from an annual incident particle flux equivalent to 10^{15} neutron hits per cm^{-2}. This causes p-type doped silicon to turn to n-type, effectively reversing the diode junctions at the heart of the devices. Further radiation damage can lead to increases in the detector current,

Figure 10. *A H $\rightarrow \gamma\gamma$ event simulated in the ATLAS detector, for $m_H = 100$ GeV at a luminosity of 10^{34}cm^{-2}s^{-1}.*

finally causing thermal runaway as the heat dissipation increases, and leading to the destruction of the detector. To cope with these problems, the cooling system is based on the evaporation of fluorocarbons, as used in domestic and industrial refrigeration units. A mixed-phase flow of liquid and vapour will remove large quantities of heat using the latent heat of evaporation, and maintain a constant temperature until the last of the liquid is used up. These properties make evaporative cooling a more attractive option than the use of a pure liquid.

Beyond the radius of the SCT, the occupancy continues to drop, and so a new technology is employed, with a lower density of readout channels. The transition radiation detector (TRT) makes use of the electromagnetic radiation created when a charged particle crosses the boundary between two materials with differing dielectric constants. The energy lost into radiation is given by $I = \alpha\gamma\hbar\omega_p/3$ where α is the fine structure constant, γ is the boost of the incident particle, and ω_p is the plasma frequency in the material. For $\gamma > 1000$ the emission is in the soft X-ray range (2-20 keV). Electrons will radiate for momenta > 0.5 GeV, while pions will only radiate for momenta > 140 GeV. Hence the X-ray signature can be used to separate electrons from pions. Since the emission angle is $\sim 1/\gamma$, the X-rays follow the particle track. The X-ray emission probability is $\sim 1\%$ per transistion, so many bonudaries are needed to obtain a significant signal.

These considerations lead to the TRT design concept. A radiator consisting of many foils of polypropolene create the X-rays. Drift tubes are used to record the ionization from the charged particles, and also to convert the X-rays to e^+e^- pairs using a heavy noble gas (Xenon). The additional ionization from the pairs is added to that of the parent particle, creating a signal that can discriminate between incident electrons and pions.

In the TRT, 4 mm diameter drift tubes are used, with a drift time measurement giving the position of the incident track with a precision of 170 μm in the coordinate perpendicular to the tube. The device provides 36 points per track, in contrast to the 6 very high precision points provided by the pixels and SCT together. The occupancy in the TRT is around 20% but the high density of points allows the pattern recognition to function. The transition radiation signature allows electrons to be selected with 90% efficiency, while rejecting 99% of incident pions.

Figure 12 shows the expected ATLAS performance in detecting the associated production channel $pp \to t\bar{t}HX$, $H \to b\bar{b}$. This channel is favoured because of the large H-top coupling. The full power of the ID is needed to find the b-vertices and to reconstruct the top signature. The events with tagged b-quarks show a resonance above the expected SM background.

3.3.3 $H \to ZZ \to llll$

The 4 lepton channels are the so-called "golden" channels for the LHC. The signature stands out very well from the SM background, and the mass reconstruction of the Higgs is straightforward. The main difficulties are with the acceptance and mass resolution for a low mass Higgs. Since the Higgs is produced by two quarks with unequal momenta, the final state is boosted along the beam axis, with the boost being potentially large for the low Higgs mass case. This means that the probability of one of the final state leptons being lost into the beam pipe is significant. For this reason, the ATLAS tracking system is designed to cover a pseudo-rapidity range of ±2.5 units. Also, for $m_H < 2m_Z$, one of the Z's is virtual, and so the Z mass cannot be used as a constraint, placing a premium on good mass resolution. In the case of electrons, the LAr calorimeter can provide the required performance. The ATLAS muon system makes use of an air-core toroid to avoid multiple scattering of the muons as they pass through the system. This enhances the resolution for these lower energy decays. The field is created by eight 25 m long superconducting coils forming the barrel system, and two endcaps, each with 8 smaller coils. Three layers of drift chambers provide the main momentum measurement, while faster resistive plate chambers, cathode strip chambers and thin gap chambers are required to form a sufficiently fast trigger. Figure 13 shows a simulation of an event in the $H \rightarrow e^+e^-\mu^+\mu^-$, where the two muons can be seen passing through the barrel muon system. Only the chambers hit by tracks are displayed, allowing the large coils of the barrel toroid magnet to be seen. The cryostat containing the forward muon toroid is also cutaway, showing the smaller coils creating the field in the forward region. The two electrons can also be seen in the barrel LAr calorimeter. Figure 14 shows a $H \rightarrow e^+e^-\mu^+\mu^-$ event in the inner detector. The high transverse momentum tracks from the Higgs decay are reconstructed in spite of the large number of softer tracks looping in the solenoid field.

The performance of the detector in the various four lepton channels varies a little, owing to the differing resolution of the muon and calorimeter systems. Figure 15 shows the reconstructed Higgs mass for $e^+e^-e^+e^-$, $e^+e^-\mu^+\mu^-$ and $\mu^+\mu^-\mu^+\mu^-$ events, for a Higgs mass of 130 GeV. The best mass resolution, of 1.4 GeV, is obtained for the 4 muon mode, where the muon system makes an excellent measurement of the track momenta. The modes with electrons show a worse resolution, with tails to low masses, caused by energy loss from the electrons as they pass through the tracker material. This worsens the mass resolution to 1.5 GeV for $e^+e^-\mu^+\mu^-$ events, and to 1.6 GeV for $e^+e^-e^+e^-$ events. Of course, for higher transverse momentum particles the position is reversed: the resolution of the muon system gets worse as the track sagitta reduces, while the calorimeter resolution improves as $1/\sqrt{E}$ until the systematic limit is reached. This means that the electron signature is more powerful for high mass Higgs and other objects such as Z' resonances.

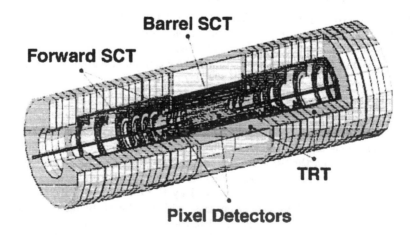

Figure 11. *The layout of the ATLAS Inner Detector.*

3.3.4 $H \to ZZ \to ll\nu\nu$ and $H \to WW \to l\nu jj$

At high mass, the production rate of Higgs bosons falls, as the available parton luminosity falls with the parton distribution function. Beyond 700 GeV, too few events are available in the four lepton channels to make a discovery. However, the rates for $H \to ZZ \to ll\nu\nu$ and $H \to WW \to l\nu jj$ remains high, since all 3 flavours of neutrino are available. The final state cannot be reconstructed, but the SM background is low at high mass. Figure 16 shows the signal, and signal plus background distributions expected for the reconstructed mass in the WW channel, for Higgs masses of 800 GeV and 1 TeV. The signal emerges clearly from the falling SM background as the Higgs mass rises, extending the discovery reach to 1 TeV.

3.4 Summary of the LHC Higgs search reach

The LHC search reach is summarised in Figure 17, which shows the significance of the Higgs signal, defined at $S/\sqrt{B}$ where S is the number of signal events, and B the expected

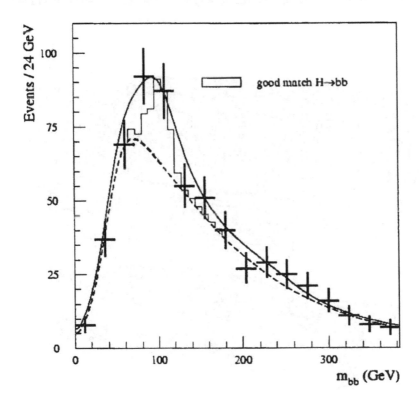

Figure 12. *The expected signal from $H \to b\bar{b}$ decays, produced in association with $t\bar{t}$ in ATLAS, for $m_H = 100$ GeV (points). The signal events with matched b-quarks are shaded. The SM background is shown as a dashed line.*

background, after 1 year of running at nominal luminosity (ie an integrated luminosity of 30 pb^{-1}). The significance of the channels contributing at each mass is added. The full Higgs mass range is covered, up to 1 TeV, with over 5σ significance, guaranteeing a discovery if the Higgs exists. After 3 years of running, the signal significance rises to over 10σ over the full mass range.

The ability of the LHC detectors to measure the mass and width of the Higgs has also been investigated. Figure 18 shows the precision expected on these parameters. Below 500 GeV, the Higgs mass can be obtained to almost 0.1% precision, with the resolution deteriorating rapidly at higher mass. The Higgs width is hard to measure at low mass (where the intrinsic width is small), but a 1% precision can be obtained for masses above 300 GeV.

Finally, an enormous amount of work has been devoted to understanding the search reach for the various Higgs bosons present in supersymmetric models. Figure 19 shows the excluded regions for various channels in the plane of m_A vs $\tan\beta$ where m_A is the mass of the CP-odd Higgs, and β is the ratio of the vacuum expectation values of the two Higgs doublet fields. The full power of ATLAS and CMS will be needed to perform the search, but with 300 pb^{-1} of integrated luminosity the whole plane can be excluded.

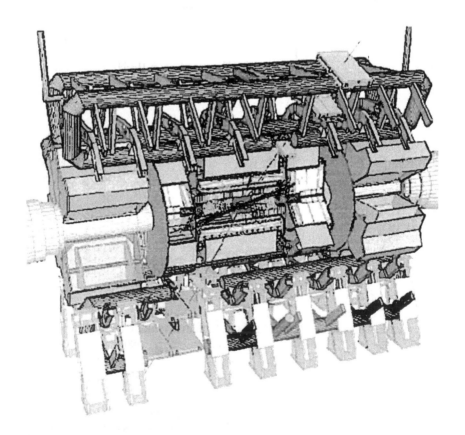

Figure 13. *A simulation of a $H \to e^+e^-\mu^+\mu^-$ decay in the ATLAS detector.*

4   Supersymmetry

The theory of supersymmetry has been developed over many years by many authors. In these lectures, I have used the excellent review by Martin (Martin S.P. 1997) as the source of the theoretical remarks. This contains comprehensive references to the original literature.

4.1   The Higgs mass and the hierarchy problem

The need for a Higgs field, or some other new physics can be seen by inspecting the cross-sections for processes such as $e^+e^- \rightarrow W^+W^-$ in the SM. If the Higgs is not included, this process, which is a pure $J = 1$ transition, has a differential cross-section of

$$\frac{d\sigma}{d\Omega} = G_F^2 \frac{E^2 \sin^2 \Theta}{8\pi^2} \, ,$$

corresponding to a partial wave amplitude of $f = G_F E^2/6\pi$. Unitarity demands that $f \leq 1$, and this limit is violated for $E > \sqrt{6\pi/G_F} = 1.3$ TeV.

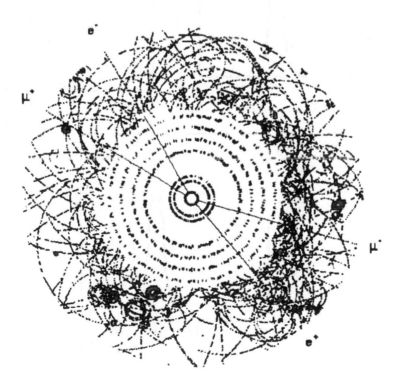

Figure 14. *A simulation of a $H \rightarrow e^+e^-\mu^+\mu^-$ decay in the ATLAS inner detector.*

In the SM, unitarity is guaranteed by extra diagrams involving Higgs exchange, which cancel those involving fermions. This guarantees acceptable high energy behaviour, as long as the Higgs couplings are as predicted in the SM: the fermion-Higgs coupling given by $gm/2M_W$, and the W-Higgs coupling by gM_W, where g is the coupling constant, m is the fermion mass, and M_W is the W mass. These couplings arise naturally in the SM. Since Higgs exchange is used to cure the unitarity problem, the Higgs mass cannot be allowed to rise too high, or the propagator will suppress the amplitude for this process, and the cancellation will not occur. From this argument, one can conclude that in the SM framework, $m_H < 780$ GeV. This limit does not apply directly in models beyond the SM, but unitarity still imposes a bound in the TeV region.

It is not possible to calculate the Higgs mass within the SM. The first order term

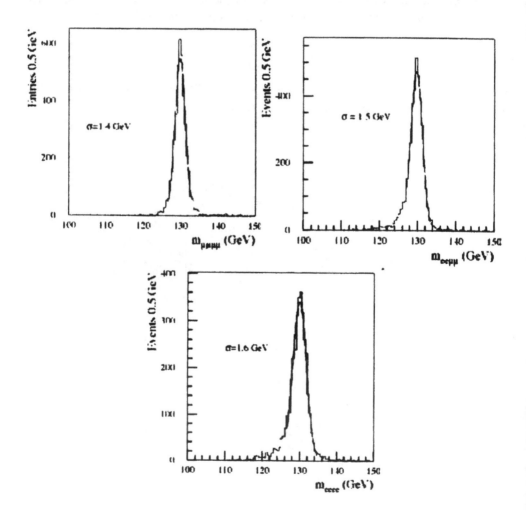

Figure 15. *Reconstructed Higgs masses for the 4 lepton decay modes: a) $\mu\mu\mu\mu$, b) $\mu\mu ee$, c) eeee.*

is given by the shape of the Higgs potential, but this is overwhelmed by contributions from fermion loop diagrams. These diagrams give a divergent contribution as there is no natural limit to the momentum that can circulate in the loop. If the Higgs exists, new physics must therefore occur at some point, stabilising its mass by some mechanism. It is possible to argue that the SM is a low-energy effective theory, and so a cut-off should be imposed on these internal momenta, corresponding to the scale of new physics. If this is done, the Higgs mass rises to the chosen scale, which must be around 1 TeV to be consistent with unitarity. Unfortunately, the only known physics above the electroweak scale is gravity, which would lead to a Higgs mass near the Planck mass, some 15 orders of magnitude too high! This is known as the "hierarchy problem" - it is very difficult to maintain two widely separated scales in the theory in the presence of quantum loop effects.

The search is therefore on for new physics: either the Higgs does not exist, in which case some other mechanism mimics the SM to provide the W and Z masses; or there is a

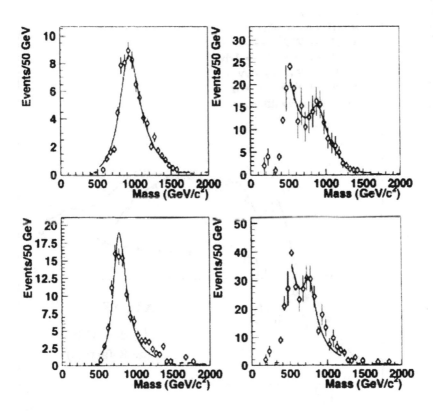

Figure 16. *The signal (left) and signal+background (right) for the reconstructed Higgs mass in the channel $H \to WW \to l\nu jj$ for $m_H = 1$ TeV (top) and $m_H = 800$ GeV (bottom).*

Higgs, in which case new physics is required to stabilise its mass. In both cases, the new physics is expected in the TeV region.

4.2   The Minimal Supersymmetric Standard Model (MSSM)

The best class of new physics would completely eliminate the quantum loop effects that cause the hierarchy problem, without introducing any arbitrary parameters into the model. It must also preserve the successes of the SM, and not violate any known facts, such as the measurements of the proton lifetime, the electric dipole moment of the neutron, or $g - 2$.

This can be achieved using supersymmetry. The fermion loops contribute a correction to the Higgs mass of

$$\Delta m_H^2 = \frac{|\lambda_f|^2}{16\pi^2}[-2\Lambda^2 + 6m_f^2 \ln(\Lambda/m_f) + ...],$$

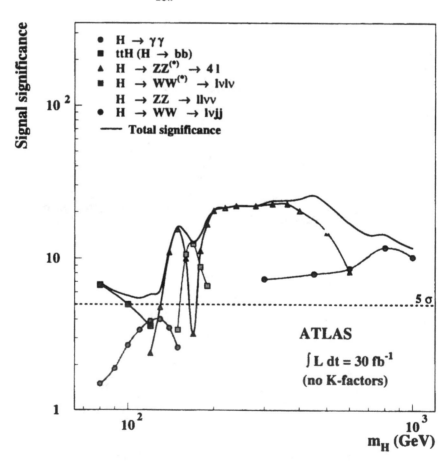

Figure 17. *The combined significance of the Higgs signals at the LHC, as a function of Higgs mass.*

while scalar particles contribute

$$\Delta m_H^2 = \frac{\lambda_s}{16\pi^2}[\Lambda^2 - 2m_s^2 \ln(\Lambda/m_s) + ...] ,$$

where m_f, m_s are the fermion and scalar masses, λ_f, λ_s are their couplings to the Higgs and Λ is the scale of new physics. The opposite signs on the leading terms are due to the spins of the particles. It is clear that if the theory were to contain two scalars for every fermion, with $\lambda_s = |\lambda_f|^2$, then the leading term would be cancelled exactly. At first sight, this appears to require the introduction of a vast number of arbitrary parameters, but this is not the case. The supersymmetry transformation links fermions to bosons, and since each Dirac spinor contains 4 degrees of freedom (particle and antiparticle in two helicity states), two scalars are generated per fermion. Furthermore, the couplings are related in exactly the way required for the cancellation.

Exact supersymmetry would require that the scalar masses were degenerate with the fermions, but this cannot be case, since no super-partners have been observed, and a scalar electron would be impossible to miss. However, the non-leading terms in the Higgs mass have only a logarithmic dependence on the fermion and scalar masses, so supersymmetry breaking does not spoil the cancellation. However, as expected, a SUSY mass scale in the TeV range is required to avoid violating the unitarity bound. SUSY cannot be exiled to a very high mass and still solve the hierarchy problem.

The simplest set of particle states based on the SM fields is known as the minimal supersymmetric standard model or MSSM. The particles involved are given in Table 1.

The only addition to the SM, beyond the doubling of states brought about by the SUSY operation itself, is the requirement that there be two Higgs doublets. This is necessary to cancel anomalies which arise in the theory. This only introduces on a new parameter above that in the SM.

The MSSM states then mix to form the physical mass eigenstates. This produces

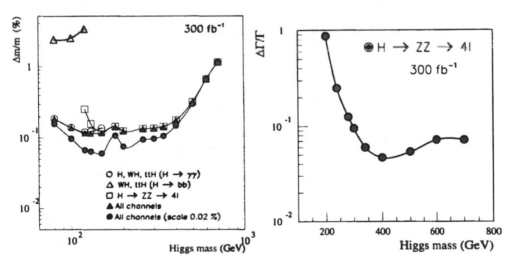

Figure 18. *Expected precision for measurements of the Higgs mass (left) and Higgs width (right).*

five Higgs bosons (after the electroweak symmetry breaking, which requires 3 degrees of freedom to give masses to the W and Z). These are two CP-even neutral states, h^0 and H^0, a CP-odd neutral state, A^0, and two charged states $H^\pm$. The gaugino and higgsino states mix to create four neutralinos, $\tilde{\chi}_1^0, \tilde{\chi}_2^0, \tilde{\chi}_3^0, \tilde{\chi}_4^0$ (in order of increasing mass), and four charginos $\tilde{\chi}_1^\pm, \tilde{\chi}_2^\pm$. The exact mixing depends on the mass spectrum in the model.

The states can be labelled with a multiplicative quantum number called R-parity: $R_p = (-1)^{3(B-L)+2s}$, where B and L are baryon and lepton number, and s is the spin quantum number. All SM particles have $R_P = +1$ and all the superpartners have $R_P = -1$. R-parity conservation has the effect of suppressing proton decay in SUSY model to negligible levels, by removing diagrams involving SUSY particles mediating between quarks in the initial state and SM particles in the final state. It also prevents any mixing between SM and SUSY particles. These properties make it much easier to construct viable SUSY models, but nonetheless models do exist in which R-parity is violated.

R-parity conservation also has important consequences for experimental searches:

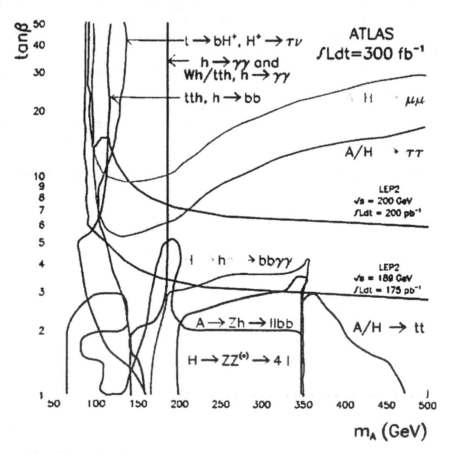

Figure 19. *The excluded regions for a variety of signatures in the plane of m_A vs $\tan\beta$ for 300 pb^{-1} of integrated luminosity.*

Multiplet name	Bosons	Fermions
Q	$(\tilde{u}_L, \tilde{d}_R)$	(u_L, d_R)
$\bar{u}$	$\tilde{u}_R^*$	$u_R^\dagger$
$\bar{d}$	$\tilde{d}_R^*$	$d_R^\dagger$
L	$(\tilde{\nu}_L, \tilde{e}_L)$	(ν, e_L)
$\bar{e}$	$\tilde{e}_R^*$	$e_R^\dagger$
H_u	(H_u^+, H_u^0)	$(\tilde{H}_u^+, \tilde{H}_u^0)$
H_d	(H_d^0, H_d^-)	$(\tilde{H}_d^0, \tilde{H}_d^-)$
Gauge bosons/	$W^\pm, W^0$	$\tilde{W}^\pm, \tilde{W}^0$
gauginos	B^0	$\tilde{B}^0$
	g	$\tilde{g}$

Table 1. *The particle content of the Minimal Supersymmetric Standard Model.*

- any initial state must have $R_p = +1$ and so SUSY particles must be created in pairs. This requires energies of twice the SUSY mass scale.

- any SUSY particle decay must be to a state with $R_P = -1$, and so each final state contains another SUSY particle.

- the lightest SUSY particle (LSP) must therefore be stable.

- a stable LSP (unless very heavy) must be electrically neutral and weakly interacting to have escaped detection. It is therefore a potential dark matter candidate.

4.3 Supersymmetry breaking

The SUSY lagrangian is very highly constrained, since the fermion and boson couplings are linked, only one new parameter is introduced to cope with the second Higgs doublet, and all other parameters are known from the SM. However, unbroken SUSY could not possibly have escaped detection, with mass-degenerate SM/SUSY partner pairs providing a wealth of experimental signatures. Hence if SUSY exists in nature, it must be a broken symmetry in the current vacuum ground state, giving SUSY particles masses large enough to evade current experimental bounds.

Soft SUSY breaking is usually used to create the SUSY particle masses. This works by introducing terms into the Lagrangian that only involve the SUSY particles, explicitly breaking the symmetry. The mechanism generating these terms is a matter for debate (see below). Only certain terms are allowed in a renormalisable theory:

- gaugino masses M_1, M_2, M_3 for $\tilde{B}^0, \tilde{W}^\pm, \tilde{g}$.

- 3×3 matrices of scalar couplings $\mathbf{a}_u, \mathbf{a}_d, \mathbf{a}_e$.

- 3×3 matrices for the $\tilde{q}$ and $\tilde{l}$ masses.

- three parameters for the Higgs sector, $m_{H_u}^2, m_{H_d}^2, b$.

Each matrix covers the 3 families for each multiplet.

This means that SUSY breaking introduces a total of 105 new parameters into the theory, including masses, mixing angles and phases, all of which have to be determined from experiment. At first sight, this is a disaster, but the assumption is that once the underlying mechanism of SUSY breaking is understood, the parameters will be revealed to have a natural pattern. Each of the mechanisms proposed so far vastly reduces the number of free parameters.

4.4 SUSY breaking mechanisms

Soft SUSY breaking merely describes, in a general way, the possible parameters introduced by the broken SUSY symmetry. What is required is a mechanism that produces the SUSY breaking in a natural way. A form of dynamic symmetry breaking, like the Higgs mechanism, is the preferred solution, in which the symmetry breaking is generated by the interactions of the fields in the theory. However, the fields in the MSSM do not produce SUSY breaking. The hypothesis is therefore that new physics at some high scale is responsible, via the presence of some new fields in a "hidden sector". Symmetry breaking occurs in this sector, with the appearance of a vacuum expectation value. The hidden sector interacts weakly with the MSSM fields, and this interaction communicates the symmetry breaking to the SUSY particles.

Since many of the unknown parameters appear in the 3×3 matrices that describe the 3 families of particles, the number of parameters is greatly reduced if the interactions are "flavour-blind". For example, in supergravity models (SUGRA), the hidden sector communicates with the MSSM fields via gravitational interactions, giving the desired flavour independence. In the simplest version of these models, we then have

$$M_1 = M_2 = M_3 = m_{1/2}$$
$$a_u = a_d = a_e = A_0 y$$
$$m_{H_u}^2 = m_{H_d}^2 = m_0^2$$
$$b = B_0 \mu$$

This reduces the parameter set to $m_{1/2}, m_0, A_0, b$ and μ. Conventionally, b and μ are replaced by $\tan \beta$ (the ratio of the vacuum expectation values of the two Higgs doublets) and μ, the Higgs mass parameter. Supergravity models also include a gravitino with mass $m_{3/2} \sim m_{Soft}$ the scale of soft SUSY breaking. This gravitino is a good dark matter candidate.

A second example is gauge mediated SUSY breaking (GMSB). In these models the interaction between the hidden sector and the MSSM fields takes place via ordinary gauge interactions, carried by "messenger" particles. These could be states from a larger gauge group such as SU(5). In these models, the gravitino mass in not related to m_{Soft}, and it is expected to be very small, in the keV range. It is thus the LSP, and all final states would contain gravitinos.

GSMB models have 6 parameters:

- the SUSY breaking scale, $F_m \sim (10^{10}$ GeV$)^2$

- the messenger mass scale, M_m

- the number of messenger 5-plets, N_5

- the ratio of the Higgs vacuum expectation values, $\tan\beta$

- the coupling of the gravitino to other fields, C_{grav}

All other SUSY particles ultimately decay to final states including gravitinos, but the coupling strength is expected to be weak, leading to potentially long lifetimes for the next to lightest SUSY particle (NLSP). Two interesting possibilities include

- the neutralino is the NLSP, decaying via $\tilde{\chi}_1^0 \to \tilde{G}\gamma$, leading to a signature with high energy photons, which do not point back to the primary vertex if the lifetime is long.

- the slepton is the NSLP, decaying via $\tilde{l}_R \to \tilde{G}l$, giving a signature from massive sleptons passing through the detector, if the lifetime is long.

Another possibility, anomaly mediated SUSY breaking will not be discussed here.

5 Supersymmetry searches at the LHC

The presence of R-parity conserving supersymmetry should be easy to detect at the LHC, as long as the SUSY mass scale is in the accessible range, since the final states will all contain two LSPs, creating a missing energy signature. The R-parity violating case is discussed later. However, given the large range of parameter space, even in models like SUGRA, understanding the SUSY spectrum is not trivial. In this section, the methods proposed to detect and unravel the SUSY states are described, following the detailed studies in (ATLAS 1999). The general method is first to measure the SUSY mass scale using inclusive signatures which are insensitive to the details of the SUSY model. Next, detailed measurements of exclusive modes are used to extract kinematic information from the end-points of decay chains. The information on particle mass differences gleaned from these studies are then input into global fits to extract model parameters. On studying the problem, it immediately becomes clear that the discovery of SUSY will be easy - understanding what has been discovered will be very hard!

5.1 The LHC SUSY discovery reach

The LHC community initially chose some points in SUGRA parameter space to check the discovery potential. For each point, the mass spectrum and decay modes were calculated and events generated for study. The points chosen are shown in Table 2.

	m_0	$m_{1/2}$	A_0	$\tan\beta$	$\mathrm{Sign}\mu$
1	400	400	0	2	+
2	400	400	0	10	+
3	200	100	0	2	-
4	800	200	0	10	+
5	100	300	300	2.1	+
6	200	200	0	45	-

Table 2. *Points in SUGRA parameter space studied by the LHC collaborations.*

In these lectures, we shall look at the study of Point 5 in some detail. The first step is to estimate the SUSY mass scale. This can be done with a variety of variables sensitive to the high transverse momentum decay products characteristic of the decays of massive SUSY particles. For example, the effective mass

$$M_{eff} = E_T^{miss} + \sum_{i=1,4} p_T^i$$

takes the scalar sum of the missing energy and the four highest transverse momentum jets or leptons. The M_{eff} distribution for point 5 is shown in Figure 20, together with the SM background. It also shows the position of the peak of the distribution, plotted against the lower of the gluino and squark masses, that is the lightest strongly produced SUSY particle. The plot was made for a large set of randomly chosen SUSY parameters. There is a strong correlation evident, showing that M_{eff} can be used to determine the SUSY scale without further reconstruction of events.

For any given set of SUGRA parameters, the SUSY mass spectrum can be predicted by evaluating the relevant renormalisation group equations numerically, from the highest scale, down through the SUSY and electroweak symmetry breaking transitions. Figure 21 shows the squark and gluino masses in the $m_{1/2}, m_0$ plane, for $\tan\beta = 2$ and $\tan\beta = 10$, and for positive and negative μ.

Armed with predictions for the SUSY masses, one can simulate and study inclusive signatures, such as combinations of leptons, jets and missing energy, without attempting to reconstruct the underlying process in detail. The procedure used by ATLAS was:

- choose a set of SUGRA parameters and simulate a sample of events;

- separate the events according to signature;

- record those points in parameter space at which an excess in M_{eff} can be detected;

- plot contours in parameter space showing the reach for each signature.

Figure 22 shows the results. The search limit is defined by the region with significance $S/\sqrt{B} > 5$, and more than 10 signal events, for an integrated luminosity of 10 pb^{-1}. The LHC expects to deliver around 30 pb^{-1} in the first year of running, so this is a conservative estimate of the reach. The signatures shown are jets with missing energy and no leptons

(0l); jets, missing energy and one lepton (1l); same sign dileptons (SS); opposite sign dileptons (OS); trileptons (3l); dileptons with jet veto (2l0j); and trileptons with jet veto (3l0j). It can be seen that the missing energy signature alone is sufficient to give a good reach, although the inclusion of a single lepton (which reduces the QCD background) can extend the reach, since many SUSY decay chains include lepton production. The events with multiple leptons have lower background, but the reach is restricted by a lower rate, since these processes depend largely on the production of weakly interacting particles. The reach does not depend greatly on the other parameters ($\text{sign}(\mu)$ and $\tan\beta$).

5.2 Reconstructing SUSY particle masses

The problem with reconstructing SUSY masses at the LHC is that in each event, two LPSs are undetected. This means that complete event-by-event reconstruction is impossible, and even the use of variables such as transverse mass will not pick out a single particle mass. The key method used to determine SUSY masses at the LHC relies on identifying particular decay chains, in order to measure kinematic edges. These edges measure mass differences between particles, which can then be fitted to models. In some cases, it is possible to uniquely determine the masses of all the particles in the chain.

The best example is the chain $\tilde{q}_L \rightarrow q\tilde{\chi}_2^0 \rightarrow q l^\pm \tilde{l}^\mp \rightarrow q l^\pm l^\mp \tilde{\chi}_1^0$, which occurs at Point 5. This is a good experimental signature, with missing energy, and a lepton pair providing background reduction. Each step in the chain is a two body decay. This leads to kinematic edges in the distributions of two body mass. For example, Figure 23 shows the distribution of dilepton masses. A clear endpoint is above the small background, most of which comes from other SUSY processes. The edge position is given by

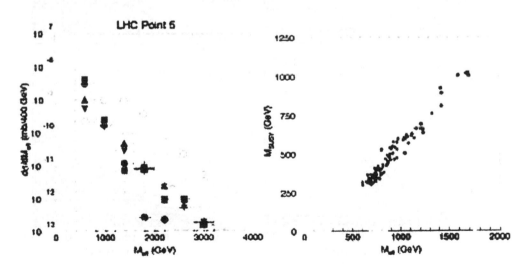

Figure 20. *Left: the distribution of M_{eff} at Point 5 for the SUSY signal (open circles), together with the total SM background (histogram), made up of $t\bar{t}$ (solid circles), W+jets (triangles), Z+jets (downwards triangles) and QCD (squares). Right: the peak of the M_{eff} distribution as a function of $M_{SUSY} = Min(m_{\tilde{g}}, m_{\tilde{u}_R})$ for various models.*

$$M_{ll}^{max} = M(\tilde{\chi}_2^0)\sqrt{1 - \frac{M^2(\tilde{l}_R)}{M^2(\tilde{\chi}_2^0)}}\sqrt{1 - \frac{M^2(\tilde{\chi}_1^0)}{M^2(\tilde{l}_R)}}$$

and the position of the edge can be measured to ~ 0.5 GeV with 30 pb^{-1} of data, giving 5800 signal events, 880 SUSY background, and only 120 SM background.

The full chain $\tilde{q}_L \to q\tilde{\chi}_2^0 \to ql^{\pm}\tilde{l}^{\mp} \to ql^{\pm}l^{\mp}\tilde{\chi}_1^0$ contains 4 unknown masses, and the dilepton edge gives one constraint. Figure 23 also shows the distribution of masses of the lq pair from this chain, where l is the first lepton to be produced. The fit includes the expected mass resolution, which is far worse than for the dilepton edge because of the jet energy measurement required. This edge adds a second constraint to the calculation of the SUSY masses. Proceeding in the same way, two edges can be extracted from the masses of the llq system, one a maximum, and the other a threshold. This gives a total of 4 constraints from this chain, on the four unknown masses. In addition, information can be obtained from the process $\tilde{q}_L \to \tilde{\chi}_2^0 q \to \tilde{\chi}_1^0 qh$, followed by $h \to b\bar{b}$. Hence it is possible,

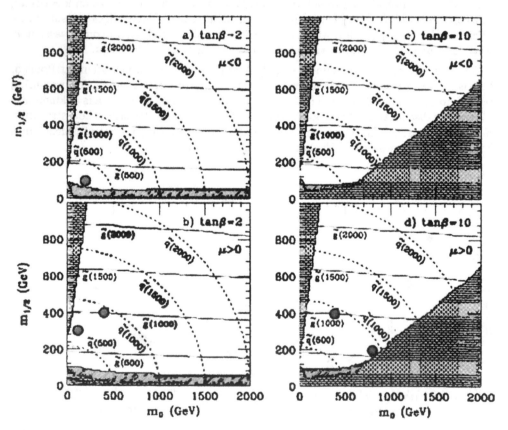

Figure 21. *Contours showing gluino and sqark masses in the $m_{1/2}, m_0$ plane for $A_0 = 0$, $\tan\beta = 2$ and $\tan\beta = 10$, and for positive and negative μ. The shaded regions are excluded by experiment, or an absence of electroweak symmetry breaking in the model (this is model dependent). The dots show the positions of Points 1 to 5.*

rather surprisingly, to completely solve for all the unknown SUSY masses, including that of the unobserved $\tilde{\chi}_1^0$.

Figure 24 shows the results in this case, for the squark and $\tilde{\chi}_1^0$ masses. The fractional errors obtained on the squark, $\tilde{\chi}_2^0$, slepton and $\tilde{\chi}_1^0$ masses are ± 3, ± 6, ± 9 and ± 12 % respectively. The masses obtained are of course highly correlated. This method demonstrates that it is possible in some cases to obtain fairly complete information on the SUSY particle masses. However, the particular example shown relies both on the existence of a favourable chain of two body decays, and sufficient rate so that the chain can be extracted from the background. At other SUSY points, there is no guarantee that this will remain true.

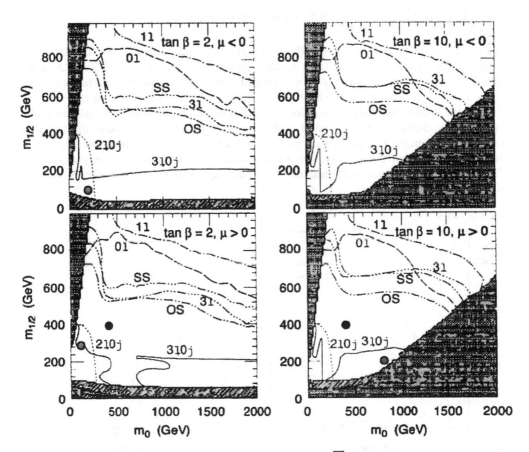

Figure 22. *Contours showing the search reach for $S/\sqrt{B} > 5$ in the $m_{1/2}, m_0$ plane for $A_0 = 0$, $\tan\beta = 2$ and $\tan\beta = 10$, and for positive and negative μ. The shaded regions are excluded by experiment, or an absence of electroweak symmetry breaking in the model (this is model dependent). The dots show the positions of Points 1 to 5.*

h mass	92.9 ± 1.0 GeV
Dilepton endpoint	108.92 ± 0.50 GeV
lq endpoint	478.1 ± 11.5GeV
Ratio of lq/llq endpoints	0.865 ± 0.060
(llq endpoint)	(271.8 ± 14.0 GeV)
hq endpoint	552.5 ± 10.0 GeV
hq threshold	346.5 ± 17.0 GeV

Table 3. *Input measurements for the SUGRA parameter fit at Point 5, with their errors, for a run of 30 pb^{-1}.*

5.3   Extracting the SUSY model parameters

Given the various measurements possible at Point 5, a fit can be performed to extract the underlying parameters of the SUGRA model. This of course assumes that SUGRA is the correct framework - the fit will not be meaningful if Nature has chosen a different solution. Table 3 shows the measurements that can be input, together with their expected errors, while Table 4 shows the output measurements. In the case of the lq and llq endpoints, the ratio is used, since they are correlated, and the scale error and other systematics cancel in the ratio. The output parameters are given both for a low and high integrated luminosity, illustrating the improvement in precision which could be obtained after 3 years of running. The main parameters are well measured, and the sign of μ is also determined, but the data is insensitive to A_0.

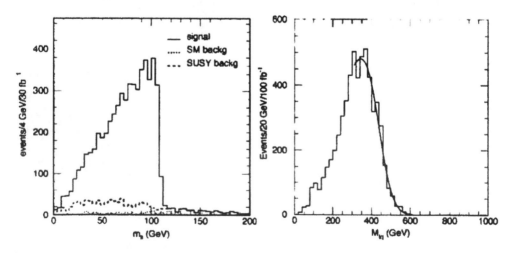

Figure 23.   *Left: The signal from the dilepton edge at Point 5 (histogram) with the background from SUSY processes (dashed) and from the SM (dotted). Right: the signal from the lq edge (histogram) with a fit including the expected resolution.*

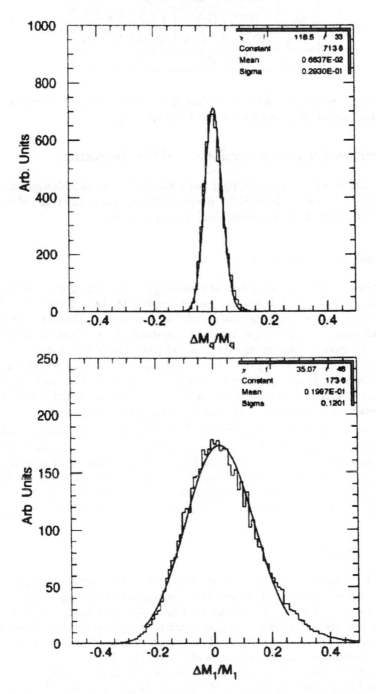

Figure 24. *Top: The mass resolution for the $\tilde{q}_L$. Bottom: the mass resoution for the $\tilde{\chi}_1^0$*

	Low Luminosity	High Luminosity
m_0	$100^{+4.1}_{-2.2}$ GeV	100.0 ± 1.4 GeV
$m_{1/2}$	300.0 ± 2.7 GeV	300.0 ± 1.7 GeV
$\tan \beta$	2.00 ± 0.1	2.00 ± 0.09

Table 4. *Output parameters for the SUGRA fit at Point 5, with their errors, for a run of 30 pb^{-1} (low luminosity) and 100 pb^{-1} (high luminosity).*

5.4 Searches for gauge mediated SUSY breaking

In these models, the gravitino is very light and is the LSP. The available signatures depend on the identity of the next lightest SUSY particle (NLSP), and the strength of its coupling to the gravitino. Four distinct possibilities have been considered.

5.4.1 Rapid decays of the $\tilde{\chi}^0_1$

If the $\tilde{\chi}^0_1$ is the NLSP, it will decay via the channel $\tilde{\chi}^0_1 \to \tilde{G}\gamma$. The decay rate is controlled by the parameter C_{grav}, which determines the coupling to the gravitino. For $C_{grav} \sim 1$ the decay is prompt, giving a signature with two high energy photons in every event, coupled with large missing energy from the gravitino. In addition, decays of heavier SUSY particles produce high energy leptons and jets. The SM background is therefore small, and the signature is very distinctive, making this case easy to distinguish from other SUSY models. Some 2% of the $\tilde{\chi}^0_1$ decays occur in the Dalitz mode, $\tilde{\chi}^0_1 \to \tilde{G}e^+e^-$. In these cases, the position of the decay vertex can be reconstructed from the electron tracks, allowing the lifetime of the $\tilde{\chi}^0_1$, and hence C_{grav} to be measured.

The decay chain $\tilde{\chi}^0_2 \to \tilde{l}^\pm l^\mp \to \tilde{\chi}^0_1 l^+ l^- \to \tilde{G} l^+ l^- \gamma$ can be used for mass reconstructions. Edges occur in the dilepton mass, the masses of the dilepton pair with each of the photons, and the masses of the photons with each lepton. If the assumption is made that the gravitino mass is negligibly small, it is possible to extract the masses of the $\tilde{l}_R$, $\tilde{\chi}^0_2$ and $\tilde{\chi}^0_1$ from this chain. The squark and gluino masses can also be found from the chain $\tilde{q} \to \tilde{g}q \to \tilde{\chi}^0_2 q\bar{q}q$, using events in which the $\tilde{\chi}^0_2$ has been reconstructed as above.

5.4.2 Slow decays of the $\tilde{\chi}^0_1$

If $C_{grav} >> 1$, the decay to gravitinos occurs very slowly. For example, if $C_{Grav} = 10^3$ then the decay length is 1.1 km. This means that the missing energy signature is essentially identical to that which occurs in SUGRA models. However, in many events, high energy photons will be detected, which do not point back to the primary vertex. This would be a smoking gun for the presence of GMSB. To establish this signature, the calorimeter must be capable of measuring the impact angle of the photon. Figure 25 shows the expected distribution in the ATLAS LAr EM calorimeter, for the angle of the photon compared to the line to the primary vertex. 82% of the photons miss the vertex by more than 5σ. After cuts on the photon energy, the overall signal efficiency is 52%. With 30 pb^{-1} of data, decay lengths as high as 100 km can be detected, corresponding to $C_{grav} = 10^8$.

5.4.3 Rapid decays of sleptons

In GMSB models with $N_5 > 1$, the NLSPs are right-handed sleptons, which decay promptly to gravitinos via the mode $\tilde{l}_R \rightarrow l\tilde{G}$. There are many signatures with leptons and missing energy, and it is simple to detect SUSY in these cases. Many mass measurements are possible. For example, Figure 26 shows the dilepton mass spectrum in one such model. Two edges are visible, one at 52.1 GeV given by $\sqrt{m^2(\tilde{\chi}_1^0) - m^2(\tilde{l}_R)}$ and one at 175.9 GeV given by $\sqrt{m^2(\tilde{\chi}_2^0) - m^2(\tilde{l}_R)}$, from the decays $\tilde{\chi}_{1,2}^0 \rightarrow \tilde{l}^{\pm}l^{\mp} \rightarrow \tilde{G}l^{\pm}l^{\mp}$.

5.4.4 Slow decays of sleptons

In the case that $C_{grav} \gg 1$, the slepton lifetime becomes long, with decay lengths reaching the kilometer range. The detectable signature is therefore the presence of two quasi-stable particles in each event. Because the slepton mass is high, time of flight can be used to measure the mass, taking advantage of the 10m path length to the muon system, and the 1ns time resolution of the muon chambers. Figure 27 shows the reconstructed mass for a model with a stau mass of 102.2 GeV, in which the stau decays via $\tilde{\tau} \rightarrow \tau\tilde{G}$. The

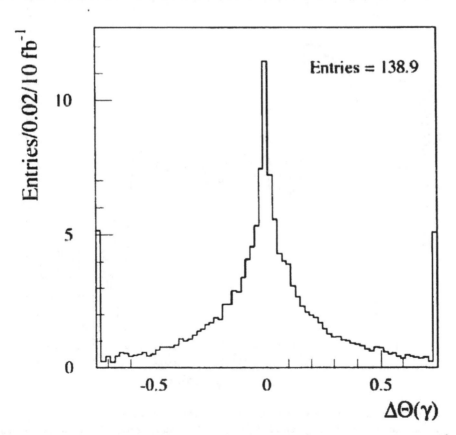

Figure 25. *The non-pointing angle for photons from $\tilde{\chi}_1^0$ decays, with $C_{grav} = 10^3$.*

mass is measured with a resolution of 3.8 GeV. Since the sleptons are observed directly, neutralino decays can be fully reconstructed in the channel $\tilde{\chi}_i^0 \to \tilde{l}_R l$. Figure 27 shows the spectrum of masses of slepton-lepton pairs, showing clear peaks at the masses of the $\tilde{\chi}_1^0, \tilde{\chi}_2^0$ and a weaker signal at the $\tilde{\chi}_4^0$. The $\tilde{\chi}_3^0$ has different decays due to its mixture of higgsino and gaugino components, and is not seen.

5.4.5 Extracting the parameters of GMSB models

In most cases, the main parameters of GMSB models can be extracted by a fit to the measurements outlined above. One year of running (an integrated luminosity of 30 pb^{-1}) is sufficient to determine Λ, N_5, $\tan\beta$ and the sign of μ with reasonable precision. However, in some cases, the effects of Λ and N_5 are not easily separated.

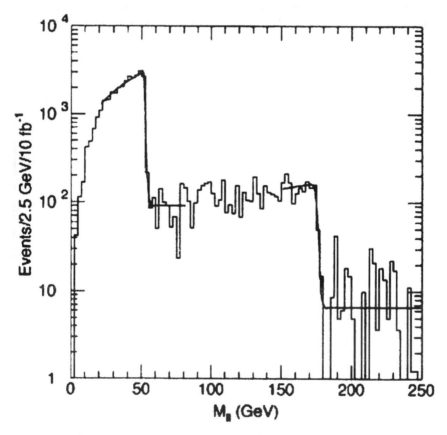

Figure 26. *The dilepton mass distribution for a GMSB model with $N_5 > 1$ and $C_{grav} = 1$. The curve shows a fit to the expected form of the edges, convoluted with the Gaussian resolution.*

5.5 R-parity violating models

R-parity provides a simple method to avoid proton decay in supersymmetric models, but the existence of R-parity itself is not motivated within the MSSM framework, and the symmetry has to be imposed by hand. There may be ways within larger grand unified frameworks to impose R-parity conservation naturally, but experiments must prepare for the possibility that R-parity is violated at some level.

R-parity can be broken by three distinct interaction terms, with couplings:

- $\lambda \neq 0$. This term violates lepton number with processes such as $\tilde{\chi}_1^0 \rightarrow l^+ l^- \nu$.

- $\lambda' \neq 0$. This term also violates lepton number, but through channels such as $\tilde{\chi}_1^0 \rightarrow q\bar{q}\nu$ or $q\bar{q}l$.

- $\lambda'' \neq 0$. This term violates baryon number, via channels such as $\tilde{\chi}_1^0 \rightarrow qqq$.

Taken together, these processes will mediate proton decay at a high rate. Therefore it is necessary to assume that all the coupling values are small or zero, which raises questions of fine tuning, and naturalness. Most experimental analyses of the search reach have assumed that only one coupling is non-zero, which effectively suppresses proton decay.

The value of the coupling strength has a strong influence on the phenomenology. For values $\lambda < 10^{-6}$ the LSP decay length is greater than the size of the detectors, and the experimental signatures are unchanged. For values in the range $10^{-6} < \lambda < 10^{-2}$, the LSP decays inside the detector. This means that the missing energy signature is suppressed, and in the case of $\lambda'' \neq 0$, removed altogether. This is a potential disaster for the LHC discovery potential, and the only case which has been studied in detail. For $\lambda > 10^{-2}$

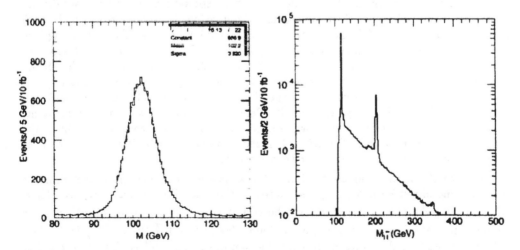

Figure 27. *Left: the slepton mass reconstructed from its time of flight to the muon system. Right: the mass of slepton-lepton pairs. Both plots are for a GMSB model with quasi-stable sleptons.*

all SUSY decay chains are grossly modified, with SUSY particles decaying promptly to normal matter. This case is not theoretically favoured.

Studies so far have found that signatures can be detected for each of the R-parity violating scenarios, with moderate strength couplings. The lepton number violating terms tend to give signatures with a large number of leptons, as well as missing energy from neutrinos. Some channels, such as $\tilde{\chi}_1^0 \to q\bar{q}l$ can be fully reconstructed; in other cases various endpoints and mass combinations allow the SUSY masses to be extracted. The most difficult case is for $\lambda'' \neq 0$, where the missing energy signature is totally absent, and the final state contains six quarks from the two LSP decays. The reconstruction of these multiquark states is difficult, given the high combinatorics, but studies indicate that a search exploiting the decay chain $\tilde{\chi}_2^0 \to \tilde{l}^{\pm}l^{\mp} \to \tilde{\chi}_1^0 l^+l^- \to l^+l^- qqq$ can reveal both the $\tilde{\chi}_2^0$ and $\tilde{\chi}_1^0$ by plotting the qqq mass against the $qqql^+l^-$ mass. The combinatoric and other backgrounds are suppressed by looking for the two resonances simultaneously.

6   Extra space dimensions

The conventional method of curing the quantum instabilities in the Higgs mass, outlined above, uses supersymmetry to cancel the offending fermion loop diagrams. This would be an extremely elegant and economical way to solve the problem, were it not for the difficulty of the supersymmetry breaking, which can in principle introduce 105 new parameters. It is entirely possible that the SUSY breaking mechanism will emerge naturally from some high-scale new physics and thus fix these parameter. However, at present the predictivity of SUSY is much reduced.

Interest in alternative ways to solve the hierarchy problem was raised by the publication of the ADD model (Arkani-Hamed *et al* 1998). This paper noted that in a Universe with extra space dimensions, the volume of the 'bulk' space-time with $3+n$ space dimensions could be much larger than the volume of the visible 3-dimensional space. In such a case, the application of Gauss' theorem to an isolated mass shows that the strength of gravity in the bulk is much larger than it appears in 3-D, since only a small fraction of the field generated by the mass appears in the 3-D subspace. The origin of the hierarchy problem is the large diffence between the strength of gravity (parameterized by the Planck mass), and the strength of the other forces, (characterized by the weak scale). If bulk gravity is actually stronger, the bulk Planck mass will be lower, and the hierarchy problem can be removed from the Higgs sector. The new physics required to stabilize the Higgs mass can simply be bulk gravity. This requires that the bulk Planck mass be of the order of the weak scale, and no larger than a few TeV.

In extra-dimensional models, the bulk Planck mass in $4+n$ dimensions (including the normal time dimension), $M_{Pl(4+n)}$, is related to the usual Planck mass M_{Pl} by $M_{Pl}^2 = M_{Pl(4+n)}^{(2+n)} R^n$, where R is the radius of the compactified extra dimensions. If we then require that $M_{Pl(4+n)} \sim M_W$, R can be found from $R \sim 10^{\frac{30}{n}-17} \text{cm} (\frac{1\text{TeV}}{m_W})^{1+2/n}$. This relation shows that the hierarchy problem has not in fact been solved, but transferred to the geometry of space - it is now necessary to explain why the n extra dimensions have a different scale from the 3 normal ones.

Setting $n = 1$ leads to $R = 10^{13}$ cm, which is the scale of planetary orbits, and clearly excluded. However, for $n = 2$, $R \sim 100$ μm - 1 mm, which is not immediately excluded. The presence of any extra dimension on this scale would be obvious, if the normal SM fields, such as fermions or photons were able to travel in them. Amongst other effects, it would be possible to send matter and light around the extra dimensions, and also the force laws for gauge interactions would be modified. The highly precise measurements of the electroweak interaction at scales of $1/M_W$ rule this out for any macroscopic scale. However, it is much more difficult to test the interaction strength of gravity at scales lower than 1 mm, and at the time of the publication of the ADD model, no measurements existed. The hypothesis is therefore made that only gravity propagates in the bulk, while the SM fields are confined to the 3D subspace, known as a brane.

Two very different approaches can be considered for direct searches for extra dimensions of this scale. The most staightforward way is to search for a deviation from the inverse square law of gravity. The other is to seek for effects at accelerators, produced by the gravitational effects in the bulk. These effects are observable because of the low scale of the Planck mass in these models, which makes the strength of gravity comparable to that of the other forces. In addition to these methods, indirect limits can be set from astronomical effects.

6.1 Experimental limits on extra dimensions from gravity experiments and supernovae

The Newtonian $1/r^2$ dependence of the gravitational force appears from Gauss' theorem as an immediate consequence of the surface area of a sphere increasing with r^2. This is a property of spheres in 3D spaces, and so if one rederives the force law in a 4+n dimensional space, the dependence changes to $1/r^{n+2}$. However, the spaces under consideration here consist of 3 infinite space dimensions, and n compacitified to a radius R. This means that a Gaussian surface of radius a will completely enclose the extra dimensions if $a > R$, and the Newtonian $1/r^2$ dependence will be seen. Only if $a < r$ does the Gaussian surface intersect the extra dimensions, and the new dependence becomes evident. Hence, it is necessary to seek for deviations from Newtonian gravity for masses separated by distances below the millimetre scale.

Gravity experiments normally present their results in terms of a Yukawa interaction rather than as r^{-n}. This gives a potential of the form

$$V(r) = - \int dr_1 \int dr_2 \frac{G\rho_1(r_1)\rho_2(r_2)}{r_{12}} \left[1 + \alpha \exp\left(\frac{-r_{12}}{\lambda}\right) \right] ,$$

where $\rho(r_1), \rho(r_2)$ are two mass distributions at r_1 and r_2 respectively, separated by r_{12}. The range of the Yukawa interaction is given by λ, and the strength relative to gravity is given by α. In the case of extra dimensions, we expect $\alpha \sim 1$, and λ of order of the extra dimension radius R.

A recent review (Long J.C. and Price J.C. 2003) shows the current limits on deviations from Newtonian gravity, from laboratory experiments, as well as the expected limits from planned experiments. They conclude that the existing measurements constrain the size of extra dimensions in the ADD model to $R < 200$ μm, or < 150 μm for $n = 2$. This restricts the range of parameter space for the model, but a reduced R can be compensated by a higher Planck scale.

Limits on large extra dimensions can be set from supernova data, since the rate of cooling and emission of neutrinos will be affected by energy loss into the bulk. The observation of neutrino emission from SN1987A provides direct information on the neutrino flux, and hence a key validation of the models of supernova collapse. However, the limits which can be obtained from supernova data are not clear: for example (Cirelli M. 2003) argues that limits from supernovae can be relaxed if R is not constant for all the dimensions, or if mixing occurs with sterile neutrino states in the bulk.

The existing constraints are therefore not sufficient to rule out the ADD model, and direct searches at colliders are needed to elimate a larger region of the parameter space.

6.2 Searches for large extra dimensions at colliders

The compactification of the extra dimensions gives a periodic boundary condition on the wavefunction of any state propagating in them. This leads to a series of excited states, equally spaced in energy, known as a "tower of Kaluza-Klein modes" after the famous theorists who first proposed a unification of EM and gravity using an extra dimensional model. In the ADD model, the large value of R leads to a small spacing between the states, which can be treated as a continuum. The only field propagating in the bulk is the graviton, and so massive gravitons can be excited by collider experiments, with masses up to the kinematic limit. This leads to a missing energy signature, as the gravitons disappear into the bulk space.

The potential for the LHC to discover such a signature has been studied by (Vacavant L., Hinchliff I., 2000) and (Kabachenko, V., Miagkov A., Zenin A., 2001). The missing energy signature can be detected in various channels, including channels with jets ($gg \rightarrow gG, qg \rightarrow qG, qq \rightarrow gG$) and with photons ($qq \rightarrow \gamma G$). In addition, virtual graviton exchange can modify the cross-sections for diphoton and dilepton production ($pp \rightarrow \gamma\gamma X, pp \rightarrow llX$).

The cross-section for real graviton production at a hadron collider is given by

$$\frac{d^4\sigma}{dm_G^2 dp_T^2 dy dy_G} = \frac{m_G^{n-2}}{2} \frac{S_{n-1}}{M_D^{n+2}} \frac{d\sigma_m}{dt} \sum_{i,j} \frac{f_i(x_1)}{x_1} \frac{f_j(x_2)}{x_2} ,$$

where m_G, y_G are the graviton mass and rapidity, p_T, y are the jet or photon transverse momentum and rapidity, S is a factor depending on the geometry of the extra dimensional space, M_D is the bulk Planck mass, $d\sigma_m/dt$ is the underlying differential cross-section as a function of the Mandelstam variable t, and the sum runs over the contributing parton distributions f of the incident partons. It is clear from this formula that the mass scale and n do not separate. This means that at fixed beam energy, it is not possible to extract them separately.

It should also be noted that at energies above M_D, the quantum theory of gravity comes into play. Since the physics of this is not understood, these calculations can only be valid at energies well above M_D, where an effective theory can be used. This can be a problem for the LHC, since its parton collision energy spans the TeV range, above and below the M_D values favoured in some models.

The dominant backgrounds to this search are from $pp \rightarrow Z + jets \rightarrow \nu\bar{\nu} + jets$, $pp \rightarrow W + jets \rightarrow \tau\bar{\nu} + jets$ and $pp \rightarrow W + jets \rightarrow e\bar{\nu} + jets$. The backgrounds from

n	M_D^{Min} (TeV)	M_D^{Max} (TeV)	R
2	~ 4	7.5	10 μm
3	~ 4.5	5.9	300 pm
4	~ 5	5.3	1 pm

Table 5. *The search reach for large extra dimensions at the LHC in the ADD scenario*

the W can be largely suppressed by the use of a lepton veto. The background from the Z cannot be removed. However, it can be calibrated using the $Z \rightarrow e^+e^-$ decays. This measures the $\nu\bar{\nu}$ contribution directly, apart from a small correction for the difference in acceptance. However, the relative branching ratios between electrons and neutrinos means that the calibration sample has only a third of the statistics of the background itself, limiting the statistical precision which can be achieved.

The 5σ discovery limits were set by requiring a missing energy above 1 TeV, for an integrated luminosity of 100 pb^{-1}. Because the theoretical predictions are not valid near M_D, the limits are only valid above a scale M_D^{Min}. While the LHC would certainly see something near M_D, there are no reliable predictions upon which to base a search strategy. The results are presented in Table 5, which includes the value of M_D^{Min}, as well as the search limit M_D^{Max}, and the corresponding value of R for various values of n. It can be seen that the search limit drops as n increases, and that for $n > 4$ there is no region of parameter space in which safe predictions can be made. The ultimate reach of the LHC appears to be for extra dimensions of a scale of 1 pm.

The single photon signal can also be used, but the search reach is more limited. This channel would only be useful as a confirmation if n and M_D are both small.

As noted above, the form of the cross section does not allow n and M_D to be extracted separately at fixed energy, since a change in one can be compensated by the other. However, because the reach in M_D is limited by the maximum available parton energy, a change in beam energy can separate the two effects. If the LHC beam energy is reduced, the highest accessible M_D is reduced, while n is fixed. The change in the total cross section can therefore separate the two parameters. For example, the ratio of the cross sections for a beam energy of 10 TeV, compared to the nominal 14 TeV is 12% for $n = 2$, 9% for $n = 3$ and 7% for $n = 4$. This means that the cross section ratio needs to be measured to around 1% to achieve a good sensitivity. The main challenge in such a measurement would be to control the systematic errors on the luminosity, while running with different beam energies.

The diphoton and dilepton channels can also be used to glean information since virtual graviton exchange diagrams interfere with the SM processes, and hence modify the angular distribution. This can give a similar reach to the direct graviton production signal for the LHC. This method has been used by the D0 and CDF collaborations (Olivier B. 2001 and Ferbel T. 2001) to set limits at the Tevatron. The signal is searched for in the plane of diphoton (dilepton) invariant mass and centre of mass production angle, giving a better separation from the SM background. A limit of $M_D > 1.44$ TeV is obtained for $n = 3$ falling to $M_D > 0.97$ TeV for $n = 7$.

Finally, low scale quantum gravity opens up the exciting possibility of black hole production in the TeV range (Dimopoulos S., Landsberg G. L. 2001). Such quantum sized holes decay by Hawking radiation with very short lifetimes. This means that there is no possibility of them ingesting nearby masses, like the detectors! Indeed, since cosmic rays continuously bombard the upper atmosphere with energies higher than the TeV scale, the Earth would not have survived if stable black holes could be created in this way. The decay of the black holes creates a spectacular signature with typical multiplicities of $\mathcal{O}(10)$ partons above 500 GeV. The production cross-section cannot be calculated reliably without a theory of quantum gravity, but a semiclassical argument leads to predictions of the order of 10,000 events per year at the LHC. The temperature of the black hole is given by $T_H = (n+1)/4\pi r_H$, where r_H, the hole radius, is given by $r_H \sim \frac{\hbar}{M_D c}(\frac{m_H}{M_D})^{1/(n+1)}$ and m_H is the black hole mass. This suggests a strategy where the final state energy spectrum is used to measure T_H and direct reconstruction of m_H from the decay products allows n to be determined. However, this method is complicated by the fact that the emission spectrum itself depends on n via the so-called "grey-body" factors which specify the rate of emission of each particle species compared to a black body spectrum, and also by the fact that the black hole mass drops, and its temperature rises during the decay. Nonetheless, it seems likely that a great deal could be learnt about the underlying theory from such a spectacular high-rate signature.

6.3   Small extra dimensions

In the ADD model, the extra dimensions need to be of macroscopic size, in order to provide enough volume of bulk space to allow a large gravitational coupling in the bulk to be sufficiently diluted in 3-D space for the hierarchy problem to be solved. ADD used flat space dimensions. However, it is possible to consider non-factorizable space-time geometries (Randall L. and Sundrum R. 1999) instead. A particular example contains one extra dimension, with the metric

$$ds^2 = e^{-2kr_c\phi}\eta_{\mu\nu}dx^\mu dx^\nu + r_c^2\phi^2 ,$$

where k is a parameter of order of the Planck mass, x^μ are the coordinates of the normal 4-D space-time, $0 < \phi < \pi$ is the coordinate of the extra dimension whose compactified radius is r_c.

Such a space-time can contain 2 branes at $\phi = 0$ and $\phi = \pi$. Because of the exponential factor (called the warp factor), the space-time volume in 4-D changes exponentially when moving between the branes. This allows sufficient space time volume to be created to solve the hierarchy problem, even if r_c is of order of the Planck length. This creates the appealing possibility that a single Planck scale can be used throughout the theory, which simultaneously creates the gravitational scale and the weak scale, with a ratio given by the warp factor. This can be achieved with a value of $kr_c \sim 10$, so little fine tuning is required - all the characterstic scales are the same.

There are still considerable theoretical obstacles to a fully worked out version of this scenario, but the experimental consequences are clear. Since the r_c value is so small, the separation between the masses of graviton resonances is large, of the order of new Planck scale (ie $\mathcal{O}(\text{TeV})$). Since the graviton couples to all SM states, such resonances would appear as peaks in the invariant mass distributions of any pair produced state. The favoured final state to consider is e^+e^- which has low background at the LHC, and benefits from the excellent mass resolution of the EM calorimeters at high energy. This signal was studied for the original model in (Allanach *et al* 2000) which showed that the signal could be observed up to a graviton mass of 2.1 TeV, while the resonance could be established as spin 2 up to a graviton mass of 1.7 TeV. A more general analysis (Allanach *et al* 2002) explored the search limits in the plane of graviton mass and the parameter $\Lambda_\pi = \bar{M}_{Pl}e^{-kr_c\pi}$ which is inversely proportional to the graviton coupling strength, for various signal channels. Figure 28 shows the limits in for the e^+e^- channel. The theoretically favoured region is below the line for $k/\bar{M}_{Pl} = 0.01$. The reach is presented as contours of the relative error in the production cross-section times branching ratio $\sigma\dot{B}$. An error of 20% corresponds to a 5σ discovery. In most of the plane, the graviton width is too small to be measured, apart from inside the shaded region at small Λ_π.

7 Conclusions

The Standard Model is a triumph of both theory and experiment, matching existing data to excellent precision. However, it is clear that it cannot be the final theory of particles and fields. The instability of the Higgs mass under quantum corrections points to new physics in the TeV range. The most favoured scenario is supersymmetry, which has the potential to solve the hierarchy problem and link to a higher theoretical framework in which gravity is included, perhaps within a supersymmetry breaking mechanism.

Alternative models based on extra dimensions have not yet reached the level of theoretical development, or experimental understanding, of supersymmetry, but offer a novel alternative picture.

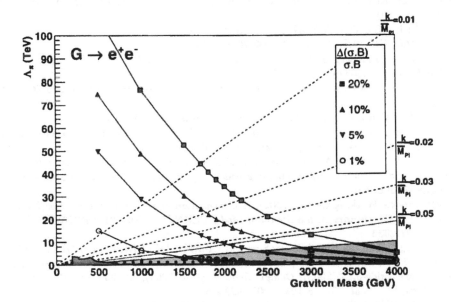

Figure 28. *Search limits for graviton resonances in the e^+e^- channel at the LHC, in models with warped extra dimensions.*

In both cases, a rich variety of new physics signatures are predicted which could be detected at the Tevatron, and will be found or excluded by the LHC. After one year of LHC running, the landscape of the TeV world should be clear to see.

Acknowledgments

I would like to thank the organizers for their kind invitation to speak at this school.

References

Allanach B.C. *et al* (2000) *Journal of High Energy Physics* 09 (2000) 019.

Allanach B.C. *et al* (2002) *Journal of High Energy Physics* 12 (2002) 039.

Arkani-Hamed, N, Dimopoulos, S and Dvali, G. (1998) *Physics Letters* B249 263.

ATLAS Collaboration (1999) *Detector and Physics Performance Technical Design Report*, CERN/LHCC/99-15, May 1999.

M Carena, J.S. Conway, H.E. Haber, J. Hobbs *et al* (2000), *Report of the Tevatron Higgs Working Group*, hep-ph/0010338.

Cirelli M. (2003) *Neutrinos in Extra Dimensions and Supernovae*, hep-ph/0305141.

CMS Collaboration,
  CMS Technical Design Reports (1997-2002), http://cmsdoc.cern.ch/docLHCC.shtml;
  CMS ECAL TDR (CERN/LHCC 97-33, Dec 1997),
  CMS HCAL TDR (CERN/LHCC 97-31, June 1997),
  CMS MAGNET TDR (CERN/LHCC 97-10, May 1997),
  CMS MUON TDR (CERN/LHCC 97-32, Dec 1997),
  CMS TRACKER TDR (CERN/LHCC 98-6, Apr 1998) and Addendum (CERN/LHCC 2000-016, Feb 2000),
  CMS TRIGGER TDR (CERN/LHCC 2000-38, Dec 2000),
  CMS DAQ TDR (CERN/LHCC/2002-26, Dec 2002).

Dimopoulos S., Landsberg G. L. (2001), *Black Holes at the LHC*, *Phys. Rev. Lett.* 87 (2001) 161602, hep-ph/0106295.

Ferbel T., (2001) *Search for "large" extra dimensions at the Tevatron*, hep-ex/0103009; 8 Mar 2001.

Hagiwara, K et al [Particle Data Group] (2002) *Physical Review* D66 010001.

James, F. (1986) *CERN Academic Training Program.* See also section on *Statistics* in Hagiwara, K *et al* [Particle Data Group] (2002), *Physical Review* D66 010001.

Kabachenko, V., Miagkov A., Zenin A., (2001) *Sensitivity of the ATLAS detector to extra dimensions in di-photon and di-lepton production processes*, ATL-PHYS-2001-012, Geneva, CERN, 21 Jun 2001.

LEP Electroweak Working Group (2002), *A Combination of preliminary Electroweak Measurements and Constraints on the Standard Model*, CERN-EP/2002-091, LEPEWWG/2002-02, hep-ex/0212036

LEP Working Group for Higgs Boson Searches (ALEPH, DELPHI, L3 and OPAL Collaborations), *Physics Letters* B 565 (2003), 61-75. Also as CERN-EP/2003-011, hep-ex/0306033.

Long J.C. and Price J.C. (2003), *Compt. Rend. Phys.* 4 (2003), 337-346, hep-ph/0303057. See also Long J.C. *et al*, *Nature* 421 (2003) 922-925.

Martin S.P. (1997) *A Supersymmetry Primer*, hep-ph/9709356.

Olivier B., (2001) *Search for Large Extra Dimensions at the Tevatron*, hep-ex/0108015; LPNHE-2001-01 - 7 Aug 2001.

Randall L. and Sundrum R., *An Alternative to Compactification, Phys. Rev. Lett.* **83** (1999) 4690, hep-th/9906064.

Spira M., Djouadi A., Graudenz D., Zerwas P.M. , *Higgs Production at the LHC, Nuclear Physics* B **453** (1995) 17-82.

Vacavant L., Hinchliff I., (2000) *Model Independent Extra-dimension signatures with ATLAS,* LBNL-45198, ATL-PHYS-2000-016, hep-ex/0005033, May 2000.

B physics

Valerie Gibson

University of Cambridge, UK

1   Introduction

Since the discovery of the b quark, its properties and interactions have been studied in great detail. One major observation is that the b quark has a very long lifetime, $\mathcal{O}(10^{-12})$ ps, thereby implying a hierarchy in the mixing between quark generations. This, together with the discovery of CP violation, has led to the main goal of the current generation of B physics practitioners; namely to understand the structure of quark mixing and its rôle in CP violation.

The physical importance of CP violation is demonstrated by the fact that it is one of the necessary ingredients to explain the dominance of matter over antimatter in our universe (Sakharov 1967). Although the Standard Model naturally accommodates CP violation through the complex quark mixing matrix (Kobayashi and Maskawa 1972), it alone can not explain the large asymmetry observed (Shaposhnikov 1986). However, baryogenesis at the electro-weak scale is still possible with various extensions to the Standard Model which introduce additional sources of CP violation (Nir 1999).

Experimentally, CP violation was unexpectedly discovered in the decay of neutral kaons in 1964 (Christenson 1964). Since then, the kaon system remained the only place CP violation had been observed until 2001 when it was observed in the b quark system for the first time by the B factory experiments, BaBar (BaBar 2001) and Belle (Belle 2001). In the B meson system, CP violation is expected to occur in many different decay channels and for some the Standard Model can make precise predictions. Therefore, the B system is considered an ideal place to test the Standard Model predictions and to hunt for the necessary new physics beyond.

These lectures focus on how B decays can be used to obtain a better understanding of the Standard Model description of quark mixing and CP violation, to search for new physics beyond the Standard Model and to provide a strategy to disentangle cleanly the new physics and the Standard Model contributions.

2   Mixing and CP violation

We start with a gentle introduction to the phenomenology of neutral meson mixing and CP violation. There are many excellent review articles on this topic; see for example (BaBar 1998 and Nir 2001). Here we describe a general approach that is applicable to any neutral meson system.

2.1   Neutral meson mixing

The phenomenon of oscillations between neutral meson particle and antiparticle is well established and is possible as a result of the non-conservation of flavour in the weak interaction. Consider the time development of an arbitrary neutral meson state, X^0, and its antiparticle, $\overline{X^0}$,

$$a(t)\,|X^0\rangle + b(t)\,|\overline{X^0}\rangle, \tag{1}$$

which is governed by the time-dependent differential equation

$$i\frac{\partial}{\partial t}\begin{pmatrix} a \\ b \end{pmatrix} = H\begin{pmatrix} a \\ b \end{pmatrix} = \begin{pmatrix} M_{11} - \frac{i}{2}\Gamma_{11} & M_{12} - \frac{i}{2}\Gamma_{12} \\ M_{12}^* - \frac{i}{2}\Gamma_{12}^* & M_{22} - \frac{i}{2}\Gamma_{22} \end{pmatrix}\begin{pmatrix} a \\ b \end{pmatrix}. \tag{2}$$

Here M_{ij} and Γ_{ij} are elements of the so-called mass and decay matrices respectively and are associated with measurable quantities. A consequence of the invariance of the combined symmetry operation CPT is that particle and antiparticle masses and lifetimes are identical, namely $M_{11} = M_{22} \equiv M$ and $\Gamma_{11} = \Gamma_{22} \equiv \Gamma$.

The eigenstates of H, $|X_1\rangle$ and $|X_2\rangle$, can be written as a linear combination of the particle and antiparticle states,

$$\begin{aligned} |X_1\rangle &= p\,|X^0\rangle + q\,|\overline{X^0}\rangle \\ |X_2\rangle &= p\,|X^0\rangle - q\,|\overline{X^0}\rangle, \end{aligned} \tag{3}$$

where p and q represent the amount of meson state mixing and are complex numbers with the normalization condition $|p|^2 + |q|^2 = 1$. The eigenvalues of $|X_{1,2}\rangle$, $E_{1,2}$, are then obtained from the characteristic equation,

$$|H - EI| = 0, \tag{4}$$

where I is a 2×2 unit matrix. The eigenvalues obtained are

$$\begin{aligned} E_{1,2} &= M_{1,2} - \frac{i}{2}\Gamma_{1,2} = M - \frac{i}{2}\Gamma \pm \sqrt{(M_{12} - \frac{i}{2}\Gamma_{12})(M_{12}^* - \frac{i}{2}\Gamma_{12}^*)} \\ &= \left(M \mp \frac{\Delta M}{2}\right) - \frac{i}{2}\left(\Gamma \mp \frac{\Delta\Gamma}{2}\right) \end{aligned} \tag{5}$$

where $M_{1,2}$ and $\Gamma_{1,2}$ are the masses and decay widths of the two eigenstates and $\Delta M = M_2 - M_1$ and $\Delta\Gamma = \Gamma_2 - \Gamma_1$. The relationship between p and q can be extracted from the eigenvector equations,

$$(H - EI)\begin{pmatrix} p \\ \pm q \end{pmatrix} = 0, \tag{6}$$

where

$$\frac{q}{p} = \sqrt{\frac{M_{12}^* - \frac{i}{2}\Gamma_{12}^*}{M_{12} - \frac{i}{2}\Gamma_{12}}}. \tag{7}$$

The time evolution of the initial particle and antiparticle states, $|X^0(t)\rangle$ and $|\overline{X^0}(t)\rangle$, obtained by rearranging equations (3), and substituting for the time evolution of the states $|X_{1,2}\rangle$,

$$|X_{1,2}(t)\rangle = |X_{1,2}\rangle e^{-i(M_{1,2} - \frac{i}{2}\Gamma_{1,2})t}, \tag{8}$$

are

$$|X^0(t)\rangle = f_+(t)|X^0\rangle + \frac{q}{p}f_-(t)|\overline{X^0}\rangle \tag{9}$$

$$|\overline{X^0}(t)\rangle = f_+(t)|\overline{X^0}\rangle + \frac{p}{q}f_-(t)|X^0\rangle,$$

where $f_\pm(t) = \frac{1}{2}\left[e^{-i(M_1 - \frac{i}{2}\Gamma_1)t} \pm e^{-i(M_2 - \frac{i}{2}\Gamma_2)t}\right]$. Hence, the probabilities of finding an $|X^0\rangle$ and an $|\overline{X^0}\rangle$ at time t from an initially pure $|X^0\rangle$ state are

$$P(X^0 \rightarrow X^0 : t) = |\langle X^0|X^0(t)\rangle|^2 = |f_+(t)|^2 \tag{10}$$

$$P(X^0 \rightarrow \overline{X^0} : t) = |\langle \overline{X^0}|X^0(t)\rangle|^2 = |\frac{q}{p}f_-(t)|^2$$

with

$$|f_\pm(t)|^2 = \frac{1}{4}\left[e^{-\Gamma_1 t} + e^{-\Gamma_2 t} \pm 2e^{-\overline{\Gamma} t}\cos(\Delta M t)\right] \tag{11}$$

and the average decay width $\overline{\Gamma} = (\Gamma_1 + \Gamma_2)/2$. The last term in equation (11) describes the oscillation character of the X^0 and $\overline{X^0}$ content in an initial pure X^0 beam. The oscillations depend crucially on the size of the oscillation parameter, defined as $x = |\Delta M|/\overline{\Gamma}$. Shown in Figure 1 are the probabilities (assuming $|q/p| = 1$) for finding an X^0 and an $\overline{X^0}$ from an initial X^0 state as a function of time. When $x = 1$, there is some probability that the meson will oscillate before it decays, and when $x = 25$, the meson exhibits very rapid oscillations before decaying. Figure 1a) corresponds approximately to $K^0 - \overline{K^0}$ mixing, in which the difference in lifetimes and hence decay widths of the two physical eigenstates (K_S^0 and K_L^0) is large due to the domination of $K_S^0 \rightarrow \pi\pi$ final states,

$$\frac{\tau_L}{\tau_S} = \frac{51.7 \pm 0.4 \text{ ns}}{0.08927 \pm 0.00009 \text{ ns}} \sim 580, \tag{12}$$

and the mass difference,

$$\Delta M = M_L - M_S = (3.483 \pm 0.009) \times 10^{-15} \text{ GeV}, \tag{13}$$

is very small (Particle Data Group 2002). Figure 1b) and 1c) correspond approximately to $B_d^0 - \overline{B_d^0}$ and $B_s^0 - \overline{B_s^0}$ meson mixing respectively. In contrast to the kaon system, the overwhelming majority of final states expected for B^0 decays are the same as for the $\overline{B^0}$ decays and hence the lifetimes of the physical eigenstates are expected to be approximately the same (Particle Data Group 2002),

$$\left(\frac{\Delta\Gamma}{\overline{\Gamma}}\right)_d \sim 0.5\%, \quad \text{and} \quad \left(\frac{\Delta\Gamma}{\overline{\Gamma}}\right)_s \sim 15\%. \tag{14}$$

However, the difference between the masses is large (HFAG 2003),

$$\Delta M_d = (3.30 \pm 0.04) \times 10^{-13} \text{ GeV} \quad \text{and} \quad \Delta M_s > 9.5 \times 10^{-12} \text{ GeV (95\% } cl), \qquad (15)$$

and when combined with the measured B_d^0 and B_s^0 lifetimes, $\tau_d = (1.542 \pm 0.016)$ ps and $\tau_s = (1.461 \pm 0.057)$ ps (Particle Data Group 2002), result in oscillation parameters of $x_d = 0.77 \pm 0.02$ and $x_s > 21$ respectively.

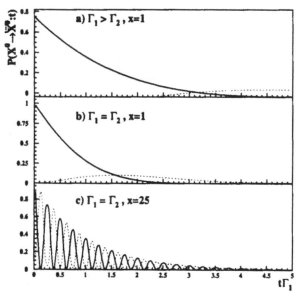

Figure 1. *The probability of finding an X^0 meson (solid curve) and an $\overline{X^0}$ meson (dashed curve) as a function of $t\Gamma_1$ from an initial pure X^0 state for the cases a) $\Gamma_1 > \Gamma_2$ with $x = 1$, b) $\Gamma_1 = \Gamma_2$ with $x = 1$ and c) $\Gamma_1 = \Gamma_2$ with $x = 25$.*

2.2   CP violation in meson decays

We now consider the time dependent decay rates for the neutral mesons X^0 and $\overline{X^0}$ to decay into the same final state f,

$$\Gamma_f(t) \equiv \Gamma\left(X^0(t) \to f\right) \qquad (16)$$

$$= |A_f|^2 \left[|f_+(t)|^2 + \left(\frac{q}{p}\frac{\overline{A_f}}{A_f}\right)^2 |f_-(t)|^2 + 2\text{Re}\left\{\frac{q}{p}\frac{\overline{A_f}}{A_f} f_+^*(t)f_-(t)\right\} \right]$$

and

$$\overline{\Gamma_f}(t) \equiv \Gamma\left(\overline{X^0}(t) \to f\right) \qquad (17)$$

$$= |A_f|^2 \left[\left|\frac{\overline{A_f}}{A_f}\right|^2 |f_+(t)|^2 + \left(\frac{p}{q}\right)^2 |f_-(t)|^2 + 2\left(\frac{p}{q}\right)^2 \text{Re}\left\{\frac{q}{p}\frac{\overline{A_f}}{A_f}^* f_+^*(t)f_-(t)\right\} \right],$$

where, $A_f = \langle f|H|X^0\rangle$ and $\overline{A_f} = \langle f|H|\overline{X^0}\rangle$ are the instantaneous decay amplitudes. Any difference between these two rates is clear proof of CP violation.

The first terms in the rates Γ_f and $\overline{\Gamma_f}$ show that CP violation is generated if $|A_f| \neq |\overline{A_f}|$. This is called *CP violation in the decay amplitudes*. The second terms of Γ_f and $\overline{\Gamma_f}$ show that CP violation is generated if $|q/p| \neq 1$, even if there is no CP violation in the decay amplitudes. From equations (10), it is clear that if $|q/p| \neq 1$, then the oscillation rate for $X^0 \to \overline{X^0}$ is different from that for $\overline{X^0} \to X^0$. This type of CP violation is called *CP violation in the mixing* and results in two physical eigenstates which are different from the CP eigenstates. Finally, if there is no CP violation in the mixing or decay amplitudes, the first and second terms of Γ_f and $\overline{\Gamma_f}$ are identical. However, CP violation can still arise due to the third term which can be expanded using

$$\text{Re}\left\{\left(\frac{q\,\overline{A_f}}{p\,A_f}\right)^{(*)} f_+^*(t) f_-(t)\right\} = \text{Re}\left\{\frac{q\,\overline{A_f}}{p\,A_f}\right\} \text{Re}\left\{f_+^*(t) f_-(t)\right\}$$

$$\mp \text{Im}\left\{\frac{q\,\overline{A_f}}{p\,A_f}\right\} \text{Im}\left\{f_+^*(t) f_-(t)\right\}. \qquad (18)$$

In this case, if

$$\text{Im}\left\{\frac{q\,\overline{A_f}}{p\,A_f}\right\} \neq 0, \qquad (19)$$

then Γ_f and $\overline{\Gamma_f}$ differ resulting in CP violation. Since the process involves the decays of $X^0(\overline{X^0})$ from the initial $X^0(\overline{X^0})$ and the decays of the $\overline{X^0}(X^0)$ oscillated from the initial $X^0(\overline{X^0})$ into a common final state, this type of CP violation is commonly referred to as *CP violation in the interference between the mixing and decay amplitudes*.

Complementary to this classification of CP violation is the widely used notation of *direct* versus *indirect* CP violation. It was motivated historically by the hypothesis of a new superweak interaction that was supposed to account for CP violation in $K_L^0 \to \pi^+\pi^-$ decays (Wolfenstein 1964). *Indirect* CP violation is any CP violating effect that can be entirely attributed to CP violation in the neutral meson mass matrix and is independent of the final state. Conversely, any effect that can not be described in this way and explicitly requires CP violating phases in the decay amplitude itself and hence depends on the final state is called *direct CP violation*. It follows that CP violation in the mixing represents indirect CP violation, CP violation in the decay is direct CP violation and CP violation in the interference contains aspects of both indirect and direct CP violation.

In the B system, where $\Delta\Gamma$ is small, it is convenient to derive the time dependent decay rates from the particle-antiparticle base,

$$\Gamma_f(t) = \frac{|A_f|^2}{2} e^{-\Gamma t} [I_+(t) + I_-(t)] \qquad (20)$$

and

$$\overline{\Gamma_f}(t) = \frac{|\overline{A_f}|^2}{2|\lambda|^2} e^{-\Gamma t} [I_+(t) - I_-(t)], \qquad (21)$$

where the two time dependent functions $I_\pm(t)$ are given by

$$I_+(t) = \left(1 + |\lambda|^2\right) \cosh\left(\frac{\Delta\Gamma t}{2}\right) + 2\text{Re}\{\lambda\} \sinh\left(\frac{\Delta\Gamma t}{2}\right) \qquad (22)$$

$$I_-(t) = \left(1 - |\lambda|^2\right) \cos(\Delta M t) + 2\text{Im}\{\lambda\} \sin(\Delta M t). \qquad (23)$$

The complex quantity λ,

$$\lambda = \frac{q}{p}\frac{\overline{A}_f}{A_f}, \tag{24}$$

has observable real and imaginary parts that quantify CP violation. Similarly, the corresponding decay amplitudes for the CP conjugated final states $\overline{f}$ are given by

$$\Gamma_{\overline{f}}(t) = \frac{|A_f|^2}{2|\lambda|^2}e^{-\Gamma t}\left[\overline{I}_+(t) - \overline{I}_-(t)\right] \tag{25}$$

and

$$\overline{\Gamma_{\overline{f}}}(t) = \frac{|\overline{A}_{\overline{f}}|^2}{2}e^{-\Gamma t}\left[\overline{I}_+(t) + \overline{I}_-(t)\right], \tag{26}$$

where $\overline{I}_\pm(t)$ are the same as $I_\pm(t)$ but with λ replaced by

$$\overline{\lambda} = \frac{p}{q}\frac{A_{\overline{f}}}{\overline{A}_{\overline{f}}}. \tag{27}$$

We will now consider the three categories of CP violation in turn.

2.2.1   CP violation in the mixing

The test of CP invariance through meson state mixing is whether the magnitudes of the off-diagonal elements of the mass and decay matrices are the same, namely,

$$|M_{12} - \frac{i}{2}\Gamma_{12}| \stackrel{?}{=} |M_{12}^* - \frac{i}{2}\Gamma_{12}^*|. \tag{28}$$

If they are not the same, it physically corresponds to a difference in the mixing rates between $X^0 \to \overline{X^0}$ and $\overline{X^0} \to X^0$. Equation (7) implies that if

$$\left|\frac{q}{p}\right| \neq 1 \tag{29}$$

then CP is violated in the meson state mixing. A difference in the mixing rates will occur when there is a phase difference between M_{12} and Γ_{12}. The standard convention is to keep Γ_{12} real and describe any observed CP violation by a small imaginary part of M_{12}.

CP violation in the mixing has been measured in the K^0 semi-leptonic charge decay asymmetry (Particle Data Group 2002),

$$A_{sl} = \frac{\Gamma(K_L^0 \to \pi^-\ell^+\nu) - \Gamma(K_L^0 \to \pi^+\ell^-\nu)}{\Gamma(K_L^0 \to \pi^-\ell^+\nu) + \Gamma(K_L^0 \to \pi^+\ell^-\nu)} = 0.327 \pm 0.012\% = \frac{1 - |q/p|^2}{1 + |q/p|^2}. \tag{30}$$

We can interpret this result in terms of the phase difference between M_{12} and Γ_{12}, if we define the phase ϕ_{12} according to

$$\frac{M_{12}}{\Gamma_{12}} = -\left|\frac{M_{12}}{\Gamma_{12}}\right|e^{i\phi_{12}} \tag{31}$$

and knowing that the CP violating effects in the K system are small ($\phi_{12} << 1$). We find that $\phi_{12} \approx 6.5 \times 10^{-3}$.

In principle, the semi-leptonic decay asymmetry can also be measured using B^0 decays,

$$A_{sl} = \frac{\Gamma\left(\overline{B^0} \to \ell^- \nu X\right) - \Gamma\left(B^0 \to \ell^+ \nu X\right)}{\Gamma\left(\overline{B^0} \to \ell^- \nu X\right) + \Gamma\left(B^0 \to \ell^+ \nu X\right)} = \frac{1 - |q/p|^4}{1 + |q/p|^4}. \tag{32}$$

However, CP violation in the mixing is expected to be small in the B system due to GIM suppression and the smallness of $\Delta\Gamma$, $\mathcal{O}\left(10^{-3}\right)$ for B_d^0 and $\mathcal{O}\left(10^{-5}\right)$ for B_s^0, and the theoretical interpretation is difficult due to large hadronic uncertainties in Γ_{12}.

2.2.2 CP violation in the decay

CP violation in the decay is due to an interference among decay amplitudes and results in a difference between the partial decay rate $X \to f$ and the charge conjugate process $\overline{X} \to \overline{f}$. The two decay amplitudes are written as

$$A \equiv \langle f | H | X \rangle, \qquad \overline{A} \equiv \langle \overline{f} | H | \overline{X} \rangle; \tag{33}$$

and if several amplitudes contribute to A and $\overline{A}$, then

$$A = \sum_i A_i e^{i\phi_i} e^{i\delta_i} \quad \text{and} \quad \overline{A} = \sum_i A_i e^{-i\phi_i} e^{i\delta_i}, \tag{34}$$

where A_i are real. Two types of phases appear in A and $\overline{A}$. ϕ_i are parameters that violate CP and appear in A and $\overline{A}$ with opposite signs. Usually they arise in the electroweak sector and hence are known as *weak phases*. δ_i are *strong phases* that arise in the scattering or decay amplitudes. They do not violate CP and appear in A and $\overline{A}$ with the same sign.

If a meson can decay by at least two different decay mechanisms with different amplitudes, having different weak and strong phases, there is an interference between the decay amplitudes resulting in

$$\left|\frac{\overline{A}}{A}\right| = \left|\frac{\sum_i A_i e^{-i\phi_i} e^{i\delta_i}}{\sum_i A_i e^{i\phi_i} e^{i\delta_i}}\right| \neq 1 \tag{35}$$

and corresponds to CP violation in the decay.

In principle, CP violation in the decay can be measured from the asymmetry in charged decays

$$A_{+-} = \frac{\Gamma\left(X^+ \to f\right) - \Gamma\left(X^- \to \overline{f}\right)}{\Gamma\left(X^+ \to f\right) + \Gamma\left(X^- \to \overline{f}\right)} = \frac{1 - |\overline{A}/A|^2}{1 + |\overline{A}/A|^2}. \tag{36}$$

However, the theoretical prediction for A_{+-} is difficult due to hadronic uncertainties in the calculation of the strong phase difference. CP violation in the decay has, however, been observed through the measurement of $\mathrm{Re}\{\epsilon'\}$ in $K^0 \to \pi^+\pi^-$ decays (see Section 2.2.4).

2.2.3   CP violation in the interference

CP violation in the interference between the mixing and decay amplitudes can be measured from the time dependent CP asymmetry,

$$
\mathcal{A}(t) = \frac{\Gamma_f - \overline{\Gamma}_f}{\Gamma_f + \overline{\Gamma}_f} = \frac{I_-(t)}{I_+(t)} \tag{37}
$$

$$
= \frac{(1 - |\lambda|^2)\cos(\Delta M t) + 2\mathrm{Im}\{\lambda\}\sin(\Delta M t)}{(1 + |\lambda|^2)\cosh(\Delta \Gamma t/2) + 2\mathrm{Re}\{\lambda\}\sinh(\Delta \Gamma t/2)}. \tag{38}
$$

This asymmetry can also be written so as to separate the direct CP violating contribution, $\mathcal{A}_{\mathrm{CP}}^{dir}$, from that describing CP violation in the interference between the mixing and decay amplitudes, $\mathcal{A}_{\mathrm{CP}}^{mix}$,

$$
\mathcal{A}(t) = \mathcal{A}_{\mathrm{CP}}^{dir}\cos(\Delta M t) + \mathcal{A}_{\mathrm{CP}}^{mix}\sin(\Delta M t). \tag{39}
$$

In the B system, where $\Delta\Gamma$ is small,

$$
\mathcal{A}_{\mathrm{CP}}^{dir} = \frac{(1 - |\lambda|^2)}{(1 + |\lambda|^2)} \quad \text{and} \quad \mathcal{A}_{\mathrm{CP}}^{mix} = \frac{2\mathrm{Im}\{\lambda\}}{(1 + |\lambda|^2)}. \tag{40}
$$

Furthermore, for decays dominated by a single CP violating phase, $\mathcal{A}_{\mathrm{CP}}^{dir}$ is negligible and the CP asymmetry reduces to

$$
\mathcal{A}(t) = \mathrm{Im}\{\lambda\}\sin(\Delta M t). \tag{41}
$$

In this case, $\mathrm{Im}\{\lambda\}$ is the phase difference between the mixing amplitude and the phase of the relevant decay amplitude and can be cleanly interpreted in terms of purely electroweak parameters.

CP violation in the interference has been observed in K^0 decays and has approximately the same magnitude, $\mathcal{O}(10^{-3})$, as CP violation in the mixing (Nakada 2002). Most recently, CP violation in the interference has been observed in the B system by the B factory experiments where a relatively large CP asymmetry has been measured (see Section 5.3).

2.2.4   The parameters ε and ε'

The parameters ε and ε' are normally used to describe CP violation in K^0 decays (Nakada 2002) (Particle Data Group 2002). However, the underlying processes producing CP violation in the K^0 and B^0 system are the same. Therefore, it is a useful exercise to translate the language of ε and ε' into the λ notation we have used here.

In the K^0 system, the CP observables

$$
\eta_{+-} \equiv \frac{A(K_L^0 \to \pi^+\pi^-)}{A(K_S^0 \to \pi^+\pi^-)} \quad \text{and} \quad \eta_{00} \equiv \frac{A(K_L^0 \to \pi^0\pi^0)}{A(K_S^0 \to \pi^0\pi^0)}, \tag{42}
$$

are related to λ by

$$
\eta_f = \frac{1 - \lambda_f}{1 + \lambda_f}, \tag{43}
$$

and are therefore affected by all 3 types of CP violation. In principle, we could define similar parameters for the B system, such as,

$$\eta_{+-} \equiv \frac{A(B_H \rightarrow \pi^+\pi^-)}{A(B_L \rightarrow \pi^+\pi^-)}, \quad \eta_{00} \equiv \frac{A(B_H \rightarrow \pi^0\pi^0)}{A(B_L \rightarrow \pi^0\pi^0)} \tag{44}$$

$$\text{and} \quad \eta_{J/\psi K_S^0} \equiv \frac{A(B_H \rightarrow J/\psi K_S^0)}{A(B_L \rightarrow J/\psi K_S^0)} \quad etc$$

where B_H and B_L are the B^0 mass eigenstates (see Section 3.1).

Instead of η_{+-} and η_{00}, the parameters ε and ε' are defined such that any direct CP violating effects are isolated into ε',

$$\varepsilon \equiv \frac{1}{3}(\eta_{00} + 2\eta_{+-}), \quad \varepsilon' \equiv \frac{1}{3}(\eta_{+-} - \eta_{00}). \tag{45}$$

In the two pion final state, two isospin terms ($I = 0, 2$) contribute to the decay amplitudes,

$$A_I = \langle(\pi\pi)_I|H|X^0\rangle \quad \text{and} \quad \overline{A}_I = \langle(\pi\pi)_I|H|\overline{X^0}\rangle. \tag{46}$$

Since the measured amplitude ratio $|A_2/A_0| << 1$, it follows that to first order in A_2/A_0,

$$\varepsilon \approx \frac{1 - \lambda_0}{1 + \lambda_0}, \tag{47}$$

where $\lambda_0 = \frac{q}{p}\frac{\overline{A}_0}{A_0}$. Therefore, since λ_0 depends only on a single final state, ε does not contain CP violation in the decay. We can show that

$$\text{Re}\{\varepsilon\} \approx \frac{1}{2}\left[1 - \left|\frac{q}{p}\right|\right] \approx \frac{\phi_{12}}{4} \tag{48}$$

and

$$\text{Im}\{\varepsilon\} \approx \frac{1}{2}\text{Im}\left\{\frac{q}{p}\frac{\overline{A}_0}{A_0}\right\} \approx \frac{\phi_{12}}{4}. \tag{49}$$

Hence, $\text{Re}\{\varepsilon\}$ and $\text{Im}\{\varepsilon\}$ correspond to CP violation in the mixing and interference respectively with approximately the same order of magnitude, $\mathcal{O}(10^{-3})$.

From the definitions of ε and ε' it can be shown that

$$\eta_{+-} \approx \varepsilon + \varepsilon' \tag{50}$$

which leads to

$$\lambda_{+-} \approx 1 - 2\varepsilon - 2\varepsilon' \tag{51}$$

and

$$|\lambda_{+-}|^2 \approx 1 - 4\text{Re}\{\varepsilon\} - 4\text{Re}\{\varepsilon'\}. \tag{52}$$

Using equation (48), we find

$$\text{Re}\{\varepsilon'\} = \frac{1}{4}\left[1 - \left|\frac{\overline{A}_{+-}}{A_{+-}}\right|^2\right] \tag{53}$$

and

$$\text{Im}\left\{\varepsilon'\right\} = \left[\frac{1}{2}\text{Im}\left\{\lambda_{+-}\right\} + \text{Im}\left\{\varepsilon\right\}\right]. \tag{54}$$

Hence, $\text{Re}\left\{\varepsilon'\right\}$ and $\text{Im}\left\{\varepsilon'\right\}$ correspond to CP violation in the decay and CP violation in the interference respectively. The $\text{Re}\left\{\varepsilon'\right\}$ has been measured in $K^0 \to \pi\pi$ decays and, after 37 years of precision experiments, has culminated in a world average (Particle Data Group 2003) of

$$\text{Re}\left\{\varepsilon'\right\} = (16.7 \pm 2.3) \times 10^{-4} \tag{55}$$

which is clear proof that CP violation in the decay exists in the kaon system. In a similar way, we could define

$$\varepsilon'_B = \frac{1}{3}(\eta_{+-} - \eta_{00}) \quad \text{or} \quad \varepsilon'_B = \frac{1}{3}(\eta_{+-} - \eta_{J/\psi K_S^0}) \tag{56}$$

for the B^0 system. Indeed, any measurement of a non-zero value of ε'_B would be evidence for direct CP violation.

3   Weak decays in the standard model

In the Standard Model with three fermion families, the weak charged current can be written as

$$J_\mu = (\bar{u}, \bar{c}, \bar{t})_L \, \gamma_\mu \, V_{\text{CKM}} \begin{pmatrix} d \\ s \\ b \end{pmatrix}_L \tag{57}$$

where V_{CKM} is the unitary 3×3 Cabibbo-Kobayashi-Maskawa (CKM) quark mixing matrix (Cabibbo 1963) (Kobayashi and Maskawa 1972) and describes the rotation between the weak eigenstates (d', s', b') and the mass eigenstates (d, s, b),

$$\begin{pmatrix} d' \\ s' \\ b' \end{pmatrix} = V_{\text{CKM}} \begin{pmatrix} d \\ s \\ b \end{pmatrix}. \tag{58}$$

The CKM matrix can be written explicitly as

$$V_{\text{CKM}} = \begin{pmatrix} V_{ud} & V_{us} & V_{ub} \\ V_{cd} & V_{cs} & V_{cb} \\ V_{td} & V_{ts} & V_{tb} \end{pmatrix} \tag{59}$$

where V_{ij} is the matrix element coupling the ith up-type quark to the jth down-type quark. The magnitudes of seven of the nine matrix elements, $|V_{ud}|$, $|V_{us}|$, $|V_{ub}|$, $|V_{cd}|$, $|V_{cs}|$, $|V_{cb}|$ and $|V_{tb}|$ are determined from nuclear β decay and pion, kaon, hyperon, D meson, B meson and top quark decays assuming first order (tree-level) weak interactions. The remaining two elements, $|V_{td}|$ and $|V_{ts}|$, can only be accessed through box diagrams or through second order weak interactions such as the QCD penguin diagrams which generate rare B hadron decay modes. Examples of Feynman diagrams for tree-level, box and penguin B decays are shown in Figure 2.

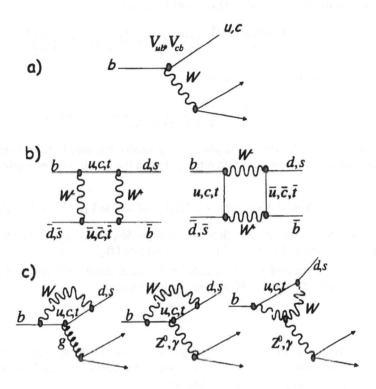

Figure 2. *The Standard Model B decay diagrams. a) is an example of a first order (tree-level) B decay, b) are the B^0-$\overline{B^0}$ box diagrams and c) are examples of penguin diagrams mediated by a gluon (QCD penguin) and Z^0/γ (electro-weak penguin).*

3.1 B mixing in the standard model

In the Standard Model, B mixing is dominated by short range interactions described by box diagrams (Particle Data Group 2002). Due to the large top quark mass and the structure of the CKM matrix, the top quark contribution in the loop is dominant. The dispersive part of the B^0-$\overline{B^0}$ oscillation amplitude is calculated to be

$$M_{12}^q = -\frac{G_F^2 m_W^2 \eta_{B_q^0} m_{B_q^0} B_{B_q^0} f_{B_q^0}^2}{12\pi^2} S(m_t^2/m_W^2) \left(V_{tq}^* V_{tb}\right)^2 \tag{60}$$

for quarks $q = d$ or s. G_F, m_W and m_{B_q} are the Fermi coupling constant, W boson mass and the B_q^0 meson mass. The function $S(m_t^2/m_W^2)$ is determined from the ratio of the top and W boson masses, and the QCD correction factor, $\eta_{B_q^0}$, can be calculated reliably with perturbative methods. The B meson decay constant, $f_{B_q^0}$, and the bag parameter $B_{B_q^0}$, have never been measured directly. The most promising theoretical approach to obtain $f_{B_q^0}$ and $B_{B_q^0}$ is through Lattice QCD calculations (Abbaneo 2003).

In calculating the absorptive part of the oscillation amplitude the contribution from the c and u quarks in the loop also needs to be considered,

$$\Gamma_{12}^q = \frac{G_F^2 m_b^2 \eta'_{B_q^0} B_{B_q^0} f_{B_q^0}^2}{8\pi} \left[\left(V_{tq}^* V_{tb}\right)^2 + \frac{8}{3} V_{tq}^* V_{tb} V_{cq}^* V_{cb} \left(\frac{m_c^2}{m_b^2}\right) + \cdots \right]. \tag{61}$$

For both B_d^0 and B_s^0 mesons, we can derive

$$\left|\frac{\Gamma_{12}}{M_{12}}\right| \approx \frac{3\pi}{2} \frac{m_b^2}{m_W^2} \frac{1}{S\left(m_t^2/m_W^2\right)} \approx 5 \times 10^{-3}, \tag{62}$$

which shows that the absorptive part is very small compared to the dispersive part. With this approximation, the mass difference between the two B mass eigenstates, B_H and B_L, can be derived,

$$\Delta M = M_H - M_L = 2|M_{12}| \quad \text{and} \quad \Delta\Gamma = \Gamma_L - \Gamma_H = 2|\Gamma_{12}|. \tag{63}$$

Note that, due to the phase difference between M_{12} and Γ_{12}, the heavy mass eigenstate (B_H) decays slower than the light mass eigenstate (B_L).

Using the measured values of $\Delta M = (0.502 \pm 0.006)\,\mathrm{ps}^{-1}$ and the average lifetime $\tau_d = (1.542 \pm 0.016)$ ps, for B_d^0 mesons, it follows that

$$\frac{\Delta\Gamma}{\Gamma} \approx 0.5\% \tag{64}$$

and that $\Delta\Gamma$ can be neglected in the B_d^0 system. For the B_s^0 system, only a lower limit of ΔM has been measured to be $14.4\,\mathrm{ps}^{-1}$ with 95% confidence level. Using the measured lifetime, $\tau_s = (1.461 \pm 0.057)$ ps, it follows that

$$\frac{\Delta\Gamma}{\Gamma} > 11\%, \tag{65}$$

and the effect of $\Delta\Gamma$ can no longer be neglected in the B_s^0 decay time distributions.

Finally, in the B system, CP violation in the mixing is given by

$$\mathrm{Im}\left\{\frac{\Gamma_{12}}{M_{12}}\right\} = 1 - \left|\frac{q}{p}\right|^2 \tag{66}$$

$$= -4\pi \frac{m_b^2}{m_W^2} \frac{1}{S(m_t^2/m_W^2)} \mathcal{O}\left(\frac{m_c^2}{m_b^2}\right) \mathrm{Im}\left\{\frac{V_{tq}^* V_{tb} V_{cq}^* V_{cb}}{|V_{tq}^* V_{tb}|^2}\right\}. \tag{67}$$

Assuming $m_b = 4.25\,\mathrm{GeV/c^2}$, $m_W = 80\,\mathrm{GeVc^2}$ and $m_t = 174\,\mathrm{GeV/c^2}$, this is approximately -6×10^{-4} for B_d^0 and 2×10^{-5} for B_s^0, showing that CP violation in the mixing is small for both the B_d^0 and B_s^0 systems.

3.2   CP violation in the standard model

CP violation in the Standard Model is due to the complex phases of the CKM matrix elements. For example, the $b \to uW$ vertex and its CP conjugate are related by the replacement $V_{ub} \overset{CP}{\to} V_{ub}^*$. For three families of quarks and leptons, 3 angles and 1 complex phase are required to define the CKM matrix. These 4 numbers are fundamental constants of nature, just like the Fermi coupling constant, G_F, or the electromagnetic coupling constant, α_{em}, and need to be determined from experiment.

Hence, the unitary 3×3 CKM matrix can be parameterized by four parameters. One possible choice (Particle Data Group 2002) is to write it in terms of 3 angles, θ_{12}, θ_{23}, θ_{13}, and one complex phase, δ, such that

$$V_{CKM} = R_{23} \times R_{13} \times R_{12},\qquad(68)$$

where

$$R_{12} = \begin{pmatrix} c_{12} & s_{12} & 0 \\ -s_{12} & c_{12} & 0 \\ 0 & 0 & 1 \end{pmatrix} \quad R_{23} = \begin{pmatrix} 1 & 0 & 0 \\ 0 & c_{23} & s_{23} \\ 0 & -s_{23} & c_{23} \end{pmatrix} \quad R_{13} = \begin{pmatrix} c_{13} & 0 & s_{13}e^{-i\delta} \\ 0 & 1 & 0 \\ -s_{13}e^{i\delta} & 0 & c_{13} \end{pmatrix}$$

$$(69)$$

and $s_{ij} = \sin\theta_{ij}$, $c_{ij} = \cos\theta_{ij}$ and i and j are generation labels.

Another very popular parameterization of the CKM matrix is the perturbative form suggested by Wolfenstein (Wolfenstein 1983) which reflects the hierarchy of the strengths of the quark transitions through charged current interactions as shown in Figure 3. The

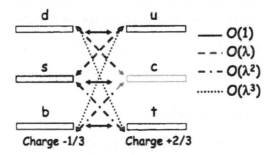

Figure 3. *The hierarchy of quark transitions parameterized by the Wolfenstein parameter* $\lambda \equiv \sin\theta_{12} \sim 0.22$.

Wolfenstein parameterization can be derived by introducing

$$\lambda = s_{12}, \quad A = \frac{s_{23}}{s_{12}^2}, \quad \rho = \frac{s_{13}\cos\delta}{s_{12}s_{23}}, \quad \eta = \frac{s_{13}\sin\delta}{s_{12}s_{23}} \qquad (70)$$

where $\lambda \equiv \sin\theta_{12} \sim 0.22$, θ_{12} is the Cabibbo angle and η represents the CP violating phase in the Standard Model. The matrix is expanded in terms of λ and is often approximated to $\mathcal{O}(\lambda^3)$,

$$V_{CKM} = \begin{pmatrix} 1 - \frac{1}{2}\lambda^2 & \lambda & A\lambda^3(\rho - i\eta) \\ -\lambda & 1 - \frac{1}{2}\lambda^2 & A\lambda^2 \\ A\lambda^3(1 - \rho - i\eta) & -A\lambda^2 & 1 \end{pmatrix} + \mathcal{O}(\lambda^4). \qquad (71)$$

In the LHC era, next-to-leading order corrections in λ will play an important role and we need to consider the CKM matrix to $\mathcal{O}\left(\lambda^5\right)$,

$$
V_{\text{CKM}} = \begin{pmatrix}
1 - \frac{1}{2}\lambda^2 + \frac{1}{4}\lambda^4 & \lambda & A\lambda^3(\rho - i\eta) \\
-\lambda + \frac{1}{2}A^4\lambda^4 - A^2\lambda^5(\rho + i\eta) & 1 - \frac{1}{2}\lambda^2 + \frac{1}{4}\lambda^4(1 - 2A^2) & A\lambda^2 \\
A\lambda^3(1 - \tilde{\rho} - i\tilde{\eta}) & -A\lambda^2 + A\lambda^4\left(\frac{1}{2} - \rho - i\eta\right) & 1 - \frac{1}{2}A^2\lambda^4
\end{pmatrix}
$$
$$
+ \mathcal{O}\left(\lambda^6\right), \tag{72}
$$

where $\tilde{\rho}$ and $\tilde{\eta}$ are given by $\tilde{\rho} = \rho(1 - \lambda^2/2)$ and $\tilde{\eta} = \eta(1 - \lambda^2/2)$.

The requirements for CP violation to be present in the Standard Model can be elegantly summarised by the formula (Jarlskog 1986)

$$
\left(m_t^2 - m_c^2\right)\left(m_t^2 - m_u^2\right)\left(m_c^2 - m_u^2\right) \times \left(m_b^2 - m_s^2\right)\left(m_b^2 - m_d^2\right)\left(m_s^2 - m_d^2\right) \times J_{\text{CP}} \neq 0. \tag{73}
$$

The first ingredient involves the quark masses and is related to the fact that the CP violating phase could be removed through an appropriate unitary transformation of quark fields, if any 2 quarks of the same charge have equal masses. Consequently, the origin of CP violation is related also to the hierarchy of quark masses. The second ingredient is the Jarlskog determinant, $J_{\text{CP}} = \left|\text{Im}\left\{V_{i\alpha}V_{j\beta}V_{i\beta}^*V_{j\alpha}^*\right\}\right|$, which arises from the unitarity of the CKM matrix and can be interpreted as a measure of the strength of CP violation in the Standard Model. Using the parameterizations of the CKM matrix, J_{CP} can be written as

$$
J_{\text{CP}} = s_{12}s_{13}s_{23}c_{12}c_{23}c_{13}\sin\delta = \lambda^6 A^2\eta = \mathcal{O}\left(10^{-5}\right) \tag{74}
$$

and gives concrete meaning to the fact that CP violation is small in the Standard Model.

3.3   The unitarity triangles

The unitarity of the CKM matrix implies that there are six orthogonality conditions between any pair of columns or any pair of rows of the matrix:

$$
\begin{array}{llcr}
V_{ud}V_{ub}^* + V_{cd}V_{cb}^* + V_{td}V_{tb}^* & = & 0 & \quad (db) \quad (75) \\
V_{us}V_{ub}^* + V_{cs}V_{cb}^* + V_{ts}V_{tb}^* & = & 0 & \quad (sb) \quad (76) \\
V_{ud}V_{us}^* + V_{cd}V_{cs}^* + V_{td}V_{ts}^* & = & 0 & \quad (ds) \quad (77)
\end{array}
$$

and

$$
\begin{array}{llcr}
V_{ud}V_{td}^* + V_{us}V_{ts}^* + V_{ub}V_{tb}^* & = & 0 & \quad (ut) \quad (78) \\
V_{cd}V_{td}^* + V_{cs}V_{ts}^* + V_{cb}V_{tb}^* & = & 0 & \quad (ct) \quad (79) \\
V_{ud}V_{cd}^* + V_{us}V_{cs}^* + V_{ub}V_{cb}^* & = & 0 & \quad (uc) \quad (80)
\end{array}
$$

where the quark pair in parenthesis (ij) represents the condition between the $i'th$ and $j'th$ column or row. Each of the orthogonality conditions requires the sum of three complex numbers to vanish and so can be represented geometrically in the complex plane as a triangle. These are known as the *unitarity triangles* and are shown in Figure 4 using the current experimental values of the various matrix elements $|V_{ij}|$ (Particle Data Group 2002). All six triangles have the same area, equal to $J_{\text{CP}}/2$, which is a measure of CP

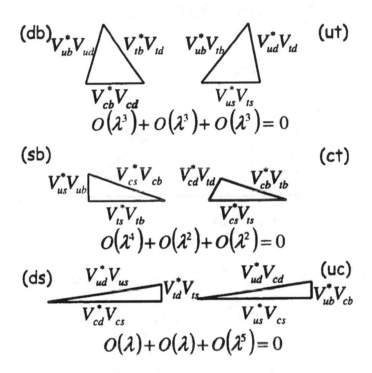

Figure 4. *The six unitarity triangles showing their relative sizes using the current experimental measurements of the CKM matrix elements.*

violation in the Standard Model. It can be seen from Figure 4 that two of the unitarity triangles, (*db*) and (*ut*), have all 3 sides of comparable magnitude. These two triangles are redrawn in Figure 5 by choosing a phase convention such that $V_{cd}V_{cb}^*$ is real and dividing the lengths of all the sides by $|V_{cd}V_{cb}^*| = A\lambda^3$. The two triangles are identical to $\mathcal{O}(\lambda^3)$ and differ only by $\mathcal{O}(\lambda^5)$ corrections. The unitarity triangle (*db*) is commonly referred to as *the* unitarity triangle.

The angles of the triangle (*db*) are denoted by α, β and γ;

$$\alpha \equiv arg\left(-\frac{V_{td}V_{tb}^*}{V_{ud}V_{ub}^*}\right), \quad \beta \equiv arg\left(-\frac{V_{cd}V_{cb}^*}{V_{td}V_{tb}^*}\right), \quad \gamma \equiv arg\left(-\frac{V_{ud}V_{ub}^*}{V_{cd}V_{cb}^*}\right), \tag{81}$$

with $\alpha + \beta + \gamma = \pi$. The higher order terms in the CKM matrix introduce a phase in the matrix element V_{ts} such that the relationship between the angles β and γ in the (*db*) triangle and the angles β' and γ' in the (*ut*) triangle,

$$\beta' \equiv arg\left(-\frac{V_{ts}V_{us}^*}{V_{td}V_{ud}^*}\right), \quad \gamma' \equiv arg\left(-\frac{V_{tb}V_{ub}^*}{V_{ts}V_{us}^*}\right), \tag{82}$$

is given by

$$\beta' = \beta + \chi \quad \text{and} \quad \gamma' = \gamma - \chi. \tag{83}$$

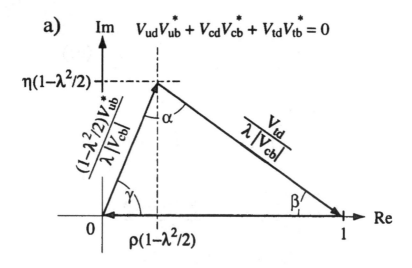

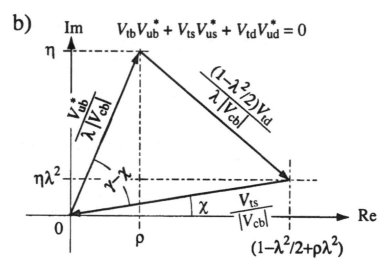

Figure 5. *The unitarity triangles (db) and (ut). The phase convention of the unitarity triangle (db) is chosen such that $V_{cd}V_{cb}^*$ is real and all sides are normalized to $|V_{cd}V_{cb}^*|$.*

The Wolfenstein parameterization of the CKM matrix implies that

$$arg\{V_{td}\} = -\beta, \quad arg\{V_{ub}\} = -\gamma, \quad arg\{V_{ts}\} = \chi + \pi, \tag{84}$$

and that all the other elements are real. Hence, including the angles in the Wolfenstein parameterization to $\mathcal{O}(\lambda^5)$, we find

$$V_{CKM} = \begin{pmatrix} 1 - \frac{1}{2}\lambda^2 + \frac{1}{4}\lambda^4 & \lambda & -|V_{ub}|e^{-i\gamma} \\ -\lambda + \frac{1}{2}A^4\lambda^4 - A^2\lambda^5(\rho + i\eta) & 1 - \frac{1}{2}\lambda^2 + \frac{1}{4}\lambda^4(1 - 2A^2) & A\lambda^2 \\ -|V_{td}|e^{-i\beta} & |V_{ts}|e^{-i\chi} & 1 - \frac{1}{2}A^2\lambda^4 \end{pmatrix} + \mathcal{O}(\lambda^6) \tag{85}$$

where β, γ and χ are related to the Wolfenstein parameters λ, ρ and η through

$$\beta = \tan^{-1}\left[\frac{\bar{\eta}}{1-\bar{\rho}}\right], \quad \gamma = \tan^{-1}\frac{\eta}{\rho}, \quad \text{and} \quad \chi = \eta\lambda^2. \tag{86}$$

In these lectures, we make allowance for the possibility of new physics beyond the Standard Model description by defining

- the B_d^0 mixing phase, $\phi_d = arg\{M_{12}^d\} = 2\beta + \phi_d^{NP}$ and
- the B_s^0 mixing phase, $\phi_s = arg\{M_{12}^s\} = -2\chi + \phi_s^{NP}$

where $\phi_{d,s}^{NP}$ is the phase of any new physics in the $B^0 - \overline{B^0}$ box diagrams. We also refer to γ as the *weak decay phase* which can be measured using $b \to u$ transitions but can also be affected by new physics in loop diagrams.

4   B physics facilities

Since the b quark was discovered at Fermilab in 1977, many measurements of its confined properties and decay properties have been made at a variety of accelerators and their corresponding experiments. A summary of the parameters of relevant past and present machines is given in Table 1. The design of the experiments and their B physics capabilities are influenced by the different production mechanisms of b quarks which are shown in Figure 6. At high energy hadron colliders, the dominant $b\bar{b}$ production mechanism is expected to be due to gluon fusion processes. At the centre-of-mass energy of the $\Upsilon(4s)$ machines, only B_d^0 and $B^\pm$ mesons are produced. At larger centre-of-mass energies, for example at the e^+e^- colliders LEP/SLC and the hadron colliders Tevatron/LHC, all species of B hadrons are produced in the approximate ratio $B^\pm : B_d^0 : B_s^0 :$ b-baryon $: B_c^\pm \approx 39 : 39 : 11 : 12 : 1\%$. In addition, B hadrons are produced with a significant boost, traveling a few millimetres before they decay. A big step therefore in the study of B physics is the deployment of high precision silicon detectors that can reconstruct B decay vertices with an accuracy of a few 10's of μm.

Some highlights of B physics over the last 25 years include:

1977 the discovery of the Υ (Fixed target at Fermilab),

1983 the measurement of the inclusive b lifetime (PEP and PETRA),

1984 direct observation of B_d^0 and $B^\pm$ mesons (CLEO and ARGUS),

1987 the observation of B mixing (UA1, ARGUS and CLEO),

1992 evidence for the Λ_b baryon and B_s^0 meson (LEP),

1993 the first observation of $B_d^0 \to K^*\gamma$ and $B_d^0 \to \pi^+\pi^-$ (CLEO),

1994 evidence for the Ξ_b baryon and B^{**} hadrons with non-zero orbital angular momentum (LEP),

Year	Machine	Energy (GeV)	σ (b)	b fraction	Luminosity $(cm^{-2}s^{-1})$	Total number b pairs $(\times 10^6)$
Symmetric e^+e^- colliders at the $\Upsilon(4s)$						
79-	CESR	10.8	1.05 nb	0.25	1.3×10^{33}	~ 10
73-	DORIS	10.8	1.05 nb	0.23	$\sim 10^{31}$	~ 0.5
Asymmetric e^+e^- colliders at the $\Upsilon(4s)$						
00-	PEP-II	10.8	1.05 nb	0.25	6.6×10^{33}†	~ 100
00-	KEK-B	10.8	1.05 nb	0.25	1.0×10^{34}†	~ 100
e^+e^- colliders in the continuum						
80-90	PEP	29	0.4 nb	0.09	3.2×10^{31}	
78-86	PETRA	35	0.3 nb	0.09	1.7×10^{31}	
e^+e^- colliders at the Z^0						
89-00	LEP	91	9.2 nb	0.22	2.4×10^{31}	~ 2
89-00	SLC	91	9.2 nb	0.22	3.0×10^{30}	~ 0.09
$p\bar{p}$ hadron colliders						
92-96	Tevatron Run-I	1800	100 μb	0.002	3.0×10^{31}	$5k$ produced
01-	Tevatron Run-II	1960	100 μb	0.002	$1 - 2 \times 10^{32}$‡	

Table 1. *Past and present experimental facilities.* † *maximum achieved by July 2003.* ‡ *expected.*

1998 the discovery of the $B_c^\pm$ meson (CDF), and

2001 the discovery of CP violation in B_d^0 decays (BaBar and Belle).

By the time LEP and SLC produced their first collisions in 1989, the inclusive b lifetime was known with about a 20% accuracy. The relatively long b lifetime provided the first evidence for a small V_{cb} and the branching ratios of B_d^0 and $B^\pm$ decays of $\mathcal{O}(10^{-3})$ had been measured. In recent years, before the start of the B factories, the main contributors to B physics studies have been the e^+e^- colliders operating at the $\Upsilon(4s)$ and at the Z^0 resonance and also the Tevatron $p\bar{p}$ collider. By the start of the 21st century, B physics was promoted to the domain of precision physics.

4.1   BaBar and Belle

In 2000, the PEP-II and KEK-B high luminosity e^+e^- colliders began operation. These machines have separate e^- and e^+ magnet rings so that they can operate at asymmetric energies. The e^+ beam runs at the lower energy of 3.1/3.5GeV (BaBar / Belle) and the e^- beam runs at the higher energy of 9/8GeV (BaBar / Belle). This gives the centre-of-mass of the collision a boost of $\beta\gamma \sim 0.6$ along the beam axis. As a result the decay positions of the two B mesons is typically separated by 250μm. The required luminosity, $\geq 10^{33}\,cm^{-2}s^{-1}$, is set by the scale of the branching ratios for B meson decays to the experimentally accessible CP eigenstates, the need to determine the flavour of the other B in the event and the time difference between the two B decays.

The BaBar and Belle detectors, shown in Figure 7, are broadly similar. Each has a silicon vertex detector surrounded by a wire tracking chamber with a helium based gas.

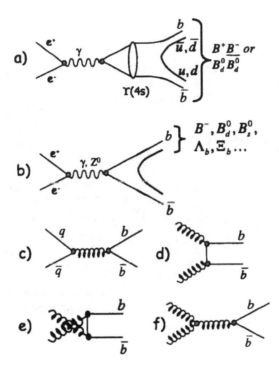

Figure 6. *b quark production mechanisms: a) $e^+e^- \rightarrow \Upsilon(4s) \rightarrow b\bar{b}$, b) $e^+e^- \rightarrow \gamma/Z^0 \rightarrow b\bar{b}$, c) and d)-f) are first order $\mathcal{O}(\alpha_s^2)$ quark-quark annihilation and gluon fusion processes respectively.*

They both have particle identification, a CsI(Tl) calorimeter, a superconducting coil and detectors interspersed with iron flux return for K_L^0 and μ identification. BaBar and Belle differ in their choice of silicon detector layout and particle identification. BaBar uses a 5 layer silicon detector with 2 layers at large radii and Belle has a 3 layer silicon detector. Both experiments exploit the dE/dx information in their tracking chambers for particle identification. In addition, BaBar uses long quartz bars to generate Cherenkov light from passing particles and to transport it via total internal reflection to a water-filled torus. Belle uses a combination of TOF and aerogel threshold Cherenkov counters for their particle identification.

The total integrated luminosity collected by Babar and Belle by July 2003 is 130.7 fb^{-1} and 158.7 fb^{-1} respectively (Browder 2003). The peak luminosity at Belle exceeded 10^{34} cm^{-2}s^{-1} in May 2003.

4.2 CDF and D0 Run-II

The Tevatron $p\bar{p}$ collider at Fermilab has undergone a major upgrade since the end of Run-I in 1996. The centre-of-mass energy has been increased to 1.96 TeV with a design luminosity of 5-8 $\times 10^{32}$ cm^{-2}s^{-1} for a final integrated luminosity target of 2 fb^{-1}. Both

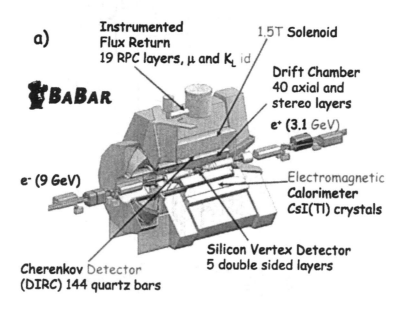

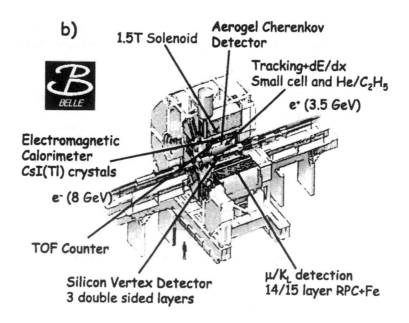

Figure 7. *The BaBar and Belle experiments.*

the CDF and D0 experiments, shown in Figure 8, have undergone major upgrades to operate at high luminosity and shortened bunch spacing (132 ns).

CDF has installed some new components for Run-II, namely a TOF detector to provide hadron particle identification and a hadronic track trigger. The track trigger finds high p_T charged tracks and links them to silicon tracks thereby enabling a cut on the impact parameter to be applied at the trigger level. This is the first time that such a trigger has been used in a hadron collider environment and is very important for the study of B physics. CDF has also updated the silicon tracker, replaced the tracking system electronics and improved the muon coverage. D0 has installed a 2 T super-conducting magnet, a silicon tracker, a fibre tracker and will also benefit from a silicon track trigger. D0 has also improved its calorimeter readout and upgraded its muon system.

Run-II started in March 2001 and has achieved a peak luminosity of $4.7 \times 10^{31} \mathrm{cm}^{-2} \mathrm{s}^{-1}$, a factor of 2 below the nominal (Paulini 2003). The main B physics aims of CDF and D0 are to measure the B_s^0 mixing parameter ΔM_s using $B_s^0 \rightarrow D_s^{\pm} \pi^{\mp}$ decays, the B_d^0 mixing phase ϕ_d using $B_d^0 \rightarrow J/\psi K_S^0$ decays and the weak decay phase γ from a combined analysis of $B_d^0 \rightarrow \pi^+ \pi^-$ and $B_s^0 \rightarrow K^+ K^-$ decays before the LHC starts operation.

5   B physics experimental programme

In the CKM picture of the Standard Model with 3 generations there are 4 independent parameters that must be determined from experiment. In the Wolfenstein parameterization these are λ, A, ρ and η. In this section we describe one possible scenario for an experimental programme to determine these four parameters and to over-constrain the Standard Model picture. The theoretical background for this section is described in detail in (Abbaneo 2003) and references therein. Unless otherwise stated, the results quoted are taken from (Schubert 2003).

The parameter λ is determined from precise measurements of $|V_{ud}|$ from nuclear and pion beta decay; $|V_{us}|$ from kaon, hyperon and τ decays; $|V_{cd}|$ from ν production of charm quarks; and $|V_{cs}|$ from beta decays of D mesons and decays of real W bosons. The current status from a fit to all the measurements is summarised by

$$\lambda = 0.2235 \pm 0.0033. \tag{87}$$

The remaining 3 parameters are determined from decays of B mesons. The parameter A is determined from the measurement of $|V_{cb}|$ from semi-leptonic B decays. ρ and η are extracted from the current knowledge of the unitarity triangle.

In order to check the internal consistency of the Standard Model picture of the CKM matrix and to search for new physics beyond, a dedicated B physics experimental programme is required. The programme can be sub-divided into 4 stages:

- Stage 1: determine $|V_{ub}|$ and $|V_{cb}|$ from tree-level B decays;

- Stage 2: determine $|V_{td}|$ and $|V_{ts}|$ from loop induced B decays;

- Stage 3: measure the B_d^0 mixing phase from CP violating B decays, thereby checking the consistency of the Standard Model;

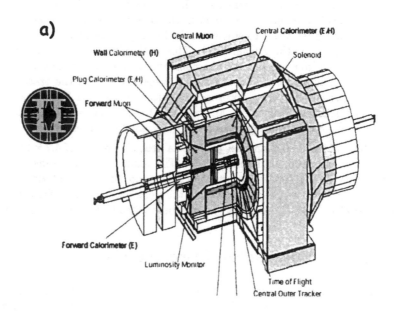

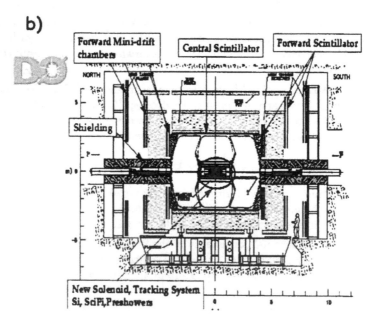

Figure 8. *The CDF and D0 experiments.*

• Stage 4: over-constrain the CKM picture and disentangle the new physics and Standard Model contributions using precision measurements of the B_s^0 mixing phase and the weak decay phase.

It is interesting to note that Stages 1 and 2, which give measurements of four CP conserving decays, tell us whether CP violation ($\eta \neq 0$) is predicted in the Standard Model.

5.1   Stage 1: V_{cb} and V_{ub}

The first step has been made by ARGUS and CLEO at the symmetric e^+e^- colliders and the four LEP experiments. BaBar and Belle at the asymmetric e^+e^- colliders are improving the precision on the determinations.

The matrix elements, $|V_{cb}|$ and $|V_{ub}|$, are measured using both inclusive and exclusive semi-leptonic decays of B mesons. Since these are tree-level decays, the values extracted are insensitive to new physics outside of the Standard Model and are therefore universal fundamental constants.

Key rôles in the determination of $|V_{cb}|$ are played by the exclusive decay mode $B_d^0 \rightarrow D^*\ell\nu$ and heavy quark effective theory (HQET). HQET is based on the idea that QCD is flavour independent such that, in the limit of infinitely heavy quarks, the quark transition occurs with a unit form factor $\mathcal{F}(1) = 1$ when the quarks are moving with the same invariant four-velocity. Experimentally, the aim is to measure the differential decay rate,

$$\frac{d\Gamma}{d\omega} = \frac{G_F^2}{48\pi^3}K(\omega)\mathcal{F}^2(\omega)|V_{cb}|^2, \tag{88}$$

as a function of ω (the product of the four-velocity of the D^* and the B^0). G_F is the Fermi coupling constant, $K(\omega)$ is a known kinematic factor and $\mathcal{F}(\omega)$ are hadronic form factors. In order to extract $\mathcal{F}(1)|V_{cb}|$, the decay rate is extrapolated to $\omega = 1$ where the D^* is at rest in the B^0 rest frame. To perform this extrapolation, assumptions have to be made about the shape of the form factor $\mathcal{F}(\omega)$. In addition, QCD corrections to $\mathcal{F}(1) = 1$ are required to take into account the effects of the finite mass of the heavy quark. Results obtained from the various QCD approaches yield an average value of

$$\mathcal{F}(1) = 0.91 \pm 0.04, \tag{89}$$

where the error is expected to reduce in the future with progress in unquenched Lattice QCD calculations and improved statistics. The measurement of $|V_{cb}|$ using exclusive decays has been performed by several groups and an example from Belle is shown in Figure 9a) for the reaction $\overline{B^0} \rightarrow D^{*+}\ell^-\bar{\nu}$.

An alternative way to determine $|V_{cb}|$ is by comparing the measurement of the semi-leptonic decay width for inclusive $b \rightarrow c\ell\nu$ decays to the prediction in the context of the Operator Product Expansion (OPE). Experimentally, the semi-leptonic decay width is determined from the branching ratio and lifetime and is related to $|V_{cb}|$ by

$$\Gamma(B \rightarrow X_c\ell\nu) = \frac{Br(B \rightarrow X_c\ell\nu)}{\tau_B} = \frac{G_F}{192\pi^3}m_b^5 f\left(\frac{m_c}{m_b},\frac{m_\ell}{m_b}\right)\gamma_c \tag{90}$$

where f is a known phase space factor, depending on the b and c quark masses. γ_c represents the perturbative and non-perturbative corrections which can be expressed using

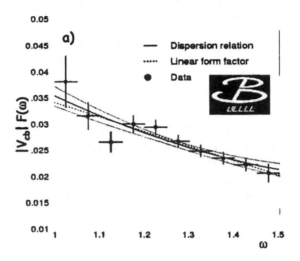

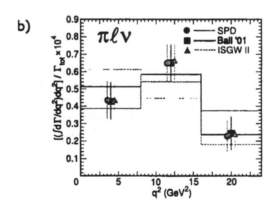

Figure 9. *a) Belle measurement of ω versus $\mathcal{F}(\omega)|V_{cb}|$ using $\overline{B^0} \to D^{*+}\ell^-\bar{\nu}$ decays. b) CLEO measurements of the q^2 distribution for $B \to \pi\ell\nu$ decays compared with various models.*

OPE in terms of a double expansion in $1/m_b$ and α_s. At $\mathcal{O}(1/m_b^3)$, the OPE introduces 9 parameters, that can be constrained using measurements of other inclusive kinematic variables, such as the photon energy spectrum in $B \to X_s\gamma$ decays and moments of the lepton and hadronic mass spectra in $B \to X_c\ell\nu$ decays. A comparison of the parameters obtained from the different measurements also provides a consistency test of the theoretical assumptions, notably quark-hadron duality. At the present level of sensitivity, the results show no evidence of any violation effects.

The present status of the combined exclusive and inclusive measurements of $|V_{cb}|$ is

$$|V_{cb}| = A\lambda^2 = 0.0415 \pm 0.0011. \tag{91}$$

The matrix element $|V_{ub}|$ is determined from semi-leptonic $b \to u\ell\nu$ transitions either through exclusive decays such as $B_d^0 \to \rho, \pi\ell\nu$ or using the event kinematics to separate out the $b \to u$ from the $b \to c$ contributions in inclusive $B \to X\ell\nu$ decays. For the exclusive decays modes, the weak decay involves a heavy to light quark transition so that HQET can not be used directly to extrapolate the form factors. Therefore, the form factors are calculated using a variety of methods based on Lattice QCD, QCD sum rules, perturbative QCD and quark models. At present, none of the methods provides a fully model-independent determination of $|V_{ub}|$. There is also a large spread in the predictions and, in particular, in the q^2 differential distribution. Experimentally, there are two important goals in studying the exclusive charmless semi-leptonic decay modes. The first is to lower the minimum lepton momentum cut and the second is to devise analysis methods that minimize the q^2 dependence of the form factors. A recent CLEO result, uses a low momentum cut of $1.5\,\text{GeV}$ and has the feature that the branching fractions are measured in three q^2 intervals. This is shown in Figure 9b) for $b \to \pi\ell\nu$ decays together with the predictions from the various models.

In principle, a clean determination of $|V_{ub}|$ using inclusive semi-leptonic decays can be obtained along same lines as $|V_{cb}|$. However, the main problem here is the large background from $b \to c$ decays. Typical strategies to separate the $b \to u$ from the $b \to c$ contributions involve applying severe kinematic cuts which significantly restrict the available phase space and exaggerate the theoretical errors. The three main cuts that have been used by the experiments include a cut on the lepton momentum, a cut on the hadronic invariant mass and a cut on the leptonic invariant mass. A reduction in the theoretical uncertainties can be achieved by using more complex kinematic cuts or, by measuring the universal structure function, which describes the distribution of the light-cone component of the residual momentum of the b quark, in some other process, such as the γ spectrum in $b \to s\gamma$ decays.

The present status of the combined exclusive and inclusive measurements of $|V_{ub}|$ is

$$|V_{ub}| = (3.80^{+0.24}_{-0.13} \pm 0.45) \times 10^{-3}. \tag{92}$$

5.2   Stage 2: V_{td} and V_{ts}

The matrix elements $|V_{td}|$ and $|V_{ts}|$ are extracted from the measurement of ΔM defined by equations (60) and (63) using a fit to the mixing asymmetry,

$$\mathcal{A}_{mix} = \frac{P(B^0 \to B^0) - P(B^0 \to \overline{B^0})}{P(B^0 \to B^0) + P(B^0 \to \overline{B^0})} = \cos \Delta M t \tag{93}$$

or fraction of mixed events,

$$\mathcal{F}_{mix} = \frac{P(B^0 \to \overline{B^0})}{P(B^0 \to B^0) + P(B^0 \to \overline{B^0})} = \frac{(1 - \cos \Delta M t)}{2} \tag{94}$$

using many different analysis methods. Experimentally, these measurements require the knowledge of the proper time of the B^0 and the flavour of the B^0 or $\overline{B^0}$ at production and decay. The extraction of $|V_{td}|$ or $|V_{ts}|$ requires the knowledge of the weak decay constant, f_{B_q}, and the bag parameter, B_{B_q}. Recent QCD sum rules and partially quenched Lattice calculations give

$$f_{B_d}\sqrt{B_{B_d}} = 223 \pm 33 \pm 12 \,\text{MeV} \quad \text{and} \quad \xi = \frac{f_{B_s}\sqrt{B_{B_s}}}{f_{B_d}\sqrt{B_{B_d}}} = 1.24 \pm 0.04 \pm 0.06. \tag{95}$$

Examples of the fits for ΔM_d are shown in Figure 10. The current world average value for ΔM_d (HFAG 2003) is

$$\Delta M_d = (0.502 \pm 0.006)\,\text{ps}^{-1} \tag{96}$$

and translates into

$$|V_{td}||V_{tb}| = (9.2 \pm 1.4 \pm 0.5) \times 10^{-3}. \tag{97}$$

No B_s^0 oscillation signal has yet been observed. However, a limit is placed on the B_s^0 oscillation frequency using an *amplitude* method. This is illustrated in Figure 10 where the world average of all the B_s^0 oscillation measurements are presented. The current limit on ΔM_s is given by

$$\Delta M_s > 14.4\text{ps}^{-1} \tag{98}$$

giving

$$|V_{ts}||V_{tb}| > 0.033. \tag{99}$$

Alternatively, $|V_{ts}|$ can be extracted from a measurement of the branching fraction of $b \to s\gamma$ decays. The present measurements result in

$$|V_{ts}||V_{tb}| = 0.047 \pm 0.008. \tag{100}$$

CDF is also making progress in the measurement of B_s^0 mixing using the exclusive decay mode $B_s^0 \to D_s^\pm \pi^\mp$ (Harr 2003). They currently have 84 ± 11 signal events from an integrated luminosity of 119 pb^{-1}, reconstructed using the $D_s^\pm \to \phi\pi^\pm$ decay channel. The B_s^0 invariant mass spectrum is shown in Figure 11. However, in order to observe Standard Model B_s^0 oscillations ($\Delta M_s \sim 20\text{ps}^{-1}$) at the 5σ level, they need a ten-fold increase in statistics.

5.3   Stage 3: The B_d^0 Mixing Phase

In the Standard Model, the B_d^0 mixing phase, ϕ_d, is given by the CKM angle β and is an example of CP violation in the interference which can be measured using the CP asymmetry given in equation (41). The *golden* decay channel for this measurement, $B_d^0 \to J/\psi K_S^0$, is dominated by the tree level Feynman diagram. The weak phase of the penguin contribution, $arg\{V_{ts}V_{tb}^*\}$ is similar to the weak phase of the tree decay but does not affect the parameter λ defined in equation (24). However, the effect of K^0-$\overline{K^0}$ mixing has to be taken into account and therefore

$$\begin{aligned}
\lambda(B_d^0 \to J/\psi K_S^0) &= \left(\frac{q}{p}\right)_{B_d^0}\left(\frac{q}{p}\right)_{K^0}\left(\frac{\overline{A}}{A}\right)_{J/\psi K_S^0} \\
&= -\left(\frac{V_{tb}^* V_{td}}{V_{tb} V_{td}^*}\right)\left(\frac{V_{cs} V_{cd}^*}{V_{cs}^* V_{cd}}\right)\left(\frac{V_{cb} V_{cs}^*}{V_{cb}^* V_{cs}}\right)
\end{aligned} \tag{101}$$

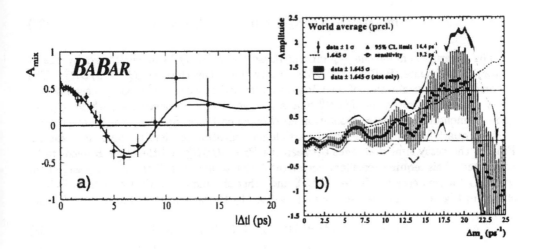

Figure 10. *a) An example of the ΔM_d fit to the mixing asymmetry A_{mix} using fully reconstructed $B_d^0 \to \overline{D}^{(*)}X$ events from BaBar (BaBar 2002) and b) the combined measurements of the B_s^0 oscillation amplitude as a function of ΔM_s (HFAG 2003).*

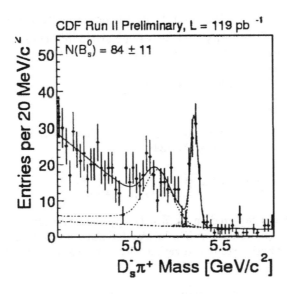

Figure 11. *The $D_s^\pm \pi^\mp$ candidate mass spectrum for $D_s^\pm \to \phi\pi^\pm$ from CDF. The curve is a fit to the signal shapes for $\overline{B_s^0} \to D_s^+\pi^-$ (narrow peak) and $\overline{B_s^0} \to D_s^{*+}\pi^-$ (wide peak), where the photon from the D_s^{*+} decay is not observed.*

where the minus sign takes into account that $J/\psi K_S^0$ is a CP odd state, and leads to

$$\text{Im}\,\{\lambda(B_d^0 \rightarrow J/\psi K_S^0)\} = \text{Im}\,\{-e^{-2i\beta}\} = \sin 2\beta. \tag{102}$$

The measurement of the B_d^0 mixing phase has recently been performed by the B factory experiments, BaBar and Belle , using the decay channel $B_d^0 \rightarrow J/\psi K_S^0$. The experimental determination of the CP asymmetry and hence the B_d^0 mixing phase involves 3 key elements. Firstly, the $J/\psi K_S^0$ final state needs to be reconstructed, which requires good momentum and energy resolution for charged particles. Secondly, the determination of the b quark flavour requires good particle identification to identify kaons and leptons. Finally, the decay time difference between the $B_d^0 \rightarrow J/\psi K_S^0$ and the other B needs to be computed. This requires excellent vertex resolution to measure distance. A measurement of the CP asymmetry for $B_d^0 \rightarrow J/\psi K_S^0$ and other charmonium/K^0 decays from Belle is shown in Figure 12. The latest results presented by BaBar and Belle (Browder 2003) are

$$\sin \phi_d = 0.741 \pm 0.067 \pm 0.033 \quad \text{BaBar} \quad 81\,\text{fb}^{-1} \tag{103}$$
$$\sin \phi_d = 0.733 \pm 0.057 \pm 0.028 \quad \text{Belle} \quad 140\,\text{fb}^{-1} \tag{104}$$
$$\sin \phi_d = 0.736 \pm 0.049 \quad \text{Average} \tag{105}$$

where ϕ_d is identical to 2β if only Standard Model box processes contribute to B_d^0 oscillations. The average value for this direct measurement of $\sin \phi_d$ can be compared to the prediction of $\sin 2\beta$ obtained from the currently known values of $|V_{cb}|$, $|V_{ub}|$, $|V_{td}|$ and $|V_{ts}|$,

$$\sin 2\beta_{side} = 0.695 \pm 0.055. \tag{106}$$

This shows that the Standard Model currently gives a very consistent picture.

The current status of the unitarity triangle is summarised in Figure 13. The measurements of $|V_{ub}|$ and $|V_{cb}|$ are related to ρ and η through

$$\sqrt{\bar{\rho}^2 + \bar{\eta}^2} = \left(1 - \frac{\lambda^2}{2}\right)\frac{1}{\lambda}\left|\frac{V_{ub}}{V_{cb}}\right| \tag{107}$$

and result in a circular band centered on $(\bar{\rho}, \bar{\eta}) = (0,0)$. The measurement of $|V_{td}|$ and the limit on $|V_{ts}|$ are related to ρ and η through

$$\sqrt{(1 - \bar{\rho})^2 + \bar{\eta}^2} = \frac{1}{\lambda}\left|\frac{V_{td}}{V_{cb}}\right| \approx \frac{1}{\lambda}\left|\frac{V_{td}}{V_{ts}}\right| \tag{108}$$

and result in a circular band centered on $(\bar{\rho}, \bar{\eta}) = (1,0)$. The direct measurement of $\sin \phi_d$ is related to ρ and η via equation (86) when interpreted in terms of the Standard Model. The best fit to these B physics results, together with the measurement of $\varepsilon \propto \bar{\eta}(1 - \bar{\rho})$ from kaon decays, gives

$$\bar{\rho} = 0.21 \pm 0.08 \pm 0.05 \tag{109}$$
$$\bar{\eta} = 0.34 \pm 0.04 \pm 0.02 \tag{110}$$

and is consistent with CP violation in the Standard Model.

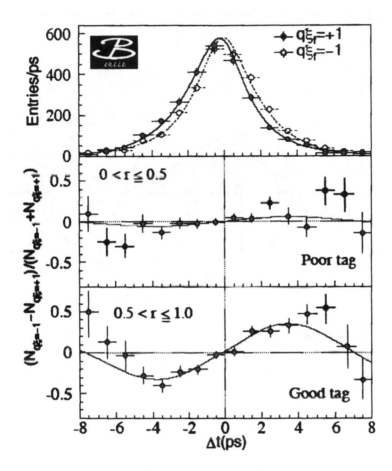

Figure 12. *The Belle proper time distribution for events with a tagged B_d^0 (open circles) and $\overline{B_d^0}$ (closed circles) meson in the event for $B_d^0 \rightarrow J/\psi K_S^0$ decays and other charmonium/K_S^0 final states. Also shown is the CP asymmetry for poor and good tags.*

5.4   Stage 4: Future

A fully comprehensive study of CP violation that includes the separation of possible new physics contributions from those of the Standard Model requires ultra-high statistics for B meson decays and the availability of the B_s^0 meson. This will be carried out by the next generation of B physics experiments and are the subject of the remainder of these lectures.

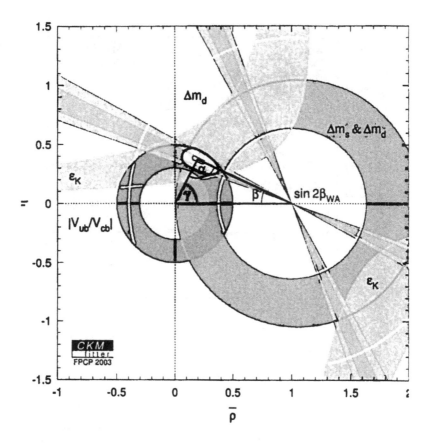

Figure 13. *The current status of the unitarity triangle from constraints on $\bar{\rho}$ and $\bar{\eta}$ using the values of $|V_{ub}|$, $|V_{cb}|$, the measurement of B_d^0 mixing, the upper limit on B_s^0 mixing, the measurement of the B_d^0 mixing phase and ϵ from kaon decays (Charles 2003).*

6   Next generation B physics experiments

The main B physics aims of the next generation of experiments, namely the dedicated B physics experiments, LHCb at the LHC and the BTeV experiment at the Tevatron collider, as well as the general purpose experiments ATLAS and CMS at the LHC, are

- to make precise measurements of CKM parameters with small model dependence;
- to search for new physics via CP violating phases; and
- to search for new physics via rare decays;

as well as covering a broad program of heavy flavour physics. A summary of the running parameters of the LHC and Tevatron colliders and experiments are given in Table 2. The main advantages of studying B physics at hadron colliders compared to e^+e^- colliders are

- a large $b\bar{b}$ cross-section, $\sigma_{b\bar{b}} \sim 100\text{-}500\,\mu\text{b}$ for $\sqrt{s} = 2\text{-}14\,\text{TeV}$, corresponding to $10^{11}\text{-}10^{12}$ $b\bar{b}$ produced per year at a luminosity of $2 \times 10^{32}\,\text{cm}^{-2}\text{s}^{-1}$;

- the production of all species of B hadrons ($B^{\pm}$, B_d^0, B_s^0, $B_c^{\pm}$ and b-baryons); and

- a large number of primary particles to determine the b production vertex.

An experimental difficulty arises from the large inelastic cross-section, $\sigma_{b\bar{b}}/\sigma_{inelastic} \sim 0.2\text{-}0.6\%$ for $\sqrt{s} = 2\text{-}14\,\text{TeV}$. The hadron collider experiments therefore require excellent triggering capabilities to reduce the minimum-bias background. The main disadvantages for performing B physics studies at hadron colliders are that many particles are produced which are not associated with B hadrons; and that the B hadrons do not evolve coherently and hence B^0 mixing dilutes the flavour tagging.

	Tevatron	LHC	
Energy/collision mode	2.0 TeV $p\bar{p}$	14.0 TeV pp	
$b\bar{b}$ cross-section	$\sim 100\mu$b	$\sim 500\mu$b	
Inelastic cross-section	~ 50mb	~ 80mb	
Ratio $b\bar{b}$/inelastic	0.2%	0.6%	
Bunch spacing	132 ns	25 ns	
Bunch length	~ 30cm	~ 5cm	
	BTeV	LHCb	ATLAS / CMS
Status	Pending finance	In construction	In construction
Start date	2009	2007	2007
Detector configuration	Single-arm forward	Single-arm forward	Central detector
Running luminosity	$2 \times 10^{32}\,\text{cm}^{-2}\text{s}^{-1}$	$2 \times 10^{32}\,\text{cm}^{-2}\text{s}^{-1}$	$1 \times 10^{33}\,\text{cm}^{-2}\text{s}^{-1}$ (10^{34})
$b\bar{b}$ events per 10^7 sec	$2 \times 10^{11} \times accep$	$1 \times 10^{12} \times accep$	$5 \times 10^{12} \times accep$
(Interactions/crossing)	~ 1.6	~ 0.5 (30% single int.)	~ 2.3 (20)

Table 2. *Comparison of the LHC and Tevatron collider running parameters and experiments.*

The dedicated B experiments, LHCb and BTeV, take full advantage of a forward detector geometry which is based on the fact that both the b- and $\bar{b}$-hadrons are predominantly produced in the same forward (or backward) direction at high energies. This is demonstrated in Figure 14 where the polar angles of the b- and $\bar{b}$-hadrons are shown in events generated with the PYTHIA simulation programme. The polar angle is defined with respect to the beam axis in the pp centre-of-mass system. Detecting both b- and $\bar{b}$-hadron at the same time is essential for the flavour tag. Further advantages of a forward geometry are

- The b-hadrons produced are faster than those in the central region. Their average momentum is about 80 GeV/c, corresponding to a mean decay length of ~ 7mm. Therefore, a good decay time resolution can be obtained for reconstructed B-mesons.

- The spectrometer can be built in an open geometry with an interaction region which is not surrounded by all the detector elements. This allows the vertex detector system to be built with sensors that can be moved away from the beam during

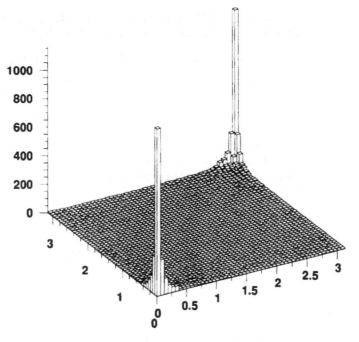

Figure 14. *Polar angles of the b- and b̄-hadrons*

injection. During data taking, the sensors are positioned closer to the beam in order to achieve a good vertex resolution.

- In the forward region, momenta are mainly carried by the longitudinal components. Therefore, the threshold value for the p_t trigger can be set low for electrons, muons and hadrons without being constrained by the detector requirements. This makes the p_t trigger more efficient than in the central region.

- The momentum range required for particle identification is well matched to Ring Imaging Cherenkov detectors.

- The open geometry allows easy installation, maintenance and possible upgrades.

A disadvantage of the forward geometry is that the minimum-bias background also peaks forward resulting in high occupancy and high track density and therefore requiring a finer segmentation of the detector components. Both LHCb and BTeV choose to run at a luminosity of $2 \times 10^{32}\,\mathrm{cm^{-2}s^{-1}}$. This optimal luminosity is chosen in a region where one inelastic interaction per bunch crossing dominates simplifying the event reconstruction and reducing radiation levels.

In order to exploit the full potential of the LHC, the experiments require the following capabilities

- a trigger sensitive to both leptonic and hadronic final states;

- a particle identification system capable of identifying p, K, π, μ and e within the required momentum range;

- a vertex detector able to reconstruct primary and B vertices very precisely;

- a tracking system with good momentum resolution; and

- an electromagnetic calorimeter capable of reconstructing π^0 decays.

We will now discuss to what extent each of the experiments matches these requirements.

6.1 LHCb

The LHCb experiment (Amato 1998) is a single-arm forward spectrometer designed to run at a lower luminosity (2×10^{32} cm^{-2}s^{-1}) than the nominal LHC luminosity (10^{34} cm^{-2}s^{-1}) and will perform its full physics programme from the start of LHC operation. The detector layout, recently optimized to reduce the material budget of the experiment and to improve the trigger performance (Antunes Nobrega 2003), is shown in Figure 15. It consists of the beam-pipe, a silicon strip Vertex Locator (VELO), a dipole magnet, a tracking system, two Ring Imaging Cherenkov detectors with three radiators, a calorimeter system and muon system. The pp interaction point is contained within the VELO detector and the 1.1 T dipole magnet is displaced $\sim$ 5 metres upstream. The LHCb trigger is designed to distinguish minimum-bias events from events containing B hadrons through the presence of particles with a large transverse momentum p_T and the existence of secondary vertices. The fringe field from the magnet, in the region between the VELO and RICH-1, is used to provide p_T information for tracks with large impact parameter.

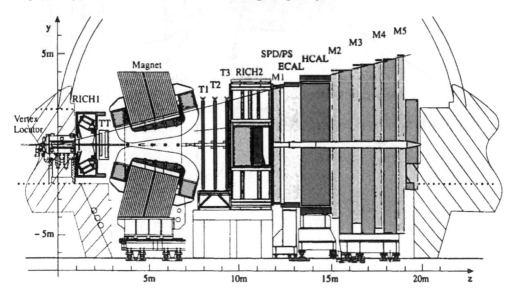

Figure 15. *The LHCb experiment.*

6.2 BTeV

BTeV is a single-arm forward spectrometer (Drobychev 2002), shown in Figure 16, and will run at the Tevatron $p\bar{p}$ collider from 2009. A key design feature of BTeV is the silicon pixel detector, which is centered on the interaction region and surrounded by a 1.5 T magnet. The main advantages of using a silicon pixel detector compared to a silicon strip detector are a better signal to noise ratio, a 5-10 μm spatial resolution, a lower occupancy and radiation hardness. The main disadvantages are the large radiation length ($\sim 1X_0$) and the number of readout channels. A unique feature of the BTeV silicon pixel detector is that it will be used in the lowest level trigger system. Other features of the experiment include a tracking system, a single RICH detector with 2 radiators, a lead-tungstate calorimeter and a muon detector.

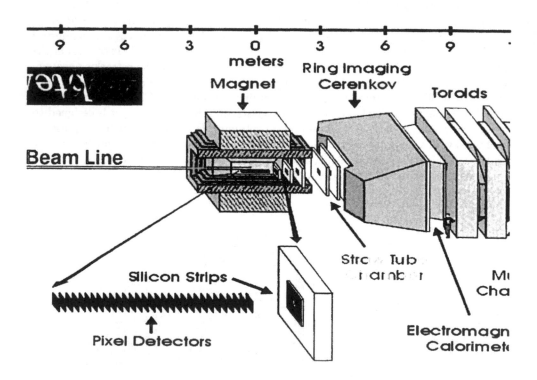

Figure 16. *The BTeV experiment.*

6.3 ATLAS and CMS

ATLAS and CMS are general purpose collider detectors designed to perform high p_T physics at the LHC (Virdee 2003). During the first three years of running, the LHC will operate at a low luminosity, 1×10^{33} cm^{-2}s^{-1}, enabling ATLAS and CMS to pursue a B physics programme. The B capabilities of the experiments are achieved through the employment of silicon pixel and microstrip detectors and specialist B triggers by reducing the lepton p_T thresholds to a minimum. Both ATLAS and CMS have excellent muon and electron detection capabilities, and a good vertex resolution, allowing them to collect a high statistics sample of decays such as $J/\psi K_S^0$ and $J/\psi \phi$. However, their first level high p_T lepton trigger is insensitive to purely hadronic final states, which can only be triggered by the presence of a semi-leptonic decay of the other b quark. The two experiments also have no $p/K/\pi$ separation capability in the relevant momentum range, although the energy loss, dE/dx, can be used for particle identification at low momenta. In particular, they cannot separate kaons from pions in the high momentum region relevant for two-body decay modes of B mesons such as $\pi^+\pi^-$ and $D_s K$. For these reasons, it is not expected that ATLAS and CMS will be able to study CP violation with all the final states necessary to perform a model-independent analysis. They will, however, have a large acceptance for very rare B decays, such as $B \to \mu^+\mu^-$.

7 B physics at the LHC

7.1 B production

The physics of generic B production (Ahmadov 2000) enters in many ways in the context of B physics in the LHC era. It affects the possibilities of B decay studies, it can be used for QCD tests and is an important background for several processes of interest. B production at hadron colliders was first observed by UA1 at the $Sp\bar{p}S$ and has subsequently been studied at the Tevatron collider. Studies include the shape of the p_T distribution for b production and the correlations between the b and $\bar{b}$. Phenomenologically, these are reasonably well explained by next-to-leading order QCD (NLO QCD). However, the observed b production cross-section measured at the Tevatron is larger than expected and is underestimated by a factor of ~ 2.4. There are 3 possible explanations for this discrepancy (Salam 2002). The first explanation is that the absolute normalisation of the cross-section is not correctly predicted due to the possible presence of large higher order terms (NNLO QCD). The second explanation is that the shape of the p_T distribution is distorted by some perturbative or non-perturbative effects, such as fragmentation. Finally, it has been suggested that the discrepancy could be due to the production of light gluinos with a mass of ~ 14 GeV. Studies of generic B production physics are under investigation by the next generation experiments. One example is the study by ATLAS of $b\bar{b}$ production correlations using the J/ψ from exclusive $B_d^0 \to J/\psi K_S^0$ or $B_s^0 \to J/\psi \phi$ decays and the muon from the semi-leptonic decay of the other B in the event (Ohlsson-Malek 2003).

The physics of B production can also be interpreted as the study of the production and properties, e.g. mass and lifetime, of the individual B species. For example, a copious number of $B_c^\pm$ mesons, $\sim 10^9$, will be produced at the LHC in one year of running. A study by the LHCb experiment of the decay mode $B_c^\pm \to J/\psi \pi^\pm$, which has an expected visible branching ratio of 6.8×10^{-4}, results in an annual yield of 14k events with a background to signal ratio of < 0.8 (Antunes Nobrega 2003). The $B_c^\pm$ meson can also be used for CP violation studies, for example in the decay modes $B_c^\pm \rightarrow J/\psi D^\pm$ and $B_c^\pm \rightarrow D_s^\pm D^0$, although several years of running will be required to provide sufficient statistics.

7.2   B Decay

We now turn our attention to the study of CP violation and the search for new physics in B decays in the LHC era. We start by considering a few examples of how the B_s^0 mixing frequency ΔM_s, the B_d^0 mixing phase ϕ_d, the B_s^0 mixing phase ϕ_s, and the weak decay phase γ can be measured at the LHC. The CKM angle α is notoriously difficult to measure, mainly because the decay mode $B_d^0 \rightarrow \pi^+\pi^-$ is plagued by large penguin contributions, and will not be discussed here. For the sake of clarity, we also will restrict our discussions to the recent studies performed by the LHCb experiment (Antunes Nobrega 2003). We will then discuss the interpretation of these measurements in the presence of new physics and how the contributions from new physics and the Standard Model can be separated. Finally, we will review the physics of very rare B decay modes.

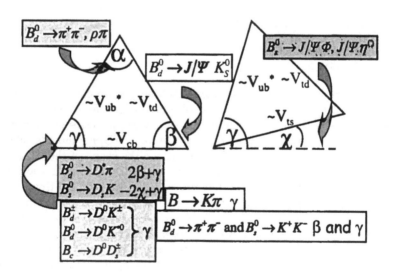

Figure 17. *A summary of the B decay modes and the relevant CKM phases they probe discussed in these lectures.*

B decays can be subdivided into three main categories, neutral (B_d^0, B_s^0) decays, charged ($B^\pm$, $B_c^\pm$) decays and decays of b-baryons (Λ_b, Ξ_b etc). Here, we shall focus on the physics of neutral B decays, which can be further categorized according to their final state:

- flavour specific final state

- flavour non-specific final state, of which there are

  - CP eigenstates
  - mixed CP eigenstates and
  - non-CP eigenstates.

At the LHC, the measurement of ΔM_s is studied using a flavour-specific final state whereas the study of CP violation uses all types of flavour non-specific final states. The B decay modes that will be discussed in these lectures and the CKM parameters they probe are shown in Figure 17.

7.2.1 B_s^0 mixing

At the LHC, the B_s^0 mixing frequency will be measured using the $B_s^0 \rightarrow D_s^{\pm}\pi^{\mp}$ decay channel, which has a large visible branching ratio ($\mathrm{Br}(B_s^0 \rightarrow D_s^{\pm}\pi^{\mp}) = 120 \times 10^{-6}$) and is an example of a flavour-specific final state in which only a single tree diagram contributes. It is therefore considered as a *gold-plated* channel for the study of B_s^0 mixing. The generic decay formulae, equations (20) and (21), can be applied with the constraint $\lambda = 0$ ($|q/p| = 1$) and a flavour asymmetry can be defined,

$$\mathcal{A}_{flav} = \frac{\Gamma_{\overline{B}\rightarrow f} - \Gamma_{B\rightarrow f}}{\Gamma_{\overline{B}\rightarrow f} + \Gamma_{B\rightarrow f}} = -D\frac{\cos \Delta M_s t}{\cosh \Delta \Gamma_s t} \tag{111}$$

where D is a dilution factor due to wrong tagging and experimental resolution. From this asymmetry, the B_s^0 oscillation frequency ΔM_s and, optionally, the decay width difference $\Delta \Gamma_s$ can be determined.

A study of B_s^0 mixing by LHCb using $B_s^0 \rightarrow D_s^{\pm}\pi^{\mp}$ decays where the D_s is fully reconstructed in the $D_s^{\pm} \rightarrow K^+K^-\pi^{\pm}$ final state has recently been performed. The analysis benefits from excellent mass (14 MeV) and decay time (44 fs) resolutions. The expected $B_s^0 \rightarrow D_s^{\pm}\pi^{\mp}$ decay rate as a function of proper time t for two different values of ΔM_s is shown in Figure 18. The expected ΔM_s sensitivity for one year of data taking is $\sim 0.01\,\mathrm{ps}^{-1}$ and results in a $> 5\sigma$ measurement up to values of $\Delta M_s = 68\,\mathrm{ps}^{-1}$. The equivalent sensitivities for ATLAS and CMS are $36\,\mathrm{ps}^{-1}$ and $30\,\mathrm{ps}^{-1}$ respectively (Ohlsson-Malek 2003).

7.2.2 The B_d^0 mixing phase

We have already discussed the extraction of the B_d^0 mixing phase, in the context of the current generation of experiments (see Section 5.3). At the LHC, the *gold-plated* decay mode $B_d^0 \rightarrow J/\psi K_S^0$, an example of a flavour non-specific CP eigenstate, will be also central to the measurement of the B_d^0 mixing phase. The current results from the B factory experiments are consistent with the Standard Model expectations, $\mathcal{A}^{dir} = 0$ (i.e. $|\lambda| = 1$) and $\mathcal{A}^{mix} = \mathrm{Im}\{\lambda\} = \sin 2\beta$. However, precision measurements are important as new physics could affect these parameters at the level of precision reached in the LHC era. In addition, we shall see in Section 7.2.4 that a precise measurement of ϕ_d is required for the extraction of the weak decay phase γ from analyses such as the combined study of the $B_d^0 \rightarrow \pi^+\pi^-$ and $B_s^0 \rightarrow K^+K^-$ decay modes. An example of a fit to the $B_d^0 \rightarrow J/\psi K_S^0$ CP

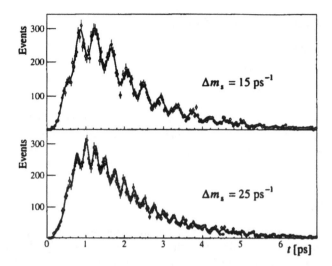

Figure 18. *The proper time distribution of simulated $B_s^0 \rightarrow D_s^{\pm}\pi^{\mp}$ candidates for one year of data at LHCb and two values of ΔM_s. The B_s^0 mesons have been flavour tagged as not oscillated.*

asymmetry is shown in Figure 19. The statistical sensitivities on the parameters $\sin\phi_d$ and $|\lambda|$ expected in one year of data taking are 0.022 and 0.023 respectively. In 5 years at the LHC, the expected precision of $\sin\phi_d$ using all the data from LHCb, ATLAS and CMS is expected to reach 0.005.

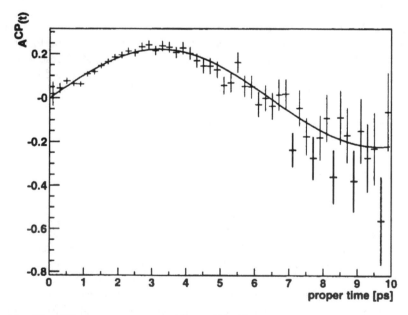

Figure 19. *The CP asymmetry using $B_d^0 \rightarrow J/\psi K_S^0$ decays in one year of data at LHCb.*

7.2.3 The B_s^0 mixing phase

The B decay channels, $B_s^0 \rightarrow J/\psi\phi$ and $B_s^0 \rightarrow J/\psi\eta^{(\prime)}$, are the SU(3) analogues of $B_d^0 \rightarrow J/\psi K_S^0$. They are both flavour non-specific final states and can be used to determine the B_s^0 mixing phase, the $J/\psi\phi$ being a mixed CP eigenstate and the $J/\psi\eta^{(\prime)}$ a pure CP eigenstate. If only Standard Model processes contribute to B_s^0 oscillations, the CP asymmetry is given by equation (41) with

$$\lambda(B_s^0 \rightarrow J/\psi\phi(\eta^{(\prime)})) = \left(\frac{q}{p}\right)_{B_s^0} \left(\frac{\overline{A}}{A}\right)_{J/\psi\phi(\eta^{(\prime)})} \tag{112}$$

$$= \frac{1-r}{1+r} \left(\frac{V_{tb}^* V_{ts}}{V_{tb} V_{ts}^*}\right) \left(\frac{V_{cb} V_{cs}^*}{V_{cb}^* V_{cs}}\right)$$

where r is the ratio of the B_s^0 decay amplitudes into the CP=-1 and CP=+1 states. This leads to

$$\mathrm{Im}\left\{\lambda(B_s^0 \rightarrow J/\psi\phi(\eta^{(\prime)}))\right\} = \mathrm{Im}\left\{e^{2i\chi}\right\} = \sin 2\chi. \tag{113}$$

The Standard Model CKM picture predicts that the B_s^0 mixing phase should be small, $\chi = \eta\lambda^2 \approx -0.04$. Therefore, the observation of a large CP asymmetry in this decay channel would be a striking signal for physics beyond the Standard Model. Similar to the B_d^0 mixing phase, the B_s^0 mixing phase is needed to extract the weak decay phase γ from analyses such as $B_s^0 \rightarrow D_s^{\pm} K^{\mp}$ and the combined analysis of $B_d^0 \rightarrow \pi^+\pi^-$ and $B_s^0 \rightarrow K^+ K^-$ decays.

Compared to $B_d^0 \rightarrow J/\psi K_S^0$ decays, the $B_s^0 \rightarrow J/\psi\phi$ and $B_s^0 \rightarrow J/\psi\eta^{(\prime)}$ decay channels present several challenges:

- The mixing frequency ΔM_s of the time-dependent CP asymmetry is very large, requiring excellent proper-time resolution.

- The B_s^0 meson is expected to have a non-negligible decay-width difference, $\Delta\Gamma_s/\Gamma_s$.

- The $B_s^0 \rightarrow J/\psi\eta^{(\prime)}$ decay mode, where the η decays predominantly to an all neutral particle final state, requires excellent neutral particle identification and reconstruction.

- Due to the fact that both J/ψ and ϕ are vector mesons, there are three distinct amplitudes contributing to the $B_s^0 \rightarrow J/\psi\phi$ decay: two CP even and one CP odd. The two components can be disentangled on a statistical basis by taking into account the distribution of the so-called transversity angle, θ_{tr}, defined in Figure 20.

The results of a study by LHCb using a fit to the transversity angle and B_s^0 mass distribution for $B_s^0 \rightarrow J/\psi\phi$ decays indicates that a statistical precision on the B_s^0 mixing phase of $\sigma(\sin\phi_s) \sim 0.06$ and on the decay width difference of $\Delta\Gamma_s/\Gamma_s \sim 0.018$ can be achieved in one year of data taking. A study by ATLAS using the $B_s^0 \rightarrow J/\psi\eta^{(\prime)}$ decay mode results in a precision of $\sigma(\phi_s) \sim 0.17$ for one year of data taking at the low luminosity of $10^{33}\mathrm{cm}^{-2}\mathrm{s}^{-1}$ (Ohlsson-Malek 2003).

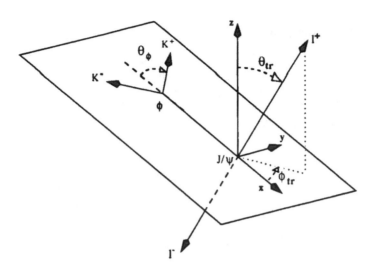

Figure 20. *Definition of the transversity angle used in the LHCb analysis of $B_s^0 \to J/\psi\phi$ decays.*

7.2.4 The weak decay phase

Many methods have been devised to extract the weak decay phase γ and an excellent summary can be found in reference (Fleischer 2002). In general, the methods can be divided into four main categories:

- Method I: time dependent asymmetries,

- Method II: time integrated amplitude relations,

- Method III: isospin symmetry relations, and

- Method IV: Uspin relations.

Methods I and II are mainly model independent as they involve only tree-level decays, whereas III and IV depend on hadronic assumptions and any isospin or Uspin breaking effects. Here, we will illustrate the measurement of the weak decay phase γ in three different ways.

The first way uses the decay mode $B_s^0 \to D_s^\pm K^\mp$, in which the final state is a non-CP eigenstate, and is an example of Method I. The decay can proceed through two tree decay diagrams and their CP conjugated processes, the interference of which gives access to the sum of the weak decay phase and the B_s^0 mixing phase, $\gamma + \phi_s$. In turn, this provides a measurement of γ, if ϕ_s is determined by some other means, for example using $B_s^0 \to J/\psi\phi$ decays. The decay formulae, equations (20) and (21), can be applied with $|\lambda| = |\bar{\lambda}| \approx 0.5$ and $arg\{\lambda\} = \Delta + (\gamma + \phi_s)$ and $arg\{\bar{\lambda}\} = \Delta - (\gamma + \phi_s)$ where Δ denotes the strong phase difference between the two tree diagrams. Experimentally, a potentially large background arises due to the $B_s^0 \to D_s^\pm \pi^\mp$ decay mode which has more than 10 times the branching fraction of $B_s^0 \to D_s^\pm K^\mp$. The LHCb RICH detectors, therefore, provide a powerful means to separate out the $B_s^0 \to D_s^\pm \pi^\mp$ and $B_s^0 \to D_s^\pm K^\mp$ contributions as illustrated in Figure 21. In the LHCb study, a simultaneous fit is performed to the $B_s^0 \rightarrow D_s^\pm \pi^\mp$ and $B_s^0 \rightarrow D_s^\pm K^\mp$ decays to determine the parameters ΔM_s, $|\lambda| = |\bar{\lambda}|$, $arg\{\lambda\}$, $arg\{\bar{\lambda}\}$ and the mistag fraction. ΔM_s and the mistag fraction are effectively determined from the flavour asymmetry in the $B_s^0 \rightarrow D_s^\pm \pi^\mp$ event sample, while being applied at the same time to measure the CP asymmetry in the $B_s^0 \rightarrow D_s^\pm K^\mp$ sample. An example of the $B_s^0 \rightarrow D_s^\pm K^\mp$ asymmetries obtained after 5 years of data taking are shown in Figure 22. The value of the weak phase $\gamma + \phi_s$ is extracted from the difference of the fitted parameters $arg\{\lambda\} - arg\{\bar{\lambda}\}$. The expected statistical uncertainty on $\gamma + \phi_s$ for one year of data is $\sigma(\gamma + \phi_s) \sim 14$-$15$ degrees for values of $\gamma + \phi_s$ ranging between 55 and 105 degrees.

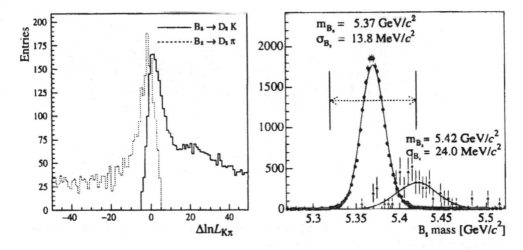

Figure 21. *The difference in log-likelihood between the RICH hypotheses for the K and π coming directly from the B_s^0 in $B_s^0 \rightarrow D_s^\pm K^\mp$ and $B_s^0 \rightarrow D_s^\pm \pi^\mp$ decays (left) and the B_s^0 mass distribution for selected $B_s^0 \rightarrow D_s^\pm K^\mp$ candidates (right). Also shown (right) are the misidentified $B_s^0 \rightarrow D_s^\pm \pi^\mp$ events normalised to the visible branching ratios.*

Our second illustration to extract γ is an example of Method II and relies on the measurement of 3 time-integrated decay rates, $B_d^0 \rightarrow \bar{D}^0(K^+\pi^-)K^{0*}$, $B_d^0 \rightarrow D_{CP}^0(K^+K^-)K^{0*}$, $B_d^0 \rightarrow D^0(\pi^+K^-)K^{0*}$, and their CP conjugates. $D_{CP}^0 = (D^0 + \bar{D}^0)/\sqrt{2}$ denotes the CP even eigenstate of the D^0-$\bar{D}^0$ system and $K^{0*} \rightarrow K^+\pi^-$. The method is theoretically clean as the decays involve tree-level diagrams, which interfere due to the presence of $D^0 - \bar{D}^0$ mixing. The value of γ extracted through this method is therefore sensitive to any new physics in $D^0 - \bar{D}^0$ oscillations. The relations between the six decay rates are used to extract the weak decay phase γ as well as the strong phase difference between $B_d^0 \rightarrow D^0 K^{0*}$ and $B_d^0 \rightarrow \bar{D}^0 K^{0*}$ decays. An advantage of this analysis is that no proper time information or tagging is required. However, the visible branching ratios are low, for example $Br(B_d^0 \rightarrow D_{CP}^0(K^+K^-)K^{0*}) \sim 0.2 \times 10^{-6}$, resulting in ~ 0.6k events reconstructed per year. The expected statistical precision on γ after one year of data taking is $\sigma(\gamma) = 9$-7 degrees for values of γ in the range 55 to 105 degrees. The sensitivity is independent on the value of the strong phase difference, but decreases by approximately 1 degree for a change in background level of 20%.

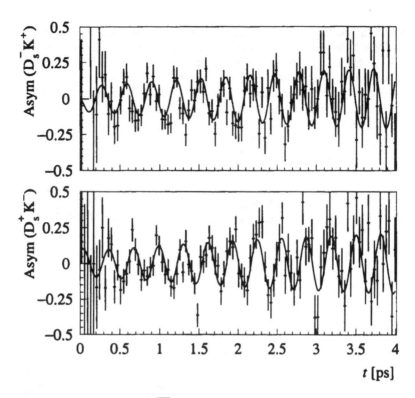

Figure 22. *Time dependent B_s^0-$\overline{B_s^0}$ asymmetry of $D_s^- K^+$ (top) and $D_s^+ K^-$ (bottom) events for $\Delta M_s = 20\,\mathrm{ps}^{-1}$ and 5 years of data.*

Our final illustration is an example of Method IV which uses Uspin symmetry (replacement of all d quarks with s quarks) to relate B_d^0 and B_s^0 decays. The decay modes we shall consider are $B_d^0 \rightarrow \pi^+\pi^-$ and $B_s^0 \rightarrow K^+K^-$. Under the assumption that the strong interaction dynamics remains invariant under this interchange, the relative contributions of the penguin process with respect to the tree process are identical in the two decay modes and γ can be determined from the time-dependent CP asymmetry for B_d^0 and $\overline{B_d^0}$ decaying into $\pi^+\pi^-$ and that for B_s^0 and $\overline{B_s^0}$ decaying into K^+K^-. Therefore, the value of γ extracted from this method is sensitive to any new physics in the penguin diagrams. Uspin symmetry can be also tested by using the values of ϕ_d and ϕ_s obtained from the CP asymmetries measured with $B_d^0 \rightarrow J/\psi K_S^0$ and $B_s^0 \rightarrow J/\psi\phi$ decays. Experimentally, a crucial requirement for the analysis of $B_d^0 \rightarrow \pi^+\pi^-$ and $B_s^0 \rightarrow K^+K^-$ decays is the ability to suppress other B hadron decays with the same two track topology and which may exhibit their own CP asymmetries. In LHCb, the rejection of these backgrounds relies on the performance of the RICH particle identification and on the invariant mass resolution. An example of the invariant mass distribution for selected and triggered $B_s^0 \rightarrow K^+K^-$ decays obtained by LHCb is shown in Figure 23. In order to extract the weak decay phase γ, a fit is performed to the $B_d^0 \rightarrow \pi^+\pi^-$ and $B_s^0 \rightarrow K^+K^-$ CP asymmetries to measure the four observables $\mathcal{A}_{CP}^{dir}(B_d^0 \rightarrow \pi^+\pi^-)$, $\mathcal{A}_{CP}^{mix}(B_d^0 \rightarrow \pi^+\pi^-)$, $\mathcal{A}_{CP}^{dir}(B_s^0 \rightarrow K^+K^-)$ and $\mathcal{A}_{CP}^{mix}(B_s^0 \rightarrow K^+K^-)$. These four observables depend on the B_d^0 mixing phase ϕ_d, the B_s^0 mixing phase ϕ_s, the weak decay phase γ and the amplitudes and phases of the penguin to tree ratios. In the limit of exact Uspin symmetry of the strong interaction, the amplitude (d) and phase (θ) of the penguin to tree ratio in $B_d^0 \to \pi^+\pi^-$ and the equivalent quantities in $B_s^0 \to K^+K^-$ decays are the same. In this case, the measurements of the four CP asymmetry observables provide a determination of γ, if ϕ_d and ϕ_s are taken from elsewhere. The expected sensitivity on γ for one year of data is $\sigma(\gamma) \sim 5$ degrees assuming the Standard Model values of $\Delta\Gamma_s/\Gamma_s = 0.1$ and $\Delta M_s = 20\,\mathrm{ps}^{-1}$, and $d = 0.3$ and $\theta = 160$ degrees. An example of the confidence regions in the (d,γ) plane for one year of data, together with the one-dimensional experimental probability density functions is shown in Figure 24.

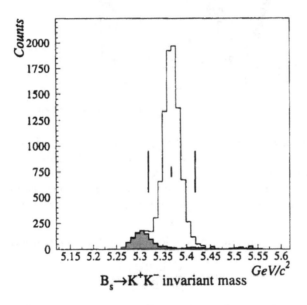

Figure 23. *Invariant mass distribution for selected $B_s^0 \to K^+K^-$ candidates. The light-shaded (yellow) region is the signal and the dark (red) one represents background from B decays with the same two charged track topology.*

7.3 Tests for new physics

There are many ways of looking for new physics outside of the Standard Model description of the CKM matrix (Stone 2001). A generic approach is to look for inconsistencies between the different kinds of measurements used to determine the apex of the unitarity triangle. For example, new physics outside the Standard Model could reveal itself as discrepancies among the values of ρ and η derived from independent determinations using the magnitude measurements ($|V_{ub}/V_{cb}|$ and $|V_{td}/V_{ts}|$), or from the different CP violation measurements in B_d^0, B_s^0 and K_L^0 decays. Another interesting generic approach is to express the CKM matrix in terms of four phases instead of the Wolfenstein parameters and to perform

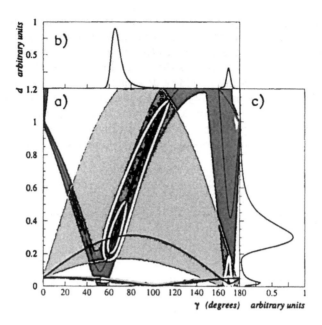

Figure 24. *a) Confidence regions in the (d, γ) plane; the darker (red) bands indicate the region obtained from $B_d^0 \to \pi^+\pi^-$; the lighter (cyan) bands indicate the region obtained from $B_s^0 \to K^+K^-$ and the black solid curves represent the exact analytical calculations. The deformed ellipses enclose the 68% and 95% confidence regions obtained by using all constraints. b) and c) are the probability density functions for γ and d respectively.*

checks between them. Silva and Wolfenstein have shown that the relationship

$$\sin \chi = \frac{\lambda^2 \sin \beta \sin \gamma}{\sin(\beta + \gamma)} \qquad (114)$$

is a critical check of the Standard Model. The final approach is to compare CP phases extracted from different exclusive B decay channels. For example, a recently observed difference by the Belle collaboration between the B_d^0 mixing phase obtained using $B_d^0 \to J/\psi K_S^0$ decays (tree diagram) and that from $B_d^0 \to \phi K_S^0$ decays (pure penguin diagram) may be an indication that new physics is just around the corner (Belle 2003).

At the LHC, the B_s^0 meson plays an essential rôle to disentangle new physics from that of the Standard Model. Let us consider the three ways we have discussed in these lectures to extract the weak decay phase γ. The value of γ measured using $B_s^0 \to D_s^\pm K^\mp$ decays will not be affected by the possible existence of new particles. However, the combined analysis of $B_d^0 \to \pi^+\pi^-$ and $B_s^0 \to K^+K^-$ decays makes explicit use of the penguin processes where new particles can contribute to the loops. Equally, new physics in D^0-$\overline{D^0}$ mixing could affect the value of γ extracted from the analysis of $B_d^0 \to D^0 K^{0*}$ decays. From these three measurements, it will be possible to determine γ and, together with the measurement of $|V_{ub}|$, extract the CKM parameters A, ρ and η even in the presence of new physics. In addition, it will be possible to extract the contribution of new physics to the oscillations and penguins. A possible scenario after one year of data taking at LHCb is shown in Figure 25. The measured value of γ could disagree with that obtained from the current Standard Model analysis of the CKM parameters using processes generated by tree and box diagrams. This would be a clear signal for the existence of new physics in the B^0-$\overline{B^0}$ oscillations and would allow the separation of the new physics from that of the Standard Model.

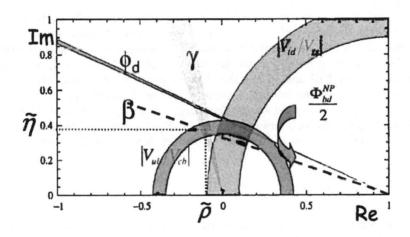

Figure 25. *A possible scenario for the allowed region in the $\bar{\rho}$-$\bar{\eta}$ plane after one year of data taking at LHCb. The contribution from new physics to the B^0-$\overline{B^0}$ oscillation is indicated.*

7.4 Rare B decays

Very rare B decays are an excellent place to search for new physics (Abbaneo 2003) (Nir 1999). Figure 2c) shows the basic Standard Model diagram for these processes which, in general, tend to have small branching ratios, for example $\mathrm{Br}(B_d^0 \to K^*\gamma) \sim 4 \times 10^{-5}$, $\mathrm{Br}(B_d^0 \to K^{0*}\mu^+\mu^-) \sim 10^{-6}$ and $\mathrm{Br}(B_d^0 \to \mu^+\mu^-) \sim 10^{-10}$. However, in the presence of new physics, new fermion-like particles can replace the quarks or new gauge-like particles can replace the W, thereby increasing the branching fractions significantly. The exclusive processes are also sensitive to new physics in observables other than the branching fractions, such as photon polarization, forward-backward asymmetries and CP asymmetries. Their branching fractions also depend on the CKM matrix elements $|V_{ts}|$ and $|V_{td}|$, thereby providing another consistency check of the unitarity of the CKM matrix. The main impact in the LHC era on the study of rare decays will be to observe rare B_s^0 decays and to provide a radical increase in statistics. $B \to \mu^+\mu^-$ decays will only be observed before the LHC era if it is enhanced drastically by the presence of new physics. The study of rare B decays in the LHC era is at a preliminary stage, mainly due to estimating the size of the background when the branching ratios are so small. However, it is expected that all the next generation experiments will have significant statistics for decay channels that have two leptons in the final state, such as $B \rightarrow \mu^+\mu^-$, $B_d^0 \rightarrow K^{0*}\mu^+\mu^-$ and $B_s^0 \rightarrow \phi\mu^+\mu^-$ and for the radiative rare B decays, such as $B_d^0 \rightarrow K^*\gamma$ and $B_s^0 \rightarrow \phi\gamma$ (Ahmadov 2000).

8   Conclusions

An exciting time for B physics has just commenced. The first experimental phase has recently provided the first observation of CP violation in the B system by BaBar and Belle. These experiments, together with CDF and D0 at the Tevatron collider, will measure the CKM parameters with unprecedented precision. The next generation of B experiments (LHCb, BTeV and the general purpose experiments, ATLAS and CMS) will start to become operational in 2007 and will be the ultimate source for B physics studies. Their goal is to thoroughly test the internal consistency of the Standard Model, to search for the necessary new physics beyond and to disentangle the new physics from the Standard Model physics. I certainly look forward to the next 10-15 years of B physics results and hope that new physics will reveal itself first in the virtual world of the B sector.

Acknowledgments

I thank the organizers of the school for inviting me to give these lectures and for their hospitality. T.Nakada and R.Thorne are deeply acknowledged for many useful comments.

References

Abbaneo, D. *et al* (2003), *The CKM Matrix and the Unitarity Triangle*, editors M.Battaglia *et al*, hep-ph/0304132v1.

Ahmadov, A. *et al* (2000), *Proceedings of the Workshop on Standard Model Physics (and more) at the LHC*, CERN 2000-04.

Amato, S. (1998), *LHCb Technical Proposal*, CERN/LHCC/98-4.

Antunes Nobrega, R. (2003), *LHCb Reoptimized Detector Design and Performance*, CERN/-LHCC/2003-030.

BaBar Collaboration [Aubert, B. *et al*] (1998), *The BaBar Physics book*, editors P.F.Harrison and H.R.Quinn, SLAC-R-504.

BaBar Collaboration [Aubert, B. *et al*] (2001), *Phys. Rev. Lett.* **87**, 091801.

BaBar Collaboration [Aubert, B. *et al*] (2002), *Phys. Rev. Lett.* **88**, 221803.

Belle Collaboration [Abe, K. *et al*] (2001), *Phys. Rev. Lett.* **87**, 091802.

Belle Collaboration [Abe, K. *et al*] (2003), *submitted to Phys. Rev. Lett.*, hep-ex/0308035.

Browder, T. (2003), *Proceedings of the XXI International Symposium on Lepton and Photon Interactions at High Energies*, Fermilab August 2003.

Cabibbo, N. (1963), *Phys. Rev. Lett.* **10**, 531.

Charles, J. *et al* [CKMFitter] (2003), http://ckmfitter.in2p3.fr/.

Christenson, J.H. *et al* (1964), *Phys. Rev. Lett.* **13**, 138.

Drobychev, G.Y. (2002), *Update to Proposal for an Experiment to Measure Mixing, CP Violation and Rare Decays in Charm and Beauty Particle Decays at the Fermilab Collider - BTeV*, BTeV-doc-316-v3.

Fleischer, R. (2002), *Physics Reports* **370**, 537.

Harr, R. (2003), *Proceedings of Beauty 2003, 9th International Conference on B Physics at Hadron Machines*, Pittsburgh October 2003.

HFAG [Heavy Flavour Averaging Group] (2003), http://qqq.slac.stanford.edu/xorg/hfag/.

Jarlskog, C. (1986), *Proceedings of the 1986 Recontre de Moriond on "Progress in Electroweak Interactions"* editor J.Tran Thanh Van 389.

Kobayashi, M. and Maskawa, K. (1972), *Prog. Theor. Phys.* **49**, 282.

Nakada, T. (2002), *Surveys in High Energy Physics* **17**, 3.

Nir, Y. (1999), *XXVII SLAC Summer Institute on Particle Physics*, hep-ph/9911321.

Nir, Y. (2001), *55th SUSSP on Heavy Flavour Physics*.

Ohlsson-Malek, F. (2003), *Proceedings of Flavour Physics and CP Violation Conference* Paris June 2003.

Particle Data Group [Hagiwara, K. *et al*] (2002), *Phys. Rev. D* **66**, 1.

Particle Data Group [Hagiwara, K. *et al*] (2003), http://pdg.lbl.gov/.

Paulini, M. (2003), *Proceedings of the Physics at the LHC Conference* Prague July 2003.

Sakharov, A.D. (1967), *JETP Lett.* **5**, 24.

Salam, G.P. (2002), *Proceedings of the X International Workshop on Deep Inelastic Scattering* Cracow Poland July 2002, hep-ph/0207147.

Schubert, K. (2003), *Proceedings of the XXI International Symposium on Lepton and Photon Interactions at High Energies*, Fermilab August 2003.

Shaposhnikov, M.E. (1986), *JETP Lett.* **44**, 364.

Stone, S. (2001), hep-ph/0111313 and references therein.

Virdee, T. (2003), *this volume*.

Wolfenstein, L. (1964), *Phys. Rev. Lett.* **13**, 562.

Wolfenstein, L. (1983), *Phys. Rev. Lett.* **51**, 1945.

Heavy ion physics at the LHC

Berndt Müller

Duke University, USA

1 Overview

These lectures[1] are meant to be an introduction into the physics opportunities for relativistic heavy ion collisions at the LHC. I will begin with a review of our present knowledge of the properties of hadronic matter at high temperature, where the gas of hadrons converts into a quark-gluon plasma characterized by the absence of quark confinement and spontaneous chiral symmetry breaking. In the next section I will discuss possible experimental probes of this new form of matter. Following a brief outline of our current ideas concerning the structure of nuclei at high energies, but small values of the momentum transfer, I will describe how hot, equilibrated matter is formed in nuclear collisions and how it disassembles into individual hadrons, which are observed in experiments (Harris and Müller 1996). I will then present an overview of the insights gathered in the first three years of the RHIC heavy ion programme. This will lead us, finally to the special opportunities for relativistic heavy ion physics at the LHC (Fries and Müller 2003)

2 QCD at high temperature: the quark-gluon plasma

There are two ways of producing matter under extreme conditions. By squeezing slowly, one can produce cold dense matter; by squeezing rapidly, one produces hot dense matter, where the term "dense" refers to the energy density of the matter created. The first approach is much more difficult to realize than the second one. The only way cold matter at densities higher than normal nuclear density can be produced in nature is in the collapse of a burnt-out star to a neutron star, where up to about ten times normal nuclear density ($\rho_0 \approx 0.15$ fm^{-1}) is reached in the inner core. We do not know how to create such matter in the laboratory, but if we could, it would be extremely interesting, because cold ultradense matter is predicted to have very unusual characteristics, such as being a transparent color superconductor (Alford 2001).

---

[1]The lecture slides can be downloaded from:
http://www.ippp.dur.ac.uk/sussp57/LectureNotes/Mueller.ppt (PowerPoint format) or
http://www.ippp.dur.ac.uk/sussp57/LectureNotes/Mueller.pdf (PDF format).

Hot matter of extremely high energy density ($\epsilon > M_N \rho_0$, where M_N is the nucleon mass) filled the entire universe at times less than 20 μs after the big bang. These are the conditions we expect to recreate in high energy collisions of heavy nuclei. This field of research has a history of almost 30 years, beginning at the Bevalac (Lawrence Berkeley National Laboratory, LBNL), continued at the SIS (GSI, Darmstadt), AGS (Brookhaven National Laboratory, BNL), SPS (CERN), and presently culminating at RHIC (also at BNL). The LHC will extend this research to a range of much higher energy densities.

The theory of ultradense matter is, in principle, quite simple, being described by the QCD Lagrangian (for introductions see: Shuryak 1980, Müller 1985):

$$L_{QCD} = -\frac{1}{4}\sum_a G^a_{\mu\nu}G^{a\mu\nu} + \sum_f \bar{\psi}_f\gamma^\mu(\partial_\mu + g\sum_a A^a_\mu t^a)\psi_f + \sum_f m_f\bar{\psi}_f\psi_f. \tag{1}$$

At equilibrium, characterized by nonzero temperature T and baryon chemical potential μ_b, the equation of state of this matter depends only on a small number of parameters: T, μ_b, g, m_f ($f = 1,\ldots,N_F$). In fact, the strong coupling constant g and the quark massses m_f are themselves functions of T and μ_b, as dictated by the renormalization group equations of QCD. At high temperature and/or high baryon potential these equations predict that g is small, permitting the use of perturbative techniques to calculate many of the properties of the dense matter. In the extreme high temperature limit (for $\mu_b = 0$) all interactions can be neglected compared with the kinetic energy of the particle excitations, leading to an ultrarelativistic equation of state $\epsilon \approx P/3$, with

$$\epsilon = \nu\int\frac{d^3p}{(2\pi)^3}E(e^{-E/T} \pm 1)^{-1} \rightarrow \nu\frac{\pi^2}{30}aT^4 \tag{2}$$

the energy density, $E = \sqrt{p^2 + m^2} \approx |p|$ and $a = 1$ for gluons, $a = 7/8$ for quarks. ν counts the number of elementary degrees of freedom: $\nu = 2(N_C^2 - 1) = 16$ for gluons and $\nu = 4N_C N_F = 12N_F = 36$ for three quark flavours.

At very high temperatures, the corrections to the energy density can be calculated perturbatively, but at practically accessible temperatures and in the interesting temperature range, where the matter undergoes a transition from hadron gas to a plasma of quarks and gluons, one needs more powerful calculational approaches. The present tool of choice are numerical simulations of QCD lattice gauge theory. The best existing calculations, including dynamical quarks and using improved lattice actions, show that the effective number of degrees of freedom $\nu_{eff} \sim 30\epsilon/\pi^2T^4$ exhibits a steep rise near $T \approx 160$ MeV from a value suggestive of a pion gas ($\nu \approx 3$) to a value slightly below a free gas of quarks and gluons ($\nu \approx 16 + 10.5N_F$). The temperature at which this transition occurs is usually called T_c (see Figure 1, reprinted from Karsch F. 2002; see also Karsch and Laermann 2003).

Whether the transition is smooth or discontinuous, corresponding to a phase transition, depends on the number of light quark flavours N_F and the numerical values of the quark masses m_f. For the physical values of the m_u, m_d, and especially m_s, the transition appears to be smooth but steep. The "critical" temperature T_c can be identified by the location of the peaks in various susceptibilities, such as the susceptibilities of the scalar quark density $\bar{\psi}\psi$, the Polyakov line, or the specific heat, which all yield the same value $T_c \approx 160$ MeV. The critical temperature depends on the baryon chemical potential, tracing out a line in the $T - \mu_b$ plane. As μ_b grows, the transition sharpens and eventually

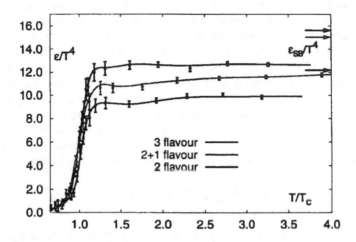

Figure 1. *Energy density in QCD with 2 and 3 quark flavors, showing a transition temperature T_c between a pion gas and a free gas of quarks and gluons. Reprinted figure with permission from (Karsch F. 2002). Copyright (2002) Elsevier B.V.*

becomes discontinuous at a critical point (T_c, μ_c). Very preliminary calculations on small lattices have located this point near $\mu_c \approx 700$ MeV, outside the range accessible in heavy ion collisions (see Figure 2, reprinted from Fodor and Katz 2002; see also de Forcrand and Philipsen 2002; Fodor, Katz and Szabo 2003). However, because the transition is so steep even at $\mu_b = 0$, the transition from a quark-gluon plasma to a hadron gas may be quite abrupt under conditions probed at heavy-ion colliders. As we will discuss later, the experimental evidence from RHIC strongly points to such a scenario.

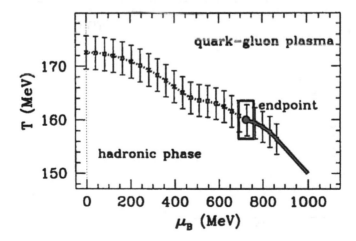

Figure 2. *The μ dependence of the transition temperature for (2+1) flavor QCD. Reprinted figure with permission from (Fodor and Katz 2002). Copyright (2002) Institute of Physics Publishing.*

Maybe the most characteristic properties of the high-temperature phase of QCD matter is that the color force is screened at short distances. This phenomenon can be derived within the framework of standard diagrammatic perturbation theory at finite temperature (Kapusta 1989), but it is illustrative to derive it in a more intuitive manner that is similar to the derivation of Debye screening in electrostatics. Consider a color capacitor charged up to some color potential ϕ^a. If the space between the capacitor plates is filled with a medium composed of mobile color charges with density ρ^a, the equation for the color potential between the plates, in lowest order of the strong coupling constant g, is given by $\nabla^2 \phi^a = -g\rho^a$. In the absence of an external color potential ϕ^a, the color charge density vanishes; in presence of a nonzero potential a nonvanishing color density is induced. In the linear response approximation, the induced color density is found to be

$$\rho^a(\phi) = \nu \mathrm{Tr}\left[t^a \frac{d^3p}{(2\pi)^3}(e^{-(E+g\phi^b t^b)/T} \pm 1)^{-1}\right] \approx -\mu^2 \phi^a, \qquad (3)$$

where μ is a screening mass, which is given by $\mu^2 = (gT)^2$ for gluons and $\mu^2 = (gT)^2/6$ for each flavor of quarks. As a result, a static pointlike color charge generates a screened potential

$$g\phi^a(r) = t^a \frac{g^2}{4\pi r}e^{-r\mu}. \qquad (4)$$

Numerical simulations show that the potential becomes rapidly screened as T exceeds T_c. The screening mass is well represented by its perturbative value given above for $T \geq 2T_c$, but rapidly disappears as $T \to T_c$, where the density of free color charges vanishes (see Figure 3, reprinted from Kaczmarek et al. 2000). This suggests that the critical temperature for color deconfinement should be related to the QCD scale parameter Λ by $g(T_c)T_c \sim \Lambda$, which is consistent with the numerical results and $g(T_c) \approx 2$.

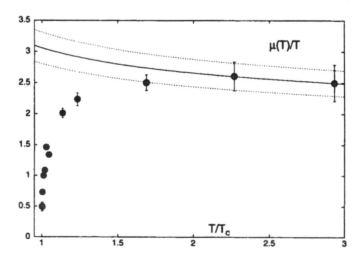

Figure 3. *Fit results for the screening mass $\mu(T)$ as a function of temperature. Reprinted figure with permission from (Kaczmarek et al. 2000). Copyright (2000) by the American Physical Society.*

The color screening by the plasma also affects the dispersion relation of the quasiparticle modes. At high temperature, where perturbation theory applies, the interactions of gluon or quark excitations with the thermal medium induce effective masses, or more correctly self energies, of order gT. These dynamically generated masses cure many of the infrared divergences, which otherwise would plague a plasma of color-charged, massless particles. They also reduce the energy density and pressure of the plasma at a given temperature and thus explain the result that the effective number of degrees of freedom in the quark-gluon plasma falls short of the value expected for free quarks and gluons.

Figure 4 (courtesy Peter Braun-Munzinger, GSI; see also Fries and Müller 2003) presents a comprehensive picture of our current state of knowledge about the QCD phase diagram and shows which region has been explored by means of heavy ion collisions. The data points are derived from chemical equilibrium fits to the hadron abundances observed in collisions of heavy nuclei at the SIS, AGS, SPS, and RHIC, ranging from about 1 GeV/nucleon on a fixed target (at SIS) to collisions at $\sqrt{s} = 200$ GeV per nucleon pair (at RHIC). The prediction from lattice QCD calculations for the phase boundary between hadron gas (HG) and quark-gluon plasma (QGP) is shown by the green line. The inital and final conditions reached at RHIC and expected for the LHC are also shown.

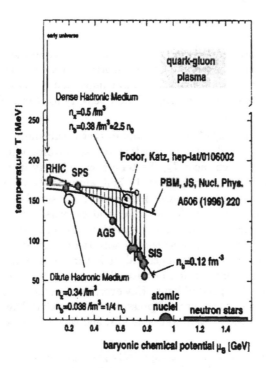

Figure 4. *Schematic QCD phase diagram in the T - μ_B plane (courtesy Peter Braun-Munzinger, GSI).*

3   Probes of ultradense matter

Many observables have been proposed as experimental probes for the quark-gluon plasma. An incomplete list certainly includes (Müller 1995):

- Effects of the "latent heat" in the (E, T) relation,

- Effects of large specific heat C_V on thermal fluctuations,

- Net charge and baryon number fluctuations,

- Enhancement of s-quark production,

- Thermal lepton pair $(\ell^+\ell^-)$ and photon (γ) radiation,

- Disappearance of light hadrons (ρ^0),

- Dissolution of heavy quark (Ψ and Υ) bound states,

- Large energy loss of fast partons (jet quenching),

- Bulk hadronization,

- Collective vacuum excitations (DCC).

In addition, there are several observables which are sensitive to the formation of an equilibrated system and thus can probe whether "matter" with well-defined thermodynamic properties is created in the collisions. The most important of these probes measure collective flow properties, such as the "elliptic" flow velocity v_2. In the following, we discuss some of the QGP probes listed above.

The steeply rising ratio ϵ/T^4 near T_c implies that a substantial amount of energy is required to "unthaw" new degrees of freedom (those corresponding to free color). As energy is added to the system near T_c, the temperature barely increases, until all of the additional degrees of freedom are liberated. Following van Hove (van Hove 1982), one would thus expect that the temperature of the created matter plateaus at a certain beam energy, remains roughly constant for a while, and eventually rises again as the beam energy is further increased. A convenient experimental measure of the temperature is the exponential slope of the transverse energy spectra of emitted hadrons. It is now well understood that this slope does not, in general, reflect the true temperature of the matter, but rather a blue-shifted temperature, because the matter flows collectively outward with a velocity close to half the speed of light. However, the argument still holds, because the flow is in itself a measure of the internal pressure of the matter. During the unthawing of new degrees of freedom the pressure remains low, and the flow velocity does not increase much.

It has been argued that kaons are especially good probes for this signature, because they are too heavy to be produced copiously at low temperatures and decouple rather early from the hadronic gas. The collision energy dependence of the spectral slopes for K^+ and K^- mesons, covering the full range accessible from the AGS to RHIC, reveals a plateau covering the whole SPS energy range (see Figure 5, reprinted from Gorenstein et al. 2003). This is a tantalizing result, which one would like to see confirmed by similar data using other particles, such as hyperons and ϕ-mesons, and by an analysis of the thermal and flow contributions to the measured slopes.

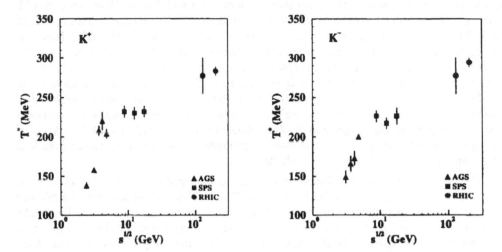

Figure 5. *The collision energy dependence of the inverse spectral slope parameter T^* for K^+ and K^- mesons, covering the full range accessible from the AGS to RHIC. Reprinted with permission from (Gorenstein et al. 2003). Copyright (2003) Elsevier B.V.*

The elliptic flow has recently emerged as a valuable tool allowing to probe the collective response of the created matter at early times (Ollitrault 1994). In the hydrodynamical model, the quadrupole component v_2 of the collective flow is generated by the anisotropic shape of the volume of hot matter created in noncentral collisions of two nuclei, resulting in a geometric anisotropy of the pressure gradient. Experimentally, v_2 is measured either as the quadrupole component of the azimuthal distribution of emmitted hadrons with respect to the reaction plane, or as the second Fourier coefficient of the azimuthal two-particle correlation function. Because of its dependence on the existence of a geometric anisotropy, which rapidly dilutes as the matter expands, the elliptic flow must be generated early, roughly within the first 3 fm/c of a collision (Kolb et al. 2001). v_2 is a function of the transverse momentum p_T, and a comparison between different hadrons can test whether the assumption of collectivity is really true.

The large difference between the constituent and current masses of u, d, and s quarks is due to the spontaneous breaking of chiral symmetry in the QCD vacuum. At temperatures above T_c this difference vanishes, because the quark condensate rapidly disappears above T_c. As a result, even the strange quark becomes "light" ($m_s < T$) in this domain, and s-quarks are abundantly created. Moreover, the liberation of gluons as dynamical degrees of freedom opens up an efficient mechanism for $s\bar{s}$ production, which is expected to lead to a rapid chemical equilibration of all light flavours (Rafelski and Müller 1982). This phenomenon is observed in Pb + Pb collisions at the SPS. Relative to hadrons composed solely of u and d valence quarks, baryons with valence s-quarks are enhanced in proportion to their net strangeness. The triply strange Ω^- hyperon is produced 15 times more copiously than expected from yields measured in $p + p$ collisions (Anderson et al. 1999).

The screening of the color force at high temperature results in the disappearance of all bound states of quarks. Because of their large size, bound states of light quarks disappear first, followed by $c\bar{c}$ bound states (charmonia) and, finally, $b\bar{b}$ bound states (Υ mesons). The dissolution of these states would thus signal the presence of color screening at sufficiently short distances (Matsui and Satz 1986, Karsch and Satz 1991). However, the most recent lattice QCD calculations of the spectral function in the $c\bar{c}$ channel suggest that the charmonium ground states J/Ψ and η_c survive as resonances up to about $1.5T_c$, which would delay the onset of this QGP signature (Datta et al. 2002, Asakawa et al. 2003). Furthermore, it has been pointed out in recent years that mesons composed of heavy quarks can be regenerated by coalescence of a $c\bar{c}$ pair when the quark-gluon plasma hadronizes (Braun-Munzinger and Stachel 2000, Thews and Rafelski 2002). At LHC energies, where heavy quarks are abundantly produced, this mechanism might even lead to an enhancement of J/Ψ and Υ production.

The NA50 experiment has observed a substantial suppression of J/Ψ and ψ' production in Pb + Pb collisions at the SPS (Abreu et al. 2001). This suppression can be explained by the inelastic interaction of produced $c\bar{c}$ pairs with the nucleons in the colliding nuclei and with produced light hadrons. Whether there is an unexplained additional suppression has been hotly debated without firm conclusion. The first, very preliminary data from Au + Au at RHIC do not indicate a much stronger suppression than that observed at the SPS.

A most interesting probe of matter containing a density of color nonsinglet excitations is the suppression of high-p_T hadrons. The mechanism responsible for this "jet quenching" is the energy loss of a hard parton as it propagates through the dense medium. The dominant energy loss process is gluon radiation after scattering in the medium, similar to the radiative energy loss of a light, energetic particle traversing ordinary matter (Gyulassy and Plümer 1990, Wang et al. 1995). The peculiarity of QCD is that the radiated gluon itself rescatters easily in the medium, leading to a L^2 dependence of the energy loss, where L is the thickness of the medium (Baier et al. 2000). In the spirit of perturbative QCD factorization, the energy loss can be expressed as a medium modification of the jet fragmentation function $D_{p \to h}(z, Q^2)$. As in QED, multiple scattering leads to a suppression in the radiation of soft gluons (the nonabelian Landau-Pomeranchuk-Midgal, or LPM effect) with frequency $\omega < \omega_c = \hat{q}L^2/2$, where

$$\hat{q} = \rho \int q^2 dq^2 \frac{d\sigma}{dq^2} \tag{5}$$

measures the scattering power of the medium. For a steeply falling parton spectrum the energy loss can either be described as a shift of the spectrum to lower momenta p_T or as a suppression of the spectrum by a quenching factor $Q(p_T) < 1$ (Baier et al. 2001). For a large nucleus and strong quenching one can show that $Q(p_T)$ is proportional to R^{-1}, where R is the nuclear radius (Müller 2003). In other words, the jet cross section grows like $A/R \sim R^2$, instead of like $A \sim R^3$, implying that jet production effectively becomes a surface effect. Jets originating in the core of the colliding nuclei lose so much energy that they are overshadowed by the more abundant jets of lower energy originating in the nuclear surface.

In practice, it is difficult to measure jets in the high multiplicity environment of heavy ion collisions. Instead, one usually just observes the leading hadron, which is proportionally suppressed in energy. The suppression factor of the hadron spectrum in a nuclear collision, compared with a $p+p$ collision, is called R_{AA}. Obviously, R_{AA} depends on the impact parameter; with $R \to 1$ in very peripheral collisions. Since the measured high-p_T hadrons originate mostly from the surface oriented toward the detector, the away-side jet has to traverse the nuclear volume to emerge on the opposite side. This leads to an additional suppression of high-p_T hadrons in the opposite azimuthal direction. Finally, jet quenching contributes to the elliptic flow, because the energy loss is a function of the azimuthal emission angle in noncentral collisions.

4 Structure of nuclei at small x

When probed at high energy with sufficiently fine resolution, an atomic nucleus looks like a collection of weakly interacting quarks and gluons, called partons. These parton distributions are not calculable, but they have been measured with great accuracy. They are functions of two kinematic variables: the longitudinal momentum fraction x, measured in units of the momentum of one of the nucleons constituting the nucleus, and the resolution, defined by the invariant de Broglie wavelength Q^2 at which the nucleus is probed. The variation of the parton distributions $f_i(x, Q^2)$ with Q^2 is determined by the renormalization group equations of QCD, called DGLAP equations in this context. The distribution of partons in transverse momentum is not well known, but many QCD processes require the assumption of an intrinsic k_T-distribution with a width of several hundred MeV/c up to 1 GeV/c. Geometrically, the parton distributions follow the density distribution of the nucleus, which is highly Lorentz contracted when the nucleus moves at high energy (by about a factor 2800 for a ^{208}Pb nucleus at the LHC).

The parton distributions grow rapidly for small values of x. This is a general property of an interacting quantum field theory; small-x gluons represent soft components of the color field, and these are very abundant. Experiments have shown that the parton distributions grow like $xf(x) \sim x^{-\lambda}$ where $\lambda \approx 0.5$; the precise numbers depend on the resolution scale. The larger Q^2, the more rapid is the growth of the parton distributions at small x. In perturbation theory this is understood as *parton splitting*, described quantitatively by the DGLAP equations. The probability for the emission of a gluon with momentum fraction $x' = xz$ ($z < 1$) by a quark or gluon with momentum fraction x grows like z^{-1} as $z \to 0$, leading to a proliferation of gluons at small x.

Obviously, this growth cannot continue without bound. At some point, the density of gluons becomes so high that the probability for the reverse process, *parton fusion*, becomes significant and needs to be taken into account in the evolution equations for the parton distributions. Eventually, the gluon density must saturate. The saturation condition can be written in the form (Gribov et al. 1983)

$$\frac{\alpha_s x G_A(x, Q^2)}{\pi R_A^2 Q^2} \sim 1, \tag{6}$$

where $G_A(x, Q^2)$ is the gluon distribution in a nucleus with mass A and radius R_A. The momentum scale Q^2, at which this condition is satisfied for a given x, is called the *saturation* scale Q_{sat}^2. Extensive measurements of the parton distributions in the proton at HERA have shown that the gluon distribution grows as a power at small x, before saturation effects set in, yielding the following scaling law for the saturation momentum:

$$Q_{sat}^2 \sim A^{1/3} x^{-\lambda}. \tag{7}$$

This relation suggests that saturation effects can be probed in heavy nuclei at larger values of x (by about a factor 30) than in isolated nucleons (Blaizot and Mueller 1987). Or put the other way around, at a given value of x, the saturation momentum for a heavy nucleus is about 2.5 times as large as for a proton. Since the concept of weakly interacting partons only applies to the perturbative domain of QCD, i.e. to sufficiently large values of the momentum scale Q, this result implies that parton saturation effects can be studied in nuclei at much larger values of x. This provides one of the motivations for the construction of an *electron-ion collider* (EIC, eRHIC).[2]

Parton saturation suggests a picture of quasiclassical glue fields at small x, where $gA \sim k$, and the perturbation expansion is in powers of α_s, but not in powers of gA (McLerran and Venugopalan 1994). In the extreme limit ($x \to 0$) the probability distribution of classical color fields is thought to be random and completely determined by the saturation scale:

$$P[A] \sim \exp\left(-\int d^2 x_\perp g^2 A^2(x_\perp)/Q_{sat}^2\right). \tag{8}$$

This ensemble of glue fields is called the *color glass condensate*. The term "condensate" is to be understood loosely. Strictly speaking, the color field strength itself has a vanishing expectation value, but the average is to be performed at the level of probabilities, not amplitudes. The saturation scale evolves with x and satisfies a nonlinear renormalization group equation, a generalization of the BFKL equation, which is universal, i.e. independent of the hadron species (meson, baryon, or nucleus), in the limit $x \to 0$ (Iancu et al. 2001, Weigert 2002).

One important question is, how many partons are "liberated" from the coherent parton cloud when two nuclei collide. Theoretical arguments show that in the classical limit ($s \to \infty$ and $x \to 0$) virtually *all* partons get liberated by interactions (Kovchegov 2001), but microscopic transport calculations suggest that the liberation probability drops significantly at small x for a fixed collision energy s. Extremely simplified model calculations of the particle multiplicity, assuming a one-to-one correspondence between liberated partons and emitted hadrons and a simple scaling law for the saturation scale with x, are in good agreement with the particle rapidity distributions measured in Au + Au collisions at RHIC (see Figure 6, reprinted from Kharzeev and Levin 2001).

An interesting aspect of the parton liberation concept is the rapidity distribution of net baryons (baryons minus antibaryons). If hadron rapidity distributions are determined to a significant degree by the initial state parton distributions, then the net baryon number at rapidity y should reflect the net baryon number distribution in the nucleons of a Au nucleus at the momentum fraction

$$x(y) = \frac{\langle k_T \rangle}{M_N} \exp(y - y_b) \tag{9}$$

where y_b denotes the beam rapidity, M_N the nucleon mass, and $\langle k_T \rangle \approx Q_{sat}$ is the average momentum transfer in the interactions that liberate the partons. If all partons were

---

[2]See http://www.phenix.bnl.gov/WWW/publish/abhay/Home_of_EIC for details of the physics opportunities and technical challenges of an electron-ion collider.

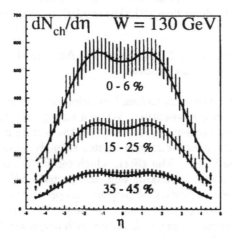

Figure 6. *Pseudo-rapidity dependence of charged hadron production as a function of centrality, measured in Au + Au collisions in the PHOBOS experiment at RHIC. Reprinted figure with permission from (Kharzeev and Levin 2001). Copyright (2001) Elsevier B.V.*

liberated, the net baryon number distribution originating from one nucleus would be given by

$$dB/dy = \frac{1}{3}Ax(y)\sum_q (f_q(x) - f_{\bar{q}}(x)).$$ (10)

For Au + Au at the maximal RHIC energy, this corresponds to a prediction $(dB/dy)_{y=0} \approx 25$, slightly below the measure value of about 30. Dynamical transport calculations in the parton cascade model (Bass et al. 2003) indicate that the liberation factor is less than unity, but they also predict a considerable amount of baryon number transport toward midrapidity by multiple parton-parton collisions, filling up the gap to the measured value. These results suggest that no novel mechanism of baryon transport is needed to explain the surprisingly large baryon asymmetry observed at RHIC. The analogous prediction for the baryon asymmetry at the maximal LHC energy is $(dB/dy)_{y=0} \approx 5 - 10$.

5   Matter formation and breakup

The canonical picture of the stages of formation, evolution, and decay of hot, dense matter in relativistic heavy ion collisions is as follows:

1. *Hard scattering:* Occurs on a very short time scale $\Delta t \sim 1/Q < 1/Q_{sat}$ and is described by pQCD.

2. *Equilibration:* Can be described by multiple parton scattering in the parton cascade model or by chaotic dynamics of classical color fields. The characteristic time scale is $\Delta t \sim 1/g^2 T$.

3. *Expansion:* Is described by relativistic fluid dynamics. The simplest scenario is boost invariant longitudinal flow (Bjorken model), which predicts the cooling law $T(\tau) \sim \tau^{-1/3}$.

4. *Hadronization:* Occurs when T falls below T_c. An unresolved question is whether hadronization is a slow or fast process.

For various reasons the interest on hadron formation has focused at RHIC on hadrons with several GeV of transverse momentum. While a strong universal suppression of the hadron yield is observed at momenta $p_T > 5$ GeV/c, the picture is more complex in the $2-5$ GeV/c range. While meson emission is also strongly suppressed at these lower momenta, baryon emission is not. The effect, which results in anomalously large p/π^+ and $\bar{p}/\pi^-$ ratios of order unity, is striking and unexpected (Adler et al. 2003a). The explanation may be that recombination of partons, as opposed to parton fragmentation, contributes to hadron formation in this momentum range. Because baryons contain three valence partons, recombination competes effectively with fragmentation up to larger values of p_T, leading to an enhanced baryon-to-meson ratio in an intermediate momentum range.

Relatively model independent calculations are possible at hadron momenta that are sufficiently high, so that details of the internal parton wave functions of the hadrons are not important. Transport calculations tell us that the single parton spectra before hadronization are nearly exponential at low momenta (up to about $1.5-2$ GeV/c), transitioning to a power-law shape at higher momenta.

Denoting the quark phase-space distribution by $w_\alpha(p)$ (α denotes color, flavor, and spin degrees of freedom) and employing light-cone coordinates along the direction of the emitted hadron, recombination predicts the following equations for meson and baryon spectra, respectively:

$$E\frac{dN_M}{d^3P} = \int d\Sigma \frac{P \cdot u}{(2\pi)^3} \sum_{\alpha\beta} \int dx\, w_\alpha(xP^+)\bar{w}_\beta((1-x)P^+)|\phi_{\alpha\beta}^{(M)}(x)|^2, \tag{11}$$

$$E\frac{dN_B}{d^3P} = \int d\Sigma \frac{P \cdot u}{(2\pi)^3} \sum_{\alpha\beta\gamma} \int dx\, dx'\, w_\alpha(xP^+)w_\beta(x'P^+)w_\gamma((1-x-x')P^+)|\phi_{\alpha\beta\gamma}^{(B)}(x,x')|^2. \tag{12}$$

Here Σ stands for the freeze-out hypersurface, u^μ for the collective flow velocity of the partonic matter, and capital letters P denote hadron coordinates. For a thermal parton distribution, $w(p) = \exp(p \cdot u/T)$, the partonic factors in both equations combine to a single thermal distribution for the hadron, $w(P^+)$, yielding trivial integrations over the momentum fractions x, x'. The consequence is that hadrons are emitted in thermal abundances. This result naturally explains the observed large baryon-to-meson ratios. On the other hand, parton fragmentation predicts a hadron spectrum of the form

$$E\frac{dN_h}{d^3P} = \int d\Sigma \frac{P \cdot u}{(2\pi)^3} \sum_\alpha \int dz\, z^{-3}\, w_\alpha(P^+/z)D_{\alpha \to h}(z), \tag{13}$$

where $z < 1$ denotes the momentum fraction carried by the hadron. It is easy to see that, for large values of the momentum P^+, recombination always wins over fragmentation, because $w(P^+/z)$ falls off faster than $w(P^+)$. On the other hand, fragmentation dominates over recombination at large P^+, when the parton distribution follows a power law. An excellent fit to all hadron spectra measured in Au + Au collisions at RHIC is obtained with a parton spectrum, which is a superposition of a thermal spectrum with a blue-shifted temperature $T_{\text{eff}} = 350$ MeV and a LO pQCD spectrum down-shifted by 2.2 GeV, to account for the energy loss of the partons in the dense medium (see Figure 7 reprinted from Fries et al. 2003).

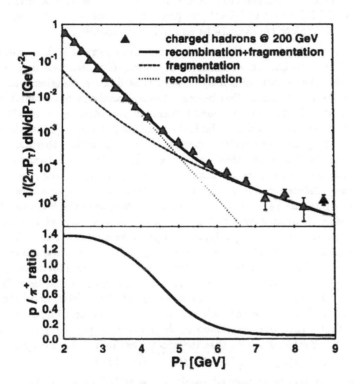

Figure 7. *Top: Inclusive p_T spectrum of charged hadrons measured in central Au+Au collisions at $\sqrt{s_{NN}} = 200$ GeV in the PHENIX detector at RHIC. Bottom: Ratio of protons to π^+ as a function of P_T. The region below 4 GeV/c is dominated by recombination, the region above 6 GeV/c by parton fragmentation. Reprinted with permission from (Fries et al. 2003). Copyright (2003) American Physical Society.*

6 The first three years at RHIC

The Relativistic Heavy Ion Collider (RHIC) can accelerate heavy nuclei as well as protons. So far, RHIC has produced collisions of Au + Au at c.m. energies of 130 GeV and 200 GeV per nucleon pair, $p + p$ collisions at 200 GeV c.m. energy, and $d + Au$ collisions also at 200 GeV in the nucleon-nucleon c.m. frame. The design luminosity is 2×10^{26} cm^{-2}s^{-1} for Au + Au and 2×10^{32} cm^{-2}s^{-1} for $p + p$. So far, about half of these values have been reached. Physics measurements at RHIC are made by four major detectors, called BRAHMS, PHENIX, PHOBOS, and STAR.[3] BRAHMS is a conventional two-arm spectrometer, PHENIX a complex system with two central detector arms and a pair of muon spectrometers in the beam direction, PHOBOS is a "table-top" all-silicon detector with a full-coverage multiplicity array and two small-acceptance spectrometer arms, and STAR consists of a solenoidal TPC covering the central two units of rapidity and several other components, such as a vertex detector and electromagnetic calorimeter. The completion of PHENIX and STAR has proceeded in steps; in the early runs data were taken with incomplete configurations and reduced capability.

One important result of the early runs was that the observed hadron spectra at low momenta, as well as the hadron abundances at midrapidity, are well described by the thermal model (for Au + Au collisions). There is still some disagreement about the best parameters for the thermal distribution. Several analyses have found a *chemical* temperature $T_{ch} \approx 175$ MeV and a baryon chemical potential $\mu_b \approx 40$ (30) MeV for 130 (200) GeV c.m. energy, describing the hadron abundance ratios (Braun-Munzinger et al. 2001, Cleymans 2002). The hadron spectra are best fit with a thermal distribution of temperature $T_{th} \approx 100 - 120$ MeV and an average collective outward flow velocity $\langle \beta_T \rangle \approx 0.5$, corresponding to an expansion with $\beta_T \approx 0.8$ at the surface (Burward-Hoy et al. 2003). The large difference between T_{ch} and T_{th} is consistent with a picture of a strongly interacting, expanding and rapidly cooling hadron gas following the hadronization of the quark-gluon plasma. Multiply strange baryons (Ξ, Ω) do not quite fit into this picture; their spectra are better described by a higher temperature of about 145 MeV and a lower flow velocity (Adams et al. 2003b). This is usually "explained" by the small cross sections of these baryons in hadronic matter, which lead to their early thermal decoupling. Other analysis groups insist that the abundance ratios and the spectra can be well described by a single temperature parameter in the range of 150 MeV (Rafelski and Letessier 2003, Broniowski et al. 2003). Since the different results are due to differences in analysis methodology, rather than large error bars in the data, it is presently not clear how to resolve this dispute. Everyone agrees that the spectra are well described by a blue-shifted temperature parameter $T_{slope} = T\sqrt{(1 + \beta_T)/(1 - \beta_T)} \approx 300$ MeV.

Perhaps the most highly publicized result of the RHIC experiments to date is the observed suppression of hadron production at high transverse momentum ($5 < p_T < 10$ GeV/c). "Suppression" here means that the yield falls below the expectation of a scaling of the yield measured in $p + p$ collisions by the number of independent nucleon-nucleon collisions in Au + Au at a given impact parameter b. The ratio

$$R_{AA}(b) = \frac{(dN/dp_T)_{\text{AuAu}}}{N_{\text{coll}}(b)\,(dN/dp_T)_{\text{pp}}} \tag{14}$$

is identified as the suppression factor and usually given as a function of *centrality* rather than impact parameter. The data clearly show that hadron emission is unsuppressed in the most peripheral Au + Au collisions and increasingly suppressed for more central collisions. The suppression factor reaches a value of 4−5 for the most central collisions (see Figure 8, Adcox et al. 2003, Adams et al. 2003a). The data also show correlations among high-p_T hadrons emitted in the same direction, characteristic of jet-like phenomena, but demonstrate the disappearance of such jet-like correlations for back-to-back emission in collisions of increasing centrality. In effect, high-p_T hadron emission becomes an outward

---

[3]See http://www.bnl.gov/RHIC/ for details on the RHIC detectors.

directed surface phenomenon, due to the combined action of the depth dependent energy loss and the steeply falling primary spectrum.

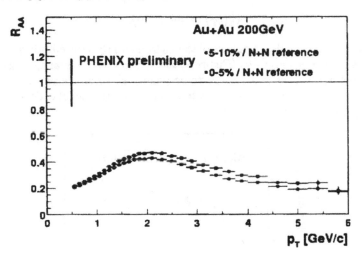

Figure 8. *Nuclear suppresion factor R_{AA} for Au-Au interactions at PHENIX in RHIC as a function of centrality (0-5% and 5-10% centrality selections). See (Adcox et al. 2003).*

As already discussed, the suppression of baryon emission sets in at significantly higher transverse momenta. This is seen for hyperons as well as protons, and naturally explained by the contribution of parton recombination up to momenta around 5 GeV/c. This effect results in a strong p_T-dependence of R_{AA} for all charged hadrons between 2 and 5 GeV/c, which is not seen for identified neutral pions. The conclusion that the suppression is caused by final state interactions, rather than an initial state effect due to nuclear modifications of the parton distributions, has been confirmed by the recent data from $d + Au$ collisions, which do not show such a suppression at midrapidity (Adams et al. 2003c, Adler et al. 2003b).

Another important result of the RHIC experiments is the strong azimuthal anisotropy of the collective flow with respect to the nuclear collision plane. This effect is measured in terms of the second Fourier component of the angular distribution of emitted particles, the so-called *elliptic flow* velocity

$$v_2 = \langle \cos 2\phi \rangle, \tag{15}$$

where ϕ is the azimuthal emission angle with respect to the pre-determined collision plane. v_2 is a function of impact parameter (the elliptic flow vanishes at $b = 0$) and of transverse momentum. The remarkable finding is that v_2 saturates the value expected from a hydrodynamic expansion, assuming very rapid build-up of transverse pressure (Ackermann et al. 2001). This result provides strong evidence for a rapid thermalization of the medium created in the collisions.

The elliptic flow velocity initially grows with p_T until leveling off above 1 GeV/c. An interesting aspect of the data is that v_2 saturates at a higher value for baryons than for mesons, and the saturation occurs later. The recombination mechanism provides a natural explanation for this phenomenon, because baryons contain more valence quarks than mesons. The model suggests a simple scaling law (Molnar and Voloshin 2003):

$$v_2^h(p_T) \approx n v_2^q(p_T/n), \tag{16}$$

where n is the number of valence quarks of the hadron and v_2^q is the elliptic flow of partons. Indeed, a plot of v_2/n versus p_T/n (Figure 9) nicely confirms the scaling, suggesting that the elliptic flow is a property of the partonic phase. v_2^q saturates at the value 0.08, in good agreement with the elliptic flow pattern caused by the angular dependence of the parton energy loss in peripheral collisions

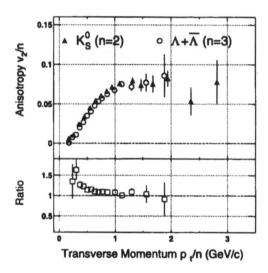

Figure 9. *Scaling of v_2/n versus p_T/n, suggesting an elliptic flow of partons (courtesy of Paul Sorensen, STAR at LBNL).*

7 Heavy ions at the LHC

What are the differences of the physics we can expect to explore with heavy ion collisions in the LHC? A fundamental advantage of the much higher energies available at the LHC is that the experimental conditions have a much higher "dynamic range":

- Matter will be produced at a much higher energy density ϵ_0 at an earlier time τ_0.

- Jet physics can be probed to momenta $p_T > 100$ GeV/c.

- b and c quarks become abundant probes of the quark-gluon plasma.

- The increased lifetime of QGP phase ($10 - 15$ fm/c) makes initial state effects less important.

- The QGP is even more dominant over final-state hadron interactions than at RHIC.

By employing the parton saturation concept, it is possible to use NLO pQCD calculations of gluon production to estimate the multiplicity and transverse energy that will be produced in Pb + Pb collisions at the full LHC energy (5.5 TeV in the NN center of mass). The saturation scale here serves as a dynamic infrared cutoff for the divergent differential cross section. The charged multiplicity dN_{ch}/dy is expected to grow from about 650 at RHIC to slightly above 2000 at the LHC. Because of the concomitant increase in average transverse momentum, this may correspond to an increase in the energy density of the created matter by a factor of 4 − 5 at the same proper time τ of a comoving observer.

The increased gluon density ρ_g leads to predictions of a significantly larger energy loss of hard partons at the LHC. Three effects contribute to this increase: The density of scatterers increases, the average momentum transferred in a collision with a medium particle increases (because the color force is screened at shorter distances), but the scattering cross section on a medium particle decreases (for the same reason). The increased energy loss has a somewhat less dramatic effect on the suppression factor R_{AA}, because the parton spectrum falls less steeply at the higher LHC energy. Theoretical estimates suggest that R_{AA} could be in the range 0.05 − 0.1 for 10 GeV/c hadrons, rising to about 0.5 at 100 GeV/c. The pQCD theory of parton energy loss predicts a strong energy dependence of the suppression factor, which can be explored over a range of a factor 10 in p_T, allowing for a test of the underlying theory that is impossible at RHIC.

Making some reasonable assumptions about the collective transverse flow and the parton energy loss, it is possible to extrapolate the calculation of hadron spectra, including both parton fragmentation and recombination, to the top LHC energy (see Figure 10, reprinted from Fries and Müller 2003). Depending on the amount of collective flow, the transition between the regions of dominance of the two hadron production mechanisms shifts to higher p_T, maybe up to 7 GeV/c; the region of anomalously large baryon-to-meson ratios is predicted to extend accordingly.

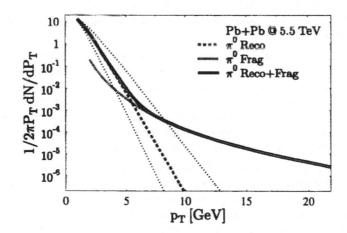

Figure 10. *Transverse momentum of π^0 for central Pb+Pb collisions at 5.5 TeV. Reprinted figure with permission from (Fries and Müller 2003). Copyright (2003) Springer-Verlag.*

Because of the feasibility to measure jets with transverse energies of 100 GeV or more, it will be practical at the LHC to identify jets over the multiparticle background even in Pb + Pb collisions. This makes it possible to not only measure the integrated parton energy loss, but to study the medium modification of the fragmentaton function through measurements of the jet profile, allowing for an additional test of the theoretical basis of the jet quenching mechanism. Another experimental approach to the same problem is to search for jets opposite to a hard photon, which is produced by the QCD Compton process $(q + g \rightarrow q + \gamma)$. Since the photon escapes without energy loss, its energy can be used as a calibration of the total transverse energy of the opposite-side jet.

Photons in the intermediate p_T range around 10 GeV/c can be copiously produced by secondary interactions of hard scattered partons with thermal medium quanta, such as quark-gluon Compton scattering or (anti-)quark annihilation. The pQCD cross sections for these reactions are peaked at backward angles, corresponding to the "conversion" of a fast quark into a photon, which inherits almost the full energy of the incident quark. This effect will allow for an independent determination of the density of scatterers in the dense medium.

Three detectors will study heavy ion collisions at the LHC. One of these, the ALICE detector,[4] is a dedicated heavy ion experiment; the other two, ATLAS and CMS,[5] are primarily designed for the $p+p$ program but also have the capability to measure interesting observables in Pb + Pb collisions. The design luminosity of 10^{27} cm^{-2}s^{-1} corresponds to an event rate of 8000 minimum bias collisions per second. The current beam use plan calls for one month of Pb + Pb collisions each year.

The ALICE detector (Figure 11) combines the concepts of the STAR and PHENIX detectors. Its central rapidity design is built around a large TPC combined with a Si pixel vertex detector. The TPC is surrounded by time-of-flight, Cherenkov, and photon detectors for particle identification as well as an electromagnetic calorimeter. ALICE also has a muon detector in the forward direction to measure decays of heavy quark bound states. ALICE will be able to carry out virtually the same physics program as STAR and PHENIX combined, but with much higher event rates for high-p_T phenomena and heavy quarkonia, especially the Υ states.

The CMS and ATLAS collaborations have recently performed extensive studies of the capabilities of their detectors in the high multiplicity environment of Pb + Pb collisions. Both detectors have excellent resolution for phenomena associated with jets allowing superior measurements of parton energy loss. CMS will be able to measure a loss of 5 GeV or more on a photon-tagged quark jet above 120 GeV transverse energy. CMS also has excellent resolution for the detection of heavy quarkonium states and will be able to collect more than 30k decays of Υ states during a one-month run.

In summary, ALICE, CMS, and ATLAS are nicely complementary in their physics capabilities aimed at studying hot, ultradense matter created in nuclear collisions at the LHC. Due to the higher c.m. energy hard probes (jets, heavy quarkonia) extend the range of matter probes into regimes allowing for more rigorous and reliable calculations. Hadrons containing b- and c-quarks will become abundant components in the final state of a heavy ion collision at the LHC. Soft probes are governed by significantly different

---

[4]See http://alice.web.cern.ch/Alice/AliceNew/ for a detailed description of the ALICE detector.
[5]See T. Virdee's lectures at this school.

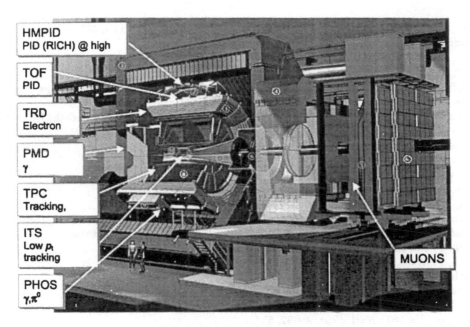

Figure 11. *ALICE detector.*

values of medium parameters, so that the theoretical conclusions drawn from the RHIC data can be put to a serious test at the LHC.

References

Abreu at al., NA50 collab. 2001 *Phys. Lett. B* **521**, 195.

Ackermann. K. H. et al., STAR collab. 2001 *Phys. Rev. Lett.* **86**, 402.

Adams, J. et al., STAR collab. 2003a *Phys. Rev. Lett.* **91**, 172302.

Adams, J. et al., STAR collab. 2003b preprint arxiv.org:nucl-ex/0307024.

Adams, J. et al., STAR collab. 2003c *Phys. Rev. Lett.* **91**, 072304.

Adler, S. S. et al., PHENIX collab. 2003a *Phys. Rev. Lett.* **91**, 172301.

Adler, S. S. et al., PHENIX collab. 2003b *Phys. Rev. Lett.* **91**, 072303.

Adcox, K. et al., PHENIX collab. 2003 *Phys. Lett. B* **561**, 82.

Alford, M. G. 2001 *Ann. Rev. Nucl. Part. Sci.* **51**, 131.

Anderson, E. et al., WA97 collab. 1999 *Phys. Lett. B* **449**, 401.

Asakawa, M. and Hatsuda, T. 2003 preprint arxiv.org:hep-lat/0308034.

Baier, R., Schiff, D. and Zakharov, B. G. 2000 *Ann. Rev. Nucl. Part. Sci.* **50**, 37.

Baier, R., Dokshitzer, Yu. L., Mueller, A. H. and Schiff, D. 2001 *JHEP* **0109**, 033.

Bass, S. A., Müller, B. and Srivastava, D. K. 2003 *Phys. Rev. Lett.* **91**, 052302.

Blaizot, J. P. and Mueller, A. H. 1987 *Nucl. Phys. B* **289**, 847.

Braun-Munzinger, P. and Stachel, J. 2000 *Phys. Lett. B* **490**, 196.

Braun-Munzinger, P., Magestro, D., Redlich, K. and Stachel, J. 2001 *Phys. Lett. B* **518**, 41.

Broniowski, W., Florkowski, W. and Hiller, B. 2003 *Phys. Rev. C* **68**, 034911.

Burward-Hoy, J. M. et al., PHENIX collab. 2003 *Nucl. Phys. A* **715**, 498.

Cleymans, J. 2002 *J. Phys. G* **28**, 1575.

Datta, S., Karsch, F. and Petreczky, P. 2002 *Nucl. Phys. Proc. Suppl.* **119**, 487.

de Forcrand, P. and Philipsen, O. 2002 *Nucl. Phys. B* **642**, 290.

Fodor, Z. and Katz, S. D. 2002 *JHEP* **0203** 014.

Fodor, Z., Katz, S. D. and Szabo, K. K. 2003 *Phys. Lett. B* **568**, 73.

Fries, R. J., Müller, B., Nonaka, C. and Bass, S. A. 2003 *Phys. Rev. Lett.* **90**, 202303; *Phys. Rev. C* **68**, 044902.

Fries, R. J. and Müller, B. 2003 *Heavy ions at LHC: Theoretical issues*, preprint nucl-th/0307043.

Gorenstein, M. I., Gazdzicki, M. and Bugaev, K. A. 2003 *Phys. Lett. B* **567**, 175.

Grandchamp, L. and Rapp, R. 2002 *Nucl. Phys. A* **709**, 415.

Gribov, L. V., Levin, E. M. and Ryskin, M. G. 1983 *Phys. Rept.* **100**, 1.

Gyulassy, M. and Plümer, M. 1990 *Phys. Lett. B* **243**, 432.

Harris, J. W. and Müller, B. 1996 *Ann. Rev. Nucl. Part. Sci.* **46**, 71.

Iancu, E., Leonidov, A. and McLerran, L. 2001 *Nucl. Phys. A* **692**, 583.

Kaczmarek, O., Karsch, F., Laermann, E. and Lütgemeier, M. 2000 *Phys. Rev. D* **62**, 034021.

Kapusta, J. I. 1989 *Finite-Temperature Field Theory* Cambridge University Press.

Karsch, F. and Satz, H. 1991 *Z. Phys. C* **51**, 209.

Karsch, F., *Nucl. Phys. A* **698**, 199 (2002).

Karsch, F. and Laermann, E. 2003 preprint arxiv.org:hep-lat/0305025.

Kharzeev, D. and Levin, G. 2001 *Phys. Lett. B* **523**, 79.

Kolb, P., Huovinen, P. and Heinz, U. 2001 *Phys. Lett. B* **500**, 232.

Kovchegov, Yu. 2001 *Nucl. Phys. A* **692**, 557.

Matsui, T. and Satz, H. 1986 *Phys. Lett. B* **178**, 416.

McLerran, L. and Venugopalan, R. 1994 *Phys. Rev. D* **49**, 2233, 3352.

Molnar, D. and Voloshin, S. A. 2003 *Phys. Rev. Lett.* **91**, 092301.

Müller, B. 1985 *Lecture Notes in Physics* **225**.

Müller, B. 1995 *Rept. Prog. Phys.* **58**, 611.

Müller, B. 2003 *Phys. Rev. C* **67**, 061901.

Ollitrault, J.-Y. 1992 *Phys. Rev. D* **46**, 229.

Rafelski, J. and Müller, B. 1982 *Phys. Rev. Lett.* **48**, 1066.

Rafelski, J. and Letessier, J. 2003 *Nucl. Phys. A* **715**, 98.

Shuryak, E. V. 1980 *Phys. Rept.* **61**, 71.

Thews, R. L. and Rafelski, J. 2002 *Nucl. Phys. A* **698**, 575.

van Hove, L. 1982 *Phys. Lett. B* **118**, 138.

Wang, X.-N., Gyulassy, M. and Plümer, M. 1995 *Phys. Rev. D* **51** 3436.

Weigert, H. 2002 *Nucl. Phys. A* **703**, 823.

Forward physics at the LHC

Albert De Roeck

CERN, Switzerland

1 Introduction

The Large Hadron Collider (LHC) experiments cover well the detection of particles produced in the central region of pp collisions, but the forward region, i.e. the region of large pseudo-rapidity, is generally less well covered. In these two lectures we will discuss the physics opportunities at the LHC using information from the forward region. One dedicated experiment, TOTEM, plans to measure the total and diffractive cross sections. Both general purpose detectors CMS and ATLAS are presently studying detector upgrades to increase the coverage in the forward region. Some of these measurements are intimately related to the luminosity determination at the LHC.

The LHC will collide protons with a total centre-of-mass (cm) system energy of 14 TeV, and will open up a new high energy frontier. First collisions are expected in 2007. The high luminosity of the LHC will be a challenge for both machine and experiments: namely $10^{33} \text{cm}^{-2}\text{s}^{-1}$ at start-up, and $10^{34} \text{cm}^{-2}\text{s}^{-1}$ for the high luminosity mode. This will lead to event samples of order 10-100 fb^{-1}/year, but also to a considerable amount of overlay events in a single bunch crossing.

The high energy and luminosity of the collider allow for new physics opportunities in the field of diffraction and low-x QCD, two typical examples of topics associated with forward physics. The parton distributions in the proton can in principle be explored for scaled parton momentum values x down to 10^{-7}. For diffractive events in which the proton beam loses 10% of its momentum or less, the resulting 'pomeron beams' have an energy of up to almost 1 TeV. Such collisions allow one to probe the structure of the pomeron down to scaled momentum values of the partons, β, of less than 10^{-3}.

Five experiments are being installed at the LHC. Two of these, CMS in IP5 (interaction point 5) and ATLAS in IP1, are general purpose experiments with an acceptance in pseudorapidity η, where $\eta = -\ln\tan\theta/2$ with θ the polar angle of the particle, of roughly $|\eta| < 3$ for tracking information and $|\eta| < 5$ for calorimeter information. Fig. 1 shows the pseudorapidity distribution of the charged particles and of the energy flow at the LHC, demonstrating that with an acceptance limited to $|\eta| < 5$ most of the energy in the collision will not be detected.

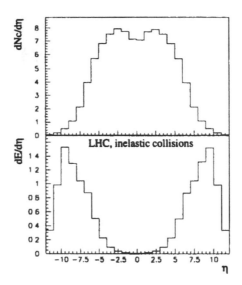

Figure 1. *Pseudorapidity distribution of the charged particles and of the energy flow at the LHC.*

The TOTEM experiment will also make use of IP5. The proposal of the TOTEM experiment can be found at http://totem.web.cern.ch/Totem/. Its main scientific program is to measure the total and elastic pp cross section at the highest energies. The experiment will use roman pot detectors to measure the low-t [1] scattered protons, and detectors for tagging inelastic events in the regions $3 < |\eta| < 5$ (T1) and $5 < |\eta| < 7$ (T2) to measure the total cross section. Both the presence of these roman pots and the inelastic event tagging detectors around CMS offer an excellent opportunity for a 'combined experiment' which will have excellent coverage of both the central and forward region. Such an experiment will allow for unique measurements as will be discussed in these lectures. In 2002 a new group was formed and approved by both collaborations which will study the physics case of combined data taking of CMS and TOTEM, and the optimization of the detectors in the forward region. More information can be found at http://agenda.cern.ch/FullAgenda.php?ida=a02824.

In these lectures we discuss opportunities for forward physics at the LHC. We start with the classical forward physics topics, such as total and elastic cross sections and general diffraction. Then we discuss the so called 'new forward physics' topics. The forward physics program at the LHC presently contains the following topics:

- Soft and hard diffraction:

    - Total cross section and elastic scattering
    - Gap survival dynamics, multi-gap events, soft diffraction, proton light cone studies (e.g. pp $\rightarrow$ $3jets + p$)

---

[1] t is the momentum transfer squared between the incoming and scattered proton: $t = (p - p')^2$.

- Hard diffraction: production of jets, $W, J/\psi, b, t$ hard photons, structure of diffractive exchange
- Double pomeron exchange events as a gluon factory
- Diffractive Higgs production (diffractive radion production)
- Supersymmetry and other (low mass) exotics; exclusive processes

- Low-x dynamics:

  - Parton saturation, BFKL/CCFM dynamics, proton structure, multi-parton scattering

- New forward physics phenomena:

  - New phenomena such as disoriented chiral condensates, incoherent pion emission, centauro events

- Strong interest from the cosmic ray community:

  - Forward energy and particle flows/minimum bias event structure

- Two-photon interactions and peripheral collisions

- Forward physics in pA and AA collisions

- QED processes to determine the luminosity to $\mathcal{O}(1\%)$ e.g. $pp \rightarrow ppee, pp \rightarrow pp\mu\mu$.

Most of these topics will be briefly discussed in the following. Many of then can be studied best with luminosities of $10^{33}\text{cm}^{-2}\text{s}^{-1}$ or lower.

In the last part of the lectures the presently considered options for forward instrumentation are discussed. Historically, the first full proposal for an experiment at the LHC that discussed forward physics was FELIX (see http://felix.web.cern.ch/FELIX/). FELIX was proposed as a dedicated experiment that could measure the full event phase space, but was put on ice in 1997. Its letter of intent is still an excellent source of information on pioneering studies for forward physics. The proceedings to a workshop on forward physics at the LHC can be found in Huiti *et al* (2000).

Roman pot detectors positioned at tens or hundreds of meters away from the interaction region are becoming a 'standard' detector to use at modern colliders. UA4 installed such detectors to measure elastic and diffractive cross sections, and these detectors were later used in conjunction with the UA2 detector by the UA8 experiment at the end of the 80's at the CERN *SppS*. HERA has roman pots installed along the proton beam line and also the experiments at the Tevatron have or had roman pot detectors in their set-up.

2 Total and elastic pp cross sections

Most proton-proton interactions are due to collisions at a large distance between the incoming protons, and the protons interact as a whole with small momentum transfer in the interaction. The particles in the final state have large longitudinal momenta but generally small transverse momenta, typically of order of a few hundred MeV. Such events are called soft events and make the large part of a so called minimum bias event sample, i.e. a sample which has as little as possible trigger bias (the exact definition of a minimum bias sample depends on the experimental acceptance and trigger). The cross section for these events at the LHC is humongous, of order of 100 mb, which is nine orders of magnitude or more larger than the Higgs cross section. The exact shape of the final state of these events at the LHC energy is not well known and the experiments will have to measure their characteristics at the LHC start-up. Even the cross section is not accurately known and needs to be measured. It is expected to be about 100 mb, with a typical uncertainty of about 20% or so.

We can easily derive how many events of this kind will be produced during a single bunch crossing at the LHC. Assume a luminosity of $10^{34} cm^{-2} s^{-1}$ which is $10^7 mb^{-1} Hz$. The non-elastic component (see below) of the pp cross section is roughly 70 mb. Thus we have an inelastic interaction rate of $7 \cdot 10^8$ Hz. The number of bunch crossings is 40 MHz, but taking into account that only about 80% of the available bunch spaces will be filled (3564/2835), one expects 23 soft overlap events per bunch crossing, at the highest luminosity, which is a huge number and will pose extra experimental challenges.

The total cross section consists of several components, the diffractive ones are shown in Fig. 2:

- The elastic cross sections, which is the reaction $pp \rightarrow pp$.

- The non-elastic diffractive cross sections (which contains single and double diffractive dissociation events). At high energies these processes are characterized by rapidity gaps, i.e regions void of particles.

- The non-diffractive inelastic cross section. These events do not have any rapidity gaps other than those compatible with fluctuations in the hadronization.

The TOTEM experiment plans to measure the total cross section and its components very accurately. In its original proposal, TOTEM planned to install three stations of roman pots on both sides of IP5. The different pots had the following purpose: the closest (90 m) were meant for high-t elastic scattering measurements, the intermediate (125 m) for low-t elastic scattering measurements and the furthest (154 m) for measuring the momentum of the scattered proton in diffractive collisions, making use of the separator magnet D2. Follow-up studies suggested three additional interesting locations, namely at $210 - 220$ m, around 420 m and between 300 and 400 m. The first are important for diffractive hard scattering studies and two-photon physics, while the latter two would be needed for the detection of diffractive Higgs production, or in general central low mass diffractive production. The most recent proposal for TOTEM now plans for two roman pot stations at both 147 and 220 m, the ones further downstream are left for consideration of a future study. The set-up is shown in Fig. 3.

Roman pots at $300 - 420$ m encounter two major difficulties: these detectors are in the cold section of the machine and therefore need special care with the integration and need to be compact. Furthermore the event buffers of CMS are about $3\mu sec$ long, thus a signal from the roman pots further than approximately 200 m distance cannot arrive in time for a trigger (91 bunch crossings) at the first trigger level: these events will need to be triggered by information in the central detectors.

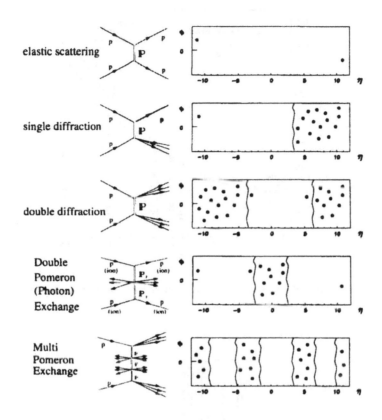

Figure 2. *Diagrams and $\eta - \phi$ pictures for various diffractive processes in pp scattering.*

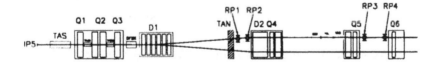

Figure 3. *Roman pot station layout for the TOTEM experiment.*

TOTEM plans to measure the total and elastic cross section during special LHC runs in so called high β^* optics mode, i.e. $\beta^* = 1000 - 1500$ m and luminosity $\sim 10^{28}$cm^{-2}s^{-1}, and medium β^*. The reach of the total cross section measurement is shown in Fig. 4.

Measuring the total cross section is naively speaking very simple: just count the amount of interactions for a given luminosity. To this end TOTEM has the inelastic event taggers, shown in Fig. 5, which allow one to record the total rate of inelastic events. However, in order to determine the cross section one needs to determine the luminosity of the data sample. TOTEM proposes to measure the total cross section to an accuracy of approximately 1%. It is as yet unclear how precise the luminosity can be determined

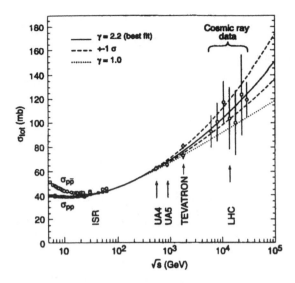

Figure 4. *Total cross section measurements with phenomenological fits added, and the region where LHC will contribute.*

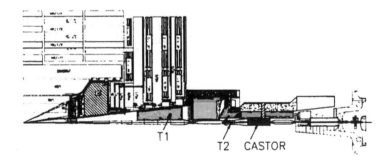

Figure 5. *Position of the inelastic event tagging detectors of TOTEM, T1, T2 and CASTOR integrated with CMS.*

at the LHC but it is likely to be more in the range of 5% using methods that involve the proton parton distributions, clearly not good enough for the aimed precision TOTEM cross section measurement. One can use, however, the so called luminosity independent method to determine the cross section whereby the optical theorem is used which relates the elastic scattering at $t = 0$ to the total cross section:

$$L\sigma_{tot}^2 = \frac{16}{1 + \rho^2} \times \frac{dN(t = 0)}{dt} . \tag{1}$$

Here L is the luminosity and ρ the ratio of the real to imaginary part of the forward scattering amplitude. The latter can be measured experimentally in principle by using elastic data in the Coulomb-nuclear interference range i.e. $|t| \sim 10^{-3}$ GeV^2/c^2 and below,

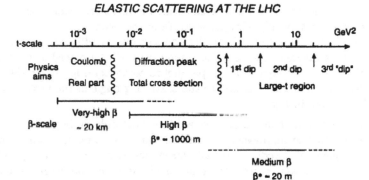

Figure 6. *Example of the reach in |t| as function of β* from the machine (for the '99 TOTEM LOI optics and detector positions).*

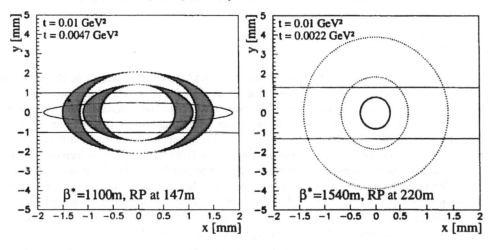

Figure 7. *Contour plots for two t-values for β* = 1100m (left) and β* = 1540 m. The 10σ_beam envelopes and the edge of a detector placed at 10σ + 0.5 mm are also shown.*

as has been done at lower energy experiments, but it may be out of reach for the LHC. The acceptance of the planned roman pots will allow one to cover $|t|$ down to a few times 10^{-3} GeV^2/c^2. The expected value of ρ at the LHC energy is in the range of $0.1 - 0.2$, so its impact on the measurement uncertainty is below the percent level. Since the number of elastic events is $L\sigma_{tot}$, the equation can be solved by knowing the number of events and the forward elastic cross section. Conversely σ_{tot} can then be used to determine the luminosity at the LHC.

Fig. 6 shows the correlation between the region that can be reached in $|t|$ and the β^*

of the machine. Values of β^* of more than $2 - 3$ km cannot be delivered by the machine. TOTEM will cover a $|t|$ range from a few times 10^{-3} to 10 GeV^2/c^2 using the high and medium β^* optics, as shown in Fig. 6. Note that during the high luminosity operation ATLAS/CMS will run with a β^* of 0.5 m.

After LHC has accelerated the beams and switched to collision optics, the roman pots will be lowered to typically $15\sigma_{beam}$ away from the beam axis, which could be as little as $1 - 2$ mm! Detectors (e.g. silicon detectors or fibers) will measure the coordinates of the elastically scattered protons. The roman pots will be placed at special locations at the machine – an optic layout called 'parallel to point focusing' – such that the measured coordinates can be directly related to the scattering angle of the proton. An example of the results is given in Fig. 7.

The measurement of the total cross section is not only an important number to know and to determine the luminosity. There are some exotic models, which, based on anti-shadowing arguments, predict unusually large cross sections, such as up to 200 mb. Note that if these models were correct the number of minimum-bias overlay events would double!

Elastic scattering is sensitive to granular structures of the proton, as proposed in some models. Here the key measurement is a precise determination of the $|t|$ distribution up to 10 GeV^2/c^2, which would allow one to distinguish between these models.

3   Diffraction

Recent data from HERA and Tevatron have re-emphasized the interest in the study of diffraction. Diffractive events can be pictured as shown Fig. 2: the characteristic is the exchange of a colorless object, termed the pomeron. After the exchange the beam particles can either stay intact or dissociate into states with (generally) a low invariant mass M according to the distribution $d\sigma/dM \sim 1/M^2$. This dissociation is termed 'diffractive dissociation'. At sufficiently high cm system energies a large rapidity gap is formed between the two dissociative states or particles. Phenomenologically the dynamics is described as the exchange of an object called the pomeron, which has quantum numbers of the vacuum and originates from Regge theory. In more modern language one tries to understand the phenomenon of diffraction and the pomeron in terms of QCD. For an overview on the work on diffraction see e.g. http://www.physics.helsinki.fi/blois_03/.

In particular the observation and study of diffractive events in ep scattering, as shown in Fig. 8, has yielded a lot of information on the diffractive exchange. It was found that the exchange could be described by partons, dominantly gluons, and within ep scattering the measured structure could be transported from e.g. inclusive deep inelastic $ep \rightarrow eX$ scattering to the production of charm or di-jets in deep inelastic diffractive scattering. An example of a structure function measurement is shown in Fig. 8. This means that the diffractive process in ep scattering is factorizable: it is the convolution of the partonic cross sections and the diffractive parton distributions, the same way as for normal parton distributions discussed in the lectures of Keith Ellis at this school.

It was however surprising that, when the diffractive PDFs are transported to the Tevatron for $p\bar{p}$ diffractive scattering, they do not give the correct predictions, i.e. factorization is broken in diffractive scattering. The data are below the predictions, which suggests that

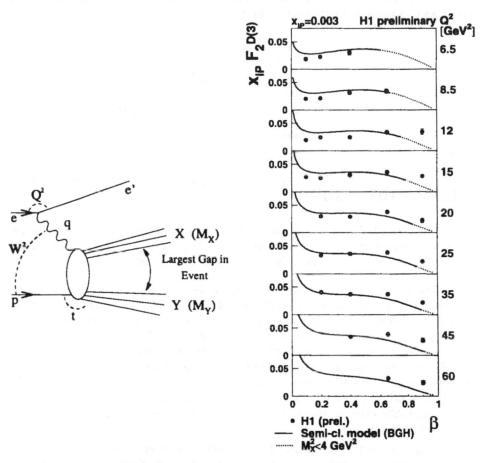

Figure 8. *Left: schematic diagram for diffractive ep scattering. Right: example of a structure function $F_2^{D(3)}$ measurement extracted from diffractive ep scattering.*

some mechanism destroys the rapidity gap. The most obvious candidate is rescattering of the proton remnants of the underlying event, and models based on this kind of dynamics can indeed describe the data. In all, diffraction is still a bizarre phenomenon and far from being understood. The LHC will contribute to its understanding.

Diffractive events at a pp collider can be experimentally selected by the observation of a rapidity gap in the event or by detecting and measuring the non-dissociated proton. The standard rapidity gap technique is expected not to be applicable at the highest luminosity at the LHC due to the large amount of overlap events per bunch crossing, as discussed before. However at the startup luminosity this selection method can be used as shown in Fig. 9. For a luminosity of 10^{33}cm^{-2}s^{-1} still in about 22% of the cases the bunch crossing will contain only one interaction. At $2 \cdot 10^{33}$cm^{-2}s^{-1} this number is reduced to 4% only. With a good control and tagging of the bunch crossings which have single collisions –

as already demonstrated to be feasible at the Tevatron – one could select and use these events for diffractive studies. At higher luminosities the usage of roman pots or similar detectors, such as microstations, will be imperative.

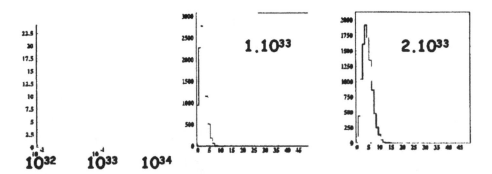

Figure 9. *Left: number of overlap events as function of luminosity. Middle: distribution of the number of overlap events for a luminosity of $1 \cdot 10^{33} cm^{-2} s^{-1}$. Right: distribution of the number of overlap events for a luminosity of $2 \cdot 10^{33} cm^{-2} s^{-1}$.*

A first study on diffraction can be made during the TOTEM runs at low luminosity. While running in high β^* mode, the acceptance of the TOTEM roman pot spectrometer is very large: it has basically a full acceptance for diffraction. The luminosity obtainable is however low: a few times 10^{29} cm^{-2}s^{-1}, up to a few 10^{30} cm^{-2}s^{-1}. Hence with e.g. two weeks of running this leads to data samples of $100 - 500$ nb^{-1}. Such samples are useful for studies of soft diffraction, such as a precise measurement of e.g. the different diffractive components (single diffraction, double diffraction, double pomeron exchange...) of the cross section, which can probably be measured with a precision of a few per cent. Some aspects of hard diffractive scattering can be studied, in those channels where the cross section is sufficiently high.

Combining CMS and TOTEM detectors will allow one to study hard diffractive processes. Two examples of such processes are given in Fig. 10: single pomeron exchange (SPE) and double pomeron exchange (DPE). The event topologies are shown in the $\eta - \phi$ plots, where ϕ is the azimuthal angle of the particles. Such measurements are very useful in studying the structure of the pomeron and the dynamics of diffraction. The program POMWIG was used to study di-jet events in SPE and DPE events. The cuts applied are $0.001 < \xi < 0.1$ and $|t| < 1$ GeV^2, and $p_T^{jet} > 10$ and 100 GeV, with $\xi = E_{pomeron}/E_{beam}$. These cuts are inspired by first calculations which indicate that the acceptance – when including roman pots as far out as 400 m – is $0.002 < \xi < 0.1$. Both jets must be observable in the CMS central detector and hence be within the acceptance of $|\eta| < 5$.

To check the sensitivity to the pomeron structure, differential distributions for several different input parton distributions are calculated. These include the QCD fits from H1 to the inclusive diffractive data (in LO: fits 4, 5 and 6; fit 4 has enforced no gluons at the starting scale of 4 GeV^2; fits 5 and 6 are different solutions with a rather large gluon content) and a simple soft $(1 - x)^5$ and hard $x(1 - x)$ parton distribution function.

The cross sections are in the order of microbarns for a p_T cut of 10 GeV, for fits 5 and 6, and 3 orders of magnitude smaller for fit 4. For a p_T cut of 100 GeV the cross section are a

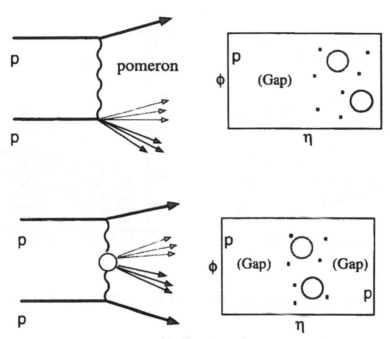

Figure 10. *Diagrams and $\eta - \phi$ pictures for single pomeron exchange (SPE) and double pomeron exchange (DPE) processes.*

factor 10^4 smaller. POMWIG is a model based on factorization, and hence the true cross sections will be a factor 10-100 smaller at the LHC, due to the factorization breaking in pp diffractive collisions mentioned before. Hence at low LHC luminosity operation (with TOTEM optics) hard diffractive scattering with two jets with a p_T of order 10 GeV/c will give several 100,000 events, but only a few tens of events with a p_T above 100 GeV/c.

For the nominal CMS high luminosity runs the number of events produced is huge: 100 events/sec and 0.1 events/sec for the two p_T cuts, respectively, certainly too much bandwidth for the trigger. Detecting and measuring the transverse energy and pseudo-rapidity of the two jets allows one to reconstruct the underlying kinematics of the hard scattering process. In LO the fractional momentum carried by the parton in the pomeron is given by $\beta = \sum_{\text{jets}} E_T e^{-\eta}/(\sqrt{s} \cdot \xi)$. The CMS/TOTEM detector combination allows one to reconstruct ξ from the roman pot information and E_T, the transverse energy, and η of the jets from the central CMS detector.

Fig. 11 shows the β spectra for the different assumptions of the parton distributions in the pomeron. The selection $p_T^{\text{jet}} > 100$ GeV emphasizes in particular the high β region. Clearly prominent differences in shape are seen for the different input parton distributions. Fig. 12 shows the η distributions of the jets. Already for this distribution the shape differences are noticeable.

The combination of the CMS central detector and LHC roman pots could open other new opportunities for diffractive studies. The diffractive cross sections and in particular the t dependence of the cross section can be considered as a source of information on the size and shape of the interaction region. It will be of interest to see how these quantities

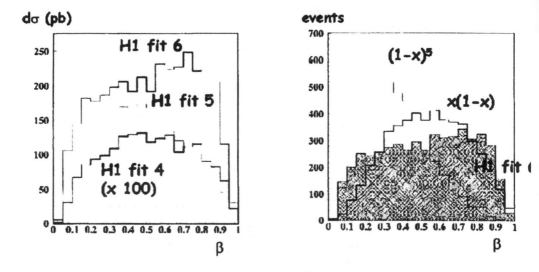

Figure 11. *β distributions calculated from di-jet events in DPE, for jets with $p_T^{jet} > 100$ GeV/c.*

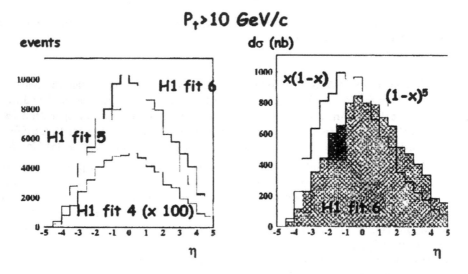

Figure 12. *η distributions calculated from di-jet events in DPE, for jets with $p_T^{jet} > 10$ GeV/c.*

evolve e.g. in the presence of a short-time perturbation which results in the production of jets in $pp \rightarrow p + \text{jet} + \text{jet} + p$ interactions.

An interesting and still somewhat unexplained phenomenon is the pomeron structure measurement by UA8. This experiment measured di-jet events at 630 GeV cm system energy in $p\bar{p}$ collisions at the $Sp\bar{p}S$ collider, for events with one proton or anti-proton tagged in the roman pots. The measured variable shown in Fig. 13 is the total longitudinal momentum of the di-jet system in the pomeron-proton cm system normalized to $\sqrt{s\xi}/2$, namely x(dijet) = x(pomeron) − x(proton), with x(pomeron) approximately the same as β defined before and x(proton) the momentum fraction of the parton in the proton. The figure shows the data compared with predictions for a hard $x(1 - x)$ and soft $(1 - x)^5$ parton distribution in the pomeron. It appears that the data are even harder than these hard PDFs. A test with POMWIG shows that the old parton distributions of H1 are very close to the hard PDFs used here and would thus also undershoot the data. In fact in order to describe the data the authors had to introduce an additional component − which they term superhard − which is basically a delta-function at $x = 1$, i.e. the parton in the pomeron takes the *full* energy of the pomeron. This result was a singularity in the field until last year when CDF made a similar (but yet unpublished) measurement, showing that the CDF data are compatible with the UA8 data. The origin of this superhard component is still mysterious

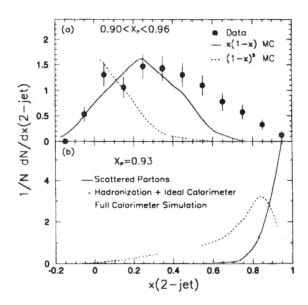

Figure 13. *(a) Observed x(dijet) distribution for UA8 events. The curves present predictions for a hard and a soft pomeron structure function after detector smearing. (b) Curves assuming that the full pomeron participates in the hard scattering, before and after simulation.*

A recent development in the study of diffractive phenomena is the revival for the search for the Higgs particle in diffractive events. There are basically two kinds of processes which are of interest here: exclusive and inclusive diffractive Higgs production, whereby the two outgoing protons do not dissociate and can be detected. This is sometimes also termed central diffractive production.

The diagram for exclusive production is shown in Fig. 14 (left). A recent calculation of the cross section for exclusive diffractive Higgs production $pp \to pHp$ gives about 3 fb

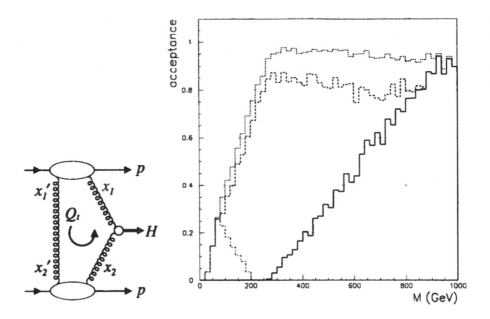

Figure 14. *Left: diagram for the exclusive production of the Higgs particle in pp interactions. Right: acceptance curves for protons from exclusive Higgs production for roman pots at positions 420+215m (dashed line), 215m alone (solid line), 420 m alone (dash-dotted line) and 420+320+215 m (dotted line).*

for a Higgs with mass of 120 GeV while for the inclusive one $pp \rightarrow p + X + H + Y + p$ it is claimed that the cross section could be as large as $100 - 300$ fb. The various processes and models are now studied in detail by several groups. Presently the spread between the different predictions is still large, more than a factor 10. A crucial point here for the exclusive models is that they should not be in conflict with the exclusive diffractive jet rate measured by CDF. Future Tevatron data will help to discriminate or tune parameters in these models, and by the time the LHC starts operating the theoretical spread should be much smaller.

The dependence of the inclusive diffractive cross section on the Higgs mass is presented in Fig. 15 (left) for one such calculation. The cross section is displayed for the Tevatron and the LHC, and is shown to fall off fast with increasing mass. The cross sections are too low for the Tevatron but large enough at the LHC, if the background can be kept under control.

With the standard TOTEM/CMS set-up it is not possible to detect both scattered protons of the exclusively produced Higgs, at low mass. This is demonstrated in Fig. 14 (right), which shows the acceptance for detectors at 215 m, at 420 m, at 420+215 m and at 420+338+215 m. The additional detectors at 300 and 400 m will be necessary for this study. Even including all detectors the acceptance for a 120 GeV Higgs will be about 40% only. For the exclusive channel $H(120) \rightarrow b\bar{b}$ with a cross section of 3 fb, the signal-to-background ratio S/B after 30 fb^{-1}, taking into account b-tag efficiencies and event selection efficiencies (at the parton level) is about 3. This is a priori not so bad

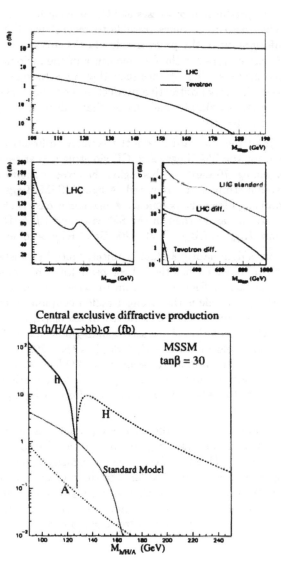

Figure 15. *Top: inclusive diffractive Higgs boson production cross section at the LHC and the Tevatron. The upper plot gives the cross sections for a low Higgs mass. The lower plots show the high Higgs mass range. The inclusive cross section at the LHC is shown as well. Bottom: the cross section times* $b\bar{b}$ *branching ratio predicted for the central exclusive production of MSSM Higgs bosons (for* $\tan\beta = 30$*) at the LHC compared with the SM result.*

if one compares with several other full inclusive channels for the Higgs discovery, which have been reviewed recently in De Roeck *et al* (2002).

Exclusive Higgs production has the advantage of the spin selection rule $J_z = 0$, which suppresses the QCD $b\bar{b}$ production processes at LO, allowing for the observability of the $H \rightarrow b\bar{b}$ decay mode. Also it becomes possible to reconstruct the mass of the Higgs particle with the information of the protons only via the missing mass method: $M_H^2 = (p_1 + p_2 - p_3 - p_4)^2$ where p_1, p_2 are the four-momenta of the incident protons and p_3, p_4 those of the scattered protons. First studies show that a resolution of about 1-2% on the mass can be achieved. Inclusive diffractive production does not have these advantages. These events can be certainly simpler and cleaner than full inclusive production but the backgrounds still have to be studied.

For scenarios beyond the standard model the exclusive diffractive event rate may be more favourable than for the SM Higgs. E.g. CP violating scenarios allow for Higgses to exist at smaller masses, eg. 70 GeV. In this region the cross sections are correspondingly larger, but a background study is still required. A light MSSM Higgs at sufficiently large $\tan\beta$ has also a promising discovery potential. A particular case is shown in Fig. 15: the cross section is a factor 10 larger than in the SM case, and the S/B can be larger than 10. Also medium mass Higgses ($M_H = 150 - 200$ GeV) may become accessible.

Since the pomeron is dominantly a gluonic object, central exclusive production could be also a way to find more exotic or other SUSY particles. An example is the radion, a scalar particle that appears in Randal-Sundrum theories. The main difference between the radion and the Higgs particle is the stronger radion coupling to gluons. First studies show that large cross sections could be obtained in exclusive pp scattering. Another exciting possibility is the production of exclusive gluino-gluino states. These could become observable if the gluino would be sufficiently light, less than $200 - 250$ GeV, with a few 10 events for the full high LHC luminosity sample, i.e. 300 fb^{-1}. Large enough samples of exclusive gluino-gluino events could allow for a precise determination of the gluino mass and properties, much like at an e^+e^- linear collider, e.g. for very light gluinos ($25-30$ GeV) as discussed in some scenarios.

Finally we like to note that, since the incoming particles in the interaction are overwhelmingly dominated by gluons (see Fig. 14, by a ratio of roughly $q/g = 1/3000$), it is basically a gluon factory. These double pomeron exchange events will be the largest clean gluon jet sample available and allow for a plethora of QCD studies. A year at the LHC would e.g. give 100,000 gluon jets with a $p_T > 50$ GeV with high purity. It is also a playground for searching for and/or studying new resonant states, like glueballs, or 0^{++} quarkonia (e.g. χ_b), in a relative background free environment.

4   Low-x

One of the most important results from HERA is the observed rise of the parton densities at small Bjorken-x, i.e. the fractional momentum of the parton with respect to the proton. Presently a debate is still ongoing whether the low-x data, reaching down to $x = 10^{-4} - 10^{-5}$ for scales Q^2 above or around 1 GeV^2, has reached the region of parton saturation, i.e. a region where the parton-parton interaction probability becomes very large. In such a region one would expect to see a reduced growth of the partons due to parton recombination and shadowing mechanisms. Perhaps saturation already occurs in a very small region around valence quarks in the proton, which leads to the formation of hotspots. A saturation region will be a new regime to study QCD where the parton

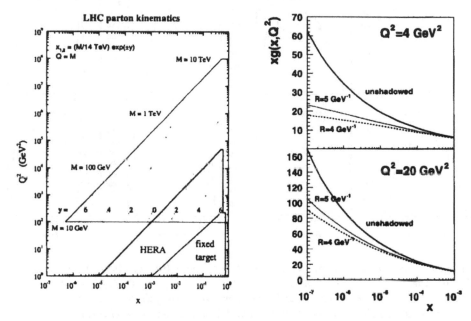

Figure 16. *Left: the kinematic plane* (x, Q^2) *and the reach of the LHC, together with that of the existing data (HERA, fixed target). Lines of constant pseudo-rapidity are shown to indicate the kinematics of the produced objects in the LHC centre of mass frame. Right: prediction of the gluon distribution at several scales, with and without saturation effects.*

densities are large but α_s is still small enough to perform perturbative calculations. Pioneering studies were done by Gribov-Levin-Ryskin (GLR) in the formulation of non-linear corrections to the Dokshitzer-Gribov-Lipatov-Altarelli-Parisi (DGLAP) equation. Low-x phenomenology is reviewed in Cooper-Sarkar *et al* (1998).

So far parton saturation has not been established and probably it can only be clearly observed at smaller x values, beyond the reach of HERA. A naive saturation limit based on geometrical scaling arguments leads to the saturation condition $xg(x) = 6Q^2$. Such values could be reached in the region $x = 10^{-6} - 10^{-7}$ for scales of a few GeV^2. In case of hotspot formation, the effects should already become visible earlier. The LHC will be the first accelerator which can access this region!

As an example we show shadowing corrections which have been estimated with GLR type of corrections to the standard parton evolution equations, using the results of triple pomeron vertex calculations. These results are shown in Fig. 16 (right), for different values of the saturation radius. The effects could be very large: at $x = 10^{-6}$, and $Q^2 = 4$ GeV^2, the effect of shadowing is as large as a factor 2. For larger Q^2 values it is reduced to a 20 − 30% effect. Estimates of saturation effects based on different models lead to similar conclusions: the effect may become strongly visible in the region of $x = 10^{-6}$.

The LHC kinematics is shown in Fig. 16 (left). It shows the x, Q^2 plane together with the region where direct DIS measurements exist. Also indicated are the lines of rapidity

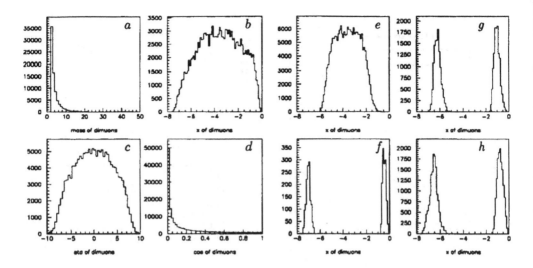

Figure 17. *Distribution of* $\ln_{10}(x)$ *for Drell-Yan production of electrons for* $M_{ee} > 4$ *GeV^2 for different η regions of the two electrons (a)* $0 < |\eta| < 5$, *(b)* $5 < |\eta| < 7$, *(c)* $7 < |\eta| < 9$, *(d)* $5.5 < |\eta| < 7.8$. *The latter corresponds roughly to a possible acceptance of microstations as calculated for ATLAS. Drell-Yan production of electrons: (e) the di-electron mass, (f)* $\ln_{10}(x)$ *values, (g) η values for the electrons, (h)* $\cos\theta$ *distribution of the electrons (normalization: number of events/pb/bin).*

in the centre of mass system of the produced heavy object (jet pair, Higgs,...) with a mass $M = Q$. The scale extends to small $x \sim 10^{-7}$, hence it is opportune to study the small x physics potential at the LHC.

In pp collisions information on parton distributions in the lowest possible x region can come from low mass

- Drell Yan production,

- direct photon production, and

- jet production.

A comprehensive study of low-x physics at the LHC has been performed for the FELIX proposal (see http://felix.web.cern.ch/FELIX/).

The Drell-Yan process $q\bar{q} \rightarrow \mu^+\mu^-$ or $q\bar{q} \rightarrow e^+e^-$ has a simple experimental signature. The $x_{1,2}$ values of the two incoming quarks relate to the invariant mass of the two–electron or muon system $M_{\mu\mu}$ as $x_1 \cdot x_2 \cdot s \simeq M_{\mu\mu}^2$, hence, when one of the $x_{1,2}$ is large ($x > 0.1$), low-x physics can be probed with low mass Drell-Yan pairs.

From Fig. 16 (left) we observe that in order to reach small masses (small scales) and low-x, one needs to probe large values of η. Hence the resulting electrons will dominantly go in the very forward direction. This is further illustrated with Fig. 17, made with the PYTHIA event generator, using MRST-LO parton distributions. Drell-Yan events were generated with a mass larger than 2 GeV. The figure shows the mass distribution, the distribution of $\ln_{10}(x)$, the pseudo-rapidity of the muons and the $\cos\theta$ distribution of the muons. Fig. 17 further shows the $\ln(x)$ distribution for the events with both muons accepted within given η ranges. It shows that in order to reach acceptances down to $10^{-7} - 10^{-6}$ in x, values of η larger than $|\eta| = 5$ should be covered.

Drell-Yan processes depend on the quark densities, in which the onset of saturation effects could be less visible. It has, however, been argued recently that the production of virtual photons, which can also produce a muon or electron pair when selected in proper kinematic regions, can be dominated by qg processes and hence be sensitive to the gluon distribution in the proton. Other processes which can be used to probe the gluon content of the proton are prompt photon (+jet) production and di-jet production. To reach the low-x area, however, processes with low scales need to be accessible, e.g. jet production with E_T^{jet} of around 5 GeV.

The kinematics for prompt photons and dijet events is very similar to the one of the Drell-Yan processes shown above. Can one within the LHC experimental environment actually extract such a low mass hard scattering signal? The event rates in any case are large. Presently CMS has no detection beyond $|\eta| = 5$. Although the T2 station of TOTEM will cover the region $5 < \eta < 7$, the instrumentation at present is modest (cathode strip chambers and RPCs), with the only task to signal the presence of a inelastic event, and does not allow the kind of study discussed here. Therefore this region is being re-examined for better detector use, as discussed in the last section.

Intimately related to low-x phenomena is the question of high energy QCD. The domain of low-x is also the domain of BFKL theory. The BFKL equation resums multiple gluon radiation of the gluon exchanged in the t channel (resums $\alpha_s \ln 1/x$ terms). It predicts a power increase of the cross section. HERA, Tevatron and even LEP have been searching for BFKL effects in the data. Presently the situation is that in some corners of the phase space the NLO DGLAP calculations indeed undershoot the QCD activity measured (e.g. forward jets and neutral pions at HERA), but BFKL has so far not been unambiguously established.

The large energy and rapidity span of the final state at the LHC may allow for a new (decisive?) attempt to establish BFKL. The 'golden measurements' identified so far are azimuthal decorrelations in the production of two jets far apart in rapidity, W-production and heavy flavour production (e.g. production of four b-quarks). In particular the di-jet measurements require detectors which cover as large a region in rapidity as is possible.

5 Forward physics and cosmic rays

The forward LHC data is also very relevant for cosmic rays analyses. The energy of the collisions at the LHC correspond to an incident energy of a proton on a fixed target proton of 10^{17} eV. This is (on a logarithmic scale) exactly in the middle between the so called knee and the ankle of the measured cosmic ray particle energy spectrum.

The development of a cosmic ray shower is depicted in Fig. 18. A primary particle penetrates the earth atmosphere and starts a shower of particles which can in turn interact. The starting point and shape of the shower depends on the incoming particle, as shown in the figure. One important method to reconstruct the incident particle type and energy is using surface measurements of the electromagnetic and muon content of the showers, and

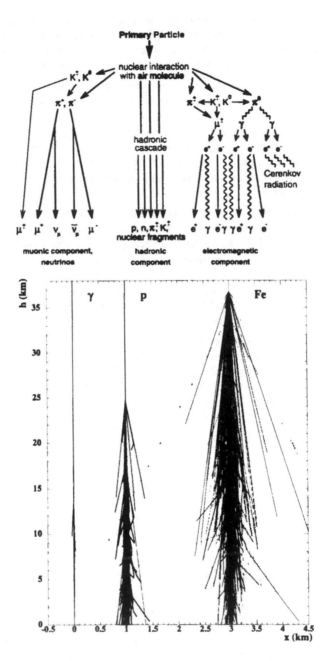

Figure 18. *Top: pictorial development of a cosmic ray air shower with its different components. Bottom: shower shapes and indicative starting points of photon, proton and iron induced showers.*

then reconstruct the shower 'bottom-up'. Such a reconstruction relies on a model for the interaction (number of particles produced, energies of the particles, etc.). In particular the forward part of the shower is of extreme importance. Also the badly known cross section decomposition in diffractive and non-diffractive events is a key ingredient in these models. Several different models are used in the cosmic ray community, tuned to as much relevant accelerator data as is available. When comparing the predictions of the models in Fig. 19 and Fig. 20 at the energy of the LHC, e.g. the momentum fraction (Feynman-x) taken by the leading particle, one finds differences larger than a factor of 2. Note that the Auger experiment will measure and reconstruct cosmic rays with energies of 10^{20} eV! Hence accurate data on particle production in the forward region at the highest energies reachable at accelerators will be very useful to constrain these models. This explains the strong interest of the cosmic ray community for this part of the program at the LHC.

A wish list produced by the cosmic ray community for measurements contains the measurements

- of the total diffractive and elastic cross section,

- leading hadron distributions, and

- general event features such as event multiplicities, correlations and low p_T jets in both the forward and central region.

Cosmic rays also form a motivation for forward physics. In very high energy cosmic rays some exotic events have been observed. Most notable are the so called centauro events, which seem to be hadronic showers with a very small or no electro-magnetic content. Such events have never been observed at an accelerator. Perhaps the energy so far was not high enough? While the signal for these kind of events is surely not unambiguous, several possible mechanisms to explain these events have been put forward: exotic extraterrestrial globs of matter, diffractive fireballs, disoriented chiral condensates, strange quark matter... Hence, forward physics at the LHC could well lead to a few surprises.

6   Two-photon physics

Two-photon physics has been traditionally studied at e^+e^- colliders because of the large fluxes of virtual photons associated with the beams. However, at the LHC the effective luminosity of $\gamma\gamma$ collisions, from photons radiated from the protons, will permit one to perform meaningful experiments. The photon spectrum can be described by the equivalent photon (or Weizsäcker-Williams) approximation (EPA). The protons can remain intact in the *elastic* production, $pp \to (\gamma\gamma \to X) \to ppX$, or one of them dissociates into a state N in the *inelastic* production, $pp \to (\gamma\gamma \to X) \to pNX$. The spectrum is strongly peaked at low photon energies ω, therefore the photon-photon center of mass energy $W \simeq 2\sqrt{\omega_1\omega_2}$ is usually much smaller than the total center of mass energy, which equals $2E = 14$ TeV. For the elastic production the photon virtuality is usually low, $\langle Q^2 \rangle \approx 0.01$ GeV^2, therefore the proton scattering angle is very small, $\simeq 20$ μrad.

As an example, in Fig. 21, assuming for each photon an integration interval of 5 GeV $< \omega < E$ and $Q^2_{min} < Q^2 < 2$ GeV^2, $S_{\gamma\gamma}$ and its integral $\int^{W>W_0} dW\, S_{\gamma\gamma}$ are shown for the

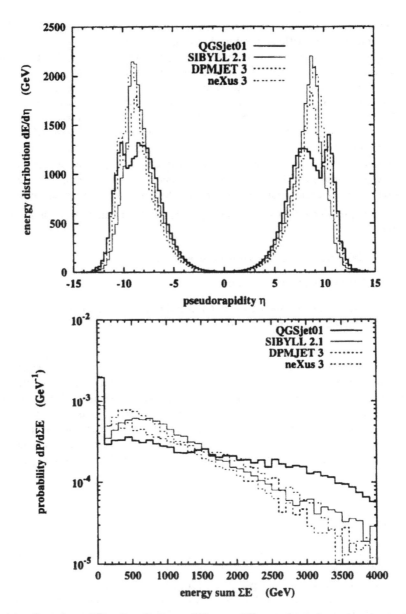

Figure 19. *Pseudorapidity distribution of the particles and total energy measurable in a forward detector ($5 < \eta < 7$) for pp interactions at 14 TeV cm energy, using different models used in cosmic ray studies.*

elastic production as a function of W and the lower integration limit, W_0, respectively. The integrated spectrum directly gives a fraction of the pp LHC luminosity available for the photon-photon collisions at $W > W_0$.

The same reaction, $pp \rightarrow ppX$, occurs also in strong interactions, via double pomeron

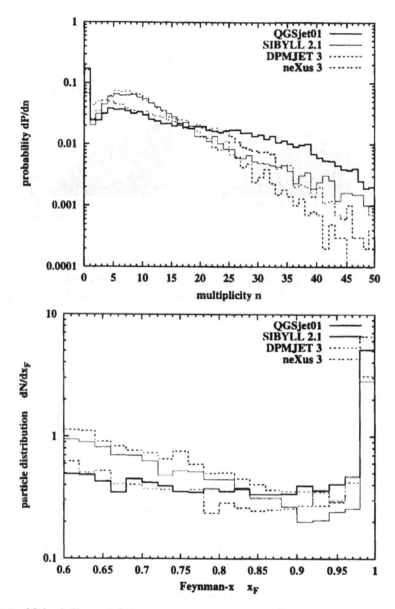

Figure 20. *Multiplicity and Feynman-x (x_F) distribution for pp interactions at 14 TeV cm energy, using different models used in cosmic ray studies.*

exchange as discussed earlier and will therefore interfere with the two-photon fusion. However, central diffraction usually results in much larger transverse momenta of the scattered protons, following the distribution $\exp(-bp_T^2)$, with the expected diffractive slope $b \simeq 4\,\mathrm{GeV}^{-2}$ at the LHC. Soft pomeron-pomeron interactions have several orders of magnitude larger cross-sections than the $\gamma\gamma$ interactions, but for the hard processes the

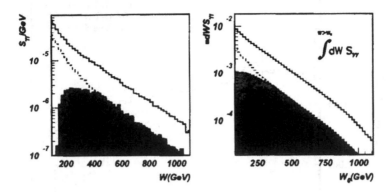

Figure 21. *Tagged photon-photon luminosity spectrum $S_{\gamma\gamma}$ and its integral $\int^{W>W_0} dW\, S_{\gamma\gamma}$ assuming double tags (shaded histograms) and single tags, for all (solid line) and only for elastic (dashed line) events; the tagging range is $70 < \omega < 700$ GeV and $Q^2_{min} < Q^2 < 2$ GeV².*

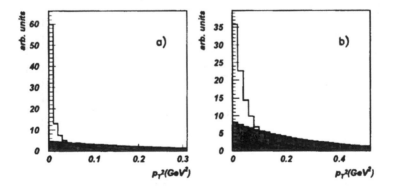

Figure 22. *a) Distribution of the transverse momenta squared of the scattered protons for the two-photon (empty histogram) and pomeron-pomeron (shaded histogram) collisions assuming the diffractive slope $b=4$ GeV^{-2}; b) the same distributions smeared by the beam divergence for the initial running conditions. Distributions have the same normalization for $p_T^2 < 2$ GeV² and correspond to a 100 GeV proton energy loss.*

cross-sections are of similar size. Therefore, the measurement of the proton p_T is vital for extracting the $\gamma\gamma$ signal. In Fig. 22, the 'true' and smeared distributions of p_T^2 are compared assuming the same cross-sections integrated over p_T^2. It shows that in such a case the two-photon signal is clearly visible and can be well extracted, and, for example, for $p_T^2 < 0.05$ GeV^2 the pomeron-pomeron contribution (neglecting interference effects) is about 20%. One should note that for the double tagged events the separation is even more powerful since one can independently use for that purpose the p_T of each proton.

The physics program for two-photon physics at the LHC contains topics such as

- the two-photon total cross section,

- QCD in two-photon physics,

- exclusive Higgs production,

- WW production, and

- SUSY particle production.

The total photoproduction cross section measurement and two-photon jet event studies could be made for W values larger than a few hundred GeV, i.e. at much larger energies than presently available. The exclusive two-photon Higgs cross section is relatively small: for a light Higgs approximately 10 events are expected for an integrated luminosity of 30 fb^{-1}, in the single tag mode (i.e. only one proton is tagged). The $\gamma\gamma \to WW$ cross section is about 200 pb, so several thousand WW pairs are produced with such an integrated luminosity. Two-photon interactions are also well suited to produce exclusive pairs of new heavy charged particles, such as charginos or slepton pairs. For a sparticle mass of 200 GeV, one expects about 20 slepton pairs or 60 chargino pairs. Note that these sparticles are produced in a very clean, exclusive channel, allowing for studies of their properties. The expected rates for sparticles versus their mass are shown in Fig. 23.

Two-photon events could also allow for a more precise determination of the luminosity at the LHC. The methods proposed so far are based on processes that involve parton distributions in the proton, which are not know better than a few per cent, hence one can hope for a precision of $4 - 5\%$ at best from e.g. measuring the luminosity using the production of W's. Two-photon processes, on the other hand, are QED processes and can be calculated at the per cent level or better (since the proton form factors have been measured sufficiently precisely). The processes $pp \to ppe^+e^-$ or $pp \to pp\mu^+\mu^-$ have been shown to be good candidates for luminosity measurements, if sufficiently clean data samples can be extracted.

In addition to two-photon physics, γp interactions also become available, and can be studied at higher energies than available at HERA.

7 Running the LHC at lower energies

The LHC also offers the possibility of studying processes at a lower centre of mass energy. Such a configuration would allow for interesting measurements which would enable one to study the energy dependences of processes or comparisons of data with the Tevatron, which produces $p\bar{p}$ collisions instead of pp collisions at the LHC.

Possible options for lower energy runs at the LHC include running at 2 TeV with a maximum luminosity of $2 \cdot 10^{32}$cm^{-2}s^{-1} and 8 TeV with $3.3 \cdot 10^{33}$cm^{-2}s^{-1}.

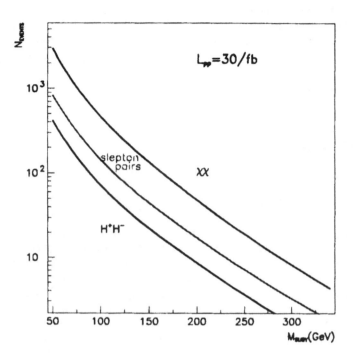

Figure 23. *Rate of sparticle pair production for 30 fb⁻¹ in two-photon collisions at the LHC.*

A first topic is remeasuring the total cross section at a cm system energy of 2 TeV. The measurements at the Tevatron from CDF and E710 show a discrepancy of several sigmas. Hence TOTEM could go back and remeasure the total cross section with a precision of 1 − 2 mb and thus referee these results. At the same time it would be a good consistency check of the TOTEM method at the LHC machine. Also the total cross section at intermediate energies can be measured.

Another important physics topic is to measure the production of Z bosons at 2 TeV and to compare the pp and $p\bar{p}$ production rates, which are sensitive to the quark and anti-quark distributions in the proton.

BFKL studies can be made in a similar way to that done recently by D0, by comparing dijet production with a jet separation in rapidity chosen such that the parton distribution effects cancel in the ratio of the rates at two different cm system energy measurements. Taking into account the acceptance of the detectors, it appears that 14 TeV and 8 TeV are the optimal energies for such measurements.

Other topics include the energy dependence of gap survival probabilities, inclusive jets, and more.

8   Detectors for forward physics

Finally we discuss briefly detectors and technologies for forward physics. The roman pots have been discussed before. A detailed design for the mechanics of these devices for TOTEM now exist. An example is shown in Fig. 24. The detector requirements include

- a high and stable efficiency near the edge facing the beam, edge sharpness better than 10 μm (guard rings at present take up to half a millimeter),

- a detector size of 3×4 cm^2,

- a spatial resolution of $\sim 20\mu$m, and

- a radiation tolerance $\sim 10^{14}$ n/cm^2 equiv.

Examples of detectors include cold silicon as developed by RD39/NA60, 3D detectors and others.

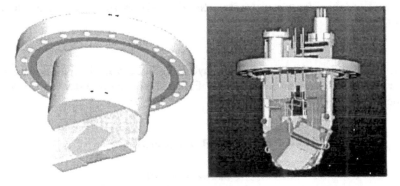

Figure 24. *Tentative design of the roman pot mechanics for TOTEM.*

Recently the Helsinki group of Orava et al. has proposed an alternative to these classical roman pot detectors. They proposed a very compact detector, named micro-station, which may make use of silicon pixel detectors or silicon strips in the prime or in a secondary vacuum. These detectors could become very important for areas where the available space is limited, e.g. in the cold magnet section at 400m and areas close to the CMS detector. However the reliability of these detectors in a real accelerator environment still has to be demonstrated.

Roman pot type detectors will be used by the TOTEM experiment and are presently being considered by the ATLAS experiment.

CMS/TOTEM will also have inelastic event tagging detectors at the location of T1/T2 as discussed in the first section. A study group is now studying upgrading the T2 region with higher performing detectors. Presently the idea is to have both a calorimeter and a tracker in this area. The tracker could be a 5 plane silicon strip tracker which would cover the η range: $5.3 < \eta < 6.5$, see Fig. 25. The present plans for a calorimeter are a tungsten absorber with a quartz fiber device of about $9\lambda_I$ length, with an electromagnetic section, covering $5.4 < \eta < 6.7$. It is expected to have 8 azimuthal sectors. The spatial resolution

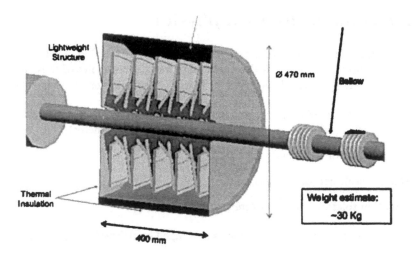

Figure 25. *A proposal for a new tracker, based on silicon strips, in the position of T2 of CMS/TOTEM.*

is poor but could perhaps be increased via layers of silicon strips in the calorimeter. The calorimeter, CASTOR, is shown in Fig. 26. Fig. 5 shows the tentative location of both the T2 tracker and CASTOR detectors.

CASTOR Calorimeter

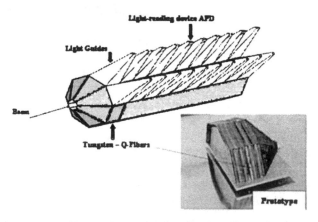

Figure 26. *The proposed CASTOR calorimeter by CMS.*

ATLAS plans a Cerenkov/quartz fiber detector called LUCID, for luminosity monitoring, which will cover the region $5.2 < \eta < 6.2$. This detector could possibly be used for the detection of rapidity gaps in an event.

9 Conclusions

- Diffraction:
  Elastic, total and soft diffractive cross sections can be measured with TOTEM in low luminosity runs. Using the combination of CMS+TOTEM allows one to study hard diffraction in detail at the highest energies. Questions on the pomeron structure and gap survival dynamics can be addressed. At these large LHC energies values of β as low as 0.001 can be reached. Diffractive Higgs production has recently gained renewed attention.

- Low-x physics:
  The LHC has the potential to reach the region in $x = 10^{-6} - 10^{-7}$, but the CMS detector would need to be upgraded in the region $5 < |\eta| < 7$. This can be achieved by redesigning the T2 detector of TOTEM, adding a calorimeter to it and include it in CMS runs. Background studies still have to be completed.

- Two-photon physics:
  A potentially interesting program for two-photon physics becomes accessible to the LHC, if proton taggers are available in the right location. Presently this is thought to be in the region of $220 - 240$ m downstream of the experiment.

- Lower energies:
  It is possible to run the LHC at smaller energies, in the range of $2 - 14$ TeV. This would allow for many studies related to QCD. Note that such low energy running is not yet planned at present.

References

Huitu K, Orava R, Tapprogge S, and Khoze V A (editors), 2000, Proceedings of the workshop *Forward physics and luminosity determination at the LHC* (Helsinki, Finland).
De Roeck A, Khoze V A, Martin A D, Orava R, Rysin M G, 2002, *Eur Phys J* **C25** 391.
Cooper-Sarkar A M, Devenish R C E, and De Roeck A, 1998, *Int J Mod Phys* **A13** 3385.

Index